The Practice of Silviculture: A Comprehensive Approach

The Practice of Silviculture: A Comprehensive Approach

Edited by Alex Edwards

MURPHY & MOORE
www.murphy-moorepublishing.com

Published by Murphy & Moore Publishing,
1 Rockefeller Plaza,
New York City, NY 10020, USA
www.murphy-moorepublishing.com

The Practice of Silviculture: A Comprehensive Approach
Edited by Alex Edwards

International Standard Book Number: 978-1-63987-532-0 (Hardback)

Cataloging-in-Publication Data

The practice of silviculture : a comprehensive approach / edited by Alex Edwards.
p. cm.
Includes bibliographical references and index.
ISBN 978-1-63987-532-0
1. Forests and forestry. 2. Trees. I. Edwards, Alex.
SD391 .S55 2022
634.95--dc23

TABLE OF CONTENTS

PREFACE

Every book is initially just a concept; it takes months of research and hard work to give it the final shape in which the readers receive it. In its early stages, this book also went through rigorous reviewing. The notable contributions made by experts from across the globe were first molded into patterned chapters and then arranged in a sensibly sequential manner to bring out the best results.

The science and art of growing and cultivating forest crops is referred to as silviculture. It involves controlling the establishment, growth, structure, and quality of forests, and management of forests stands primarily for timber production. Silviculture also focuses on ensuring that the treatment of forest stands is used to conserve and improve their productivity. Silviculture is based on the knowledge of silvics, which is the study of the life-history and general characteristics of forest trees and stands. Silviculture has applications in locality factors, nursery raising, sand dunes stabilization, mangrove forest management, watershed management, tending operation, silvicultural systems, natural regeneration, site maintenance, and wildlife conservation. This book unravels the recent studies on silviculture and its applications. It picks up individual topics and explains their need and contribution in the context of a growing economy. This book will provide comprehensive knowledge to the readers.

It has been my immense pleasure to be a part of this project and to contribute my years of learning in such a meaningful form. I would like to take this opportunity to thank all the people who have been associated with the completion of this book at any step.

Editor

1

Increasing wood mobilization through Sustainable Forest Management in protected areas of Italy

Mauro Maesano[1*], Bruno Lasserre[1] , Marco Marchetti[1]

Abstract - The European Community has long recognized the need to further promote renewable energy. Under the overall objective to support and enhance sustainable management, the promotion of the use of forest biomass could help to mitigate climate change by substituting fossil fuel, increasing carbon stock in wood products and improve energy self-sufficiency enhancing security of supply and providing job opportunities in rural areas. To what extent Italian forests can satisfy an increased wood demand, without compromising the others Ecosystem Services (ESs) remains an open question. Our aim was to assess the potential supply of woody biomass from the network of protected areas in Italy considering the felling constraints. We estimated the theoretical annual potential increment from forest inventory data performing a correlation with the Corine Land Cover 2006 at the IV level with a 1:100,000 resolution elaborated in a GIS (Geographic Information System) environment. The average annual potential increment at national level available for felling was 4.4 m^3ha^{-1}. Within the network of protected areas (EUAP and Natura 2000), the average annual increment, available to felling, was 0.98 m^3ha^{-1}, respectively 0.81 m^3ha^{-1} from coppice and 1.14 m^3ha^{-1} from non-coppice forests. Based on data obtained from this study, the availability of wood materials could be increased of almost 20 % at national level by pursuing an active management within the network of protected areas. In Italy, the actual level of resource utilization is rather low; increasing felling together with the implementation of an active management within protected areas could allow satisfying, theoretically, the Italian wood consumption.

Keywords - wood mobilisation, protected area, Sustainable Forest Management, woody biomass, supply

Introduction

In recent years, the use of forest biomass for energy purposes has gained outstanding importance due to its role in the reduction of fossil fuel dependence, the diversification of the products deriving from forests (Lamers et al. 2013), the climate change mitigation and in providing job opportunities in rural areas. Within the 2020 strategy (http://ec.europa/europe2020/) for a smart, sustainable and responsible growth, forest biomasses play a crucial role in the achievement of objectives.

The woody biomass accounts about half of Europe's renewable-energy consumption, in its various forms, from sticks to pellets to sawdust, as well as it is the more widespread renewable fuel used (Economist, April 6th 2013).

In 2010, the total supply of all woody resources in the EU 27 was about one billion cubic meters where of almost 70% came from forest and 30% from outside the forest (Mantau et al. 2010). Currently, more than half of the wood harvested from European forests is used for industrial processing purposes and future increases in the demand for timber would require forest managers to increase future management intensity (Sohngen et al. 1999). The type of management within protected areas can influence the amount of increment that may be harvested. The choice between different forest management practices is a crucial step. A Forest Management Approach (FMA) provides a structure for decision-making, including a range of silvicultural operations throughout the stand development phases (PEER 2011). Dunker et al. (2012) proposed five FMAs by placing the management goal and decisions along a gradient of intensity of intervention with the natural process. The strategic management choices of where to conserve nature, and where to produce wood are often done at the management unit level (PEER 2011).

Most of European forests are managed (MCPFE 2007) and thus offer the possibility to improve their adaptation ability by human intervention (Köhl et al. 2010). Nevertheless about 25% of the European forest area is subject to management constraints to secure the Ecosystem Services (ESs) such as nature conservation, soil protection, water supply or recreation (MCPFE 2003), but also timber production. In Italy forests account for about 10.9 million hectares corresponding to 37% of the land area and are mainly located in hill and mountain ranges. The forest area available for wood supply

[1] Department of Biosciences and Territory, University of Molise, Italy
[*] corresponding author: mauro.maesano@unimol.it

is about 8.08 million hectares (FOREST EUROPE, UNECE and FAO, 2011). According to the National Statistics Agency (Istituto Nazionale di Statistica - http://agri.istat.it - visited 07/2013), the national average harvested volume from forests amounted to almost 7.6 million cubic metres (21% of total annual volume increment) in 2001, of which about 67% was fuelwood.

Not all forest and Other Wooded Land (OWL) are available for wood supply, and this reflects the forests multifunctionality: as well as providing economic resources, the forests are important for both the environment and the community.

To what extent Italian forests can satisfy an increased wood demand, without compromising the others ESs remains an open question. Main concerns are biodiversity protection and carbon neutrality (Muys et al. 2013). Maintaining biodiversity in forested areas can be achieved through an adequate network of Protected Areas (PAs) and by the implementation of a silvicultural management, which integrates conservation and production functions (Parviainen and Frank 2003). Establishing PAs is a common conservation strategy pursued for the aim of biodiversity conservation, but it is also an important opportunity to safeguard the others ESs provided by forests (Cash et al. 2006, Turner et al. 2007, Nelson et al. 2008, Egoh et al. 2009, Pettorelli et al. 2012, Willemen et al. 2013). Biological and landscape diversity protection measures can result in a range of possible impacts on the economic use of forest resources. These PAs are subject to national and/or regional legislation that may dictate some restrictions on forest management (Verkerk et al. 2008), for example, on the future land use options (Norton-Griffiths and Southey 1995).

In Italy, the Law (MATTM 2010) protects 27.5% of the forest area, with a higher incidence in some regions of central and southern Italy. The NATURA 2000 network sites (SCI and SPA) include 22.2% of the national forest area (INFC 2005).

In this study, extensions and annual volume increment of the Italian forest area have been estimated with the aim of determining how PAs and others areas subject to different forms of protection weigh upon the annually available wood amount. The target was to identify concrete measures for wood mobilization in PAs. Extension and biomass volumes available for felling were estimated in three steps:

a. assessment of forest extensions with different types of protection and silvicultural treatments through CLC 2006 (Corine Land Cover - 2006) (ISPRA 2010) and INFC (Italian National Forest Inventory - 2005) data;
b. calculation of the annual increment of forest types through the correspondence between INFC and CLC 2006 data;
c. assessment of the potential felling amounts in the network of PAs in Italy and determination of the annually available wood within the PAs supposing the implementation of an active management.

The volume of wood that cannot be harvested from these PAs was assessed estimating the unconstrained potentials for wood fellings (i.e. disregarding all forms of protection or limitations to mobilize these resources) and by reducing this potential by applying felling restrictions to the protected areas (Verkerk et al. 2008).

Materials and Methods

General approach

The paper is based on capturing a large set of data and information on Italian forests and elaborating it in a GIS (Geographic Information System) environment in the CLC 2006 resolution (1:100,000) at the IV level (ISPRA 2010). For this study OWL classes (322, 323, 324 CLC classes) were not considered. The analysis mainly focused on the CLC classes typically considered as "Forest" (311, 312, and 313).

In the first step, the forest extensions were determined by overlapping the PAs network areas with CLC 2006 (ISPRA 2010). Each CLC 2006 forest polygon was classified according to the silvicultural system (coppice/non-coppice), as reported by the Italian National Forest Inventory 1985 (ISAFA 1988). A Protection Index (I_{PW}: Index of Protection Wood - ratio between protected forest area and total forest extension) was calculated at regional and national level for each type of protection, silvicultural system and CLC class.

In the next step, a correlation between the forestry categories listed within INFC (2003) and the IV level CLC forest classes was carried out to calculate the increments of each identified forest categories (Table 1) on the basis of INFC annual increment data. The correlation between INFC and CLC categories was performed considering the vegetation description of each category according to the inventory classification of forest vegetation (INFC 2003) and the description of the vegetation categories within CLC 2006 (http://sia.eionet.europa.eu/CLC2000/classes/index_html).

In the last step, the felling potential and restriction within the network of PAs were determined. Figure 1 reports the methodology applied to develop the objectives of the research.

Table 1 - Correlation table between the forests categories INFI and IVth level CLC.

Inventory Category (INFC 2005)	CODE INFC	CLC 2006 CLASSES	CODE CLC 2006	Correspondence with INFC code
Larch and Stone pine (L. decidua, P. cembra)	1	Holm-oak and cork-oak forests	3111	15;16;17
Norway spruce (P. abies)	2	Deciduous oak forests	3112	9;10
Fir (A. alba)	3	Mesophilous broad-leaved forests	3113	12
Scots pine and mountain pine (P. sylvestris, P. mugo)	4	Chestnut forests	3114	11
Black pines (P. nigra, P. laricio, P. leucodermis)	5	Beech forests	3115	8
Mediterranean pines (P. domestica, P. marittima, P. halepensis)	6	Igrophilous forests	3116	13
Others coniferous forest	7	Non Native broad-leaved forests	3117	14;19;18
Beech (F. sylvatica)	8	Mediterranean pine forests	3121	6;7
Temperate oaks (Q. petrea, Q. pubescens, Q. robur)	9	Mountain and oromediterranean pine forests	3122	4;5
Mediterranean oaks (Q. cerrid, Q. frainetto, Q. trojana)	10	Silver fir and spruce forests	3123	2;3
Chesnut (C. sativa)	11	Larch and Arolla pine forests	3124	1
Hornbeam and Hophornbeam (Carpinu spp., Ostrya Carpinifolia)	12	Non Native coniferous forests	3125	20
Hygrophilous forests	13	Mixed forests dominated by Holm oak	31311	25%(15;16;17)+ 75%(6; 7; 9; 10; 14)
Others deciduos broadleaved forests	14	Mixed forests dominated by deciduous oak	31312	25%(9;10)+75% (6; 7; 14; 15; 16; 17)
Holm oak (Q. Ilex)	15	Mixed forests dominated by mesophilous broad-levead	31313	25%(12)+75%(5; 7; 14)
Coark oak (Q. suber)	16	Mixed forests dominated by chestnut	31314	25%(11)+75%(5; 7; 14)
Other evergreen broadleaved forests	17	Mixed forests dominated by beech	31315	25%(8)+ 75%(3; 4; 5; 7; 14)
Poplar plantations	18	Mixed forests dominated by igrophilous species	31316	25%(13)+75%(6; 7; 14)
Other broadleaved plantations	19	Mixed forests dominated by non native broad-leaved	31317	25%(14; 18; 19) +75%(6; 7; 14)
Coniferous plantations	20	Mixed forests dominated by mediterranean pine	31321	25%(6; 7)+ 75% (7; 9; 10; 14; 15; 16; 17)
		Mixed forests dominated by mountain and oromediterranean pines	31322	25%(4; 5)+ 75%(2; 3; 7; 8; 14)
		Mixed forests dominated by silver fir and spruce	31323	25%(2; 3)+ 75%(4; 5; 7; 8; 14)
		Mixed forests dominated by larch and Arolla pine	31324	25%(1)+ 75%(2; 3; 4; 5; 8; 14)
		Mixed forests dominated by non native coniferous	31325	25%(20)+ 75%(7; 14; 19)

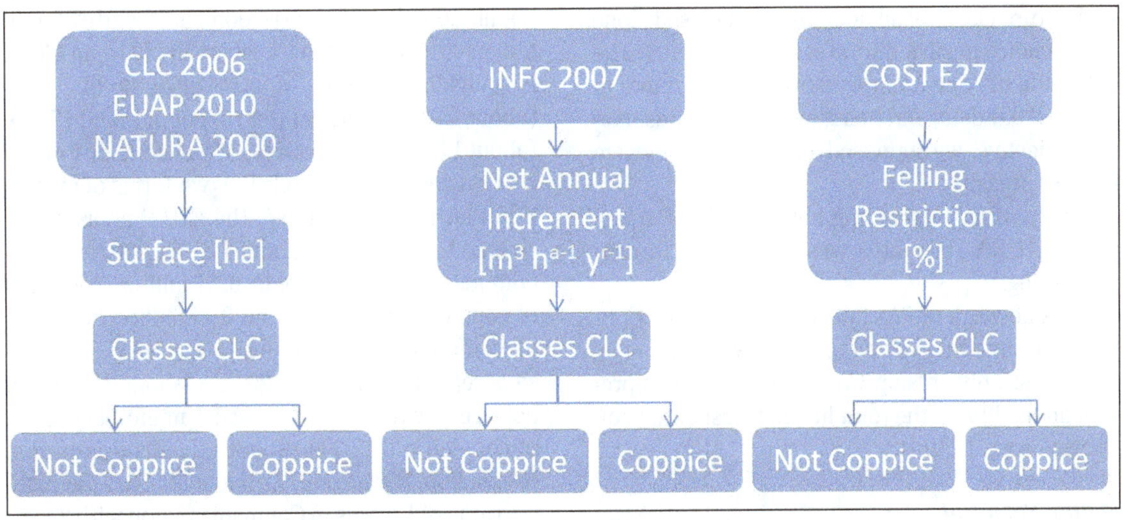

Figure 1- Schematic approach for determining the potential annual increment available and not.

Network of Protected Areas

The EUAP (Official List of the Protected Areas - MATTM 2010) areas and Natura 2000 sites made up the network of PAs used in this research. EUAP areas include 24 National Parks, 144 State Natural Reserves, 134 Regional Natural Parks, 365 Regional Natural Reserves and 171 Other Natural Protected Areas. Natura 2000 (Habitat directive 92/43/CEE) land sites (2011 Update - www.eea.europa.eu) consist in 2,549 Special Areas of Conservation (SAC) and include 2,269 Sites of Community Importance (SCI) and 600 Special Protection Areas (SPAs).

Table 2 - Comparison between the MCPFE classes, EUAP, Natura 2000, EEA and the applicable felling restrictions.

	MCPFE Classes	EUAP	Natura 2000	EEA	Restriction
Management Objective 1 "Biodiversity Conservation"	1.1 No Active Intervention	Zone A PNZ	-	A	1
	1.2 Minimum Intervention	-	-		A 2
	1.3 Conservation Through Active Management	Zone B-C-D PNZ, PNR, RNS, RNR, AANP	Natura 2000	A	3

The identified PAs network belong to the main classification of Protected Forest Areas (PFAs) at European level as they fulfilled the three general principles outlined in the MCPFE (Ministerial Conference on the Protection of Forests in Europe) guidelines (MCPFE 2002).

The MCPFE guidelines distinguish two classes of forest management in the PFAs. The first one focuses on biodiversity and is split up in three subclasses; 1.1: no active intervention; 1.2: minimum intervention; and 1.3: conservation through active management (Table 2). The second one focuses on landscapes protection and specific natural elements. The PAs network areas are included in the first class with a differentiation of the sub-classes based on the parks zonation and/or management plans.

The EUAP areas include also the CDDA (Common Database on Designated Areas) list, coordinated by EEA (EEA 2007). Regarding National Parks within EUAP, the national law established as a planning tool, the zoning of territory (Law 394/91). Based on this law, Ciancio et al. (2002), define four classes different for management degree. Zone A: the goal is the preservation regardless of the naturalness degree; zone B: preservation is realized through systemic silviculture; zone C: in addition to systemic silviculture classical silviculture can be used; zone D: the choice of silvicultural system is broader. The extension of zone A in the national parks refers to INFC (2005) data, and being evident the conservation objective, we supposed a non-coppice system in these areas.

The partial overlapping between EUAP areas and NATURA 2000 sites was considered in data processing.

This network of PAs falls within the network of High Conservation Value Forest (HCVF) defined by the Forest Stewardship Council (FSC) that represent about 40 % of the total Italian forest extension (Maesano et al. 2014).

Felling Potential

The felling potential reflects the theoretical maximum volume that is available for harvesting. There are several approaches to estimate the maximum felling potential, but the most straightforward is the Net Annual Increment (NAI) method (Verkerk et al. 2008). NAI has to be considered an upper limit, because annual felling should not exceed the annual increment to ensure long-term sustainability (MCPFE 2003). The potential increment for the whole network of PAs was calculated for each forest type, administrative region and silvicultural system according to the INFC (2005) data of the NAI.

Felling Restrictions

Data on felling restrictions in forests protected for biodiversity aims were collected from the COST Action E27 study on PFAs in Europe (Frank et al. 2007). In COST Action E27 restriction are given on a scale of 1-4, where: 1- activity is strictly prohibited; 2 - activity is usually prohibited, but with some exceptions or conditions; 3 - activity is usually allowed, but with some exceptions or conditions; 4 - activity is allowed with no restrictions. Based on the restriction scale, the limitation levels were: 1 - 0% of the potential can be harvested; 2 - 33% of the potential can be harvested; 3 - 67% of the potential can be harvested; 4 - 100% of the potential can be harvested (Verkerk et al. 2008). Table 2 shows the felling restriction applied in the network of PAs.

Results

Network of Protected Areas

The forest area inside the network of PAs was about 2.6 million hectares compared with 7.8 million hectares of Italian forest area (CLC data). This is equivalent to an I_{PW} of 33.6% made up of 15.96% (about 1.23 million hectares) of coppice and 17.64% (about 1.37 million hectares) of non-coppice forests.

Considering separately the two forms of protection that partially overlap, the EUAP areas represent an I_{PW} of 17.64% (about 1.37 million hectares) composed of 8.06% (about 0.63 million hectares) of coppice and 9.58% (about 0.74 million hectares) of non-coppice forests; while the NATURA 2000 sites represent an I_{PW} of 29.72% (about 2.3 million hectares) with 14.33% (about 1.1 million hectares) of coppice and 15.39% (about 1.2 million hectares) of non-coppice forests. At regional level, the highest value of I_{PW} was in Campania region with 3.32%, while the smallest one was in Valle d'Aosta with 0.14%. Regarding silvicultural system, in coppice forests the I_{PC} (Index of Protection Coppice) varied from 2.25% in Campania to 0% in Valle d'Aosta, while in non-coppice forests the I_{PNC} (Index of Protection Non-Coppice) ranged from 1.74% in Calabria to 0.14% in Valle d'Aosta (Tab. 3).

Considering the CLC classes, the more repre-

Table 3 - Protection indices at regional scale for the areas EUAP, Natura 2000 and EUAP+Natura 2000 . Legend: I_{PC}: Index of protection coppice; I_{PNC}: Index of protection non-coppice; I_{PW}: Index of protection wood.

	EUAP + NATURA 2000			EAUP			NATURA 2000		
Territorial District	I_{PC}	I_{PNC}	I_{PW}	I_{PC}	I_{PNC}	I_{PW}	I_{PC}	I_{PNC}	I_{PW}
Abruzzo	0.96%	1.26%	2.22%	0.73%	0.99%	1.71%	0.95%	1.24%	2.18%
Basilicata	0.61%	0.97%	1.57%	0.46%	0.90%	1.36%	0.52%	0.66%	1.18%
Calabria	1.09%	1.74%	2.83%	0.92%	1.36%	2.28%	0.77%	1.00%	1.77%
Campania	2.25%	1.06%	3.32%	1.76%	0.88%	2.64%	1.93%	0.80%	2.73%
Emilia Romagna	1.11%	0.38%	1.48%	0.46%	0.19%	0.65%	1.05%	0.36%	1.42%
Friuli V.G.	0.11%	0.79%	0.90%	0.03%	0.34%	0.37%	0.11%	0.76%	0.86%
Lazio	1.52%	1.03%	2.54%	0.58%	0.49%	1.08%	1.43%	0.86%	2.29%
Liguria	0.91%	0.34%	1.25%	0.15%	0.07%	0.22%	0.87%	0.32%	1.19%
Lombardia	0.61%	1.06%	1.67%	0.18%	0.35%	0.53%	0.59%	1.00%	1.60%
Marche	0.77%	0.24%	1.01%	0.44%	0.13%	0.56%	0.64%	0.20%	0.83%
Molise	0.44%	0.36%	0.80%	0.07%	0.13%	0.20%	0.44%	0.36%	0.80%
Piemonte	0.55%	0.99%	1.54%	0.28%	0.39%	0.67%	0.48%	0.95%	1.43%
Puglia	0.66%	0.52%	1.18%	0.43%	0.29%	0.72%	0.59%	0.50%	1.09%
Sardegna	0.63%	1.02%	1.65%	0.28%	0.22%	0.49%	0.56%	0.93%	1.49%
Sicilia	0.35%	1.25%	1.61%	0.33%	0.76%	1.09%	0.31%	1.14%	1.45%
Toscana	1.83%	1.02%	2.85%	0.62%	0.59%	1.21%	1.64%	0.94%	2.59%
Trentino Alto Adige	0.12%	1.52%	1.64%	0.04%	1.04%	1.08%	0.12%	1.34%	1.46%
Umbria	0.84%	0.25%	1.09%	0.17%	0.07%	0.24%	0.76%	0.23%	0.99%
Valle d'Aosta	0.00%	0.14%	0.14%	0.00%	0.08%	0.08%	0.00%	0.14%	0.14%
Veneto	0.60%	1.69%	2.29%	0.12%	0.32%	0.44%	0.58%	1.67%	2.25%
Italy	15.96%	17.64%	33.6%	8.06%	9.58%	17.64%	14.33%	15.39%	29.72%

sentative was the deciduous oak forest (3112), which covers 0.55 million hectares, equal to I_{PW} of 7.14%. According to silvicultural system, the classes with the highest diffusion in coppice forests were the beech forests (3115) with an I_{PC} 4.49% (0.35 million hectares) while in non-coppice forests the deciduous oak forests (3112) with an I_{PC} 2.90 % (0.23 million hectares).

Felling potential

At national level, the potential annual increment was about 35 million cubic meters, of which 42.42% derived from coppice and 57.58% from non-coppice forests. The highest annual increment of coppice forests was found in the Toscana region with 19.8%, equal to about 2.8 million cubic meters, while the Trentino Alto Adige represented the largest annual increment in non-coppice forests with 18.58% amounting to about 3.7 million cubic meters. The CLC classes with the highest annual increment were respectively the 3114 (chestnut forests) for coppice forests with 12.03% (about 4.1 million cubic meters) and the 3123 (silver fir and spruce forests) for non-coppice forests with 14.12% (about 4.9 million cubic meters) (Tab. 4).

The potential annual increment within the network of PAs amounted to about 11.7 million cubic meters, equivalent to 33.79% of the total annual increment divided in 13.53% for coppice forests (about 4.7 million cubic meters) and 20.26% for non-coppice forests (about 7 million cubic meters). The CLC classes with the highest annual increment were beech forests (3115) with 5.16% of the total annual increment (about 1.1 million cubic meters) in coppice forests, while silver fir and spruce forests (3123) with 4.17% (about 0.9 million cubic meters) in non-coppice forest (Tab. 4).

The potential annual increment distinguished by the form of protection and CLC classes was reported in Table 4.

Felling restrictions and available wood

The availability of wood in the network of PAs depends on the type of protected area and on the silvicultural system. The felling restriction affected the Zone A of the national parks, involving 1.19% (0.09 million hectares) of the forest extension and 1.09% of the national annual increment.

The annual increment unavailable within the network of PAs represents 11.88% (about 4.1 million cubic meters) of the total annual increment. However, the annual increment available through active management represents 21.9% of the total annual increment, amounting to about 7.6 million cubic meters (Tab. 5).

At national level, beech forests (3115) class was the one with the higher available annual increment (5.6% - 1,940,000 cubic meters) (Fig. 2). This is easily explainable, because the beech forest in Italy is one of the most widespread forest type representing 12% of the national forests (INFC 2005 data).

General overview of the results split up by silvicultural system showed that the annual increment was respectively 3.8 m^3ha^{-1} in coppice forests and 5.1 m^3ha^{-1} in non-coppice forests. Within the network of PAs the available annual average increment was 0.98 m^3ha^{-1} (0.81 m^3ha^{-1} in coppice and 1.14 m^3ha^{-1} in non-coppice forests). Considering the two forms of protection, the available annual average increment was 0.51 m^3ha^{-1} in EUAP areas (0.42 m^3ha^{-1} in coppice, 0.60 m^3ha^{-1} in non-coppice) while 0.9 m^3ha^{-1}in Natura 2000 sites (0.74 m^3ha^{-1} for coppice and 1.06 m^3ha^{-1} for non-coppice).

Table 4 - Share of the potential annual increment within the EUAP areas, Natura 2000 and CLC 2006. Legend: I_{CC}: Increment Current Coppice; I_{CNC}: Increment Current Non-Coppice; I_{CW}: Increment Current Wood.

	EUAP + NATURA 2000			EAUP			NATURA 2000			CLC 2006		
CLC 2006 Classess	Ic_C	Ic_{NC}	Ic_W	Ic_C	Ic_{NC}	Ic_W	Ic_C	Ic_{NC}	Ic_W	Ic_C	Ic_{NC}	Ic_W
3111	1.02%	0.90%	1.92%	0.54%	0.39%	0.93%	0.87%	0.70%	1.57%	2.22%	2.52%	4.74%
3112	2.46%	2.20%	4.66%	1.15%	1.37%	2.52%	2.23%	1.77%	4.00%	9.15%	7.49%	16.63%
3113	1.61%	1.02%	2.63%	0.70%	0.60%	1.30%	1.46%	0.94%	2.40%	6.00%	2.56%	8.56%
3114	2.60%	0.73%	3.33%	1.18%	0.49%	1.67%	2.29%	0.51%	2.80%	12.03%	2.42%	14.45%
3115	5.16%	3.36%	8.52%	3.09%	2.29%	5.38%	4.86%	3.12%	7.98%	9.86%	5.29%	15.15%
3116	0.03%	0.27%	0.31%	0.01%	0.12%	0.12%	0.03%	0.24%	0.27%	0.14%	1.16%	1.30%
3117	0.07%	0.16%	0.23%	0.05%	0.07%	0.11%	0.05%	0.15%	0.20%	0.37%	1.13%	1.50%
3121	0.00%	1.35%	1.35%	0.00%	0.68%	0.68%	0.00%	1.26%	1.26%	0.00%	2.99%	2.99%
3122	0.00%	1.06%	1.06%	0.00%	0.75%	0.75%	0.00%	0.78%	0.78%	0.00%	2.61%	2.61%
3123	0.00%	4.17%	4.17%	0.00%	1.65%	1.65%	0.00%	3.91%	3.91%	0.00%	14.12%	14.12%
3124	0.00%	1.08%	1.08%	0.00%	0.53%	0.53%	0.00%	1.05%	1.05%	0.00%	2.81%	2.81%
3125	0.00%	0.03%	0.03%	0.00%	0.01%	0.01%	0.00%	0.03%	0.03%	0.00%	0.19%	0.19%
31311	0.07%	0.20%	0.27%	0.05%	0.14%	0.18%	0.05%	0.14%	0.20%	0.14%	0.34%	0.48%
31312	0.10%	0.32%	0.42%	0.05%	0.19%	0.23%	0.08%	0.26%	0.34%	0.42%	1.08%	1.50%
31313	0.09%	0.17%	0.26%	0.03%	0.07%	0.11%	0.08%	0.15%	0.23%	0.73%	0.85%	1.57%
31314	0.08%	0.08%	0.16%	0.03%	0.04%	0.07%	0.06%	0.06%	0.12%	0.54%	0.49%	1.03%
31315	0.22%	0.64%	0.87%	0.13%	0.38%	0.50%	0.17%	0.59%	0.76%	0.71%	1.62%	2.32%
31316	0.00%	0.01%	0.01%	0.00%	0.00%	0.00%	0.00%	0.01%	0.01%	0.00%	0.02%	0.03%
31317	0.00%	0.00%	0.00%	0.00%	0.00%	0.00%	0.00%	0.00%	0.00%	0.04%	0.05%	0.09%
31321	0.00%	0.48%	0.48%	0.00%	0.25%	0.25%	0.00%	0.43%	0.43%	0.00%	1.65%	1.66%
31322	0.01%	0.96%	0.97%	0.01%	0.68%	0.69%	0.00%	0.69%	0.69%	0.02%	2.97%	2.99%
31323	0.01%	0.83%	0.84%	0.00%	0.30%	0.30%	0.01%	0.81%	0.81%	0.06%	2.55%	2.61%
31324	0.00%	0.21%	0.21%	0.00%	0.06%	0.06%	0.00%	0.21%	0.21%	0.00%	0.60%	0.60%
31325	0.00%	0.02%	0.02%	0.00%	0.01%	0.01%	0.00%	0.02%	0.02%	0.00%	0.07%	0.07%
Italy	13.53%	20.26%	33.78%	7.01%	11.04%	18.05%	12.26%	17.80%	30.06%	42.42%	57.58%	100.00%

Table 5 - Annual increments available and not within the EUAP areas and Natura 2000.

	Wood Unavailable [m3]			Wood Unavailable [%]		
	Coppice	Not Coppice	Total	Coppice	Not Coppice	Total
EAUP (PNZ - Zone A)	0	380870	380870	0.0%	1.1%	1.1%
EUAP (Zone B-C-D PNZ, PNR, RNS, RNR, AANP)	806106	1144084	1950190	2.3%	3.3%	5.6%
EUAP (Total)	806106	1524954	2331060	2.3%	4.4%	6.7%
Natura 2000	1409644	2046279	3455923	4.0%	5.9%	9.9%
EUAP + Natura 2000	1555447	2584103	4139550	4.5%	7.4%	11.9%
	Wood Available [m3]			**Wood Available [%]**		
	Coppice	**Not Coppice**	**Total**	**Coppice**	**Not Coppice**	**Total**
EAUP (PNZ - Zone A)	0	0	0	0.0%	0.0%	0.0%
EUAP (Zone B-C-D PNZ, PNR, RNS, RNR, AANP)	1636640	2322836	3959476	4.7%	6.7%	11.4%
EUAP (Total)	1636640	2322836	3959476	4.7%	6.7%	11.4%
Natura 2000	2862004	4154566	7016570	8.2%	11.9%	20.1%
EUAP + Natura 2000	3158029	4473231	7631260	9.1%	12.8%	21.9%

Discussion

At national level, 88.4% of the forest area was available for forest logging (Gasparini and Tabacchi 2011) and the share of net available annual increment was 39.5% in 1990, 33.1% in 2000 and 26.3% in 2005 (EUROSTAT 2012). The average increment potential available for felling at national level was 4.4 m^3ha^{-1}.

According to EUROSTAT data (http://epp.eurostat.ec.europa.eu/), European countries have a large margin to increase logged wood volumes. Italy has a level of wood mobilization among the lowest ones in Europe. Considering the very low level of harvesting, a large majority of harvested volumes comes from production forests outside protected areas. Therefore, there is a potential to enhance wood supply in a sustainable way, for energy and raw materials for industry, encouraging wood mobilization (Muys et al. 2013). In contrast, due to the global economic crisis there was a rising interest in fuelwood, resulting in an increase of fuelwood imports of about 26% in Italy annually. Italy is the first importer country of firewood in Europe.

Increasing the consumption of the wood sustainably produced and boost the bio-energy in countries all over Europe could lead economic, environmental and geopolitical gains (FOREST EUROPE 2010). There are many credible reasons to move away from fossil fuels, including the reduction of the dependence on foreign petroleum, maintaining forest management infrastructures, and encouraging conservation of forest work (Gunn et al. 2012). Moreover, the biomass has several other well-recognized benefits

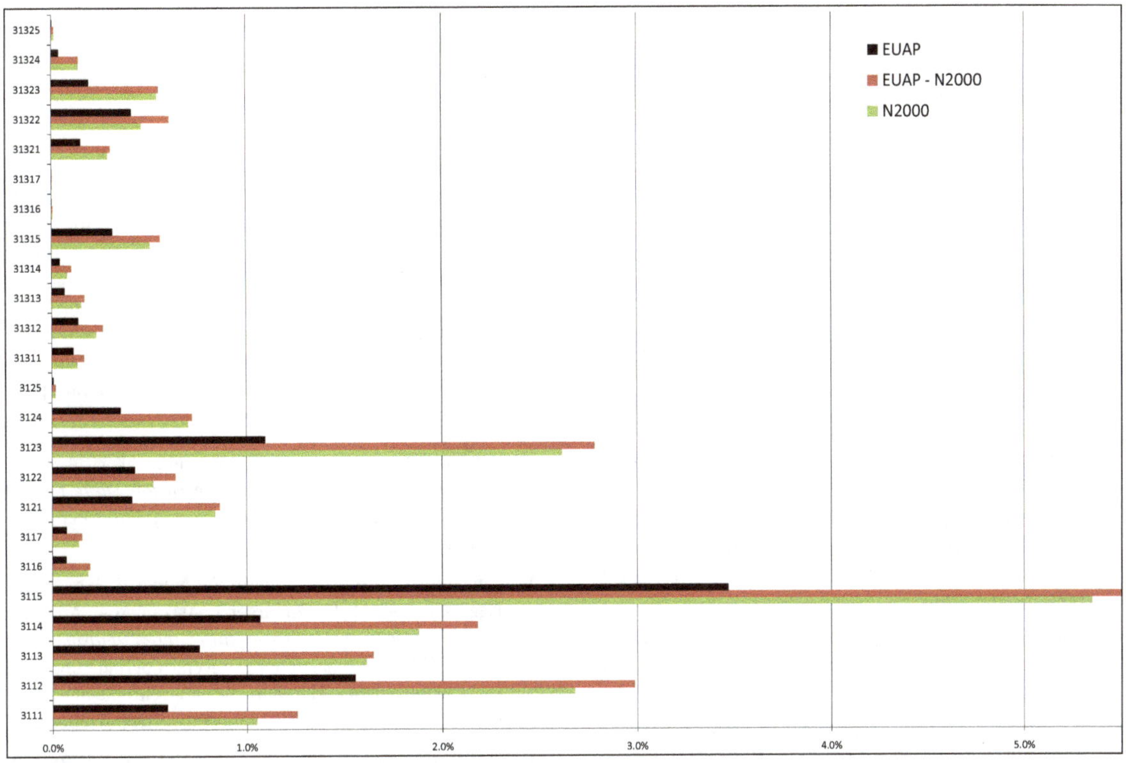

Figure 2 - Annual increments available for each CLC class within the protected areas through an active management of the territory.

such as improved security of supply, contributing to improve air quality and creation of new jobs and businesses - many of them in rural areas (Ragwitz et al. 2006).

Forests represent a combination of important ecosystems that provide habitat for numerous species, regulate water cycles, clean air and provide timber for economic use (Maes et al. 2011).

The forest growing stock, the increment and fellings could be considered directly ecosystem services indicators for sustainability of ecosystem services over time, to ensure that the long-term benefit flow of services is represented (Maes et al. 2011). Mobilizing more wood is an effort of the whole forest sector. There is a potential to enhance wood supply in a sustainable way, for energy and raw materials for industry. In line with the policy commitments all over Europe and in particular in light of the EU renewable energy targets, wood mobilization should be further encouraged in countries all over Europe (FOREST EUROPE 2010).

Conclusions

This paper assessed the amount of wood that is technically available in the protected areas by active management of the forest resources. To assess the felling potential in the network of PAs, the unbounded available wood was evaluated and divided into the two silvicultural systems (coppice, non-coppice). This upper limit does not take into account management restrictions due to biological and landscape diversity protection that may apply in PFAs and does not take into account other environmental, social and economic factors that may limit the felling potential. The research provides the dataset and suggests the rationale for wood mobilization in areas usually considered of close to strict environmental protection. Wood harvesting can be undertaken without compromising other forest functions if it is made in a sustainable way, in other words, without compacting soil, causing soil erosion, or disturbing the reproductive cycle of plants or animals.

References

Cash D.W., Adger W., Berkes F., Garden P., Lebel L., Olsson P., Pritchard L., Young O. 2006 - *Scale and cross-scale dynamics: governance and information in a multilevel world.* Ecology and Society 11: 8.

Ciancio O., Corona P., Marchetti M., Nocentini S. 2002 - *Linee guida per la gestione ecosostenibile delle risorse forestali e pastorali nei Pachi Nazionali.* Ministero dell'Ambiente, Servizio conservazione della Natura, Accademia Italiana di Scienze Forestali, Firenze, 302 p. http://www.aisf.it/monografie/linee%20guida%20parchi/Linee%20Guida%20Parchi.pdf.

Duncker P., Barreiro S., Hengeveld G.M., Lind T., Spiecker H., Mason B., Ambrozy S., Spiecker H. 2012 - *Classification of forest management approaches: a new methodological framework and its applicability to European forestry.* Ecology and Society 17 (4): 51.

EEA 2007 - *Nationally designated areas (national-CDDA).* Version 9. EEA, Copenhagen [Online] http://www.eea.europa.eu/data-and-maps/data/nationally-designated-areas-national-cdda-5.1 [11 Oct 2011]

Egoh B., Reyers B., Rouget M., Bode M., Richardson D.M. 2009 - *Spatial congruence between biodiversity and ecosystem services in South Africa.* Biological Conservation 142: 553-562.

EUROSTAT 2012 - *Forestry. Forest increment and fellings (ts-dnr520).* [Online] http://epp.eurostat.ec.europa.eu/portal/page/portal/forestry/data/main_tables [2012]

FOREST EUROPE/UNECE/FAO 2010 - *Information Document on Data Collection and Compiling the Statistics on Protected and Protective Forest and Other Wooded Land for Pan-European Reporting.* [Online] http://www.unece.org/fileadmin/DAM/timber/publications/soef2011-protected-forest.pdf [2012]

FOREST EUROPE 2010 - *Good practice guidance on the sustainable mobilisation of wood in Europe.* Publications Office of the European Union, ISBN: 978-92-79-13933-8, DOI: 10.2762/17910.

FOREST EUROPE, UNECE and FAO 2011 - *State of Europe's Forests 2011. Status and Trends in Sustainable Forest Management in Europe.*

Frank G., Parviainen J., Vandekerkhove K., Latham J., Schuck A., Little D. (eds.). 2007 - *COST Action E27 Protected forest areas in Europe-analysis and harmonization (PROFOR): results, conclusions and recommendations. Federal research and training centre for forests, natural hazards and landscape (BFW).* Vienna. 211 p.

Gasparini P., Tabacchi G. 2011 - *L'inventario Nazionale delle foreste e dei serbatoi forestali di carbonio INFC 2005. Secondo inventario forestale nazionale italiano. Metodi e risultati.* Ministero delle Politiche Agricole, Alimentari e Forestali, Corpo Forestale dello Stato; Consiglio per la Ricerca e la sperimentazione in Agricoltura, Unità di ricerca per il monitoraggio e la pianificazione forestale. Edagricole, Milano, 653 p.

Gunn S.J., Ganz D.J., Keeton W. 2012 - *Biogenic vs. geologic carbon emissions and forest biomass energy production.* Global Change Biology 4: 239-242.

INFC 2003 - *Guida alla classificazione delle vegetazione forestale.* Istituto Sperimentale per l'Assestamento Forestale e per l'Alpicoltura - ISAFA, Trento.

INFC 2005 - *Inventario Nazionale delle Foreste e dei Serbatoi Forestali di Carbonio.* Ministero delle Politiche Agricole Alimentari e Forestali, Ispettorato Generale - Corpo Forestale dello Stato. CRA - Istituto Sperimentale per l'Assestamento Forestale e per l'Alpicoltura, Trento. http://www.sian.it/inventarioforestale/jsp/home.jsp.

ISAFA 1988 - *Inventario Forestale Nazionale Italiano 1985 (IFNI 85).* Istituto sperimentale per l'Assestamento forestale e per l'Alpicoltura, Trento.

ISPRA 2010 - *La realizzazione in Italia del progetto europeo Corine Land Cover 2006.* Rapporto 131/2010 ISPRA, Istituto Superiore per la Protezione e la Ricerca Ambientale. Roma, 50 p. ISBN 978-88-448-0477-0. http://www.isprambiente.gov.it/site/_contentfiles/00008300/8329_rapporto_131_2010.pdf.

Köhl M., Hildebrandt R., Olschofksy K., Köhler R., Rötzer T., Mette T., Pretzsch H., Köthke M., Dieter M., Abiy M., Makeschin F., Kenter B. 2010 - *Combating the effects of climatic change on forests by mitigation strategies.* Carbon Balance and Management 5: 8.

Lamers P., Thiffault E., Paré D., Junginger M. 2013 - *Feedstock specific environmental risk levels related to biomass extraction for energy from boreal and temperate forests.* Biomass and Bioenergy 55: 212-226.

Maes J., Braat L., Jax K., Hutchins M., Furman E., Termansen M., Luque S., Paracchini M.L., Chauvin C., Williams R., Volk M., Lautenbach S., Kopperoinen L., Schelhaas M.J., Weinert J., Goossen M., Dumont E., Strauch M., Görg C., Dormann C., Katwinkel M., Zulian G., Varjopuro R., Ratamäki O., Hauck J., Forsius M., Hengeveld G., Perez-Soba M., Bouraoui F., Scholz M., Schulz-Zunkel C., Lepistö A., Polishchuk Y., Bidoglio G. 2011 - *A spatial assessment of ecosystem services in Europe: methods, case studies and policy analysis - phase 1.* PEER Report No 3. Ispra: Partnership for European Environmental Research. Italy, 148 p.

Maesano M., Lasserre B., Tonti D., Marchetti M. 2014 - *First Mapping of the main High Conservation Value Forests (HCVFs) at national scale: the case of Italy.* Plant Biosystem (in press).

Mantau U., Saal U., Prins K., Steierer F., Lindner M., Verkerk H., Eggers J., Leek N., Oldenburger J., Asikainen A., Anttila P. 2010 - *EUwood - real potential for changes in growth and use of EU forests.* Final Report. University of Hamburg. Centre of Wood Science, Hamburg. http://ec.europa.eu/energy/renewables/studies/doc/bioenergy/euwood final report.pdf.

MATTM 2010 - *Elenco Ufficiale delle Aree naturali Protette (EUAP) 6° Aggiornamento 2010.* Dipartimento per l'assetto dei valori ambientali del territorio, Direzione per la Conservazione della Natura, Ministero dell'Ambiente e della tutela del Territorio, Roma.

MCPFE 2002 - *Assessment Guidelines for Protected and Protective Forest and Other Wooded Land in Europe.* Annex 2 to Vienna Resolution 4. Fourth Ministerial Conference on the Protection of Forest in Europe, 28-30 April 2003, Vienna, Austria.

MCPFE 2003 - *State of Europe's forests 2003.* The MCPFE report on sustainable forest management in Europe. Jointly prepared by the MCPFE Liaison Unit Vienna and UNECE/FAO, Vienna.

MCPFE 2007 - *State of Europe's Forests 2007.* http://www.foresteurope.org/filestore/foresteurope/Publications/pdf/state_of_europes_forests_2007.pdf

Muys B., Hetemäki L., Palahi M. 2013 - *Sustainable wood mobilization for EU renewable energy targets.* Biofuels Bioproducts and Biorefining 7: 359–360.

Nelson E., Polasky S., Lewis D.J., Plantinga A.J., Lonsdorf E., White D., Bael D., Lawler J.J. 2008 - *Efficiency of incentives to jointly increase carbon sequestration and species conservation on a landscape.* In: Proceedings of the National Academy of Sciences 105: 9471-9476.

Norton-Griffiths M., Southey C. 1995 - *The opportunity costs of biodiversity conservation in Kenya.* Ecological Economics 12: 125-139.

Parviainen J., Frank G. 2003 - *Protected forests in Europe approaches-harmonising the definitions for international comparison and forest policy making.* Journal of Environmental Management 67: 27-36.

PEER 2011 - *A spatial assessment of ecosystem services in Europe: Methods, case studies and policy analysis - phase I.* European Commission Joint Research Centre Institute for Environment and Sustainability (JRC-IES).

Pettorelli N., Chauvenet A.L.M., Duffy J.P., Cornforth W.A., Meillere A., Baillie J.E.M 2012 - *Tracking the effect of climate change on ecosystem functioning using protected areas: Africa as a case study.* Ecological Indicators 20: 269-276.

Ragwitz M., Toro F., Resch G., Faber T., Haas R., Hoogwijk M., Voogt M., Rathmann M. 2006 - *Economic analysis of reaching a 20 % share of renewable energy sources in 2020.* European Commission, DG Environment. ENV.C.2/SER/2005/0080r. http://ec.europa.eu/environment/enveco/others/pdf/res2020_final_report.pdf

Sohngen B., Mendelsohn R., Sedjo R. 1999 - *Forest Management, Conservation, and Global Timber Markets.* American Journal of Agricultural Economics 81 (1): 1-13.

Turner W.R., Brandon K., Brooks T.M., Costanza R., da Fonseca G.A.B., Portela R. 2007 - *Global Conservation of Biodiversity and Ecosystem Services.* BioScience, 57: 868-873.

Verkerk P.J., Zanchi G., Lindner M. 2008 - *Impacts of Biological and Landscape Diversity Protection on the Wood Supply in Europe.* EFI Technical Report 27, European Forest Institute, 29 p.

Willemen L., Drakou E.G., Dunbar M.B., Mayaux P., Egoh B.N. 2013 - *Safeguarding ecosystem services and livelihoods: Understanding the impact of conservation strategies on benefit flows to society.* Ecosystem Services 4: 95-103.

2

Early and long-term impacts of browsing by roe deer in oak coppiced woods along a gradient of population density

Francesco Chianucci[1*], Luca Mattioli[2], Emilio Amorini[1], Tessa Giannini[1], Andrea Marcon[3], Roberta Chirichella[3], Marco Apollonio[3], Andrea Cutini[1]

Abstract - Over the last few decades, wild ungulate populations have exhibited relevant geographic and demographic expansion in most European countries; roe deer is amongst the most widespread ungulate species. The increasing roe deer densities have led to strong impact on forest regeneration; the problem has been recently recognized in coppice woods, a silvicultural system which is widespread in Italy, where it amounts to about 56% of the total national forested area.

In this study we investigated the effect of roe deer browsing on the vegetative regeneration of Turkey oak few years after coppicing, along a gradient of roe deer density. A browsing index revealed that browsing impact was high at any given roe deer density but increased at higher density, with the browsing rate ranging from 65% to 79%. We also analyzed the long-term impact of browsing six and eleven years after coppicing under a medium roe deer density. Results indicated the early impact are not ephemeral but produced prolonged impacts through time, with an average reduction in volume of -57% and -41% six and eleven years after coppicing, respectively. Based on these results we proposed integrating browsing monitoring with roe deer density estimation to allow identifying ungulate densities which are compatible with silvicultural and forest management objectives. The proposed browsing index can be regarded as an effective management tool, on account of its simplicity and cost-effectiveness, being therefore highly suitable for routine, large scale monitoring of browsing impact.

Keywords - *Capreolus capreolus*, Turkey oak, forest damage, stool, coppice forest

Introduction

The coexistence of ungulates and forest ecosystems has gaining growing attention, mainly because of the steady increase in wild ungulate populations observed in the last few decades. This is a particularly relevant issue for roe deer (*Capreolus capreolus* L.), i.e., the most widespread European deer species, with an estimated 10 million individuals occurring in the continent (Apollonio et al. 2010, Linnell et al. 1998). Roe deer population densities have been growing notably during the last few decades (Cederlund et al. 1998, Gill 1990) due to the abandonment of rural areas, changes in human land use, restocking and lack of predators (Apollonio et al. 2010). In addition, changes in silvicultural systems have entailed a general decline in wood exploitation, improving productivity, biomass accumulation and structural complexity of forest ecosystems (Cutini et al. 2013). The reduced pressure on wood also resulted in a lower human disturbance on ungulates in general and roe deer in particular.

The increasing roe deer densities have led to conflicting interests between game and forest management (Apollonio et al. 2010, Cutini et al. 2011, Gill 1992). A land-owner aiming at silviculture and wood production may see his goals frustrated by high densities of ungulates permitted on neighbouring grounds (Kramer et al. 2006). The problems of deer browsing have long been recognized in conifer species (Gill 1992, Motta 1996, Mysterud and Østbye 1999). Recently, the problem of deer impact has also been observed in coppice woods (Cutini et al. 2011), a silvicultural system which is widespread over Mediterranean countries. In Italy, coppice is the most frequently adopted silvicultural system in private forests, and it amounts to about 56% of the total forested areas in Italy (National Forest Inventory, www.infc.it). Deciduous Turkey oak (*Quercus cerris* L.) occupies the intermediate vegetation belt between sclerophyllous and mountain broadleaved forest over one million hectares (Di Matteo et al. 2014). This species represent an economically relevant species with regard to coppice management and also a potentially key food resource for roe deer (Cutini et al. 2011). Despite such a large diffusion

[1] Consiglio per la ricerca in agricoltura e l'analisi dell'economia agraria, Forestry Research Centre (CRA-SEL), Arezzo, Italy
[2] Province of Arezzo, Arezzo, Italy
[3] Department of Sciences for Nature and Environmental Resources, University of Sassari, Sassari, Italy
[*] corresponding author: francesco.chianucci@entecra.it

and importance, few studies have focused on roe deer browsing on coppice wood, probably because the problem of deer impact on coppice regeneration in Italy is relatively recent (Cutini et al. 2011). After coppicing, stems are cut down close to ground level, where the growing shoots are exposed to deer browsing, expecially during the early years following cutting. Because the browsing incidence can heavely be influenced by ungulate density (Chevrier et al. 2011) it is important to evaluate density-related effects of browsing. In addition, because browsing can delay the growth of vegetative regeneration (Motta 2003), it is important to evaluate the long-term browsing influence on coppice growth. Indeed, as browsing impact are expected to be limited to the early phase of coppice rotation, since shoots are more exposed to browsing, an important question is whether the effects of browsing are likely to be permanent, or simply ephemeral due to changes associated with stand maturity (Gill and Beardall, 2001).

The objective of this study is twofold:

i) to assess the effects of roe deer browsing on vegetative regeneration of Turkey oak (stands after coppicing, along a gradient of ungulate density;

ii) to evaluate the long-term effects of browsing through time (i.e., six and eleven years after coppicing).

Both information are nowadays strongly needed for sustainable planning of deer density and for assessing economic impact of deer on coppice (e.g., for refunding purposes).

Material and Methods

The study was carried out in the Province of Arezzo (Tuscany, Italy). Suitable study areas were individuated by integrating a series of information on roe deer densities and forest data available in the Province. Roe deer densities were obtained by means of drive censuses conducted by the Fish and Wildlife Service of the Province of Arezzo (Cutini et al. 2013, Mattioli et al. 2004, Davis et al. 2012). Data were collected from 2011 to 2013 on a network of 187 permanent sample plots (43.9 ± 26.0 S.D. ha of mean surface and 8116 ha of total area) uniformly distributed along the Province. Roe deer density at local scale was calculated by spatial interpolation using the inverse distance weighting method (Li and Heap 2008) in ARCMAP 9.2 ESRI package. The setting used was: $p = 2$ and the minimum number (n) of sampled points used for estimation was 5, within a radius of 5 Km, and the pixel size of the output was 1 ha. Roe deer density was calculated for each study area as a mean of the values of each hectare included in the area, and for each plot inside the study areas was calculated as a mean of the values in a buffer of 1 km around the centre of the plot. Data of geographical distribution and size of coppiced areas were obtained from "ARTEA" dataset of Regional Administration (www.artea.toscana.it).

Study areas were only deemed suitable if other deer species influence was absent. Also, a similar wooded and coppice coverage were considered to ensure comparability between the study areas; this was a major determinant in the choice of study areas and consequently after experimental design. Three study areas were selected representing a gradient of roe deer population densities which was representative of hilly and mountain areas of Central Italy (hereafter scenarios; Tab. 1). The observed mean roe deer densities were comparable to that indicated by the Italian National Institute for Environmental Protection and Research (ISPRA) as low (i.e., between 10 and 15 individuals per km^2), medium (i.e., between 20 and 25 individuals per km^2) and high (i.e, more than 25 individuals per km^2) roe deer densities in Apennines and Mediterranean environments.

Within each study area, a number of oak coppice stands aged 0-2 years with minimum size of 2500 m^2 were individuated. Within these plots, we randomly sampled 744 stools, of which half stools were located near the centre of each stand and half stools were located along the border of each stand. The observation period ran between October 15 and November 15, i.e. close to the end of the growing season. For each stool, we recorded the number of sprouts, top height of sprouts (i.e., the height of the tallest sprout of each stool), and number of recently browsed sprouts. The cover area of each stool was estimated by the geometric mean of the maximum and minimum diameter of projected area (assuming an elliptical shape). The browsing index was calculated as the ratio of browsed sprouts over the total number of sprouts for each surveyed stool. Browsing estimates were then averaged for each plot. Differences in oak browsing, top height, and cover area in the three different scenarios were compared by means of ANOVA. If ANOVA indicated that a significant difference existed between them in the variable of interest then Tukey's pair-wise comparison test was used to compare results from the three scenarios. We used R version 3.0.2 (R Development Core Team 2013).

To evaluate the impact of browsing on

Table 1 - Study areas

Study area	Average roe deer (n km-2)	Scenario
Valdambra	38.3 (2.6)	High
Alpe di Poti	13.9 (0.9)	Low
Alpe di Catenaia	22.6 (0.5)	Medium

Table 2 - Mean value of browsing ratio, top height and cover area in the different roe deer density scenarios.

Roe deer density	Browsing	Top height (cm)	cover area (m^2)
Low	0.65 (0.02)	110.0 (3.3)	1.28 (0.1)
Medium	0.77 (0.01)	88.1 (3.2)	1.18 (0.1)
High	0.79 (0.01)	90.2 (3.0)	0.99 (0.1)

Table 3 - Main mensurational parameters in protected (P) and non-protected (NP) Turkey oak coppiced areas.

Year	area	N (shoots)	dm (cm)	hm (m)	G (m^2ha^{-1})	V V (m^3ha^{-1})
2008	P	6768	3.8 (0.1)	3.5 (0.2)	8.83 (0.10)	27.61 (2.23)
	NP	3330	3.8 (0.1)	3.2 (0.3)	3.75 (0.07)	11.95 (1.27)
2013	P	6907	5.6 (0.1)	6.5 (0.2)	18.11 (0.22)	57.07 (5.12)
	NP	4826	6.0 (0.1)	6.2 (0.2)	14.23 (0.13)	33.76 (3.51)

the long term, additional data were collected from permanent monitoring plots previously established in the Alpe di Catenaia (Cutini et al. 2011), i.e., the medium density scenario. The permanent plots were located within three 1-ha stands dominated by Turkey oak, which were coppiced in 2002. Two sampling plots sized about 200 m^2 were established within each stand, one of which was fenced (protected, P), while the other was left accessible to browsing (non-protected, NP). Within each plot, we measured diameter at breast height and total height of sprouts. Volume was then estimated by applying a formula developed in a previous study for Turkey oak coppices in Tuscany (Amorini et al. 1998). We compared difference in growing stock (basal area and volume) between P and NP by two inventories made in 2008 (i.e., 6 years after coppicing) and 2013 (11 years after coppicing).

Results

The incidence of ungulate browsing was high in all the three different scenarios. The percentage of browsed sprouts ranged from 65% to 79% (Table 2). Damage was not uniformly distributed over all height classes of sprouts, but rather concentrated in those at the browsing height, i.e. between 20 and 130 cm (Figure 1). Moreover, browsing was negatively correlated with top height ($r=-0.46$, $p<0.001$) and cover area ($r=-0.35$, $p<0.001$).

ANOVA and Tukey's test indicated that browsing was significantly higher and height and cover area were significantly lower in the higher density scenario, compared with the others (Tukey's HSD test, $p < 0.05$). By contrast, medium and higher density scenarios did not exhibit significant differences in these attributes (Table 2).

Analysis of browsing impact 6 years after coppicing in Alpe di Catenaia (2008) revealed that growing stock in not-fenced areas was lower than that in fenced areas, with an average reduction in basal area and volume of -58% and -57% respectively, a consequence of the steady reduction in the number of shoots (Table 3). These differences were lower 11 years after coppicing (2013), even though browsing still reduced basal area and volume by- 21 % and -41 %, respectively (Table 3).

Discussion

The study revealed that browsing by roe deer has a significant impact on oak during the early years after coppicing. The observed browsing pressure was noticeably high even at lowest deer density, in which about two third of the sprouts were browsed. The browsing impact showed a density-dependency because significantly higher impacts were observed at medium and high density, compared with the low density scenario. Results are in agreement with previous studies, which reported a strict relationship between browsing and deer density (Chevrier et al. 2011). The negative relationship trend observed between browsing incidence and top height of sprouts further confirmed that the early years after coppicing are particularly critical, because the emerging shoots grow beneath the browsing height and are therefore exposed to sustained and prolonged impacts (Rooke et al. 2004).

On the other hand, we demonstrated that the effects of early browsing were not ephemeral but produced prolonged impacts through time; although the trend decreased between 2008 and 2013, the average reduction in volume observed eleven years after coppicing is noticeably high to heavily retard shoot growth, with relevant ecological as well as economic consequences. Based on the observed

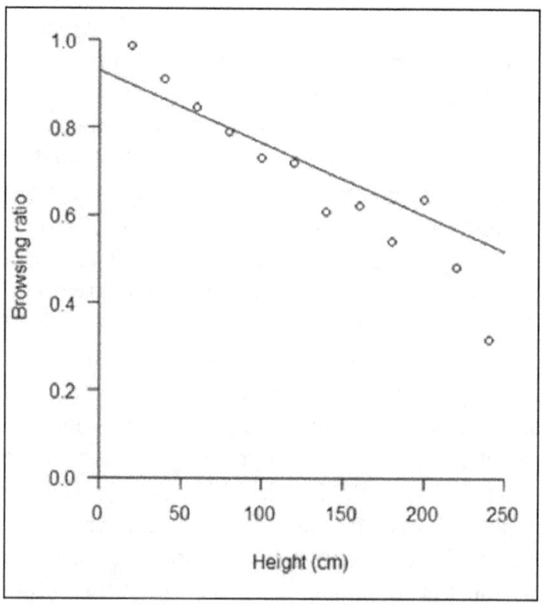

Figure 1 – Relationship between browsing ratio and top height of stools.

impacts and trends we could expect a reduction of -25% in volume at the end of the coppice rotation period, which may also imply a prolongation in the minimum rotation period (which is 18 years in Tuscany according to regional laws) to achieve a profitable harvesting, under the observed medium roe deer density.

Regardless economic considerations, both the early and long-term results indicated the browsing impacts are sustained in a range of roe deer density regarded as "normal" (ISPRA 2013) and regularly occurring in Apennines. In addition, roe deer in Italy is frequently sympatric with other deer species, e.g., fallow deer and red deer; hence, taking into account the current deer densities and their browsing pressure, we could expect that browsing intensity can seriously compromise forest dynamics in coppice woods. In these situations, the sustainable planning of ungulate densities should consider all the deer species on the whole. For example, Mattioli (2010) proposed a general ungulate density threshold based on roe-deer-equivalent conversion factors. The author individuated a maximum carrying capacity of 18 roe-deer equivalent individuals in Apennines, with red deer and fallow deer having a roe-deer-equivalent factor of 4.5 and 2.3, respectively (i.e., 1 red deer = 4.5 roe deer; 1 fallow deer = 2.3 roe deer. However, as Reimoser and Putman (2011) correctly noted, if a relationship between deer density and browsing impacts exists, the relationship is complex and not linear, making the identification of a critical ungulate density complex. Indeed, at any given deer density, sustained impact levels are affected by a wide range of other factors such as site conditions, landscape mosaic, availability and quality of alternative food resources, ungulate sympatry etc. (Ward et al. 2008, Gill and Morgan 2010, Reimoser and Putman 2011). Therefore, it is important to move beyond single-factor approach (e.g., ungulate density) to embrace the complexity of fauna-forest interactions (Weisberg and Bugmann 2003). An integrated forest-fauna management is frequently advocated by many authors as an effective way to consider the interactions among ungulates and vegetation (Bianchi et al. 2014, Cutini et al. 2013, Weisberg and Bugmann 2003). To accomplish this purpose, there is a need for rapid and objective indicators to monitor the impact of roe deer on vegetation (Reimoser et al. 1999). The browsing index proposed in this study can be regarded as an useful management tool on account of its fast, simple and cost-effective procedures, it being therefore highly suitable for large scale monitoring of the effects of deer browsing on vegetative regeneration. Integrating browsing monitoring with deer density estimation can be regarded therefore as an effective option for sustainable forest-fauna management (Cutini et al. 2013), i.e., to determine whether the relationships between ungulates and forest ecosystems are consistent with meeting silvicultural and forest management targets (Keigley and Frisina 2011, Reimoser et al. 1999). We also advocate the need to collect information on deer impact over large areas because ungulate effects on forests cannot be generalized to the spatial and temporal scales that are relevant to management. Integrating information on deer impact and deer density in e.g., regional and national forest inventory alongside with the definition of rigorous and systematic long-term monitoring programs to measure the interaction forest-ungulates can be a decisive step towards the definition of appropriate protection measures in landscape and large scale planning.

Acknowledgments

The study was supported by the Research Project "Valutazione del disturbo arrecato alla vegetazione forestale da popolazioni di capriolo di differente densità" funded by the Province of Arezzo. Field assistance by Umberto Cerofolini is greatly acknowledged.

The authors want also to thank the anonymous reviewer for his suggestions.

References

Amorini E., Brandini P., Fabbio G., Tabacchi G. 1998 – *Modelli di previsione delle masse legnose e delle biomasse per i cedui di cerro della Toscana centro-meridionale.* Annali dell'Istituto Sperimentale per la Selvicoltura 29: 41-56.

Apollonio M., Andersen R., Putman R. 2010 - *European ungulates and their management in the 21th century.* Cambridge University Press, Cambridge.

Berquist J., Orlander G. 1998 - *Browsing damage by roe deer on Norway spruce seedlings planted on clearcuts of different ages. 1. Effect of slash removal, vegetation development, and roe deer density.* Forest Ecology and Management 105: 283-293.

Bianchi L., Bartoli L., Paci M., Pozzi D. 2014 - *Impatto degli ungulati selvatici sui boschi cedui della Val di Bisenzio (Prato).* Forest@ 11: 116-124 [Online] Avalaible: http://www.sisef.it/forest@/contents/?id=efor1259-011 [2014].

Cederlund G., Bergqvist J., Kjellander P., Gill R.M.A., Gaillard J.M., Boisabert B., Ballon P., Duncan P. 1998 - *Managing roe deer and their impact on the environment: maximising the net benefits to society.* In: Andersen R., Duncan P., Linnell J.D.C. -Eds- The European Roe Deer: The Biology of Success, Scandinavian University Press, Oslo: 337-372.

Chevrier T., Saïd S., Widmer O., Hamard J. P., Saint-Andrieux C., Gaillard, J. M. 2012 - *The oak browsing index correlates linearly with roe deer density: a new indicator for deer management?.* European Journal of Wildlife Research 58 (1): 17-22.

Cutini A., Bongi P., Chianucci F., Pagon N., Grignolio S., Amorini E., Apollonio M. 2011 - *Roe deer (Capreolus capreolus L.) browsing effects and use of chestnut and Turkey oak coppiced areas.* Annals of Forest Sciences 68: 667-674.

Cutini A., Chianucci F., Chirichella R., Donaggio E., Mattioli L., Apollonio M. 2013 - *Mast seeding in deciduous forests of the northern Apennines (Italy) and its influence on wild boar population dynamics.* Annals of Forest Sciences 70: 493-502.

Davis M.L., Stephens P.A., Willis S.G., Bassi E., Marcon A., Donaggio E., Capitani C., Apollonio M. 2012 - *Prey selection by an apex predator: the importance of sampling uncertainty.* PLOS One 7 (10) : 1-10.

Di Matteo G., Tunno I., Nardi P., De Angelis P., Bertini G., Fabbio G. 2014 - *C and N concentrations in different compartments of outgrown oak coppice forests under different site conditions in Central Italy.* Annals of Forest Science (Early view): doi: 10.1007/s13595-014-0390-4.

Gill R.M.A. 1990 - *Monitoring the status of European and North American cervids.* The global environment monitoring system -GEMS- UNEP, Nairobi.

Gill R.M.A. 1992- *A review of damage by mammals in north temperate forests: 1 Deer.* Forestry 65: 145-169

Gill R.M.A., Beardall V. 2001 - *The impact of deer on woodlands: the effects of browsing and seed dispersal on vegetation structure and composition.* Forestry 74 (3): 209-218

Gill R.M.A., Morgan G. 2010 - *The effects of varying deer density on natural regeneration in woodlands in lowland Britain.* Forestry 83: 53–63.

ISPRA (2013) *Linee guida per la gestione degli ungulati.* Manuali e Linee Guida 91/2013, 221 pp.

Keigley R.B., Frisina M.R. 2011 - *Process to monitor and manage ungulate browsing pressure.* Natural Resources and Environmental Issues 16: article 29 [Online] Avalaible: http://digitalcommonsusuedu/nrei/vol16/iss1/29 [2014].

Kramer K., Groot Bruindering G.W.T.A., Prins H.H.T. 2006 - *Spatial interactions between ungulate herbivory and forest management.* Forest Ecology and Management 226: 238-247

Kuiters A.T., Slim P.A. 2002- *Regeneration of mixed deciduous forest in a Dutch forest-heatland, following a reduction of ungulate densities.* Biological Conservation 105: 65-74.

Li J., Heap A.D. 2008 - *A Review of Spatial Interpolation Methods for Environmental Scientists.* Geoscience Australia, Record 2008/23, 137 p.

Linnell J.D.C., Duncan P., Andersen R. 1998- *The European roe deer: a portrait of a successful species.* In: Andersen R., Duncan P., Linnell J.D.C (Eds). The European Roe Deer: The Biology of Success, Scandinavian University Press, Oslo: 1-22.

Mattioli L. 2010 - *La fauna come fattore di disturbo del bosco.* In: Mercurio R. "Restauro della foresta mediterranea." Clueb Ed., 368 pp.

Mattioli L., Capitani C., Avanzinelli E., Bertelli I., Gazzola A., Apollonio M. 2004- *Predation by wolves (Canis lupus) on roe deer (Capreolus capreolus) in North eastern Apennine.* Italy Journal of Zoology 264 (3): 249-258.

Motta R. 1996 - *Impact of wild ungulates on forest regeneration and tree composition of mountain forests in the Western Italian Alps.* Forest Ecology and Management 88 (1): 93-98.

Motta R. 2003 - *Ungulate impact on rowan* (Sorbus aucuparia *L) and Norway spruce* (Picea abies *(L) Karst) height structure in mountain forests in the eastern Italian Alps* . Forest Ecology and Management 181 (1): 139-150.

Mysterud A., Østbye E. 1999 - *Cover as a habitat element for temperate ungulate: effects on habitat selection and demography.* Wildlife Society Bulletin 27: 385-394.

Putman R., Langbein J., Green P., Watson P. 2011 - *Identifying threshold densities for wild deer in the UK above which negative impacts may occur.* Mammal Review 41 (3): 175-196.

R Development Core Team 2011 - *R: A language and environment for statistical computing.* R Foundation for Statistical Computing, Vienna (Austria).

Reimoser F., Armstrong H., Suchant R. 1999 - *Measuring forest damage of ungulates: what should be considered.* Forest Ecology and Management 120 (1): 47-58.

Reimoser F., Putman R.J. 2011 - *Impact of large ungulates on agriculture, forestry and conservation habitats in Europe.* In: Putman R.J., Apollonio M., Andersen R. (eds). "Ungulate Management in Europe: Problems and Practices." Cambridge University Press, Cambridge, UK: 144–191.

Rooke T., Danell K., Bergström R., Skarpe C., Hjälten J. 2004 - *Defensive traits of savanna trees – the role of shoot exposure to browsers.* Oikos 107: 161-171

Ward A.I., White P.C.L., Walker N.J., Critchley C.H. 2008 - *Conifer leader browsing by roe deer in English upland forests: Effects of deer density and understorey vegetation.* Forest Ecology and Management 256: 1333–1338.

Weisberg PJ, Bugmann H 2003 - Forest dynamics and ungulate herbivory: from leaf to landscape Forest Ecology and Management 181: 1-12.

3

Integrated forest management to prevent wildfires under Mediterranean environments

Piermaria Corona[1], Davide Ascoli[2], Anna Barbati[3], Giovanni Bovio[2], Giuseppe Colangelo[4], Mario Elia[4], Vittorio Garfì[5], Francesco Iovino[6], Raffaele Lafortezza[4], Vittorio Leone[7], Raffaella Lovreglio[8], Marco Marchetti[5], Enrico Marchi[9], Giuliano Menguzzato[10], Susanna Nocentini[9], Rodolfo Picchio[11], Luigi Portoghesi[3], Nicola Puletti[1], Giovanni Sanesi[4], Francesco Chianucci[1*]

Abstract - This review presents a multidisciplinary framework for integrating the ecological, regulatory, procedural and technical aspects of forest management for fires prevention under Mediterranean environments. The aims are to: i) provide a foreground of wildfire scenario; ii) illustrate the theoretical background of forest fuel management; iii) describe the available fuel management techniques and mechanical operations for fire prevention in forest and wildland-urban interfaces, with exemplification of case-studies; iv) allocate fire prevention activities under the hierarchy of forest planning. The review is conceived as an outline commentary discussion targeted to professionals, technicians and government personnel involved in forestry and environmental management.

Keywords - Wildfire prevention, forest fuel management, wildland-urban interface, prescribed burning

Basic definitions used in the text

(Forest) fuel management: The planned manipulation of flammable materials to adjust fire behavior (e.g., rate of spread and intensity), fire effects, and the effectiveness and costs of fire suppression (the European Glossary for Wildfires and Forest Fires, 2012, EUFOFINET;
http://www.fire.uni-freiburg.de/literature/EUFOFINET-Fire-Glossary.pdf)

(Forest) fire management: The activities concerned with the protection of people, property, and forest areas from wildfire and the use of prescribed burning for the attainment of forest management and other land use objectives, all conducted in a manner that considers environmental, social, and economic criteria (2003 Glossary of Forest Fire Management Terms;
http://bcwildfire.ca/mediaroom/Backgrounders/2003_Fire_Glossary.pdf)

(Forest) fire hazard: i) Any situation, process, material or condition that can cause a wildfire or that can provide a ready fuel supply to augment the spread or intensity of a wildfire ,all of which pose a threat to life, property or the environment;
EUFOFINET;
http://www.fire.uni-freiburg.de/literature/EUFOFINET-Fire-Glossary.pdf)
ii) fire hazard (potential): The difficulty of controlling potential wildfire. It is
commonly determined by fire behavior characteristics such as rate-of-spread,
intensity, torching, crowning, spotting, and fire persistence, and by resistance to-control
EUFOFINET;
http://www.fire.uni-freiburg.de/literature/EUFOFINET-Fire-Glossary.pdf)
iii) A fuel complex, defined by volume, type condition, arrangement, and location, that determines the degree of ease of ignition and of resistance to control

[1] Consiglio per la ricerca in agricoltura e l'analisi dell'economia agraria, Forestry Research Centre (CRA-SEL), Arezzo, Italy
[2] University of Turin, Department of Agriculture, Forest and Food Sciences, Grugliasco, Italy
[3] University of Tuscia, Department for Innovation in Biological, Agro-food and Forestry systems (DIBAF), Viterbo, Italy
[4] University of Bari, Department of Agricultural and Environmental Sciences, Bari, Italy
[5] University of Molise, Department of BioScience and Territory, Pesche (IS), Italy
[6] University of Calabria, Department of Informatics, Modeling, Electronics and Systems Engineering (D.I.M.E.S.), Arcavacata di Rende (CS), Italy
[7] University of Basilicata, Department of Crop Systems, Forestry and Environmental Sciences, Potenza, Italy
[8] University of Sassari, Department of Agriculture, Nuoro, Italy
[9] University of Florence, Department of Agricultural, Food and Forestry Systems, Firenze, Italy
[10] University Mediterranea of Reggio Calabria, Department of Agriculture, Reggio Calabria, Italy
[11] University of Tuscia, Department of Science and Technology for Agriculture, Forests, Nature and Energy, Viterbo, Italy
[*] corresponding author: francesco.chianucci@entecra.it

(Glossary of Wildland Fire Terminology 2006)

Fire-smart management: Management of forest landscapes aimed to control the fire regime by intervening on vegetation (fuel) to foster more fire-resistant (less flammable) and/or fire-resilient environments

(Forest) fire risk: The probability of a wildfire occurring and its potential impact on a particular location at a particular time. Wildfire risk is calculated using the following

equation: Fire risk = probability of occurrence x potential impact

(the European Glossary for Wildfires and Forest Fires, 2012, EUFOFINET; http://www.fire.uni-freiburg.de/literature/EUFOFINET-Fire-Glossary.pdf)

Introduction

Wildfires are major disturbances in Mediterranean ecosystems worldwide (Bond and van Wilgen 1996, Keeley et al. 2012). Wildfires affect natural resources and ecosystem services (e.g., tourism, protection, biomass production, biodiversity conservation) but also hinder the development of policies concerning the enhancement of forest landscapes (Lafortezza et al. 2008). They are complex phenomena determined by natural and anthropogenic causes. Wildfires have been historically used by humans as a land management system to modify the natural environment to adapt it to agricultural and residential activities (Cissel et al. 1999, Cochrane 2003).

Over the last few decades, the massive abandonment of rural lands and the reduced pressure on the use of forests and pastures have determined a pronounced woodland and shrubland re-colonization in many areas. These new landscape conditions, together with the increase of the wildland-urban interface, have led to increasing wildfire hazard in many Countries under Mediterranean environments (Moreira et al. 2011).

In the past, most protection programs against wildfires have focused mainly on fire suppression activities and related infrastructures with limited or null support to prevention activities (Blasi et al. 2005, Moreira et al. 2011, Fernandes et al. 2013). Current-day problems related to wildfire have also originated from this policy: fire-fighting services, fostered in periods of maximum fire frequency, are often not directly integrated with silvicultural prevention; consequently, in many cases, the problem is left unsolved. The fire-suppression policy should be turned into a more integrated "360°-view" fire management approach.

An effective and pro-active model of wildfire control, not limited to fire suppression and provision of equipment, is an integrated fire management model, which combines science and fire management (Alexander 2000) with socio-economic needs, at multiple levels of organization (Silva et al. 2010, FAO 2011). The integrated approach includes activities such as prevention, awareness raising, preparedness and restoration (Gaylor 1974). The actions required for fire-risk reduction can be thought as links in a cyclic chain (Turner 1976, Beaver 2011) (Figure 1).

The territory-scale approach of integrated fire

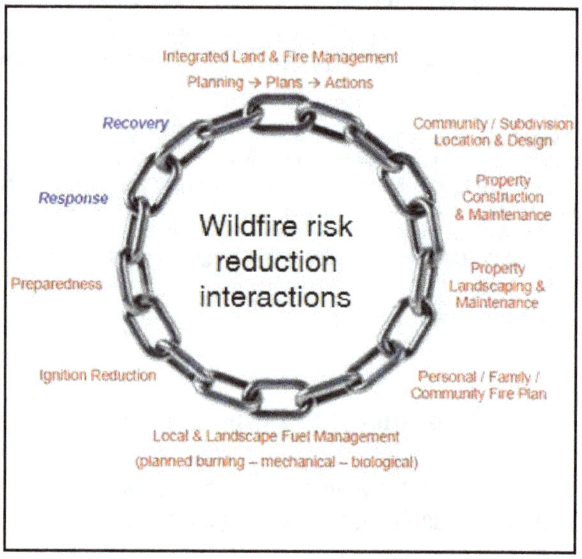

Figure 1 - Fire-risk reduction interactions (Beaver 2011, modified)

management is the "fire smart management of forest landscapes" (Fernandes 2013) which aims to control the fire regime by intervening on vegetation (fuel) to foster more fire-resistant (less flammable) and/or fire-resilient environments. It is an improvement of the "fire safe forest" concept (Agee 1996) based, at stand scale, on fuel conditions that limit surface fireline intensity.

Considering the large scattered amount of publications on the subject, this paper brings together in-depth knowledge and insights of experts, under an integrated and comprehensive framework. We start by examining the Mediterranean scenario of wildfire. Then the rationale of forest fuel management, which represents a pillar of integrated wildfire management, is described. We then move to outline the main silvicultural techniques available for forest fuel management. A specific section focuses attention on fuel management in the wildland-urban interface, which represents a significant issue in terms of fire-risk management especially in Mediterranean environments. Finally, the review deals with the integration of forest management planning and fire prevention planning.

Forest environment and wildfires

Throughout Europe ancestral fire use techniques have been developed to reach land management objectives like e.g., burning forests to open land for pasture and farming purposes (Pyne 1997, Castellnou et al. 2010, Valese et al. 2014). The disappearance of a large part of the forests in the Mediterranean region is directly related to this phenomenon (Blasi et al. 2005, Keeley et al. 2012). The continuous action of natural fires combined with human activity has underpinned the evolution of fire-adapted forest ecosystems. Most Mediterranean tree species have evolved adaptive functional traits (e.g., resprouting, thick bark, serotiny) to respond specifically to the influence of fire. Yet, adaptation involves certain kinds of fire regimes. Wildfires vary by intensity, rate of spread, size, frequency and seasonality, and their effects on the forest ecosystems vary accordingly, much depending also on the phenological stage of plant species and the structural characteristics of forest stands (Nocentini and Coll 2013).

The European scenario of wildfires is characterised by a high frequency of wildfires, mainly occurring in southern European countries. Spain, Portugal, France, Italy and Greece (the so called "fireclub") are the most fire affected countries in Europe where, in the last two decades, wildfires burned about half a billion hectares per year, with an average of 50,000 fires recorded annually (Schmuck et al. 2013). Most fires in the region range from small to fairly large (> 50 ha). The trend of very large fires, those larger than 500 ha, is shown quite stable in the last decades (San-Miguel-Ayanz and Camia, 2010) The temporal distribution of fires shows a seasonal pattern. The environmental condition that underpins fire regimes is the rhythm of wetting and drying: fires occur mainly during winter in the Alpine (Valese et al. 2014) and some continental regions and during summer in the Mediterranean ones. There are some exceptions with fire events grouped in both periods, e.g. in the Liguria region, Italy (Blasi et al. 2005).

A recurrent observations in Italy , at the regional, provincial and municipality level, is that the area burnt, as percentage of forested area, is higher where the forestry ratio (forested area vs territorial surface) is lowest (Leone et al. 2003).

Wildfires exhibit a skewed distribution with a small number of large fires (> 100 ha) accounting for a relevant part of the area annually burned : in 2012, the worst year in the last decades , fires with a burned area less than 1 ha, were 40% as number, 1,3% as burned surface; from 1 to 5 hectares about 30% as number, but less 7% as burned surface; from 5 to 50 ha about 25% as number and 37% % burned surfaces; over 100 ha burned area (less than 2% as number) are responsible of about 38 % of the annual burned surface; over 500 ha about 0.1 % as number but 9 % of burned surface (MIPAF, 2012).

Although fire regimes are clearly related to bioclimatic characteristics, fire events have also a deep interdependence with anthropogenic determinants. As mentioned in the introduction, fire events have increased greatly according to socio-economic changes observed in southern European countries (e.g., depopulation of rural areas, abandonment of farming, increase of the rural-urban interface). Moreover, landscapes of southern Europe are passing through a period of intensive recovery of autochthonous forest vegetation that entails the presence of pioneer and early stages species and structures that are highly flammable. The presence of a remarkable ladder fuel layer, especially within thermophilous forest types, creates conditions that facilitate the shift from surface to the most dangerous crown fires i.e. the passage of fire from neighboring trees (Graham and Caffrey 2003, Moreira et al. 2011).

Forest fire prevention

Fire prevention requires knowledge of the spatial distribution of the vegetation and the adaptations that vegetation can exploit to respond to fire without undergoing degradation. Incombustible forest stands do not exist, although experience leads us to believe that formations with dominance of certain tree species burn more easily than others: for instance, all the other conditions being equal, pinewoods tend to burn more easily than chestnut or beech woods (Silva et al. 2009).

The variability of dominance and diversity of tree and shrub species in a given area creates forest stands characterised by heterogeneous fuel accumulation patterns (living and dead biomass), both in quantitative and qualitative terms. Thus, fires do not pass through the same woodland in the same manner: in fact, they often leave unburnt vegetation patches. All the other environmental (meteorological, geomorphological, etc.) factors being equal, the difference of fire behaviour depends on the composition, structure and phenology of the fuel elements, on their mass, compactness and continuity, both horizontal and vertical ("fuel ladder"), and on their moisture content (Agee and Skinner 2005).

Fire prevention aims to manage fuel conditions by reducing fire hazard over a given territory, i.e. the level of both ease of ignition and fire suppression difficulty. In doing this, fire prevention affects fire severity, i.e. the magnitude of significant negative wildfire impacts on wildland systems (Keeley 2009, Simard 1991).

Fire-risk assessment includes two aspects: fire danger (ignition or propagation potential) and vulnerability (potential damage), being the total risk a product of both aspects (Bachmann and Allgöwer 1999, Bowman et al. 2009, Gorte and Bracmort 2012, Elia et al. 2012).

Fire ignition risk over large forest areas is significantly influenced by the flammability of forest fuels. It is impossible to modify these characteristics over large areas at a reasonable cost. However, as different areas exhibit different fire-risk potential, it is possible to concentrate the protection efforts in more fire-prone areas (Moreira et al. 2011, Fernandes et al. 2013), also to prevent recurrent fires that might lead to land degradation (Barbati et al. 2013). Under such a perspective, an innovative approach to map forest fuels flammability has been proposed by Corona et al. (2014), that is based on the calibration of forest type-based baseline values of forest flammability provided at the European level by Xanthopoulos et al. (2012) with a measure of the actual fire incidence by forest types (i.e. the ratio between the percentage of burned area of a given forest type compared to the total burned forest area and the percentage of the area of the forest type compared to the total forest area) (Bajocco and Ricotta 2008, Barbati et al. 2013, Corona et al. 2014). The approach can be easily applied in those countries, like e.g. Greece, Italy, Portugal and Spain, where both forest type maps and historical geospatial information about wildfire perimeters are available (Chiriacò et al. 2013): for instance, in Italy more than 40% of forest land has proven to be characterized by high forest flammability, in particular in the Southern regions. Another innovative approach to map forest fuel flammability involves use of remotely-sensed information to monitor fuel phenology; recent studies highlighted the strong potential of this method (Baiocco et al. 2010, De Angelis et al. 2012a, De Angelis et al. 2012b).

Large-scale mapping of forest flammability can result in significant improvement of fire management. It provides information about the (static) spatial pattern of fire hazard over a Country or a region. Such information is essential to define long-term priorities among forest areas. At the same time, knowledge of spatial pattern of fire hazard can be integrated with daily fire danger maps, based on meteorological conditions affecting fuel moisture content and related flammability, in order to improve the effectiveness of fire detection and allocation of fighting resources during critical days of the fire season. In short, static and dynamic mapping of wildfire dynamics and features can significantly contribute to effective wildfire management and fire prevention management.

Theoretical bases for the management of forest fuels

Fuel management is a pillar of fire smart management of forest landscapes, which demands to reduce the energy output rate of fires (Martin et al. 1979) to a point where conventional firefighting methods can be effective (Weatherspoon and Skinner 1996, Alexander 2000). Fuel treatments carried out over large surfaces can reduce both the size and severity of wildfires and their effects on communities and the environment; they can facilitate suppression by providing safe access and egress for firefighters, as well as possible counter-firing opportunities (Martinson and Omi 2003). Furthermore, fuel treatments represent valuable options for landscape management that balance societal preferences with the unavoidable recurrence of wildfires (Martinson and Omi 2003); they also significantly reduce the likelihood of crowning, abate fire severity and moderate environmental damages (Graham and Caffrey 2003, Graham et al. 2004) and redistribute fire-risk on the landscape (Cochrane et al. 2012). "Fire smart" landscapes are obtained by area-wide fuel modification and fuel conversion, rather than by fuel isolation (Fernandes 2010 and 2013, Moreira et al. 2011). As little as 1% of the landscape to be treated per year is required for substantial benefits (Finney et al. 2007) and even modest quantities of landscape fuel treatment (5%) can affect the final size of wildfires (Cochrane et al. 2012).

Fuel management involves modifications on the three layers of live and dead fuels (i.e. surface fuels, ladder fuels, crown fuels) in the structure of a stand, namely: i) removing or modifying surface dead fuels to reduce their flammability; ii) selectively removing or modifying live/dead and ladder fuels to reduce their vertical arrangement and horizontal continuity, thereby reducing the probability of crown fire. The rationale of fuel management is in the fact that three types of factors - weather, fuels (i.e. vegetal combustible materials of various size and type), and terrain (i.e. slope, exposure) - work in concert (Graham and Caffrey 2003) to give origin to fire of given characteristics, under the influence of a given casual factor. But out of such factors, only fuel abundance can be directly influenced through treatments aimed to decrease the amount of fuels in all three layers.

Byram's (1959) definition of fireline intensity (Ryan 2002, Trollope et al. 2004, Alexander and Cruz 2012) explains the close link between fire behavior and fuels: interventions aimed at decreasing the fuel load (which is the potential energy accumulated on the ground or in the canopies), reduce the intensity of a fire. Fireline intensity (I_B, kW m^{-1}) is actually the product of the fuel's net low heat of combustion

(H, kJ kg^{-1}), the mass of fuel consumed in the active flaming front (W_a, kg m^{-2}), and the linear rate of fire spread (r, m s^{-1}) (Byram 1959, Alexander 1982), under the form of the simple expression:

$$I_B = H \cdot W_a \cdot r$$

Fuels treatments, which are defined as the "act or practice of controlling flammability and reducing resistance to control of wildland fuels through mechanical, chemical, biological, or manual means, or by fire, in support of land management objectives" (Xanthopoulos et al. 2006) are designed to reduce the loading and continuity of potential fuels (Cochrane et al. 2012), therefore modifying the rate of spread and intensity of fire. The load and the arrangement of both live and dead fuels (i.e. reducing tree density and canopy cover, decreasing basal area, increasing the height-to-live crown base) can be modified through (Valliant et al. 2009):

i) horizontal, linear isolation of fuel (through firebreaks, fuel breaks and greenbelts);
ii) fuel reduction through physical removal, prescribed burning and intensive utilization;
iii) surface fuel reduction also by prescribed grazing ;
iv) change of fuel bed compactness by lopping and scattering (manually or by tractor crushing) and chipping;
v) breaking vertical continuity through pruning;
vi) change of fuel moisture content through dead fuel removal, even local irrigation (Chandler et al. 1983) and "introduction" of live understory vegetation (Agee 1996).

In order of effectiveness, the most successful methods to change potential crown fire occurrence and severity are to (Agee and Skinner 2005):

i) reduce surface fuels;
ii) increase the canopy base height (CBH);
iii) reduce canopy bulk density (CBD);
iv) reduce continuity of the forest canopy.
v) keep big trees of resistant species.

This multi-tiered approach breaks the continuity of surface, ladder and crown fuels (Van Wagner 1977, Agee et al. 2000) making fire propagation rather difficult (Valliant et al. 2009). Interventions on fuels that modify more than one component of forests or use more than one treatment type are frequently more effective (Valliant et al. 2009).

Silvicultural measures, aimed at reducing fire effects through the reduction of fuel load, have their corner stone in thinning, i.e. the planned removal of excess trees from an overstocked stand. In detail, it reduces canopy bulk density and increases canopy base height, thus making crown fires less likely (Cruz and Alexander 2010). The onset of crowning can actually occur when the intensity of a surface fire I_B meets or exceeds I_0 (kW m^{-1}), i.e. the critical surface fire intensity for crown combustion, the latter being function of CBH according to the Van Wagner (1977) equation:

$$I_0 = \left[0.01 \cdot CBH \cdot (460 + 25.9 \cdot f_{mc})\right]^{1.5}$$

where fmc is the foliar moisture content in %.

The transition of fire's spread rate from the initial onset of crowning (i.e. passive crown fire spread) to the point of active crown fire development is based on Van Wagner's (1977) concept of a critical minimum spread rate for active crowning (R_0, m min^{-1}):

$$R_0 = \frac{S_0}{CBD}$$

where S_0 is the critical mass flow rate for solid crown flame (kg m^2 min^{-1}). Van Wagner (1977) provided one estimate of S_0, namely 3.0 kg m^2 min^{-1}. On the basis of this concept, the lower the CBD the higher the critical minimum spread rate.

Unfortunately, there is no one-size-fits-all for fuel treatments: thus, managers must consider many factors when implementing a fire smart forest landscape management plan (Vaillant at al. 2009), along with the relevant effect of some environmental factors (Table 1).

Table 1 - The immediate-term effects of fuel treatments on factors that affect crowning (from Scott 1998 modified and integrated).

Fuel treatment	Surface fuel	Dead fuel load	CBH	Wind speed moisture	CBD
Coppice conversion to high forest	--	--	+	+	--
Understory removal	--	--	+	+	*
Thinning	*	--	+	+	--
Prescribed burning	--	--	+	+	--
Grazing	--	--	+	+	*
Pruning	--	--	+	+	--
Linear Infrastructure	--	--	*	+	*

: + increase,- - decrease; * no effect

Thinning practices for the management of forest fuels

Thinning, together with prescribed burning, play the most important role in the silvicultural prevention of wildfires. Basically, thinning is effective in reducing both ladder fuels and crown density (Chianucci and Cutini 2013), while prescribed burning is effective in reducing surface fuels (Agee and Skinner 2005). The decrease in stocking after thinning reduces competition for resources in the remaining trees, leading to stands with fewer but larger trees, with thicker bark (Leone and Lovreglio 2005), which are therefore more resistant to running fires and able to intercept more light and to use more water and nutrients (Cantore and Iovino 1989, Reich et al. 2006). The increased light, nutrient and water availability enhances tree growth as a result of improved resource-use efficiency and reduced drought stress (Bréda et al. 1995), which is a critical fire-risk in Mediterranean areas, mainly because of summer drought. The higher water availability also reduces the lower heating value of woody debris due to increased fuel moisture content (Dimitrakopoulos and Papaioannou 2001). According to Giordano (1981), the heating value of conifers is on average 18.84 MJ kg^{-1} when the moisture content is 5%, but it decreases to 15.28 MJ kg^{-1} (-19%) when the moisture content is 30%, reaching a minimum of 9.84 MJ kg^{-1} (-48%) when the moisture content is 100%. The biomass removal after thinning subtracts thermal energy, so decreasing the flammability and fire intensity potential, according to the equation:

$$E_{cal} = B_{abv} * HCV$$

where E_{cal} is the thermal energy (kJ m^{-2}), Babv is the aboveground biomass (kg m^{-2}) and *HCV* is the higher calorific value (kJ kg^{-1}).

Similarly, the quantity of water theoretically required for fire suppression (A_{H_2O}) can be estimated as

$$A_{H_2O} = \frac{E_{cal}}{\beta \Delta h_w}$$

where β is the efficiency coefficient, defined as the ratio between the quantity of water used for suppression over the total quantity of water provided and Δh_w is the latent heat of evaporation at 100° C (2.440 kJ kg^{-1}). Based on these simple equations, Garfì et al (2014) assumed that in a typical chestnut coppice the average reduction has been quantified as 48,505 kJ m^{-2} after an early thinning (4 years after coppicing), which represents 71% of the total energy potentially produced by burnable fuel fraction, and 44,058 kJ m^{-2} after an intermediate thinning (10 years after coppicing), which represent 27% of the total energy potentially produced by burnable fuel fraction. Such a reduction in thermal energy implies a theoretical reduction in suppression cost by decreasing the quantity of water required in fire suppression: -49 kg m^{-2} after early thinning and -45 kg m^{-2} after intermediate thinning. Similar values can be obtained from thinning conifer plantations aged between 20-50 years old (Garfì et al. 2014).

Thinning is defined by type, intensity, cycle and targeted final desired structure. Thinning type affects vertical fuel continuity and crown fire potential. It includes low (thinning from below), crown (thinning from above) and selective thinning. All the three types reduce average canopy bulk density but their influence on canopy structure differ; for example, low thinning increases the canopy base height, while crown and selective thinning do not (Agee and Skinner 2005). Thinning intensity affects directly the horizontal fuel continuity and indirectly the flammability potential (e.g., by increasing soil water content after tree removal; Compostella and Iovino 2009). Thinning cycle refers to the age of first intervention and the interval between repeated interventions. Early start of thinning is particularly important for stand subject to high wind and fire risk like conifers and very dense stands (e.g., coppices). The early intervention is recommended as younger trees react more promptly to release from competition, therefore allowing the achievement of more stable stand conditions (Cutini et al. 2009). Conversely, late start of thinning will yield a prolonged competition between trees, with resulting higher self-pruning (broadleaves) and lower branch size (conifers). Such option is often considered to reduce felling cost while increasing the economic revenues (larger trees are harvested). The desired distribution of trees is another important issue when performing thinning treatments. A uniform distribution is generally required in situations where timber production represents the priority of the management; however, this assumption may not be valid for forest fires-prevention strategy. There is evidence that a higher diversity in forest ecosystems entails a higher capacity of reducing the spread of disturbing agents, including wildfire (Nocentini and Coll 2013): for example, variable density thinning can be effective to diversify the structure of a stand by creating a diverse range of light environments (Chianucci and Cutini 2013, Gómez-Aparicio et al. 2009, González-Moreno et al. 2011), with consequences on the medium-long run on recruitment, establishment and ecophysiological behaviour of natural regeneration (Cutini et al. 2009); such option therefore should be strongly recommended in artificial plantations to progressively "re-naturalise" these forest types in order to achieve their self-organizing capacity to respond to natural disturbances, including resistance and resilience to fires.

Specific attention should be given to the treatment of coppice woods, a silvicultural system which is largely widespread in Mediterranean countries, where it covers an area of about 23 million of hectares (UN/ECE-FAO, 2000).

According to the Italian National Forest Inventory, about 42% of the total forested area consists of coppices (www.infc.it), the prevalent species being oak (either deciduous or evergreen oak), hornbeam, chestnut and beech. The key prerequisite for coppice management is the sprouting ability of coppiced tree species, which has been interpreted as a fire-tolerance strategy, i.e. a plant adaptation in ecosystem characterized by frequent crown fire occurrences, as commonly observed in Mediterranean areas (Bond and van Wilgen 1996). This fire-tolerance strategy allows plants to withstand a degree of burning and continue growing despite fire damage, making stand recovery in the coppice system more likely than in other forest types; for example, Providoli et al. (2002) demonstrated that coppicing chestnut stands immediately after fire promotes the rapid recovery of ground vegetation, which in turn reduces the rate of erosion caused by splash transport.

Fuel management and reducing fire-risk in coppice woods is currently a major challenge in Mediterranean countries, where the observed recent changes in forest management perspectives, combined with changing socio-economic conditions and the increasing use of non-renewable raw material have strongly modified the composition and structure of large forest areas previously managed as coppice. These processes resulted in the decline of overall wood exploitation and lengthening of rotation period in coppice woods (Cutini et al. 2013). The interruption of the traditional tending operations and active management has led to substantial biomass accumulation in these forest formations, threatening the conservation of the coppice system from a fire-risk perspective. The abandoning of coppice leads to a slow natural evolution of the stand towards high forest, but a more active management can foster dynamics and the achievement of more mature stand structures, reducing the risk of fire damage, and still allowing economic revenue from silvicultural interventions (Ciancio et al. 2006). Converting coppice to high forest represents an effective option to improve stand conditions and for preventing fire-risk in these forest formations. Various studies have demonstrated that applying low, moderate thinning decreases fire intensity, as they combine relatively low production of dead organic material and good decomposition conditions for it (Zagas et al. 1998, Zagas et al. 2004). It was also previously outlined that low thinning is more suitable for fire prevention than crown and selective thinning, as they are more effective in reducing canopy base height and canopy bulk density (Agee and Skinner 2005, Leone and Lovreglio 2005).

Some potential concerns related to thinning have to be considered in planning fire prevention treatments. In particular, thinning may increase fire damages and tree mortality (Raymond and Peterson 2005) because of:

- an open understory, that allows surface wind to increase and surface fuels to become drier, and thus may increase fire intensity and rate of spread;
- additional fine wood left from the thinning operation (even when the whole-tree system is applied) and grass and shrubs re-growth most likely cause higher surface fireline intensity, thus contributing to severe scorch of overstory trees.

Fuel treatments intended to minimize tree mortality will be most effective if both ladder and surface fuels are treated, i.e. thinning followed by underburning (prescribed burn) (Pollet and Omi 2002, Stephens and Moghaddas 2005, Agee and Lolley 2006).

Mechanical operations for forest-fires prevention

Mechanical treatments have the potential to recover product value so that treatment costs are partially offset. As mentioned, thinning allows controlling species selection, fuel ladders and canopy density; however, thinning can be problematic when: (i) the value of removed wood does not offset treatment costs; (ii) slash significantly increases surface fuel loads (McIver and Ottmar 2007); (iii) activities are limited by road access, slope steepness, or sensitive soils. The effectiveness of mechanical-only treatments depends on how much surface fuel remains.

While there are several approaches to treat critical surface fuels, a key constrain is the high costs of the interventions. Cutting, processing and hauling large numbers of small-volume pieces is costly and every additional operation further increases the costs (Rummer 2004). This is a particularly crucial issue in situations where a market for biomass products does not exist, while the increasing importance of forest biomass for energy uses (Berndes et al. 2003) may help in reducing the costs of silvicultural treatments.

Three harvesting methods are commonly used in forest operation: cut-to-length system (CTL); tree length system (TLS); whole-tree harvesting (WTH) (Fight et al. 1999, Hartsough et al. 1998, Windell and Bradshaw 2000). Removing the limbs and tops of trees and cutting them into merchantable lengths can occur in the forest or at the landing. In CTL, trees are felled and processed at the stump before being extracted to a landing (Spinelli et al. 2010a). In TLS, trees are felled, delimbed and topped at the stump before being extracted to a landing (Rushton et al. 2003). In WTH, trees are felled and immediately removed to a landing where they are processed, thereby reducing potential fire severity more than methods which leave slash within the stand (Hartsough et al. 2008, Marchi et al. 2014, Picchio et al. 2012a). Slash left on ground and deposition of additional fine- to medium-sized surface fuels significantly increase flame length and torching potential (Agee and Skinner 2005).

Biomass bundling is an alternative that allows recovery of forest residues for potential utilization when CTL or TLS are applied (Rummer 2004). However, a complete removal of logging residues can negatively affect post-logging soil erosion (Rice and Datzman 1981, Edeso et al. 1999) and soil fertility (Zabowski et al. 1994), and thus tree growth (Egnell and Valinger, 2003, Ampoorter et al. 2007).

In forest thinning, the main operations are felling, processing, bunching and extraction. Felling and processing can be efficiently carried out by chainsaw (Arno and Fiedler 2005, Stephens and Moghaddas 2005). Chainsaw felling requires a limited capital investment and can be applied in areas where modern forest machinery (harvester and feller) cannot work. However, chainsaw felling is more dangerous for the workers, because they are exposed to numerous risks (Montorselli et al. 2010). Modern machinery like harvester and feller provides greater comfort and safety for workers (Bell 2002). The use of harvester and feller increases productivity and reduces the costs of forest operation, when compared with motor-manual (chainsaw) felling and processing (Wang et al. 1998) (Table 2).

Ground slope and roughness are the main limiting conditions for using these modern machineries (Stampfer 1999). The optimal operating range of mechanized felling and processing is on terrain with a maximum slope of 40%, in areas with high accessibility and where roads and forest tracks are well developed (Spinelli et al. 2008).

In thinning, wood bunching and extraction can be carried out mainly by skidding, forwarding or cable yarding, with different performances in terms of productivity (Table 3). In thinning, the most affordable extraction system is by skidder with winch, due to low effective cost and flexibility. Traditionally, the winch cable is used in bunching logs or trees to skid trail (winching), so that only a limited portion of topsoil of the area is damaged by skidder trafficking (Ampoorter et al. 2012). On slightly sloping terrains, however, skidder with grapple can be used. The hydraulic grapple is operated by the skidder driver, and grabs bundles or logs without requiring further workers (Hunter et al. 2007). In order to maximize the performances of skidding systems, a strategically-planned network of forest trails is necessary to ensure good productivity and minimize the impact on forest soil (Picchio et al. 2011, 2012a, 2012b).

Table 2 - Productivity in felling and processing in forest thinning when cut-to-length (CTL), tree length (TLS), or whole-tree harvesting (WTH) systems are applied.

Harvesting system	Felling/ processing method (m³ day⁻¹)	Gross productivity	Reference
CTL	Chainsaw	2.4 – 18	Balimunsi et al. 2012
	Harvester	25 – 40	Spinelli and Stampfer 2002
TLS	Chainsaw	16-18	Baldini and Picchio 2001
WTH	Chainsaw	24-64	Marchi et al., 2012
	Feller/harvester	40 – 80	Spinelli and Nati, 2006; Bergström et al., 2010

Table 3 - Productivities in bunching-extraction in forest thinning when cut-to-length (CTL), tree length (TLS), or whole-tree harvesting (WTH) systems are applied.

Harvesting system	Bunching/ extraction	Gross productivity (m^3 day^{-1})	Reference
CTL	Skidding	4.8 – 30	Spinelli et al. 2014, Hippoliti and Piegai 2000
	Forwarding	20 – 90	Spinelli and Magagnotti, 2010, Spinelli et al. 2014
TLS	Skidding	10-40	Spinelli and Nati, 2006
WTH	Skidding	15-50	Picchio et al. 2012a, Spinelli et al. 2014
	Cable yarding	10-20	Spinelli et al. 2010c

Extraction by forwarder or by tractor and trailer is normally not recommended in the first thinning and in the selective thinning due to the considerable size and weight of these machines, while this extraction method may be applied in geometric or mixed thinning. In the last few years, mini-forwarders have been made available, with size and weight suited for their use in thinning (Spinelli et al. 2010c). In cable yarding, trees are removed via a skyline cable system and can be processed at the stump or at the landing. Cable yarding is generally applied in steep slopes, where the use of ground-based forest equipment is restricted. This extraction system has high direct cost, due to the time for assembling and disassembling the lines. Only simple cable yarder can be used in thinning when a small wood volume has to be extracted (Neri and Piegai 2007a). In WTH, trees are processed at the landing by using a processor, and the residues are usually chipped for biomass production (Neri and Piegai 2007a, 2007b).

Thinning is sometimes not effective in increasing canopy-base height, unless smaller trees and lower limbs are removed as well (Hunter et al. 2007). Low pruning is usually carried out with manual (hacksaw, loppers, etc.) or motor-manual systems (chainsaw) on branches with < 10 cm (2-4 cm on average) diameter. Pruning is a high labour-intensive and very expensive operation, especially when is not associated with thinning.

Mastication is another mechanical treatment that can be used to reduce fuels in the stand. Small standing trees can be processed using masticators, which grind trees to the stump and disperse the resulting small material on the ground. Mastication is an increasingly common fuel treatment that redistributes "ladder" fuel to the forest floor in order to reduce vertical fuel continuity, crown fire potential, and fireline intensity (Kane et al. 2009, Vitorelo et al. 2009, Kreye et al. 2014). Mastication can be used as a stand-alone treatment (Battaglia et al. 2010, Reiner et al. 2009), but may also be applied following understory thinning (Kane et al. 2009), or prior to prescribed burning (Knapp et al. 2011, Kreye et al. 2013). Masticated debris has a propensity to flame and smolder for long durations; however, improved knowledge to elucidate fire behavior and effects in these treatments is still required (Kreye et al. 2014). Fire burning in masticated fuels behaves in unexpected and contradictory ways, likely because the shredded, compact fuel created when trees and shrubs are masticated contains irregularly shaped pieces in mixtures that are quite different from other woody fuels.

In mechanical treatments for fuel management, the methodologies of felling, processing and extraction have to be planned on a large spatial and temporal scale. They cannot be localized events, and must be framed into the socio-economical and environmental context. Forest operations have to be planned during implementation and execution of the work phases. Therefore they require a schedule of forest works that integrates all activities of prevision and preparation of infrastructure and resources needed for their proper execution.

Table 4 - CO_2 emission and energetic input for some forest yards (Heinimann 2012, modified).

Reference	CO_2 emission (kg m^{-3})	Energy input (MJ m^{-3})
Ecoinvent 2012[1]	12.4	165
ProBas 2012[1]	17.9	50
Magagnotti e Spinelli, 2011[2]	-	100-200
Magagnotti e Spinelli, 2011[3]	-	9-20
Picchio et al., 2009[2,4]	-	700-900
Gonzàles-Garcìa et al. 2009[5]	28-33	175-279
Valente et al. 2011[5]	4-6	162
Magagnotti e Spinelli, 2011[6]	-	60-150
Picchio et al., 2012c[4,7]	-	850

[1]With forest road maintenance; [2]With loading on truck; [3]Bunching-Extraction by horses; [4]Gross energy requirement methodology; [5]With transport enclosed; [6]Skidding; [7]Chips production

An important aspect to be considered in forest operation planning is the impact on the environment, especially on soil and residual trees. During forest operation, the area affected by soil disturbance (compaction, rutting, soil mixing and displacement) may range between 10 and 70% of the total logged stand and the impact on the ecosystem can be substantial (Picchio et al. 2012b). A range of 4–21% of damaged residual trees has been recorded for thinning in conifer and mixed stands (Picchio et al. 2011, Spinelli et al. 2010b, Vasiliauskas 2001).

Other impacts, such as the energy input or CO_2 emissions, should be assessed when planning fire prevention thinning (Table 4). As an example, we compare the values recorded for two thinning yards in conifer stands in Italy (Figure 2– unpublished data), with the values calculated for two levels of suppression activity (low and high) in the same unthinned areas. The results highlight the importance of thinning as fuel treatment, also because the CO2 emissions caused directly by fire are not included.

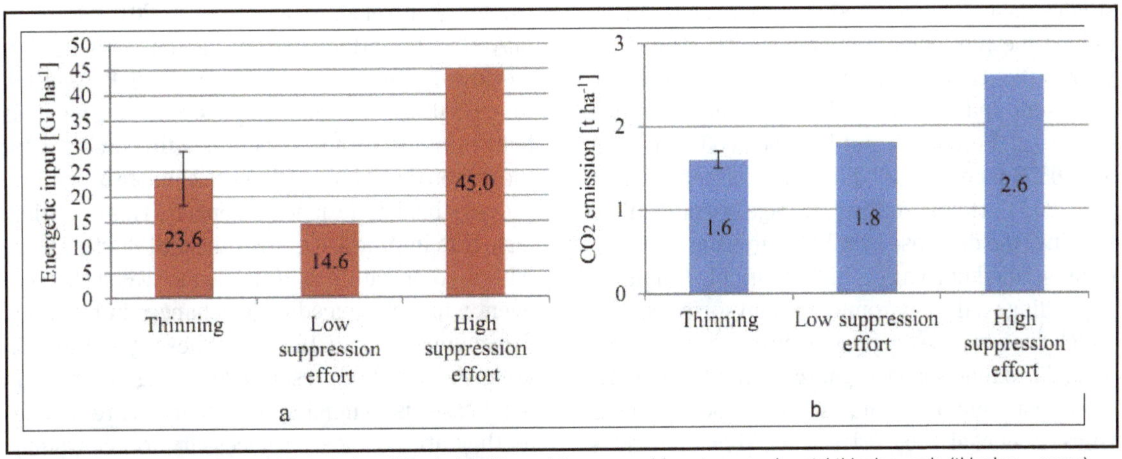

Figure 2 - Energetic input (a) and CO_2 emissions (b) calculated from data measured from two experimental thinning yards (thinning, ± range) and estimated for two levels of fire suppression activity assuming that thinning was not carried out (low and high suppression efforts).

Prescribed burning

Prescribed burning is one of the most common and effective fuel management techniques for fire hazard reduction (Fernandes and Botelho 2003, Boer et al. 2009, Rego et al. 2010, McIver et al. 2013). It is applied to maintain strategic fuel breaks (Fernandes et al. 2013) and buffer zones at the wildland-urban interface (Keeley 2002, Ryan et al. 2013), or for area wide fuel treatments in fire-prone landscapes (Russel-Smith et al. 2013, Ryan et al. 2013, van Wilgen 2013).

In forested areas, low intensity burns (i.e., between 50 and 700 kW m^{-1}), are generally carried out over treatment units of variable size, i.e., 10 - 1000 hectares (Hartsough et al. 2008, Boer et al. 2009, Fernandes et al. 2013) to reduce the load of surface fuels (upper litter, fine dead wood, grasses and low shrubs); these burns may also increase the distance between the base of the tree canopy and understory fuels (McIver et al. 2013) by scorching lower branches of trees, consuming ladder fuels (e.g., epiphytic vegetation), and by thinning elevated shrubs and young trees (Figure 3). In savannas, grasslands and shrublands, prescribed burning is generally planned over larger units in comparison to forested areas (Higgins et al. 2011, Russell-Smith et al. 2013, van Wilgen 2013), to consume accumulated dead material and remove shrub canopy fuels; in these cases, moderate to relatively intense burns (above 1000 kW m^{-1}) are needed to allow crown fire sustainment (Keeley 2002, Castagneri et al. 2013). Prescribed burning is often used in combination with other fuel management techniques such as mastication, thinning, herbicide and grazing (e.g., Kobziar et al. 2009). Mechanical treatment followed by prescribed burning is considered the most effective in reducing stand flammability (McIver et al. 2013).

The popularity of prescribed burning for fire hazard abatement is due to several aspects. Many studies have demonstrated its effectiveness in reducing the fuel complex flammability for several years after treatment, and in mitigating wildfire behaviour and severity at the stand and landscape scale (e.g., Fernandes and Botelho 2003, Finney et al. 2005, Boer et al. 2009, Kobziar et al. 2009, McIver et al. 2013). However, the "leverage" of prescribed burning (the decrease in wildfire area per unit area of prescribed burning) is highly variable (Table 5), depending on the environmental and social aspects which characterize the fire regime of the considered area, the extent, spatial organization and connectivity of treatment units, and the landscape characteristics (Boer et al. 2009, Price et al. 2012a, Fernandes et al. 2013). Prescribed burning leverage is higher in Mediterranean countries of Europe where can reach values >1 (prescribed burning decreases the total area burned) (Vilén and Fernandes 2011, Price et al. 2015). In France, Portugal and Spain this technique has been adopted since the '80s of the XX century (Fernandes et al. 2013), and recently it has been experimented also in some Italian regions (Ascoli and Bovio 2013). In these countries, prescribed burning is implemented in strategic areas relatively small in size along major pathways of recurrent anthropogenic fires with predictable spread patterns (Castellnou et al. 2010, Fernandes 2013). In addition, in several regions it has been successfully used to assist grazing management and replace the clandestine use of fire by shepherds in high fire danger periods (Coughlan 2014, Fernandes et al. 2013). The above mentioned aspects, together with a fragmented landscape, and a milder fire weather in comparison to other geographical areas (Vilén and Fernandes 2011), contribute to increase prescribed burning leverage in Southern Europe (Fernandes et al. 2013). However, also in areas where prescribed burning programs have a leverage < 1 (prescribed burning increases the total area burned) (Price and Bradstock 2011, Price et al. 2012b), several other benefits are expected in treated areas, such as a higher fire suppression capacity (Fernandes and Botelho 2003), wildfire emissions abatement (Russell-Smith et al. 2013), and more resilient landscapes (Fernandes and Botelho 2003, Boer et al. 2009, Ryan et al. 2013).

One of the main reasons which makes this technique so popular in some countries is that prescribed burning may attain multiple land management goals at the same time, such as ecosystem restoration, increased livestock production, fire-fighters training (Ryan et al. 2013), and can help to solve social conflicts related to traditional fire uses (Fernandes et al. 2013, Valese et al. 2014). In particular, historical fire regime restoration to maintain ecosystem health is a complementary target of most prescribed burning programs for fire hazard reduction (Boer et al. 2009, Russell-Smith et al. 2013, Ryan et al. 2013). Fire has played an important role in shaping many biomes worldwide (Bond and van Wilgen 1996, Pausas and Keely 2009). Consequently, prescribed burning has been the most attractive fuel-reduction practice, for the reason that it is most likely to emulate the natural process that it is designed to replace. This is particularly true in Americas, Australia and South Africa, where colonization by Europeans was a discrete event which produced major changes in historical fire regimes (Pyne 1997). In these geographical areas the target of prescribed burning programs in most cases is to mimic pre-settlement fire regimes as they are considered necessary to ecosystems health (Russell-Smith et al. 2013, Ryan et al. 2013, van Wilgen 2013).

Last, but not least, the financial costs of prescribed burning can be very competitive if compared to other fuel management techniques (Hesseln 2000), depending on the extent of the area treated, and on environmental (e.g., vegetation type, available weather windows) and operative (e.g., training level of fire personnel) aspects. In a comparative study of different fuel treatments throughout western USA, Hartsough et al. (2008) found that the gross costs of prescribed burning were the lowest, but the net costs of mechanical treatments after deducting the values of harvested products were less than those of fire. However, a comprehensive analysis of the economics of prescribed burning must take into account social costs, such as potential damage from escape, air quality, safety issues, and aesthetics (Hesseln 2000), and long term benefits, such as reduced fire control costs, wildfire carbon emission abatement (Vilén and Fernandes 2011, Russel-Smith et al. 2013), and nature conservation (Hartsough et al. 2008).

Despite its popularity, prescribed burning continues to face numerous social–ecological challenges in every geographical area it has been applied (Keeley 2002, Boer et al. 2009, Fernandes et al. 2013). Major impediments are economics, air quality concerns and environmental laws, risk of liability for escaped fires, high costs or lack of insurance availability, lack of adequate personnel and risk-averse policies (e.g., Quinn-Davidson and Varner 2012, Ascoli and Bovio 2013). Several ecological concerns exist; prescriptions reducing fire hazard may not always enhance resource values and sometimes may have a detrimental effect (Keeley 2002, Ryan et al. 2013). The combination of fire with mechanical or grazing treatments may increase the spread of exotic plant species (Lonati et al. 2009, McIver et al. 2013).

Given the complexity in assessing the cost-effectiveness of prescribed burning and its long-term implications on ecosystem services, several planning tools have been developed to assist prescribed burning (e.g., Higgins et al. 2011, Fernandes et al. 2012, Ager et al. 2013). Technology has expanded the ability to plan prescribed burning. However, to minimize undesired effects and maximize prescribed burning benefits it is necessary an understanding of ecosystem services trade-offs and an evidence-based adaptive approach, applied with scientific rigor and management practicality (Ascoli et al. 2009, Fernandes et al. 2013, Ryan et al. 2013, van Wilgen 2013).

Table 5 - Prescribed burning leverage in landscapes with different characteristics. Where leverage >1, prescribed burning treatment leads to a reduction in the total area burnt (by prescribed and wildfires); where leverage <1, treatment increases the total burnt area.

Landscape characteristics	Prescribed burning organization	Leverage	Reference
South Western USA	Spatial organization: random	0	Price et al. 2012a
Mediterranean shrublands	Annual treated area: 1.5-2.7%		
Western Australia	Spatial organization: random	0.25	Boer et al. 2009
Eucalypt temperate forests	Annual treated area: 7-9%		
South Eastern Australia	Spatial organization: random	0.33	Price and Bradstock 2011
Eucalypt temperate forests	Annual treated area: 0.5%		
Northern Australia	Spatial organization: strategic	1	Price et al. 2012b
Eucalypt tropical savannas	Annual treated area: 17.4%		
South Western Europe	Spatial organization: random	1.0-1.8	Vilén and Fernandes 2011
Mediterranean shrublands	Annual treated area: 5%		

Fuel management in the wildland-urban interface

Fuel management to reduce fire-risk is one of the most crucial issues in densely populated areas adjacent to forested landscapes. These areas, called wildland-urban interfaces (WUIs), have recently received considerable attention because of increases in both the number of structures destroyed and

Figure 3 - Short term (6 months) effects on understory fuel load and structure of prescribed burning (left side) compared to an untreated control area (right side). The burn was carried out in winter in Central Italy in a Pinus pinaster stand.

the peri-urban area burned annually by wildfire (Radeloff et al. 2005, Moreira et al. 2011). For this reason, WUIs are a significant issue in terms of fire-risk management (Cohen 2010), and it becomes relevant to identify effective fuel management in WUIs that respond to the needs of society (i.e., to protect structures and human settlements and ensure the efficiency of forest ecosystem) (Elia et al. 2012).

According to Stewart et al. (2007), the term "wildland-urban interface" is almost exclusively used in the context of wildland fire. However, there are different definitions for this term. A wildland-urban interface is an area where houses meet - or intermingle with - undeveloped wildland vegetation (Radeloff et al. 2005, Theobald and Romme 2007). When urban areas are interspersed with forests, wildfire threat to human well-being arises at the interfaces where anthropogenic development merges with wildland areas (Gill and Stephens 2009).

The interconnection space between housing facilities and vegetation is always very tight, yet considerably different from case-to-case. In this regard, numerous studies have analyzed and explored WUI growth dynamics in different countries worldwide (Kaplan e Austin 2004, Radeloff et al. 2005, Nowak e Walton 2005, Zhang e Yang 2008, Lein e Stump 2009). In the United States, the Federal Register (2001) classifies WUIs into three categories: *intermix* where the houses are mixed with dense vegetation (trees and shrubs), *interface* where the houses bordering vegetation fuel are separated by a defined demarcation line, and *occluded* where "islands" of wildland vegetation are present inside a largely urbanized area (Table 6).

Radeloff et al. (2005) estimated that in 2000 the WUIs in the United States covered a surface of 719,216 km2 and included approximately 39% of all residential units of the continental United States. Nowak and Walton (2005) have forecasted an increase of urban sprawl from 3.1% to 8.1% by the year 2050. Zhang and Yang (2008) have conducted a research in southeastern United States from 1990 to 2000 and have observed the number of patches decreased from 1,362 to 1,282 despite a 65% increase in the mean size of WUIs; this suggests that most WUIs merged forming larger patches with average surface increases from 178 km2 (1990) to 233 km2 (2000). This growth was partially explained by the gradual demographic flows of the population from urban to rural areas (Kaplan and Austin 2004). Population displacement in rural areas has led to the rapid development of peri-urban fringes, which have become an environmentally and culturally relevant local resource (Lein and Stump 2009).

In Europe, especially in the Mediterranean Basin, the urbanization issue has resulted in the expansion of size and number of WUIs. During the last few decades, the constant transformation of peri-urban areas, agricultural land abandonment and "re-naturalization" have led to substantial biomass accumulation in a context of uncontrolled residential growth (Badia et al. 2011). WUIs create an environment in which fire can easily spread from forest fuels to human settlements (Bowman et al. 2009). Therefore, it becomes essential to manage wildland areas close to human developments and, in particular, to create a "defensible space" between buildings and adjacent vegetation (Bovio et al. 2002, Lovreglio et al. 2010). This space is an area where fire can spread from houses to vegetation and vice-versa. Fuel accumulation, the improper use of fire to eliminate waste and harmful animals, and inaccurate fuel management are a few of the many factors causing wildfires in defensible space.

The WUIs are complex areas characterized by a mixture of man-made and vegetation structures and sizes (Stewart et al. 2007). In these contexts human activities are in close contact with forest fuels, resulting mostly in ignition points for potential wildfires. The delineation of WUIs by appropriate maps, at different scales, becomes an essential knowledge-base for landscape planning and fuel managing (Spyratos et al. 2007, Syphard et al. 2007).

Many studies have investigated landscape planning and fuel management in WUI using Geographic Information Systems (GIS) to process remotely

Table 6 - Definition of interface, intermix, and occluded WUIs and resulting areas from the Federal Register (2001), Stewart et al. (2007), and Theobald and Romme (2007). HU, housing units

	Description	Federal register (2001)	Stewart et al. (2003)	Theobald and Romme (2007)
Interface	Clear demarcation between structura and wildland fuels	7.5 HU/ha or >1person/ha	> 1 HU per 16 ha and < 50% vegetation	> 1 HU per 2 ha and > 10 ha patch
Intermix	Structure dispersed; continuous wildland fuels	>1 HU per 16 ha or 11<people/ha<96	> 1 HU per 16 ha and > 50% vegetation	1 HU per 2 ha to 1 HU per 16 ha
Occluded	Structures surround wildland fuel area	< 400 ha wildlands	Not considered	Not considered
Distance to untreat wildlands*	Not specified	2.4 km	0.8, 1.6 and 3.2 km	

*Untreated wildlands are areas where no fuel treatments have been carried out to mitigate wildland fire-risk in the WUI community

sensed images and lidar data (Zhu and Hejmanowska, 2013). Several authors (Caballero et al. 2007, Dumas et al. 2008, Lein and Stump 2009, Lampin-Maillet et al. 2010, Badia et al. 2011) have described and classified WUIs for wildfires prevention, beginning with the assumption that fire-risk is a function of variables related to man-made activities and the presence/absence of fuel. In particular, these studies consider two categories of variables: biophysical and human. Among biophysical variables are vegetation (fuel) types and their structural attributes (e.g., height, age of the population, diameter at breast height, etc.), topographic features (e.g., altitude, slope, aspect) and climate.

A study on the mapping of fire-risk in WUIs conducted by Lein and Stump (2009) led to the design of an ignition points map in WUIs (Ohio, U.S.) using two types of criteria: environmental, strictly related to fuel flammability, and human, related to ignition (fuel accessibility, proximity to transport infrastructures, population density, etc.). Lampin-Maillet et al. (2010) developed a method for mapping WUIs in the south of France. This method is based on the combination of four spatial classes of urban density (isolated, scattered uniformly, scattered in groups, dense) and three categories of presence/absence of vegetation (low, medium, high). Through a spatial analysis, the map of the 12 WUI types has been then intersected with the map of ignition points and the distribution of burnt areas. Badia et al. (2011) identified a method for determining the likely ignition points in two WUI areas located in Catalonia (northeast Spain); This study integrated multivariate regression analysis, to identify the most significant biophysical and anthropogenic variables affecting wildfire-risk, and the spatial analysis of these variables using data acquired from remote sensing images. Elia et al. (2014) developed a Spatial Allocation Index to determine where and what type of WUI areas may be eligible for fuel removal in southern Italy (Figure 4). The Authors integrated ecological and social factors as a product of natural (i.e., fuel load) and human system (i.e., population, housing, and roads).

The main goals of this research were to identify WUI landscapes in which fires may occur, the points with the highest probability of ignition, and the critical factors causing major events (Amatulli et al. 2007, Romero-Calcerrada and Novillo 2008). After identifying the most critical areas, the best method for reducing risk is to alter the fuel in terms of loadings and stand structure (Shang et al. 2004, Hessburg et al. 2007, Schmidt et al. 2008, Ager et al. 2010, Elia et al. 2014). Much effort has been spent on changing the conditions of WUI fuels through fuel treatments (Conard et al. 2001, Gärtner et al. 2008).

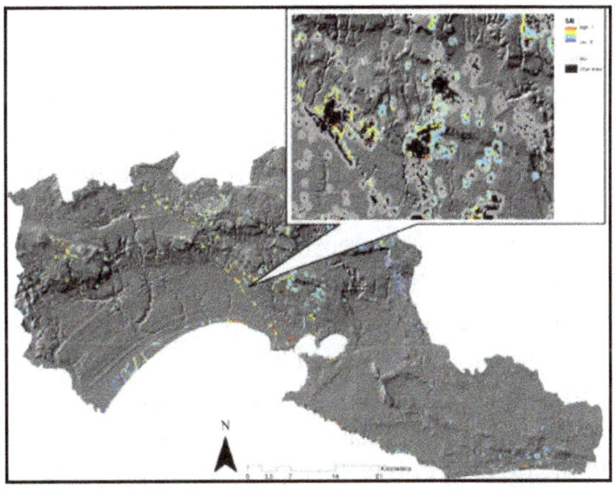

Figure 4 - Map of the Spatial Allocation Index (Elia et al. 2014) by combining 1-hour fuel load, urban density, road density, and population density. The index reveales where and what type of forest areas may be eligible for fuel removal across the WUI interface.

Physics-based models can provide fire behaviour predictions over a broad range of wildland and residential fuel types under a variety of climate and topography conditions. Various models that include sufficient experimental and field measurements could be used to support forest policy-making and fire resource management for WUI fuels and environmental conditions (Thompson et al. 2011, Scott et al. 2013, Lafortezza et al. 2013). Although such a program, especially one with large-scale field measurements, would be expensive, the cost of not undertaking it in the long run would be even higher (Whitman et al. 2013).

Forest management planning and fire prevention planning

Fire prevention involves the active management of woodlands while land abandonment promotes the spread of wildfires. Among the tools that foster active forest management, planning is a leading one in order to program and coordinate actions in a given territory. Forest planning can be implemented at various spatial scales; they will be briefly mentioned here, with reference to the Italian context, defining the specific contribution of each of them to fire prevention. Effective land planning needs spatial data and among these fire-risk maps are an indispensable tool to fulfill the fire damage prevention and mitigation objectives. The main characteristics of this important information layer will be outlined in the second part of this paragraph.

Large-scale forest plans in Italy, namely the Regional Forest Plans (RFP), are effective tool for strategic programming in the forest sector. They take into account the multifunctional role of forests and respond to the policy objectives established at national as well as international level. Elements of reference in their preparation are the criteria for sustainable management. In all the Italian administrative regions, RFRs contain directions to foster silvicultural treatments with experimented positive effects in terms of fire prevention, particularly pre-commercial thinning, thinning and pruning. Additional rules concern planning, building and maintenance of infrastructures, such as roads, skid tracks or bridges, which are important to allow for allowing a more effective forest access for vehicles and people committed to extinguish fires.

A more operational role is given to the Regional Plan Against Forest Fires (RPAFF), specifically dedicated to forecast, prevent and fighting wildfires, which is mandatory for each administrative Region. Starting from the classification and mapping of the territory of each Municipality according to the risk of fire, the plan sets out the actions needed to make fire fighting more timely and effective, also by public participation.

At a territorial scale level intermediate between the RFP plan and management plans for single forest estates, there is the Forest Landscape Plan (FLP), which generally covers forested areas between 50 and 150 thousands of hectares in homogeneous landscapes coinciding with administrative units (e.g., large protected areas, mountain Communities) and comprising both public and private forests. The objectives of this type of plan are: i) defining management guidelines for different parts of the woodland based on the predominant functions to orient forest planning at lower hierarchical level (management plan for single forest ownership); ii) establishing priorities of action also for the purposes of allocation of public funds; iii) promoting synergies and solving conflicts between forest functions also through the participation of stakeholders in decision-making.

FLP incorporates the constraints and rules on the use of forest deriving from higher hierarchical levels of land planning (e.g. regional urban plans, watershed management plans, national park planning). Among the various layers of information to be considered by FLPs, wildfire-risk maps have a special relevance. Maps included in RPAFFs can be transformed at a larger scale detail (1:10,000) if spatially distributed information on local climate, topography, forest types, live and dead biomass, roads, are readily available. Risk of fire is one of the elements to be evaluated in order to decide the priority of implementing silvicultural treatments and building infrastructure to prevent or mitigate the damage of fires. The degree of risk must be correlated to the importance of the functions assigned to the different parts of the wooded area. The attention will concentrate in areas where high fire-risk affects stands with functions of greater social (protection forests), naturalistic (e.g., priority habitats of the Natura 2000 network) and economic (stands with high stumpage prices) values.

At the lowest hierarchical level of forest planning is the management plan for a single forest estate. It regulates in time and space the application of management prescriptions to forest stands during a period of time which usually does not exceed 10-15 years. Traditional goal of the plan is the optimization of timber production and the income of the owner. Currently, the plan is being conceived as a more flexible and articulated tool than in the past as forest management is increasingly multifunctional and must confront the challenges posed to forest ecosystems by global change. For the purpose of fire prevention, directions of RFP and FLP on silvicultural treatments, their priority and urgency are translated into an action program detailing the type and intensity of interventions based on site conditions and stand structure.

At various spatial scale levels the knowledge of the potential level of fire-risk, in term of magnitude and distribution in the territory, is essential to ensure appropriate fire prevention planning (Bar-Massada et al. 2009). From a methodological point of view, the analysis of risk assessment may differ according to the type of data used and for the detail of cartographic output. In particular, it can be carried out based either only on historical data or on the assessment of predisposing and determinant factors. The analysis of historical data regards the statistics of the fires occurred in the past. Instead, the analysis of predisposing factors requires the morphological, climatic and vegetation characterization, together with the human and natural characterization, for the determinant factors analysis. From a morphological point of view, the different combinations of altitude and aspect, in relation to the slopes, define a different probability of front flame propagation, as well as the climatic characteristics. The different vegetation types may give rise to different kind of fuel loads and spatial distribution and, therefore, to different probability of propagation. Finally the anthropic factor is the main determining factor for a fire ignition (Finney 2005, Hardy 2005).

The vulnerability assessment evaluates the negative fire effects. These effects could be assessed through the , pre- and post-fire, analysis of the socio-economic values (e.g., properties, wood resources, recreational importance, carbon stocks), degradation potential (i.e., soil and vegetation conditions), and landscape value (e.g., uniqueness, conservation status (Hardy 2005, Chuvieco et al. 2010). An example of framework for fire-risk assessment, which includes the aspects mentioned above, is shown

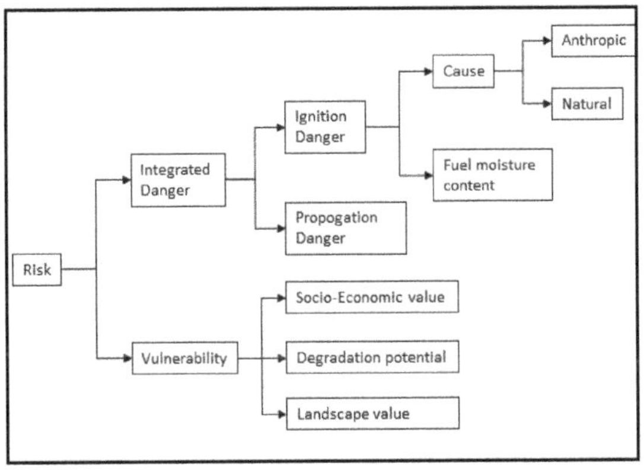

Figure 5 - Framework for fire-risk assessment (Chuvieco et al. 2010 modified).

in Figure 5.

To obtain an operational assessment of fire-risk conditions the following steps are required (Chuvieco et al. 2003, Chuvieco et al. 2010): i) generation of risk factors, using a common geographical unit; a target scale and spatial resolution needs to be defined, in relation to what sources of data are available; ii) conversion of risk factors to a common risk scale; to integrate the input risk variables, the original measurement scale of each input variable should be first converted to a common risk metric; otherwise, we could not generate synthetic indices; iii) development of criteria to integrate risk factors; the different input variables have different impacts on fire-risk conditions; they can be combined in many different ways and using a wider range of techniques: i.e. qualitative cross-tabulation, multi-criteria evaluation, regression techniques or probabilistic models.

Conclusions and perspectives

Successful forest fuel management program is gauged by (i) expansion of wildfire suppression options and effectiveness, and (ii) reduction of environmental and socio-economical impacts of wildfires (Birot 2009). Under this perspective, this review presents a multidisciplinary framework for integrating the ecological, regulatory, procedural and technical aspects of forest management for fires prevention under Mediterranean environments. Accordingly, it represents a resource text for professionals, technicians and government personnel involved in forestry and environmental management

Active human presence in the forest with appropriate and planned silvicultural interventions is distinctively important in Mediterranean environments, as it is one of the main guarantees against abandonment and consequent increased risk of fire. Therefore, public information and awareness on the importance of fuel management interventions should be promoted (Raftoyannis et al. 2014), as far as social fire science, including enhanced understanding of the cost and benefits of wildland fires and the organizational effectiveness. It is important to activate research on public interactions, to understand socioeconomic aspects and how human attitudes values and beliefs influence management options and how communities respond to risk in wildlands or rural or urban interface.

Other important and innovative perspectives for fire research concern the ability of scientist to quantify interacting effects of fire and fuel management strategies on ecology and environment, such as: on landscape, at multiple scales; understanding how fuel treatments and post fire rehabilitation affect ecosystems and their functioning; exploring opportunities in biomass utilization. Moreover, methods for new and improved fire danger assessment should be developed, investigating further physical processes that govern multi-scale fire behavior (for decisions support for treating fuels), large spatial and temporal patterns, atmospheric interactions and dynamics, including smoke forecasting systems and carbon balance. Effects of fire on soil and watershed, interactions between fire regime and invasive species, use of fire for restoration ecology should finally be studied with specific reference to Mediterranean environmental and socioeconomic conditions.

Acknowledgments

This paper has been partially produced under the framework of the project MedWildFireLab ("Global Change Impacts on Wildland Fire Behaviour and Uses in Mediterranean Forest Ecosystems, towards a « wall less » Mediterranean Wildland Fire Laboratory") funded by ERANET ForestTerra and coordinated by Francisco Castro Rego. We thank Sofia Bajocco and Marco Ottaviano for helpful comments on an earlier draft.

References

Agee J.K. 1996 - *The influence of forest structure on fire behavior.* In: Proceedings, 17th Annual Forest Vegetation Management Conference. Redding, CA. January 16-18, 1996:52-68. (http://www.fs.fed.us/rm/pubs/rmrs_gtr292/1996_agee).

Agee J.K., Bahro B., Finney M.A., Omi P.N., Sapsi D.B., Skinner C.N., van Wagtendonk J.W., Weatherspoon C.P. 2000 - *The use of shaded fuelbreaks in landscape fire management.* Forest Ecology and Management 127(1): 55-66.

Agee J.K., Skinner C.N. 2005 - *Basic principles of forest fuel reduction treatments.* Forest Ecology and Management 11: 83-96.

Agee J.K., Lolley M. R., 2006. *Thinning and prescribed fire effects on fuels and potential fire behaviour in an eastern Cascades Forest, Washington, USA.* Fire Ecology 2: 3-19.

Ager A.A., Vaillant N.M., Finney M.A. 2010 - *A comparison of landscape fuel treatment strategies to mitigate wildland fire risk in the urban interface and preserve old forest structure.* Forest Ecology and Management 259: 1556-1570.

Ager A.A., Vaillant N.M., McMahan A. 2013 - *Restoration of fire in managed forests: a model to prioritize landscapes and analyze tradeoffs.* Ecosphere 4 (2): art29.

Alexander M.E. 1982 - *Calculating and interpreting forest fire intensities.* Canadian Journal of Botany 60: 349-357.

Alexander M.E. 2000 - *Fire behavior as a factor in forest and rural fire suppression.* Forest Research, Rotorua, in association with the National Rural Fire Authority, Wellington. Forest Research Bulletin No. 197, Forest and Rural Fire Scientific and Technical Series, Report No. 5, 30 pp.

Alexander M.E., Cruz M.G. 2012 - *Interdependencies between flame length and fireline intensity in predicting crown fire initiation and crown scorch height.* International Journal of Wildland Fire 21: 95–113.

Amatulli G., Pérez-Cabello F., de la Riva J. 2007 - *Mapping lightning/human caused wildfires occurrence under ignition point location uncertainty.* Ecological Modelling 200: 321-333.

Ampoorter E., Goris R., Cornelis W.M., Verheyen K. 2007 - *Impact of mechanized logging on compaction status of sandy forest soils.* Forest Ecology and Management 241: 162–174.

Ampoorter E., Schrijver A., Nevel L., Hermy M., Verheyen K. 2012 - *Impact of mechanized harvesting on compaction of sandy and clayey forest soils: results of a meta-analysis.* Annals of Forest Science 69: 533–542.

Arno S.F., Fiedler C.E. 2005 - *Mimicking nature's fire: Restoring fire-prone forests in the west.* Island Press, Washington, D.C.

Ascoli D., Beghin R., Ceccato R., Gorlier A., Lombardi G., Lonati M., Marzano R., Bovio G., Cavallero A. 2009 - *Developing an Adaptive Management approach to prescribed burning: a long-term heathland conservation experiment in north-west Italy.* International Journal of Wildland Fire 18: 1-9.

Ascoli D., Bovio G. 2013 - *Prescribed burning in Italy: issues, advances and challenges.* iForest - Biogeosciences and Forestry 6 (1): 79-89.

Bachmann A., Allgöwer B. 1999 - *The need for a consistent wildfire risk terminology.* Joint Fire Science Conference and Workshop, Boise Idaho, 15-17 June 1999.

Badia A., Serra P., Modugno S. 2011 - *Identifying dynamics of fire ignition probabilities in two representative Mediterranean wildland-urban interface areas.* Applied Geography 31: 930-940.

Bajocco S., Ricotta C. 2008 - *Evidence of selective burning in Sardinia (Italy): which land-cover classes do wildfires prefer?* Landscape Ecology 23 (2): 241-248.

Bajocco S., Rosati L., Ricotta C. 2010- *Knowing fire incidence through fuel phenology: a remotely sensed approach.* Ecological Modelling 221: 59–66.

Baldini S., Picchio R. 2001 - *Primo diradamento con messa a punto di nuove metodologie di lavoro in una pineta dei Cimini.* Linea Ecologica 6: 47-54.

Balimunsi H., Grigolato S., Picchio R., Nyombi K., Cavalli R. 2012 - *Productivity and energy balance of motor manual harvesting in Uganda.* Forestry Studies in China 14 (4): 276-282.

Bar-Massada A., Radeloff V.C., Stewart S.I., Hawbaker T.J. 2009 - *Wildfire risk in the wildland-urban interface: a simulation study in northwestern Wisconsin.* Forest Ecology and Management 258: 1990-1999.

Barbati A., Corona P., D'Amato E., Cartisano R. 2013 - *Is Landscape a Driver of Short-term Wildfire Recurrence?* Landscape Research (early view) doi:10.1080/01426397.2012.761681.

Battaglia M.A., Rocca M.E., Rhoades C.C., Ryan M.G. 2010 - *Surface fuel loadings within mulching treatments in Colorado coniferous forests.* Forest Ecology and Management 260: 1557–1566.

Beaver A. 2011 - *Wildland-urban intermix. Disasters by Design.* In: Proceedings of WILDFIRE 2011, the 5th International Wildland Fire Conference, Sun City, South Africa, 9-13 May 2011 [Online]. Available: http://www.wildfire2011.org/material/papers/Al_Beaver.pdf. [25.07.2014]

Bell J. 2002 - *Changes in logging injury rates associated with the use of feller-bunchers in West Virginia.* Journal of Safety Research 33 (4): 463–471.

Bergström D., Bergsten U., Nordfjell T. 2010 - *Comparison of boom-corridor thinning and thinning from below harvesting methods in young dense Scots pine stands.* Silva Fennica 44: 669–679.

Berndes G., Hoogwijk M., Van den Broek R. 2003 - *The contribution of biomass in the future global energy supply: a review of 17 studies.* Biomass and Bioenergy 25: 1–28.

Blasi C., Bovio G., Corona P., Marchetti M., Maturani A. 2005 - *Fires and ecosystem complexity.From forest assessment to habitat restoration. Executive summary.* Ministero dell'Ambiente e della Tutela del Territorio, Società Botanica Italiana, Palombi & Partner, Rome.

Boer M.M., Sadler R.J., Wittkuhn R.S., McCaw L., Grierson P.F. 2009 - *Long-term impacts of prescribed burning on regional extent and incidence of wildfires-evidence from 50 years of active fire management in SW Australian forests.* Forest Ecology and Management 259 (1): 132-142.

Bond W.J., van Wilgen B.W. 1996 – *Fire and plants.* Population and Community Biology Series 14, Chapman & Hall.

Bovio G., Camia A., Marzano R. 2002 - *Metodi di indagine a scala regionale e locale. Incendi boschivi in interfaccia urbano-foresta.* Linea Ecologica 34: 34-41.

Bowman D.M.J.S., Balch J.K., Artaxo P., Bond W.J., Carlson J.M., Cochrane M.A., D'Antonio C.M., DeFries R.S., Doyle J.C., Harrison S.P., Johnston F.H., Keeley J.E., Krawchuk M.A., Kull C.A., Marston J.B., Moritz M.A., Prentice I.C., Roos C.I., Scott A.C., Swetnam T.W., van der Werf G.R., Pyne S.J. 2009 - *Fire in the Earth System.* Science 324: 481-484.

Bréda N., Granier A., Aussenac G. 1995 – *Effects of thinning on soil and tree water relations, transpiration and growth in an oak forest (*Quercus petraea *(Matt.) Liebl.).*Tree Physiology 15: 295-306.

Byram G.M. 1959 - *Combustion of forest fuels.* In: Davis, K.P. (ed.). Forest fire: control and use. McGraw-Hill, New York: 61–89.

Caballero D., Beltrán I., Velasco A. 2007 - *Forest Fires and Wildland-Urban Interface in Spain: Types and Risk Distribution.* In: Proceedings of the "IV International Wildfire Conference". Seville, Spain, 13-17 May.

Cantore V., Iovino F. 1989 - *Effetti dei diradamenti sull'umidità del suolo in popolamenti di douglasia della Catena Costiera (Calabria).* Annali Istituto Sperimentale per la Selvicoltura XX: 15-35.

Castagneri D., Esposito A., Bovio G., Mazzoleni S., Seneca U., Catalanotti A.E., Ascoli D. 2013 - *Fuel vertical structure affects fire sustainability and behaviour of prescribed burning in Spartium junceum shrublands.* Annals of Forest Science 70 (8): 863-871.

Castellnou M., Kraus D., Miralles M. 2010 - *Prescribed burning and suppression fire techniques: from fuel to landscape management.* In: Montiel C., Kraus D. (Eds) - Best practices of fire use – prescribed burning and suppression fire programmes in selected case-study Regions in Europe. European Forest Institute Research Report 24: 3-16.

Chandler C., Cheney P., Thomas P., Trabaud L., Williams D. 1983 - *Fire in Forestry. Vol. II: Forest fire management and organization.* John Wiley and Sons, New York, 298 pp.

Chianucci F., Cutini A. 2013 – *Estimation of canopy properties in deciduous forests with digital hemispherical and cover photography.* Agricultural and Forest Meteorology 168: 130-139.

Chiriacò M.V., Perugini L., Cimini D., D'Amato E., Valentini R., Bovio G., Corona P., Barbati A. 2013 - *Comparison of approaches for reporting forest fire-related biomass loss and greenhouse gas emissions in southern Europe.* International Journal of Wildland Fire 22: 730-738.

Chuvieco E., Aguado I., Yebra M., Nieto H., Salas J., Martín M. P., Vilar L., Martínez J., Martín S., Ibarra P., de la Riva J., Baeza J., Rodríguez F., Molina J. R., Herrera M. A., Zamora R. 2010 - *Development of a framework for fire risk assessment using remote sensing and geographic information system technologies.* Ecological Modelling 221: 46–58.

Chuvieco E., Allgöwer B., Salas F.J., 2003 - *Integration of physical and human factors in fire danger assessment.* In: Chuvieco E. (Ed.), "Wildland Fire Danger Estimation and Mapping. The Role of Remote Sensing Data." World Scientific Publishing, Singapore: 197–218.

Ciancio O., Corona P., Lamonaca A., Portoghesi L., Travaglini D. 2006- *Conversion of clearcut beech coppices into high forests with continuous cover: a case study in central Italy.* Forest Ecology and Management 224: 235-240.

Cissel J.H., Swanson F.J., Weisberg P.J. 1999 – *Landscape management using historical fire regimes: Blue River, Oregon.* Ecological Applications 9: 1217-1231.

Cochrane M.A. 2003 – *Fire science for rainforests.* Nature 421: 913-919.

Cochrane M.A., Moran C. J., Wimberly M. C., Baer A. D., Finney M. A., Beckendorf K. L., Eidenshink J., Zhu Z. 2012 -*Estimation of wildfire size and risk changes due to fuels treatments.* International Journal of Wildland Fire 21: 357–367.

Cohen J. 2010 - *The wildland-urban interface fire problem.* Fremontia 38: 16-22.

Compostella G., Iovino F. 1999 – *Studio sull'umidità del suolo in relazioni ai diradamenti in popolamenti di pino laricio.* L'Italia Forestale e Montana 65: 308-323.

Conard S.G., Hartzell T., Hillbruner M.W., Zimmerman G.T. 2001 - Changing fuel management strategies -*The challenge of meeting new information and analysis needs.* International Journal of Wildland Fire 10: 1553-166.

Coughlan M.R. 2014 - *Farmers, flames, and forests: Historical ecology of pastoral fire use and landscape change in the French Western Pyrenees,* 1830–2011. Forest Ecology and Management 312: 55-66.

Corona P., Ferrari P., Cartisano R., Barbati A. 2014 - *Calibration assessment of forest flammability potential in Italy.* iForest - Biogeosciences and Forestry 7: 300-305.

Cruz M.G., Alexander M.E. 2010 - *Assessing crown fire potential in coniferous forests of western North America: a critique of current approaches and recent simulation studies.* International Journal of Wildland Fire 19: 377–398.

Cutini A., Chianucci F., Chirichella R., Donaggio E., Mattioli L., Apollonio M. 2013 –*Mast seeding in deciduous forests of the northern Apennines (Italy) and its influence on wild boar population dynamics.* Annals of Forest Science 70: 493-502.

Cutini A., Chianucci F., Giannini T. 2009 - *Effetti del trattamento selvicolturale su caratteristiche della copertura, produzione di lettiera e di seme in cedui di faggio in conversione.* Annals of Silvicultural Research 36: 109-124.

De Angelis A., Bajocco S., Ricotta C. 2012a - *Phenological variability drives the distribution of wildfires in Sardinia.* Landscape ecology 27: 1535-1545.

De Angelis A., Bajocco S., Ricotta C. 2012b- *Modelling the phenological niche of large fires with remotely sensed NDVI profiles.* Ecological Modelling 228: 106–111.

Dimitrakopoulos, A.P., Papaioannou K.K. 2001 - *Flammability assessment of Mediterranean forest fuels.* Fire Technology 37: 143-152.

Dumas E., Jappiot M., Tatoni T. 2008 - *Mediterranean urban-forest interface classification (MUFIC): A quantitative method combining SPOT5 imagery and landscape ecology indices.* Landscape and Urban Plan 84: 183-190.

Ecoinvent 2012 - *Ecoinvent – Swiss centre for LCI.* www.ecoinvent.org.

Edeso J.M., Merino A., González M.J., Marauri, P. 1999 - *Soil erosion under different harvesting managements in steep forestlands from northern Spain.* Land Degradation and Development 10: 79–88.

Egnell G., Valinger E. 2003 - *Survival, growth and growth allocation of planted Scots pine trees after different levels of biomass removal in clear-felling.* Forest Ecology and Management 177: 65–74.

Elia M., Lafortezza R., Tarasco E., Colangelo G., Sanesi G. 2012 - *The spatial and temporal response of fire on insect abundance in Mediterranean forest ecosystems.* Forest Ecology and Management 263: 262-267.

Elia M., Lafortezza R., Colangelo G., Sanesi G. 2014 - *A streamlined approach for the spatial allocation of fuel removals in wildland–urban interfaces.* Landscape Ecology (early view). doi 10.1007/s10980-014-0070-7.

FAO 2011 - *Community-based fire management. A review.* FAO forestry papers 166, 99 pp.

Fernandes P.M. 2010 - *Creating fire-smart forests and landscapes.* Forêt Méditerranéenne. XXXI: 417-422.

Fernandes P.M. 2013 - *Fire smart management of forest landscapes in the Mediterranean basin under global change.* Landscape and Urban Planning 110: 175-182.

Fernandes P.M., Botelho H.S. 2003 - *A review of prescribed burning effectiveness in fire hazard reduction.* International Journal of Wildland Fire 12: 117-128.

Fernandes P.M., Davies G.M., Ascoli D., Fernández C., Moreira F., Rigolot E., Stoof C.R., Vega J.A, Molina D. 2013 - *Prescribed burning in southern Europe: developing fire management in a dynamic landscape.* Frontiers in Ecology and the Environment 11: 4-14.

Fernandes P.M., Loureiro C., Botelho H. 2012 - *PiroPinus: A spreadsheet application to guide prescribed burning operations in maritime pine forest.* Computers and Electronics in Agriculture 81: 58-61.

Fight R.D., Gicqueau A., Hartsough B.R. 1999 - *Harvesting costs for management planning for ponderosa pine plantations.* General Technical Report PNW-GTR-467. Portland, Oregon. U.S. Department of Agriculture, Forest Service, Pacific Northwest Research Station, 8 p.

Finney M.A., 2005 - *The challenge of quantitative risk analysis for wildland fire.* Forest Ecology and Management 211: 97-108.

Finney M.A., McHugh C.W., Grenfell I.C. 2005 - *Stand-and landscape-level effects of prescribed burning on two Arizona wildfires.* Canadian Journal of Forest Research 35 (7): 1714-1722.

Finney M.A., Seli R.C., McHugh C., Ager A.A., Bahro B., Agee J.K. 2007 - *Simulation of long-term landscape level fuel treatment effects on large wildfires.* International Journal of Wildland Fire 16: 712–727.

Garfì V., Iovino F., Menguzzato G., Nicolaci A., Marziliano P.A. 2014 – *Modalità applicative della prevenzione selvicolturale. Stima della quantità di combustibile eliminabile con interventi selvicolturali.* In: Bovio G., Corona P., Leone V. 2014 " Gestione selvicolturale dei combustibili forestali per la prevenzione degli incendi boschivi." Compagnia delle Foreste, Arezzo, 208 p.

Gärtner S., Reynolds K.M., Hessburg P.F., Hummel S., Twery M. 2008 - *Decision support for evaluating landscape departure and prioritizing forest management activities in a changing environment.* Forest Ecology and Management 256: 1666-1676.

Gaylor H.P. 1974 - *Wildfires: Prevention and Control.* Robert J. Brady Company (ed.), Bowie, Maryland. 319 p.

Gill A.M., Stephens S.L. 2009 - *Scientific and social challenges for the management of fire-prone wildland–urban interfaces.* Environmental Research Letters 4 034014.

Giordano G. 1981 – *Tecnologia del legno.* UTET ed., Torino.

Gómez-Aparicio L., Zavala M.A., Bonet F., Zamora R., 2009 - *Are pine plantations valid tools for restoring Mediterranean forests? An assessment along gradients of climatic conditions, stand density and distance to seed sources.* Ecological applications 19: 2124-2141.

Gonzàles-Garcìa S., Berg S., Feijoo G., Moreira M.T. 2009 - *Environmental impacts of forest production and supply of pulpwood: Spanish and Swedish case studies.* The International Journal of LCA 14 (4): 340-353.

Gonzalez-Moreno P., Quero J.L., Poorter L., Bonet F.J., Zamora R. 2011 - *Is spatial structure the key to promote plant diversity in Mediterranean forest plantations?* Basic and Applied Ecology 12: 251-259.

Gorte R.W., Bracmort K. 2012 - *Wildfire protection in the wildland–urban interface.* CRS Report for Congress, Congressional Research Service, January 23, 2012.

Graham R.T., McCaffrey S. M. 2003 - *Influence of forest structure on wildfire behavior and the severity of its effects.* Miscellaneous Publication. Washington, DC: USDA Forest Service, 12 p.

Graham R.T., McCaffrey S.M., Jain T.B. (Eds.) 2004 - *Science basis for changing forest structure to modify wildfire behavior and severity.* General Technical Report RMRS-GTR-120. Fort Collins, Colorado: U.S. Department of Agriculture, Forest Service, Rocky Mountain Research Station, 43 p.

Hardy C.H. 2005 - *Wildland fire hazard and risk: problems, definition, and context.* Forest Ecology and Management 211: 73-82.

Hartsough B.R., Abrams S., Barbour R.J., Drews E.S., McIver J.D., Moghaddas J.J., Schwilk D.W., Stephens S.L. 2008 - *The economics of alternative fuel reduction treatments in western United States dry forests: Financial and policy implications from the National Fire and Fire Surrogate Study.* Forest Policy and Economics 10 (6): 344–354.

Hartsough B.R., Gicqueau A., Fight R.D. 1998 - *Productivity and cost relationships for harvesting ponderosa pine plantations.* Forest Products Journal 48 (9): 87-93.

Hessburg P.F., Reynolds K.M., Keane R.E., James K.M., Salter R.B. 2007 - *Evaluating wildland fire danger and prioritizing vegetation and fuels treatments.* Forest Ecology and Management 247: 1-17.

Hesseln H. 2000 - *The economics of prescribed burning: a research review.* Forest Science 46 (3): 322-334.

Higgins A., Whitten S., Slijepcevic A., Fogarty L., Laredo L. 2011 - *An optimisation modelling approach to seasonal resource allocation for planned burning.* International Journal of Wildland Fire 20 (2): 175-183.

Hippoliti G., Piegai F. 2000 - *Tecniche e sistemi di lavoro per la raccolta del legno* - Compagnia delle Foreste, Arezzo.

Hunter M.E., Shepperd W.D., Lentile L.B., Lundquist J.E., Andreu M.G., Butler J.L., Smith F.W. 2007 - *A Comprehensive Guide to Fuels Treatment Practices for Ponderosa Pine in the Black Hills, Colorado Front Range, and Southwest.* General Technical Report RMRS-GTR-198. Fort Collins, CO: U.S. Department of Agriculture, Forest Service, Rocky Mountain Research Station, 93 p.

Kane J.M., Varner J.M., Knapp E.E. 2009 - *Novel fuel bed characteristics associated with mechanical mastication treatments in northern California and southwestern Oregon, USA.* International Journal of Wildland Fire 18: 686–697.

Kaplan R., Austin M. 2004 - *Out in the country: sprawl and the quest for nature nearby.* Landscape and Urban Plan 69: 235–243.

Keeley J.E. 2002 - *Fire management of California shrubland landscapes.* Environmental Management 29 (3): 395-408.

Keele J.E. 2009 - *Fire intensity, fire severity and burn severity: a brief review and suggested usage.* International Journal of Wildland Fire 18 (1): 116–126.

Keeley J.E., Bond W.J., Bradstock R.A., Pausas J.G., Rundel P.W. 2012 - *Fire in Mediterranean Ecosystems.* Cambridge University Press, 515 p.

Knapp E.E., Varner J.M., Busse M.D., Skinner C.N., Shestak C.J., 2011 - *Behaviour and effects of prescribed fire in masticated fuel beds.* International Journal of Wildland Fire 20: 932–945.

Kobziar L.N., McBride J.R., Stephens S.L. 2009 - *The efficacy of fire and fuels reduction treatments in a Sierra Nevada pine plantation.* International Journal of Wildland Fire 18 (7): 791-801.

Kreye J.K., Kobziar L.N., Zipperer W.C. 2013 - *Effects of fuel load and moisture content on fire behaviour and heating in masticated litter-dominated fuels.* Internationa Journal Wildland Fire 22: 440–445.

Kreye J.K., Brewer N.W., Morgan P., Varner J.M., Smith A.M.S., Hoffman C.M., Ottmar R.D. 2014 - *Fire behavior in masticated fuels: A review.* Forest Ecology and Management 314: 193–207.

Lafortezza R., Corry R.C., Sanesi G., Brown R.D. 2008 - *Cultural determinants of spatial heterogeneity in forest landscapes.* In: Lafortezza R., Chen J.M., Sanesi G., Crow T.R. Eds "Patterns and processes in forest landscapes: multiple use and sustainable management". Springer, Netherlands: 17-32.

Lafortezza R., Sanesi G., Chen J. 2013 - *Large-scale effects of forest management in Mediterranean landscape of Europe.* iForest - Biogeosciences and Forestry 6: 331-335.

Lampin-Maillet C., Jappiot M., Long M., Bouillon C., Morge D., Ferrier J.P. 2010 - *Mapping wildland–urban interfaces at large scales integrating housing density and vegetation aggregation for fire prevention in the South of France.* Journal of Environmental Management 91: 732-741.

Lein J.K., Stump N.I. 2009 - *Assessing wildfire potential within the wildland–urban interface: A southeastern Ohio example.* Applied Geography 29: 21-34.

Leone V., Lovreglio R. 2005 – *Pre- and post-fire treatments in Aleppo pine stands: prevention sylviculture and restoration.* In: Proceedings of the "II Conferencia Internacional sobre Estrategias de Prevención de Incendios en el Sur de Europa. Gestión forestall como herramienta para la prevención de incendios". Barcelona (Spain) 9-11 May 2005. CFC-COSE, USSE Y CTFC, 182 p.

Lonati M., Gorlier A., Ascoli D., Marzano R., Lombardi G. 2009 - *Response of the alien species Panicum acuminatum to disturbance in an Italian lowland heathland.* Botanica Helvetica 119 (2): 105-111.

Lovreglio R., Leone V., Giaquinto P., Notarnicola A. 2010 - *Wildfire cause analysis: Four case-studies in southern Italy.* iForest – Biogeosciences and Forestry 3: 8-15.

Magagnotti N., Spinelli R., 2011 - *Financial and energy cost of low-impact wood extraction in environmentally sensitive areas.* Ecological engineering 37 (4): 601-606.

Marchi E., Amatangelo G., Verani S., Picchio R. 2012 - *Productivity and energetic analysis in felling operation of pine forest thinnings.* In: Proceedings of the 45th International Symposium on Forestry Mechanization, "Forest Engineering: Concern, Knowledge and Accountability in Today's Environment" October 8 – 12, 2012 Dubrovnik (Cavtat) Croatia.

Marchi E., Picchio R., Spinelli R., Verani S., Venanzi R., Certini G. 2014 - *Environmental impact assessment of different logging methods in pine forests thinning.* Ecological Engineering 70: 429-436.

Martin R.E., Anderson H.E., Boyer W.D., Dieterich J.H., Hirsch S.N., Johnson V.J., McNab W.H. 1979 - *Effects of fire on fuels: A state-of-knowledge review.* General Technical Report WO–13. Washington, DC: USDA Forest Service.

Martinez J., Vega-Garcia C., Chuvieco E. 2009 - *Human-caused wildfire risk rating for prevention planning in Spain.* Journal of Environmental Management 90: 1241–1252.

Martinson E.J., Omi P.N. 2003 - *Performance of fuel treatments subjected to wildfires.* In: Omi P.N., Joyce L.A. editors " Fire, fuel treatments, and ecological restoration: Conference Proceedings", April 16-18, 2002. RMRS-P-29. Fort Collins, CO: U.S. Department of Agriculture, Forest Service, Rocky Mountain Research Station: 7-13.

McIver J.D., Ottmar R. 2007 - *Fuel mass and stand structure after post-fire logging of a severely burned ponderosa pine forest in northeastern Oregon.* Forest Ecology and Management 238 (1-3): 268-279.

McIver J.D., Stephens S.L., Agee J.K., Barbour J., Boerner R.E., Edminster C.B., Erickson K.L.,Farris K.L., Fettig C.J., Fiedler C.E., Haase S., Hart S.C., Keeley J.E., Knapp E.E., Lehmkuhl J.F., Moghaddas J.J., Otrosina W., Outcalt K.W., Schwilk D.W., Skinner C.N., Waldrop T.A., Weatherspoon C.P., Yaussy D.A., Youngblood A., Zack S. 2013 - *Ecological effects of alternative fuel-reduction treatments: highlights of the National Fire and Fire Surrogate study (FFS).* International Journal of Wildland fire 22 (1): 63-82.

Montorselli N. B., Lombardini C., Magagnotti N., Marchi E., Neri F., Picchi G., Spinelli R. 2010 - *Relating safety, productivity and company type for motormanual logging operations in the Italian Alps.* Accident Analysis and Prevention 42 (6): 2013–2017.

Moreira F., Viedma O., Arianoutsou M., Curt T., Koutsias N., Rigolot E., Barbati A., Corona P., Vaz P., Xanthopoulos G., Mouillot F., Bilgili E. 2011 - *Landscape e wildfire interactions in southern Europe: Implications for landscape management.* Journal of Environmental Management 92: 2389-2402.

Neri F., Piegai F. 2007a - *Biomass chipping operations: cases study in Tuscany (Central Italy).* In: Proceedings of Austro2007/FORMEC´07: "Meeting the Needs of Tomorrow Forests: New Developments in Forest Engineering." 07 – 11 October 2007, Vienna.

Neri F., Piegai F. 2007b - *Produttività e costi di trasformazione nell'utilizzazione di materiale legnoso in biomassa (chips).* L'Italia Forestale e Montana 62 : 385-398.

Nocentini S., Coll L. 2013. *Mediterranean Forests: Human Use and Complex Adaptive Systems.* In: Messier C., Puettmann K.J., Coates K.D. "Managing Forests as Complex Adaptive Systems. Building Resilience to the Challenge of Global Change." Routledge, London and New York: 214-243.

Nowak D.J., Walton J.T. 2005 - *Projected urban growth (2000–2050) and its estimated impact on the US forest resource.* Journal of Forest 103: 383-389.

Pausas J.G., Keeley J.E. 2009 - *A burning story: the role of fire in the history of life.* BioScience 59 (7): 593-601.

Picchio R., Maesano M., Savelli S., Marchi E. 2009 - *Productivity and Energy balance in conversion of a* Quercus cerris *L. coppice stand into high forest in Central Italy.* Croatian Journal of Forest Engineering 30: 15–26.

Picchio R., Neri F., Maesano M., Savelli S., Sirna A., Blasi S., Baldini S., Marchi E. 2011 - *Growth Effects of Thinning Damage in a Corsican Pine (*Pinus Laricio *Poiret) Stand in Central Italy.* Forest Ecology and Management 262 (2): 237–243.

Picchio R., Magagnotti N., Sirna A., Spinelli R. 2012a - *Improved winching technique to reduce logging damage.* Ecological Engineering 47: 83-86.

Picchio R., Neri F., Petrini E., Verani S., Marchi E., Certini G. 2012b - *Machinery-induced soil compaction in thinning two pine stands in central Italy.* Forest Ecology and Management 285: 38-43.

Picchio R., Sirna A., Sperandio G., Spina R., Verani S. 2012c - *Mechanized harvesting of eucalypt coppice for biomass production using high mechanization level.* Croatian Journal of Forest Engineering 33 (1): 15-24.

Pollet J., Omi. P. N. 2002 - *Effect of thinning and prescribed burning on crown fire severity in ponderosa pine forests.* International Journal of Wildland Fire 11: 1-10.

Price O.F., Bradstock R.A. 2011 - *Quantifying the influence of fuel age and weather on the annual extent of unplanned fires in the Sydney region of Australia.* International Journal of Wildland Fire 20: 142-151.

Price O.F., Bradstock R.A., Keeley J.E., Syphard A.D. 2012a - *The impact of antecedent fire area on burned area in southern California coastal ecosystems.* Journal of Environmental Management 113: 301-307.

Price O.F., Pausas J.G., Govender N., Flannigan M., Fernandes P.M., Brooks M.L., Bird R.B. 2015 - *Global patterns in fire leverage: the response of annual area burnt to previous fire.* International Journal of Wildland Fire (early view).

Price O.F., Russell-Smith J., Watt F. 2012b - *The influence of prescribed fire on the extent of wildfire in savanna landscapes of western Arnhem Land, Australia.* International Journal of Wildland Fire 21 (3): 297-305.

ProBas 2012 – http://www.probas.umweltbundesamt.de.

Providoli I., Elsenbeerr H., Conedera M. 2002 - *Post-fire management and splash erosion in a chestnut coppice in southern Switzerland.* Forest Ecology and Management 162: 219-229.

Pyne S.J. 1997 - *Vestal Fire: an Environmental History Told through Fire, of Europe and Europe's Encounter with the World.* University of Washington Press, Seattle, 679 p.

Quinn-Davidson L.N., Varner J.M. 2012 - *Impediments to prescribed fire across agency, landscape and manager: an example from northern California.* International Journal of Wildland Fire 21 (3): 210-218.

Radeloff V.C., Hammer R.B., Stewart S.I., Fried J.S., Holcomb S.S., McKeefry J.F. 2005 - *The wildland–urban interface in the United States.* Ecological Applications 15: 799–805.

Raftoyannis Y., Nocentini S., Marchi E., Calama Sainz R., Garcia Guemes C., Pilas I., Peric S., Amaral Paulo J., Moreira-Marcelino A.C., Costa-Ferreira M., Kakouris E., Lindner M. 2014 - *Perceptions of forest experts on climate change and fire management in European Mediterranean forests.* iForest 7: 33-41.

Raymond and Peterson 2005 - *Fuel treatments alter the effects of wildfire in a mixed-evergreen forest, Oregon, USA.* Canadian Journal of Forest Research 35: 2981–2995.

Rego F., Silva J.S., Fernandes P., Rigolot E. 2010. *Solving the Fire Paradox - regulating the wildfire problem by the wise use of fire.* In: Silva J.S., Rego F., Fernandes P., Rigolot E. (Eds.) "Towards Integrated Fire Management - Outcomes of the European Project Fire Paradox." European Forest Institute Research Report 23: 220-228.

Reich P.B., Tjoelker M.G., Machado J.L., Oleksyn J. 2006 - *Universal scaling of respiratory metabolism, size and nitrogen in plants.* Nature 439: 457-461.

Reiner A.L., Vaillant N.M., Fites-Kaufman J., Dailey S.N. 2009 - *Mastication and prescribed fire impacts on fuels in a 25-year old ponderosa pine plantation, southern Sierra Nevada.* Forest Ecology and Management 258: 2365–2372.

Rice R.M., Datzman P.A. 1981 - *Erosion associated with cable and tractor logging in northwestern California.* International Association of Science and Hydrology 132: 362–374.

Romero-Calcerrada R., Novillo C.J. 2008 - *GIS analysis of spatial patterns of human-caused wildfire ignition risk in the SW of Madrid (Central Spain).* Landscape Ecology 23: 341-354.

Rummer B. 2004 - *Technology for treating fuels and small-diameter material. One Forest Under Two Flags.* In: Proceedings of the 2004 SAF National Convention. Society of American Foresters, Bethesda, Maryland.

Rushton T., Brown S., McGrath T. 2003 - *Impact of tree length versus short - wood harvesting systems on natural regeneration.* Forest Research Report 70, Department of Natural Resources, Halifax, Nova Scotia.

Russell-Smith J., Cook G.D., Cooke P.M., Edwards A.C., Lendrum M., Meyer C.P., Whitehead P.J. 2013 - *Managing fire regimes in north Australian savannas: applying Aboriginal approaches to contemporary global problems.* Frontiers in Ecology and the Environment 11: 55–63.

Ryan K.C. 2002 - *Dynamic interactions between forest structure and fire behavior in boreal ecosystems.* Silva Fennica 36 (1): 13–39.

Ryan K.C., Knapp E.E., Varner J.M. 2013 - *Prescribed fire in North American forests and woodlands: history, current practice, and challenges.* Frontiers in Ecology and the Environment 11: 15-24.

Schmidt D.A., Taylor A.H., Skinner C.N. 2008 - *The influence of fuels treatment and landscape arrangement on simulated fire behavior, Southern Cascade range, California.* Forest Ecology and Management 255: 3170-3184.

Schmuck G., San-Miguel-Ayanz J., Camia A., Durrant T., Boca R., Libertà G., Schulte E. 2013 - *Forest Fires in Europe Middle East and North Africa 2012.* Joint Research Centre Technical Report, Publications Office of the European Union, Luxembourg. doi:10.2788/58397.

Scott J.H. 1998 - *Sensitivity analysis of a method for assessing crown fire hazard in the northern Rocky Mountains, USA.* In: Proceedings of the III International Conference on Forest Fire Research and 14th Conference on Fire and Forest Meteorology, Luso, Portugal. 16/20 November 1998, Vol. II: 2517-2532.

Scott J.H., Thompson M.P., Calkin D.E. 2013 - *A wildfire risk assessment framework for land and resource management.* General Technical Report RMRS-GTR-315. U.S. Department of Agriculture, Forest Service, Rocky Mountain Research Station, 83 p.

Shang B.Z., He H.S., Crow T.R., Shifley S.R. 2004 - *Fuel load reductions and fire risk in central hardwood forests of the United States: a spatial simulation study.* Ecological Modelling 180: 89-102.

Silva J.S., Moreira F., Vaz P., Catry F., Godinho-Ferreira P. 2009 - *Assessing the relative fire-proneness of different forest types in Portugal.* Plant Biosystems 143: 597-608.

Silva J.S., Rego F.C., Fernandes P., Rigolot E. 2010 - *Introducing the Fire Paradox.* In: J.S. Silva, F.C. Rego, P. Fernandes, E. Rigolot (Eds.) "Towards Integrated Fire Management. Outcomes of the European Project Fire Paradox." European Forest Institute Research Report 23: 2-6 and 229.

Simard A. 1991. *Fire severity, changing scales, and how things hang together.* International Journal of Wildland Fire 1 (1): 23–34.

Spinelli R., Lombardini C., Magagnotti N. 2014 - *The effect of mechanization level and harvesting system on the thinning cost of Mediterranean softwood plantations.* Silva Fennica 48 (1): 1-15.

Spinelli R., Magagnotti N. 2010 - *Performance and cost of a new mini-forwarder for use in thinning operations.* Journal of Forest Research 15: 358–364.

Spinelli R., Magagnotti N. Dellagiacoma F. 2008 - *Meccanizzazione nelle fustaie Alpine. Due diversi sistemi di lavoro.* Sherwood 147: 45-49.

Spinelli R., Magagnotti N., Picchi G. 2010a - *Deploying mechanized cut-to-length tecnology in Italy: fleet size, annual usage and costs.* International Journal of Forest Engineering 21: 23–31.

Spinelli R., Magagnotti N., Nati C. 2010b - *Benchmarking the impact of traditional small-scale logging systems used in Mediterranean forestry.* Forest Ecology and Management 260: 1997–2001.

Spinelli R., Magagnotti N., Lombardini C. 2010c - *Performance, Capability and Costs of Small-Scale Cable Yarding Technology.* Small-scale Forestry 9: 123–135.

Spinelli R., Nati C. 2006 - *A Low-Investment Fully Mechanized Operation for Pure Selection Thinning of Pine Plantations.* Croatian Journal of Forest Engineering 30 (2): 89–97.

Spinelli R., Stampfer K. 2002 - *Un harvester per i primi diradamenti.* Sherwood 75: 39-46.

Spyratos V., Bourgeron P.S., Ghil M. 2007 - *Development at the wildland-urban interface and the mitigation of forest-fire risk.* In: Proceeding of the National Academy of Science USA 104: 14272-14276.

Stampfer K. 1999 - *Influence of terrain conditions and thinning regimes on the productivity of a track-based steep-slope harvester.* In: Sessions J., Chung C. (Eds.). Proceedings of the International Mountain Logging and 10th Pacific Northwest Skyline Symposium. 20 March – 1 April 1999, Corvallis, Oregon.

Stephens S.L., Moghaddas J.J. 2005 - *Experimental fuel treatment impacts on forest structure, potential fire behavior, and predicted tree mortality in a California mixed conifer forest.* Forest Ecology and Management 215: 21-36.

Stewart S.I., Radeloff V.C., Hammer R.B., Hawbaker T.J. 2007 - *Defining the wildland-urban interface.* Journal of Forestry 105: 201-207.

Syphard A.D., Radeloff V.C., Keely J.E., Hawbaker T.J., Clayton M.K., Stewart S.I., Hammer R.B. 2007 - *Human influence on California fire regimes.* Ecological Applications 1: 1388-1402.

Theobald D.M., Romme W.H. 2007 - *Expansion of the US wildland–urban interface.* Landscape Urban Planning 83: 340-354.

Thompson M.P., Calkin D.E., Finney M.A., Ager A.A., Gilbertson-Day J.W. 2011 - *Integrated national-scale assessment of wildfire risk to human and ecological values.* Stochastic Environmental Research and Risk Assessment 25 (6): 761-780.

Trollope W.S.W., De Ronde C., Geldenhuys C.J. 2004 - *Fire Behaviour.* In: Goldammer J.G., De Ronde C. (Eds.) "Wildland Fire Management Handbook for Sub-Sahara Africa. Global Fire Monitoring Center": 27-59.

Turner B.A. 1976 - *The development of disasters – a sequence model for the origin of disasters.* Sociological Review 24: 753 – 774.

Vaillant N.M., Fites-Kaufman J.A. and Stephens S.L. 2009 - *Effectiveness of prescribed fire as a fuel treatment in Californian coniferous forests.* International Journal of Wildland Fire 18: 165–175.

Valente C., Spinelli R., Hillring B.G. 2011 - *LCA of environmental and socio-economic impacts related to wood energy production in the alpine conditions: Valle di Fiemme (Italy).* Journal of Cleaner Production 19: 1931-1938.

Valese E., Conedera M., Held A., Ascoli D. 2014 - *Fire, humans and landscape in the European Alpine region during the Holocene.* Anthropocene . DOI: 10.1016/j.ancene.2014.06.006

Van Wagner C.E. 1977 - *Conditions for the start and spread of crown fire.* Canadian Journal of Forest Research 7: 23–34.

van Wilgen B.W. 2013 - *Fire management in species-rich Cape fynbos shrublands.* Frontiers in Ecology and the Environment 11: 35-44.

Vasiliauskas R. 2001 - *Damage to trees due to forestry operation and its pathological significance in temperate forests: a literature review.* Forestry 74 (4): 319–336.

Verde J.C., Zêzere J.L. 2010 - *Assessment and validation of wildfire susceptibility and hazard in Portugal.* Natural Hazards and Earth System Sciences 10: 485-497.

Vilén T., Fernandes P.M. 2011 - *Forest fires in Mediterranean countries: CO2 emissions and mitigation possibilities through prescribed burning.* Environmental Management 48 (3): 558-567.

Vitorelo B., Han H.S., Varner J.M. 2009 - *Masticators for fuel reduction treatment: equipment options, effectiveness, costs, and environmental impacts.* In: Proceedings of the 2006 Council on Forest Engineering (COFE) meeting, Lake Tahoe, CA, 11 p.

Wang J., Greene W.D., Stokes B.J. 1998 - *Stand, harvest, and equipment interactions in simulated harvesting prescriptions.* Forest Products Society 48: 81-85.

Weatherspoon C.P., Skinner C. N. 1996 - 56. *Landscape-level strategies for forest fuel management in Sierra Nevada Ecosystem Project: Final report to Congress, vol. II, chap. 56.* Davis University of California, Centers for Water and Wildland Resources.

Whitman E., Rapaport E., Sherren K. 2013 - *Modeling fire susceptibility to delineate wildland–urban interface for municipal-scale fire risk management.* Environmental Management 52: 1427–1439.

Windell K., Bradshaw S. 2000 - *Understory biomass reduction methods and equipment.* 0051-2828-MTDC. USDA Forest Service, Technology and Development Program, Missoula, MT.

Xanthopoulos G., Caballero D., Galante M., Alexandrian D., Rigolot E., and Marzano R. 2006 - *Forest Fuels Management in Europe.* In: Andrews P.L, Butler B.W. (Eds.) Proceedings of the Conference on "Fuels Management—How to Measure Success", March 28-30, 2006, Portland, Oregon, USA. USDA Forest Services, Rocky Mountain Research Station, Fort Collins, Colorado. RMRS-P-41: 29-46.

Xanthopoulos G., Calfapietra C., Fernandes P. 2012 - *Fire hazard and flammability of European forest types.* In: Moreira F., Arianoutsou M., Corona P., De Las Heras G. (Eds.) " Post-fire management and restoration of Southern European forests." Springer, Managing Forest Ecosystems 24: 79-92.

Zabowski D., Skinner M.F., Rygiewickz P.T. 1994 - *Timber harvesting and long-term productivity: weathering processes and soil disturbance.* Forest Ecology and Management 66: 55–68.

Zagas T.D., Ganatsas P.P., Tsitsoni T.K., Hatzistathis A. 1998 - *Influence of silvicultural treatment on ecology, quality and fire resistance in Quercus ilex coppice stands, in N. Greece.* In: Proceedings of VII International Congress of Ecology (INTECOL), Florence 19-25 July 1998: 473.

Zagas T. D., Ganatsas P.P., Tsitsoni T.K., Tsakaldimi M. 2004. *Thinning effect on stand structure of holm oak stand in northern Greece.* Proceedings of the 10th MEDECOS Conference, 25th April – 1st May 2004, Rhodes, Greece.

Zhang Y., He H.S., Yang J. 2008 - *The wildland–urban interface dynamics in the southeastern U.S. from 1990 to 2000.* Landscape Urban Planning 85: 155-162.

Zhu Q., Hejmanowska B. 2013 - *Analysis of GIS-based spatial variability and risk assessment.* Journal of Chemical and Pharmaceutical Research 5: 372-380.

4

Stand structure and influence of climate on growth trends of a Marginal forest population of *Pinus nigra* spp. *nigra*

Maurizio Marchi[1], Cristiano Castaldi[1*], Paolo Merlini[1], Susanna Nocentini[2], Fulvio Ducci[1]

Abstract - The Black pine of Villetta Barrea (*Pinus nigra* ssp. *nigra* var *italica*) is a variety of the *nigra* subspecies. It is naturally distributed only in the Abruzzo Region, near the village of Villetta Barrea, with a rear-edge Marginal Population. A dendrochronological sampling of the population was implemented with the aim of studying its stand structure and the most probable interactions between growth trends and climate. Mensurational data were used to characterize the stand and, furthermore, the general correlation function (CF) and the moving correlation function (MCF), with a 30 years window, were used to assess the interrelation between the growth of the tree rings and the climate.The results indicated that the past forest management, mainly carried out with thinnings from below and selective cuttings, influenced the current structure of the forest (mean diameter) but no differences in growth trends were detected within the population. The survey on Villetta Barrea Black pine showed a positive and statistically significant correlation between the ring-width and the average temperatures of the months of December (before the ring formation - t-1), February and March; but it also showed a negative correlation with the temperatures of July, September and October of the current year (t). Moreover, the analysis with moving correlation functions suggested that, in the last decades, the population has negatively reacted to very few climate factors and, in particular, to the changes in temperatures (both minimum and maximum temperatures). This is especially true for the shifts occurred in September, the year of the ring formation.

Keywords - *Pinus nigra* ssp. *nigra* var. *italica,* Villetta Barrea, Marginal Forest Population, Abruzzo, tree-rings.

Introduction

Tree-rings analysis is becoming a common technique used for analysing forest resources, with particular regard to the marginal forest populations (Hampe and Petit 2005). It can help in: (i) investigating past endogenous and exogenous dynamics; (ii) detecting correlations between species and climate and finding pointer years (Spiecker 2002, Gallucci and Urbinati 2009); and, finally, (iii) forecasting species adaptation potential to the Global Change effects (Linares and Tiscar 2010, Amodei et al. 2012, Mazza et al. 2013). Furthermore, it is also a useful tool for the management of forest resources, because climate change may have different impacts on different forest ecosystems.

The expected changes in tree-growth might influence the competitive relationships between species, genetic pools and the geographic distribution of forest species (Lindner 2000). Following climate change effects, it is likely that forest species will migrate or will have to adapt to new environmental conditions (Parmesan 1996, Parmesan 2006).

In such a framework, and to make a better use of existing potentials and/or minimize the negative impacts of climate change on forests, conservation strategies should be developed principally in the southern Mediterranean area. A large number of "rear-edge Marginal populations" (Hampe and Petit 2005) is actually located here because the area was a glacial refugium (Petit et al. 2003, Afzal-Rafii and Dodd 2007).

In particular, the European Black pine (*Pinus nigra* J.F. Arnold 1785) can be considered a collective and Tertiary-relict species which was already present during the Pliocene, in sites where it occurs at present (Vidakovic 1974). It continued to spread during the Quaternary age (Gellini and Grossoni 2003) and, due to its wide and very discontinuous distribution, genetic and phenotypic variability is very high among populations (Isajev et al. 2004). Botanists have described at least fifteen subspecies in the course of time (Fenaroli and Gambi 1976), but today there is a general agreement on the division into six subspecies (Quézel and Médail 2003), ranging from Spain to Turkey. In spite of this, many isolated populations have been recognized, due to the fragmented range of the Black pine across Europe.

The Black pine of Villetta Barrea (*Pinus nigra* ssp. *nigra* var. *italica*) belongs to the nigra subspe-

[1] Consiglio per la ricerca in agricoltura e l'analisi dell'economia agraria, Forestry Research Centre (CREA-SEL), Arezzo, Italy
[2] University of Florence, Department of Agricultural, Food and Forestry Systems, Firenze, Italy
[*] corresponding author: cristiano.castaldi@entecra.it

cies and is native of the Abruzzo Region (Central Italy), near the small town of Villetta Barrea (41.7768 N, 13.9374 E). The species is currently distributed on the two opposite sides of the Sangro valley, on approximately 400 hectares, and is partly included in an integral reserve in the Camosciara area (Abruzzo National Park).

In order to study the relationships between growth, climate and forest management of the Villetta Barrea Black pine, we investigated the only official seed stand for the *italica* variety of *nigra* subspecies existing in Italy and in Europe (code number "ABR04" of the Regional Register of the Basic Material of Abruzzo), which is also the only managed part of this Marginal Forest Population (MaP). Mensurational data were used to determine the stand structure in relation to the site attributes and the past management, while the tree ring series were analysed to determine the influence of climate on growth trends.

Materials and methods

Study area and silvicultural history

The seed stand covers an area of about 105 hectares which is characterized by a typical Mediterranean climate (Fig. 1). It is mainly a pure-pine stand, variously mixed with beech stools, mostly in the more humid zones and at the higher elevations. The last management plan of the Municipality of Villetta Barrea (valid for the period 2002-2011), divided the seed stand into five compartments (see Tab. 1).

Many documents were found in the repository of the forest administration of Villetta Barrea and reported in the last management plan (Chapter 5 of the approved document). Even if very few documents described the past silvicultural treatment, the main practice of the past was a sort of selective felling. It is mentioned that the oldest cutting dates back to the early 1600, when the Black pine was used principally for buildings and infrastructures. A sort of selective logging was often conducted across the whole area, until the first decades of the twentieth century.

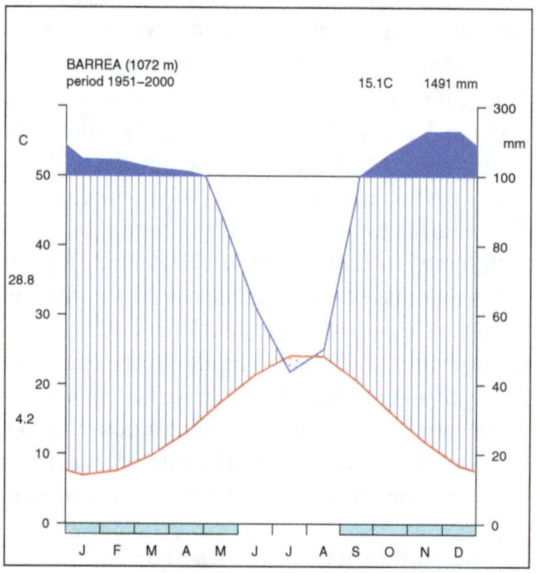

Figure 1 - Walter & Lieth diagram for the location of Barrea, the nearest meteorological station to Villetta Barrea.

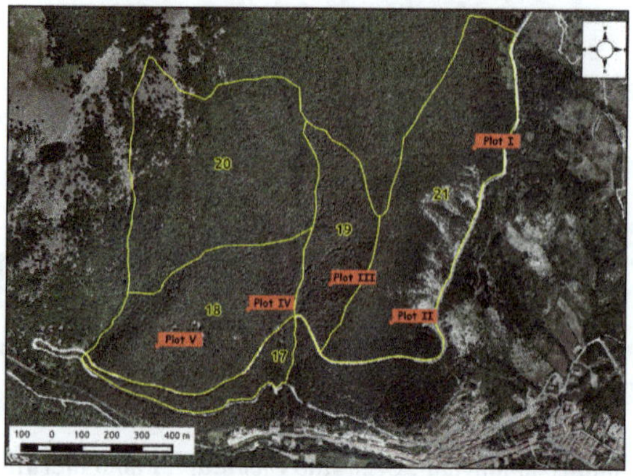

Figure 2 - Spatial distribution of survey plots (red points and labels) and forest compartments (yellow lines and numbers) across the investigated area.

Many economic transactions were registered by the forest administration, reporting the amount of the sold trees, the purchaser and the price. A few trees were also felled for the internal use of the community of Villetta Barrea.

Since 1924, the National Forest Service (office of Villetta Barrea) promoted a change, from this type of selective felling, towards an even-aged silvicultural approach which was aimed at giving more light to the soil; this condition being believed to better maintain the black pine in pure stands. Nevertheless, this practice was never really applied and pine stands have been slowly abandoned.

Survey of mensurational parameters

Five circular plots with 20 metres radius were placed in the study area to describe stand variability (Fig. 2). Plot position was initially chosen following a random but stratified criterion (one plot per compartment). However, after a first field survey, two compartments were excluded from the analysis: (a) the unit 17, because too small and without any substantial structural difference from unit 18 (Tab. 1); (b) the unit 20, mainly dominated by beech coppice black pine, being present only with isolated trees or small groups.

Consequently, 2 of the 5 plots were re-assigned to the two larger compartments (18 and 21). The position of each plot was recorded using GPS and all the sampled trees were numbered, making the plots permanent for any further analysis or survey.

In each plot, we measured the diameter at breast height (DBH) of all trees and the total height on a

Table 1 - Main characteristics of the analysed compartments and survey plots.

Compartment	Area	Elevation	Slope	Aspect	Plot
17	5.13 ha	1080 m	40%	S	-
18	22.63 ha	1175 m	50%	SE	IV ; V
19	11.64 ha	1200 m	50%	W	III
20	33.74 ha	1470 m	60%	SE	-
21	31.86 ha	1200 m	65%	SE	I ; II

representative sample of 10-15 trees (25% of total number). Also the trees volume of each species was calculated with the ForIT package (Puletti et al. 2014) of R Cran software (R Core team 2014), i.e. the implementation of the biomass and volume models worked out by Gasparini and Tabacchi (2011) and Tabacchi et al. (2011) for the 2nd Italian National Forest Inventory (INFC 2005).

A non-parametric ANOVA on DBH was performed to assess the differences between plots. Each plot was considered as a different treatment, while differences among the plots in slope and aspect were included into the error variance.

Dendrochronological sampling

The dendrochronological samples were collected following the "Principle of aggregate tree-rings" (Cook 1987) this being the more largely used in dendrochronology. The principle states that any individual tree-growth series can be "decomposed" into an aggregate of environmental factors, both "endogenous" and "exogenous", affecting the pattern of tree-growth over time. The terms "endogenous" and "exogenous" are used to differentiate forest disturbances developed by the trees themselves (endogenous), from disturbances arising from processes independent from the forest (exogenous). For instance, the gap-phase is an endogenous stand dynamics process, whilst an insect attack or a thinning implementation are exogenous processes.

The aggregate series in a fixed moment " t " can be expressed with the formula [1]:

$$R_t = A_t + C_t + D1_t + D2_t + E_t$$

where:

- R_t is the observed ring-width series;
- A_t is the age-size related trend in ring-width;
- C_t is the climatically-related environmental signal;
- $D1_t$ is the disturbance pulse caused by a local endogenous disturbance;
- $D2_t$ is the disturbance pulse caused by a stand-wide exogenous disturbance;
- E_t is the largely unexplained year-to-year variability not related to the other signals.

A_t, C_t, and E_t are assumed to be continuously present in R_t, while $D1_t$ and $D2_t$ may or may not be present depending on a disturbance occurred at a considered time or not. When the aim is to study the relationships between growth (R_t) and climate (C_t), the other factors should be minimized and, consequently, a correct sampling of the trees is very important.

Following the structure of the formula, a data collection and a subsequent pre-analysis on the seed stand of Villetta Barrea were made to minimize the effects of A, D1 and D2 and maximise C. After the forest mensuration survey, 5-7 cores at each plot (depending on the abundance of suited trees) were extracted with a 5-mm diameter increment borer, at breast height, from healthy, dominant and straight trees, to maximize the effect of "C". Each series was then prepared in the lab on a wooden support, measured with LINTAB6® and cross-dated to minimize or eliminate any probable error due to missing rings or reading mistakes. Afterwards, each tree-ring series was standardized separately with a double detrending procedure. The negative exponential curve was used to remove the trend due to the tree circumference increasing with age (A), whereas low-frequency variance and other disturbances (D1 and D2) were removed using a spline function (Cook 1981) with a 50% frequency response (cut-off) of 10 years to emphasize higher inter-annual frequency climatic variance (Cook 1981, Biondi and Visani 1993, Amodei et al. 2012).

Before any classical climatic correlation analysis, and in order to complete the information derived from ANOVA on DBH, both not-standardized and standardized tree ring series were also used, in a second ANOVA, to asses differences between plots (for the common growing period only). The analysis was performed to check dissimilarities and to verify if the plots were grouped in the same way as DBHs.

The influence of climate on tree-ring growth was investigated using the mean correlation function (CF) and the moving correlation functions (MCF) based on Pearson's correlation coefficients (Fritts 1976). CF and MCF are based on the same assumptions but they use different principles and time-periods. While CF takes into account the whole growing period of trees, MCF uses a smaller interval of time repeated for the whole period (generally 20-30-40-50 years, depending on the available data, species sensitivity and case-study). In this case, a 30-year window was used to avoid biases which could be induced by extreme events and to retain the influence of mid-frequency climate variations. Each moving window analyzes a cycle of 30 years, going forward from year to year. In both cases (CF and MCF) monthly climate variables (maximum, minimum temperatures and monthly precipitation) were sequenced from October of the previous year

Table 2 - Mean mensurational values of survey plots.

PLOT (Comp.)	M_dbh (cm)	CV_dbh	M_ht (m)	Tr/ha (m^2)	G/ha_P (m^3)	Vol/ha_P (m^2)	G/ha_B	Age
I (21)	24.5	42%	13.9	1138	46.6	321.0	1.3	124-179
II (21)	26.4	56%	14.3	923	50.6	372.5	0.2	104-145
III (19)	22.8	64%	11.8	1082	39.6	259.7	0.7	143-184
IV (18)	28.0	35%	19.5	1178	60.2	574.5	0.8	80-99
V (18)	28.3	41%	16.52	724	45.7	266.0	2.0	104-118

M_dbh = mean diameter at breast height; CV_dbh = coefficient of DBH variation; M_ht = mean height; Tr/ha = number of trees ha^{-1}; G/ha_P = Basal area of Pinus ha^{-1}; Vol/ha_P = Volume of Pinus ha^{-1}; G/ha_B= Basal area of Broadleaves ha^{-1}

(t-1) to September of the year of growth (t) and the results of the correlation functions were tested for significance using the 95% percentile range method after a bootstrap process with 1,000 replications.

The *dplR* package (Bunn et al. 2014) and the *bootRes* package (Zang and Biondi 2012) of R cran software were used for the tree ring series management and the computation of the climatic correlation.

Climatic data

Data from a weather station located in the Barrea village (41,7570 N; 13,9919 E approx. 1,000 metres a.s.l.) were initially made available. Data series were anyway discontinuous, especially regarding rainfall. To get through the problem, a complete dataset was derived from ClimateEU database (http://www.ualberta.ca/~ahamann/data/climateeu.html), an unpublished software for the interpolation of climate data from PRISM database (Daly et al. 2008). It was applied for monthly precipitations and temperatures from 1901 to 2009. ClimateEU data were compared with local data calculating the fitness of ClimateEU tested with a regression method. In addition, Mann-Kendall non-parametric test was applied to assess the presence of climatic trends, possibly influencing the analysis (Brunetti et al. 2006).

Results

Mensurational data were collected during the growing season 2012 and are summarized in Tab. 2, Fig. 3 and 4. Mean DBH ranged between 22.8 cm (plot III) and 28.3 cm (plot V) whilst maximum tree height varied from 23 m (plot IV) to 16.1 m (plot III). The highest values of tree mean height (19.5 m), number of trees (1,178 ha^{-1}), basal area per hectare (60.2 m^2 ha^{-1}) and volume (574.5 m^3 ha^{-1}) were measured in plot IV. Plot III had the lowest mean tree height (11.8 m), basal area (39.6 m^2 ha^{-1}) and volume (259.7 m^3 ha^{-1}) and the highest variation of DBHs (64%). In addition, plots III and IV were characterized by the oldest and the youngest dominant trees, respectively. Height curves are compared in Fig. 3. Plots I, II and III were shaped similarly as for slope but with different values. In general, the stand structure revealed to be very variable chiefly for mean tree height and standing volumes.

The ANOVA on DBH confirmed the high variability and, as expected, grouped plots exactly according to compartments (Tab. 3 and Fig. 4).

The dendrochronological sampling was implemented in the following year (2013) before the start of the growing season. The main statistics on tree rings series, plot chronologies and mean stand chronology are reported In Tab. 4. The age of dominant trees ranges between 80 and 184 years with a mean annual increment between 0.627 mm (Plot IV Tree 1 - P4T1) and 1,967 mm (Plot V Tree 1 - P5T1). Sensitivity was generally low whereas the first order autocorrelation was quite high, it emphasizing the high inter-annual correlation between tree-ring widths. Calculating first-order autocorrelation coefficients on standardized indices, the absolute values ranged between 0.11 and 0.22. That means that only a small amount of the low-frequency year-to-year variation was not corrected with the standardization procedure.

Similarly to DBH ANOVA, the analysis on raw tree rings series on the common growth period (1933-2012) too, showed statistically significant differences among plots. Plots were however grouped differently following the age of sampled trees (Tab. 5 and Fig. 5). As expected, ANOVA on detrended (i.e. standardized) chronologies did not show the same differences (Fig. 6).

Before the searching of climatic correlations, ClimateEU data were carefully checked and compared to local data for the common time-period available (1955-2010). Regressions were always highly significant with a very high adj-R2 for temperature, but not for precipitation due to many missing values (mainly for storage errors) in the Barrea's database. At all

Table 3 - Results of Kruskal-Wallis Rank Sum Test and relative post-hoc test (Wilcoxon Rank Sum and Signed Rank Tests).

Statistical Mean DBH	Plot (compartment)	Group
26.43	IV (18)	A
26.28	V (18)	A
23.03	II (21)	B
22.63	I (21)	B
19.19	III (19)	C

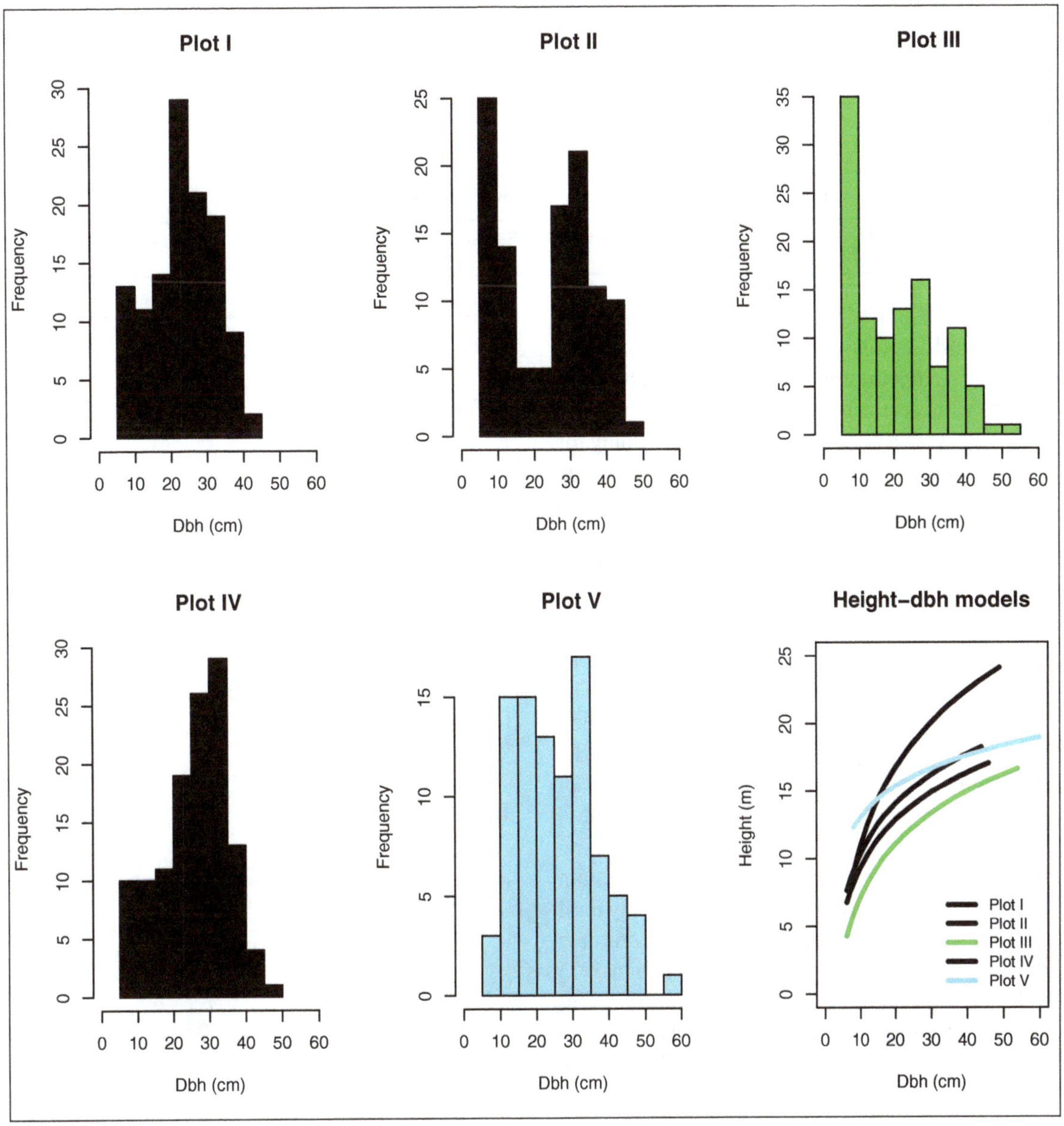

Figure 3 - Histograms of dbh values in each plot and comparison between the height-dbh models.

cases, however, the slope of functions showed a very high p-level, that means both the ClimateEU's and the Barrea's trends were statistically comparable (Tab. 6).

The Mann-Kendall test assessed the absence of climate trends, even if a growing linear trend in annual maximum and minimum temperatures was observed (Fig. 7) with +0.009°C/year for minimum and +0.004°C/year for maximum temperatures. Particular trends were detected only over the periods 1900-1940, 1941-1975 and 1976-2010. Over the first period, minimum temperatures, compared with maximum temperatures, were detected as growing more rapidly (+0.015°C/year vs. +0.006°C/year). Within the central period, minimum t. remained quite stable whilst maximum t. had a sensible decrease until 1960, reaching the same values of the early 1900. Over the third period, temperature continued to rise with close mean values (+0.029°C/year for minimum temperatures and +0.026°C/year for maximum t.). The same periods and trends were analysed for the precipitation amount. Very low adjusted-R2 (0.07-0.04) were detected, demonstrating lack of statistical relationships. Nevertheless, a general decrease between 1901 and 2010, even if relatively low (approx. -1.02 mm/year), was quite clear (Fig. 8, third plot with green line).

Correlations between species' growth and climate were then detected with CF and MCF, these showing different trends and results. With CF a significant and positive correlation was found with the minimum temperatures (Fig. 8) of the previous month of December, and the months of February and March of the same year, whilst a negative cor-

Table 4 - Main statistics of tree rings series.

Series	First	Last	Years	M	SD	SK	S1	S2	G	AR	ARS
M-P1	1834	2012	179	1.080	0.847	2.287	0.123	0.127	0.287	0.926	0.142
M-P2	1868	2012	145	1.137	1.493	1.163	0.156	0.157	0.359	0.899	0.100
M-P3	1829	2012	184	1.036	0.679	2.692	0.147	0.151	0.279	0.925	0.067
M-P4	1914	2012	99	1.745	0.785	0.698	0.165	0.162	0.249	0.872	0.060
M-P5	1895	2012	118	1.984	1.62	1.916	0.156	0.157	0.402	0.852	0.078
chrono	**1829**	**2012**	**184**	**1.578**	**0.823**	**1.165**	**0.141**	**0.146**	**0.278**	**0.873**	**0.081**

M = mean increment (mm); SD = standar deviation; SK = Skewness; S1 & S2 = sensitivity of first and second order; G = Gini coefficient; AR = Autocorrelation; ARS = Autocorrelation after standardization.

relation was found with the minimum temperatures of July and September of the current year. The maximum temperatures (Fig. 9) of the same winter months were correlated in the same way (positive), whilst a negative correlation was detected only for May within the same growing season. Concerning precipitation, the growth was significantly and negatively correlated only with the events of the previous December (Fig. 10). The analysis with MCF (Fig. 11) highlighted many periods in which we had a significant relationship between radial growth and climate. Whilst some of them were roughly stable with time, others appeared unstable throughout the period. The main correlations with temperatures were generally confirmed, even if not continuously, across the whole time-series. The minimum temperatures of August and September were negatively correlated over different decades. More evident and continuous correlations were detected with the minimum temperatures of February and March but only until 1950. Furthermore, only the minimum temperatures of March were significant, but only between 1975 and 1982. Other positive correlations were present and more continuous in the considered period and concerned maximum temperatures of the same months. As for the last decades of the concerned period, significant correlations were found only during October (minimum), December and September (maximum). In relation to the rainfall amount, very low and fragmented events were detected, especially in the last period (1973-2010) and regarding May and June of the growing season (t).

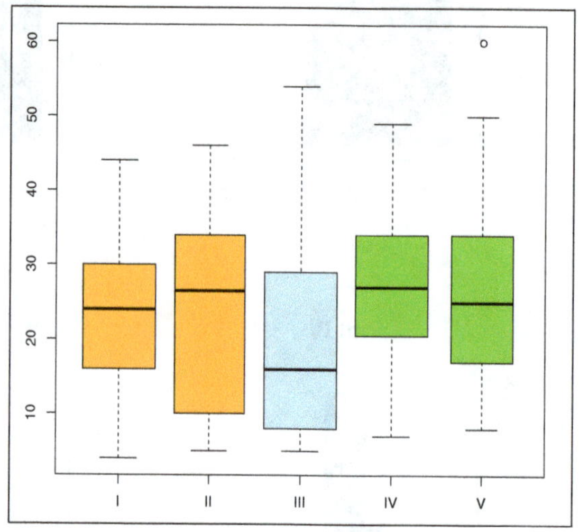

Figure 4 - Boxplot of DBH values.

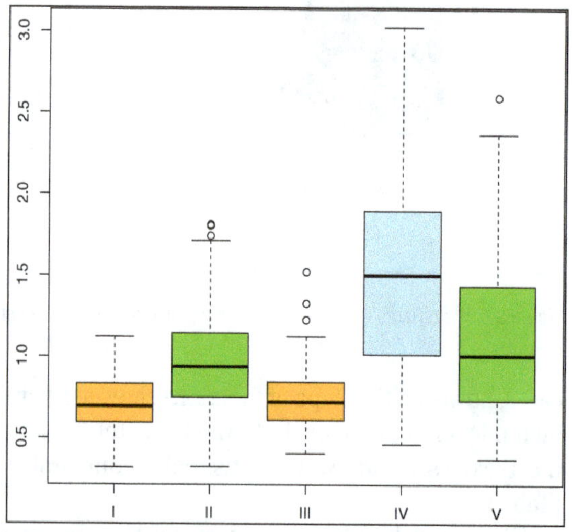

Figure 5 - Boxplot of ringwidths of raw chronologies.

Discussion

The seed stand of the MaP is a pure Black pine forest where broadleaves are quite sporadic and relatively more present where local conditions (such as ditches or cooler exposures) allowed them to be more competitive. In general, the vertical structure of stands showed two well-recognizable levels with Black pine as dominant species. Beech and ash were particularly present as stools following the selective cutting operated in the past and the animal browsing. In many areas, above all where beech was unable to compete, such as where we

Table 5 - Results of Kruskal-Wallis Rank Sum Test and relative post-hoc test (Wilcoxon Rank Sum and Signed Rank Tests) on raw chronologies on the common growth period.

Statistical Mean ringwidth	Plot (compartment)	Group
1.515 mm	IV (18)	A
1.137 mm	V (18)	B
0.973 mm	II (21)	B
0.744 mm	III (19)	C
0.710 mm	I (21)	C

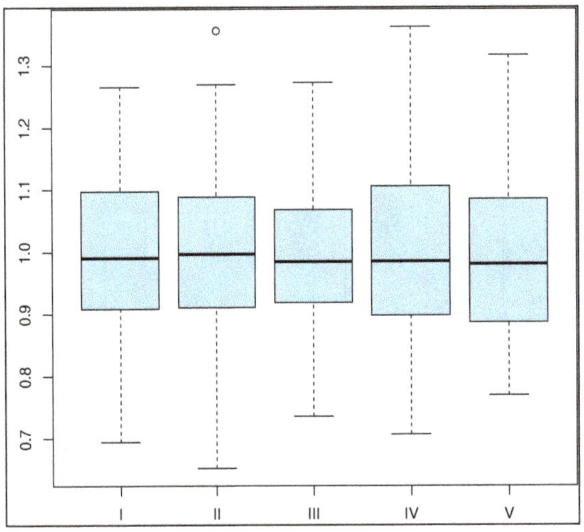

Figure 6 - Boxplot of ringwidths of detrended chronologies.

have southern exposures or shallow soils, Black pine occupied both levels.

Stand data collected in 2012 did not show any significant difference between the present time and the situation before the approval of the last management plan. In fact, any practices' implementation was reported by the management administration during the years 2001-2011 and no traces of tree felling were detected on the ground. ANOVA on DBH showed that forest structure was influenced by forest management, but also that mean radial increment was mainly influenced by stand age, site quality and ecological dynamics. According to our results, three different stand structures could be identified as follows:

- Even-aged adult stands (compartment 18 - plots IV and V): pine stands in these plots can be considered as even-aged with trees from medium to large size- and small DBH variation. The standing volume was chiefly related to the high number of medium-sized trees, and ranged from 259 m3 ha-1 to 574 m3 ha-1. The maximum age of dominant trees varied between 80 and 120 years, small-sized trees were sporadic and mainly young broadleaves. Maximum tree height was 21 m.
- Even-aged stands with old trees (compartment 21 - plots I and II): These plots can be considered as even-aged; the plot II had a high number of small-sized trees with DBH lower than 15 cm, probably due to a natural regeneration process. Standing volumes were quite similar, around 350 m3 ha-1 but the age of the bigger trees was very variable, ranging from 104 to 179 years.
- Uneven-aged stands with very old trees (compartment 19 - plot III): mean DBH and tree height

Table 6 - Results of regressions analysis based on the linear formula y = mx + q. Significance of parameters are reported with the following legend: p<0.1 (.), p<0.05 (*), p<0.01 (**), p<0.001 (***).

Variable	Adj R^2	Intercept (q)	Slope (m)
Minimum annual temperature	0.927	-0.655***	0.918***
Maximum annual temperature	0.977	2.617***	0.955***
Total annual precipitation	0.369	15.705**	1.143***

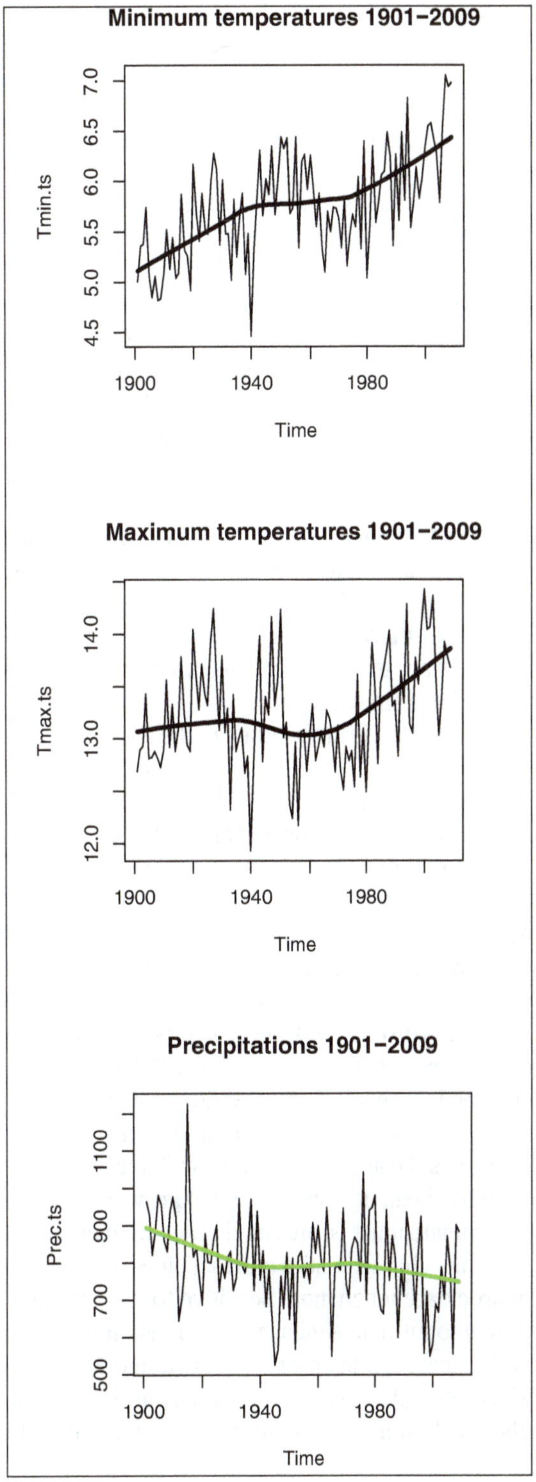

Figure 7 - Trends of minimum (blue line) and maximum (red line) annual temperatures and annual precipitations (green line) from 1901 to 2009 in Villetta Barrea (ClimateEU data).

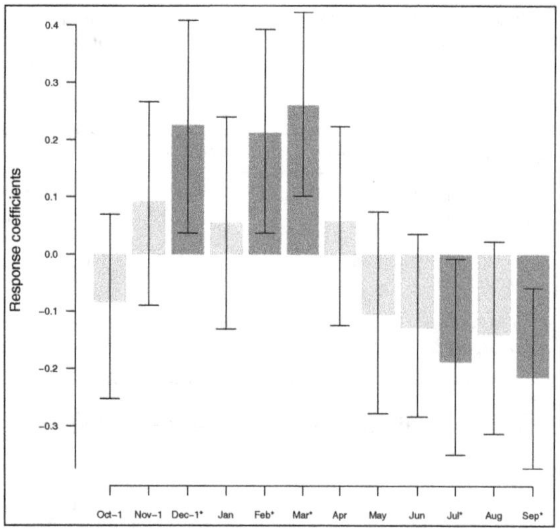

Figure 8 - Correlation function for Minimum temperatures.

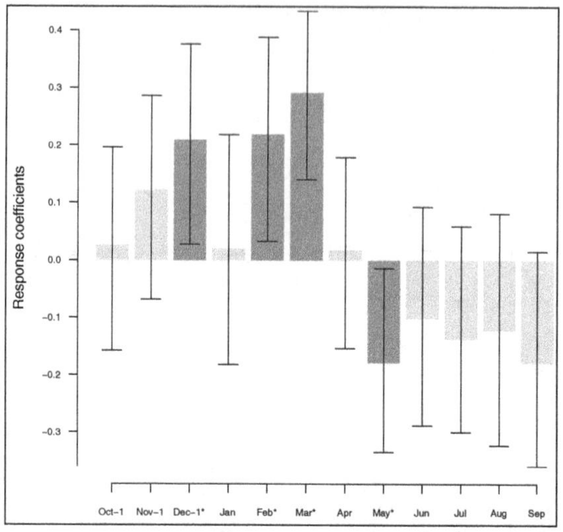

Figure 9 - Correlation function for Maximum temperatures.

were the smallest surveyed over the whole area. Tree density was very high, surely due to the high number of small-sized trees, and the coefficient of variation of DBH was high too, this indicating a very variable vertical structure. Standing volumes were the lowest measured, around 250 m3 ha-1. This stand could be considered as uneven-aged, where Black pine has established a natural regeneration dynamics. In this area the oldest (dominant) trees are aged from 143 to 184 years.

Climate influenced in the same way the growing trends, despite differences in age and size of trees. Dendrochronological analyses demonstrated that, despite a different mean increment in each of the sampled areas, mainly related to tree age, trees reacted in a similar way.

As reported in many other studies on Black pine spp., sensitivity to high summer temperatures was confirmed (Génova and Fernández 1999, Martin-Benito et al. 2010, Amodei et al. 2012). On the other hand, no continuous and recent correlation with the precipitation amount effect was detected, this probably depending on two main factors: the autecology of the Black pine of Villetta Barrea and the local soil structure. Black pine of Villetta Barrea is a bit more tolerant to drought stresses than other *Pinus nigra* subspecies (Gellini and Grossoni 2003). In addition, soil and local ecological conditions can increase or reduce water availability during different seasons as well as the amount of snow in winter.

It has been demonstrated that growth response and climate must be analysed in accordance to soil substrate characteristics and that pines on quartzite bedrock are much more sensitive to the precipitation amount than similar populations on dolomite and calcareous bedrocks (Génova and Martinez-Morillas 2002). In calcareous soils, the effect of the chemical alteration of the bedrock may offer to

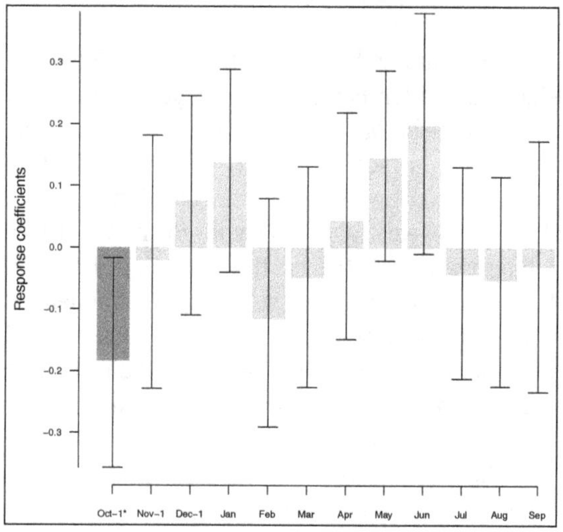

Figure 10 - Correlation function for Precipitations.

pine roots the opportunity to explore deeper levels (Amodei et al. 2012).

A positive correlation between radial growth and late-winter temperatures (February and March) in CF results was also reported by Linares and Tiscar (2010) on Salzmann pine in southern Spain. In this situation, trees may open buds earlier and, consequently, increase the length of the growing season. This is very likely to occur in Villetta Barrea, where water availability does not appear to be a problem. Results from MCF analysis pointed out few months with very unstable correlation and the lack of statistical significance over the last decades. The months detected as main driving forces with CF, proved to be relevant only for past events such as the temperatures of February and March. The endemic population of Villetta Barrea seemed to be less sensitive than a few planted stands analysed by Piermattei et al. (2014).

Some factors, not relevant with CF, turned out to be important with the MCF analysis and especially

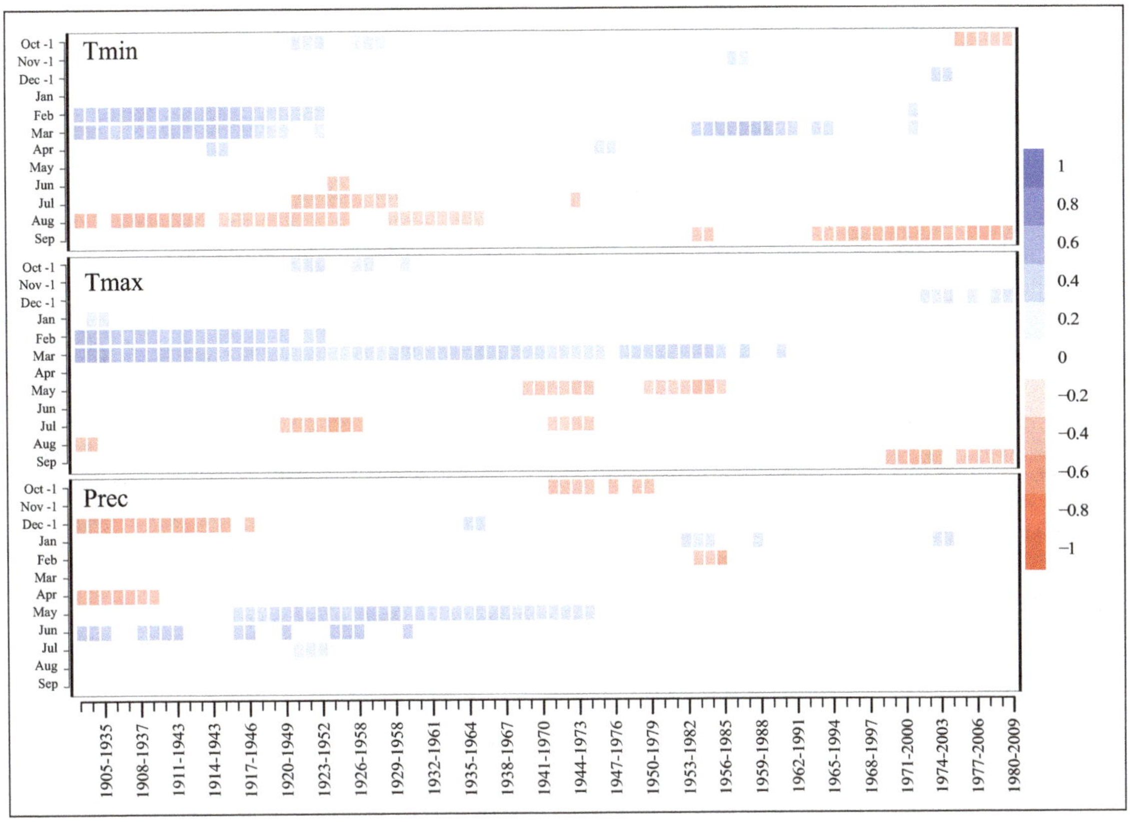

Figure11 - Result of the analysis with moving correlation function from period 1901-1930 (the first on on the left side) to 1980-2009 (the last on on the right side). Only significant correlations (p-value > 0.05) are reported.

concerning the last decades, such as the minimum temperatures of the previous October. In addition, also the shifted response from August to September was detected. In fact, the minimum temperature of August was statistically significant only in the past decades as long as the late '60s, whilst the same variable in September becomes relevant over the last decades. In this case the connection is probably linked to the cold stress that is likely to occur in late September at high elevation. High temperatures in late summer months can induce a longer growing season which can suddenly stop when an extreme event occurs. In this case, the moving windows approach of MCF appeared as a very good tool to address the study of forest species adaptability to climate change.

Definitively, the temperatures of September demonstrated to be the main driving force for the growth of Black pine of Villetta Barrea in the MaP and, in any case, the genetic provenance must also be accounted as a possible driver for variability.

Conclusions

Tree-ring analysis showed that the relationship between radial growth and climate was not affected by different stand structures and tree ages. The influence of climate factors is currently no longer active and the species is probably adapting to a new-established, changing environment. This will probably play a key role in the future and it is an issue to be duly taken into account in the management planning.

Many other questions and issues must be developed to cover all the possible analyses connected to marginal forest populations and to the management of forest genetic resources in view of the expected Global Change. For instance, the available information about tree species genetics and local adaptation progress must be carefully accounted.

The comparison between the population of Villetta Barrea and other populations and subspecies (mainly *nigra* and *laricio*) was useful to assess whether the one examined here is a case of genetic adaptation or an example of phenotypic plasticity and further similar surveys should be encouraged.

Acknowledgements

This study has been carried out as part of the ABRFORGEN project "*Implementation of forest nursery chain and organization of a modern management of forest genetic resources in Abruzzo*" a partnership between CREA-SEL and Abruzzo

Region and the COST Action FP 1202 "*Strengthening conservation: a key issue for adaptation of marginal/peripheral populations of forest trees to climate change in Europe (MaP-FGR)*". We thank also the Regional Agro-meteorological Center of Scerni for providing climatic data and scientific support.

The authors also thank the reviewers for helpful comments on an early draft.

References

Afzal-Rafii Z., Dodd R. 2007 - *Chloroplast DNA supports a hypothesis of glacial refugia over postglacial recolonization in disjunct populations of Black pine (Pinus nigra) in western Europe.* Molecular Ecology 16: 723–736. DOI: 10.1111/j.1365-294X.2006.03183.x

Amodei T., Guibal F., Fady B. 2012 - *Relationships between climate and radial growth in Black pine (Pinus nigra Arnold ssp. salzmannii (Dunal) Franco)* from the south of France. Annals of Forest Science 70 (1): 41-47.

Biondi F., Visani S. 1993 - *Cronologie dendroanulari per la Penisola Italiana [Tree-ring chronologies for the Italian Peninsula; in Italian with English abstract].* Annali della Facoltà di Agraria, Catholic University of Milan, Italy 33 (1): 3-23.

Brunetti M., Maugeri M., Monti F., Nanni T. 2006 - *Temperature and precipitation in Italy in the last two centuries from homogenised instrumental time series.* International Journal of Climatology 26: 345-81.

Bunn A., Korpela M., Biondi F., Campelo F., Mérian P., Mudelsee M., Qeadan F., Schulz M., Zang C. 2014 - *dplR: Dendrochronology Program Library* in R. R package version 1.6.0. http://CRAN.R-project.org/package=dplR

Cook E.R., Peters K. 1981 - *The smoothing spline: a new approach to standardizing forest interior tree-ring width series for dendroclimatic studies.* Tree-Ring Bull 41: 45-53.

Cook E.R. 1987 - *The decomposition of tree-ring series for environmental studies.* Tree-Ring Bulletin 47: 37-59.

Daly C., Halbleib M., Smith J. I., Gibson W. P., Doggett M. K., Taylor G. H., Curtis J. et al. 2008 - *Physiographically sensitive mapping of climatological temperature and precipitation across the conterminous United States.* International Journal of Climatology 28 (15): 2031–2064. DOI: 10.1002/joc.1688

Fenaroli L., Gambi G. 1976 - *Alberi, Dendroflora Italica. Museo Tridentino di Scienze Naturali.* Trento, Italy: 89-100.

Fritts H.C. 1976 - *Tree Rings and Climate.* New York, Academic Press, 567 p.

Gallucci V, Urbinati C. 2009 - *Dinamismi di accrescimento e sensitività climatica dell'abete bianco (Abies alba Mill.) nel SIC Alpe della Luna-Bocca Trabaria (PU).* Forest@ 6: 85-99 [Online] Available: http://www.sisef.it/forest@/. [2009-03-25].

Gasparini P., Tabacchi G. (a cura di) 2011- *L'Inventario Nazionale delle Foreste e dei serbatoi forestali di Carbonio, INFC 2005. Secondo inventario forestale nazionale italiano. Metodi e risultati.* Ministero delle Politiche Agricole, Alimentari e Forestali; Corpo Forestale dello Stato. Consiglio per la Ricerca e la Sperimentazione in Agricoltura, Unità di ricerca per il Monitoraggio e la Pianificazione Forestale. Edagricole-Il Sole 24 ore. Bologna, 653 p.

Gellini R., Grossoni P. 2003 - *Botanica forestale. I - Gimnosperme.* Padova (IT), CEDAM: 160-170. ISBN 88-13-19785-3

Génova M., Fernández A. 1999 - *Tree rings and climate of Pinus nigra subsp. salzmannii in central Spain.* Dendrochronologia 16-17: 75-85.

Génova M., Martinez-Morillas D. 2002 - *Estudio dendroecologico de Pinus nigra en Checa (Guadalajara).* Ecologia 16 :83–95.

Hampe A., Petit R.J. 2005 - *Conserving biodiversity under climate change: the rear edge matters.* Ecology Letters 8 (5): 461-467.

Isajev V., Fady B., Semerci H., Andonovski V. 2004 - *EUFORGEN Technical Guidelines for genetic conservation and use for European black pine (Pinus nigra).* International Plant Genetic Resources Institute, Rome, Italy, 6 p.

Linares J.C., Tiscar P.A. 2010 - *Climate change impacts and vulnerability of the southern populations of Pinus nigra subsp. Salzmannii.* Tree Physiology 30: 795-806.

Lindner M. 2000 - *Developing adaptive forest management strategies to cope with climate change.* Tree Physiology 20: 299–307.

Martin-Benito D., Del Rio M., Cañellas I. 2010 - *Black pine (Pinus nigra Arn.) growth divergence along a latitudinal gradient in Western Mediterranean Mountains.* Annals of Forest Science 67: 401-401. DOI:10.1051/forest/2009121

Mazza G., Gallucci V., Manetti M.C., Urbinati C. 2013 - *Tree-Ring Growth Trends of Abies alba Mill: Possible Adaptations to Climate Change in Marginal Populations of Central Italy.* The Open Forest Science Journal 6 (1: M4): 46-49.

Parmesan C. 1996 - *Climate and species' range.* Nature 382: 765-766. DOI: 10.1038/382765a0

Parmesan C. 2006 - *Ecological and Evolutionary Responses to Recent Climate Change.* Annual Review of Ecology, Evolution and Systematics 37: 637-669. DOI: 10.1146/annurev.ecolsys.37.091305.110100

Petit R. J., Aguinagalde I., de Beaulieu J.-L., Bittkau C., Brewer S., Cheddadi R., Ennos R., et al. 2003 - *Glacial refugia: hotspots but not melting pots of genetic diversity.* Science 300 (5625): 1563–1565. DOI: 10.1126/science.1083264

Piermattei A., Garbarino M., Urbinati C. 2014 - *Structural attributes, tree-ring growth and climate sensitivity of Pinus nigra Arn. at high altitude: common patterns of a possible treeline shift in the central Apennines (Italy).* Dendrochronologia 32 (3): 210-219. ISSN 1125-7865, DOI: 10.1016/j.dendro.2014.05.002.

Puletti N., Mura M., Castaldi C., Marchi M., Chiavetta U., Scotti R. 2014 - *ForIT: Functions from the 2nd Italian Forest Inventory (INFC).* R package version 1.0. http://CRAN.R-project.org/package=ForIT

Quézel P., Médail F. 2003 - *Ecologie et biogéographie du bassin méditerranéen.* Elsevier, 576 p.

R Core Team 2014 - *R: A language and environment for statistical computing.* R Foundation for Statistical Computing, Vienna, Austria. http://www.R-project.org/

Spiecker H. 2002 - *Tree rings and forest management in Europe.* Dendrochronologia 20 (1-2): 191-202. ISSN 1125-7865, DOI: 10.1078/1125-7865-00016.

Tabacchi G., Di Cosmo L., Gasparini P., Morelli S. 2011 - *Stima del volume e della fitomassa delle principali specie forestali italiane. Equazioni di previsione, tavole del volume e tavole della fitomassa arborea epigea.* Consiglio per la Ricerca e la sperimentazione in Agricoltura, Unità di Ricerca per il Monitoraggio e la Pianificazione Forestale. Trento, 412 p.

Vidakovic M. 1974 - *Genetics of European black pine (Pinus nigra Arn.).* Annales Forestales 6 (3): 57-86

Zang C., Biondi F. 2012 - *Dendroclimatic calibration in R: The bootRes package for response and correlation function analysis.* Dendrochronologia 31 (1): 68-74. DOI: 10.1016/j.dendro.2012.08.001.

5

Multi-species and multifunctional smallholder tree farming systems in Southeast Asia: timber, NTFPs, plus environmental benefits

James M Roshetko[1*], Manuel Bertomeu[2]

Abstract - The rapid increase in human population, and the corresponding worldwide enhancement of social and economical conditions, are exerting a considerable pressure to convert forests to other uses. Moreover, these phenomena raise the demand for food, fuel, wood fibers and other non-wood products, contributing to a further boost of the production pressure in the surviving forests. Simultaneously, these forests are expected to provide a diverse array of environmental services. Furthermore, smallholder forestry systems are prominent components of 'trees outside the forest' in Southeast Asia and they are primarily 'planted' systems that rehabilitate or reforest marginal lands, in order to produce tree products and services. As they traditionally are a means of producing goods for home consumption, they have become significant suppliers of products for local, national and international markets. The aim of this paper is to demonstrate that smallholder forestry systems are a viable management system which is significantly contributing to global environmental goals and local economic objectives. This paper reviews global and Asian trends of human population growth, deforestation, and demand for forest and tree products. The origin, the diversity, the adaptable management and the importance of smallholder tree-based systems are here discussed and significant details are provided on the role of smallholder tree-based systems in the mitigation of deforestation, which could be obtained by expanding regional forest resources; in supplying alternative sources of forest products and environmental benefits; and in making significant contributions to local livelihoods for rural communities.

Keywords - Smallholder, tree-farming, livelihood enhancement, timber, NTFPs, environmental benefits.

Introduction

The global human population reached 7 billion in October 2011, only 12 years after having reached 6 billion, and having doubled since 1968 (Worldometers 2011). With an annual growth rate of 75 million, the population is projected to be over 9 billion by 2050 (United Nations 2015). This rapid human population growth, and a corresponding increase in the wealth of some nations, are exerting a significant pressure capable of converting forests to agricultural, industrial, and residential uses. This also increases the demand for food, fuel, wood and non-wood products, intensifying the pressure on the surviving forest systems. Simultaneously, these forest systems are expected to provide a diverse array of environmental services. For this reason, the United Nations Millennium Development Goals are calling for a more considerable per capita wealth growth which might lead to the eradication of extreme poverty and hunger, while ensuring environmental sustainability (United Nations 2012).

Agroforestry is a dynamic, ecologically based, natural resources management system which, through the integration of trees on farms and in the agricultural landscape, diversifies and sustains the production of goods for the increased social, economic and environmental needs of land users at all levels (Mead 2004). These systems are increasingly recognized as important opportunities for smallholder livelihoods, with neutral-to-positive environmental impacts, and they have received significant research attention over the last two decades (Leakey et al. 2012).

'Agroforestation' refers to the establishment of smallholder agroforestry systems and implies land rehabilitation through the establishment of tree-farming systems and intensification of land management (Roshetko et al. 2007a). Farmers develop and manage such systems by nurturing trees on their farms, pasture lands and homesteads. These tree-farming systems are efficient agricultural and natural resources production systems. As a prominent component of 'trees outside the forest', smallholder tree-farming systems are primarily 'planted' systems that rehabilitate or reforest marginal farmlands where agricultural crop production is no longer biophysically or economically viable.

[1] World Agroforestry Centre (ICRAF) Southeast Asia Regional Program, Bogor, West Java, Indonesia
[2] University of Extremadura, UEX, Plasencia, Cáceres, Spain
[*] corresponding author: J.Roshetko@cgiar.org

These systems can also be used to reclaim degraded lands. Smallholder systems could be more productive and profitable if the common policy barriers that limited their development were removed (van Noordwijk et al. 2008).

This paper emphasizes the contribution of smallholder tree-farming systems to environmental sustainability and local livelihoods. It first reviews the trends of regional deforestation and human population growth both globally and in Asia, with an emphasis on South and Southeast Asia. Subsequently, common tree-farming systems are described and their potential to produce forest/tree products and environmental benefits are finally discussed. A particular emphasis is placed on the potential of smallholder tree-based systems to reduce the pressure on the remaining natural forests, to expand regional forest resources, to produce forest/tree products and environmental benefits as well as to make major contributions to local livelihoods for rural communities.

Forest loss, environmental degradation and loss of forest services

The rate of global forest loss in the 20 years between 1990 and 2010 was alarming: For the period 1990-2000, the global annual deforestation rate was 16 million hectares; for 2000-2010, it was 13 million hectares. In addition, this alarming rate likely underreports the damage sustained by the global forest resource, as forest degradation is not included.

Forest cover was reduced to slightly more than 4 billion hectares (30% of the global land area) and the two countries with the largest loss of forest area, over the 20 years period, were Brazil and Indonesia, which respectively lost 2.8 million and 1.2 million hectare/year, representing 0.5% and 1.1% of the annual loss of their forest area (FAO 2010). These changes primarily represent the loss of tropical forests to other land uses: conversion from diverse tropical ecosystems to annual agricultural systems, monoculture tree plantations and cleared (but left not used) landscapes.

Fortunately, the rate of forest loss in both countries and across the globe has declined, but the rate is still far from being sustainable. The rate of deforestation is somewhat offset by planting and natural forest regeneration. The total net change in global forest area was a decline of 8.3 million hectare/year in 1990-2000 and of 5.2 million hectare/year in 2000-2010, the difference with the above deforestation figures being between the areas planted or naturally regenerated (FAO 2011). Efforts to plant new forests and trees have gained momentum as the planted forests represent 7% of total global resources, with an increase of 5 million hectares in the first 10 years of the millennium (FAO 2010).

In Asia, the deforestation-afforestation trend has been mixed. On the basis of FAO data for the 1990-2000 period, the Asia-Pacific region lost forest cover at a rate of 700,000 hectare/year. However, in the subsequent 10 years the trend reversed, with the regional forest cover increasing by 1.4 million hectare/year (FAO 2011, FAO 2010). The reversal in regional deforestation was largely due to the successful tree planting programs in China, India, Vietnam and Thailand. In the 20 years under consideration, China amazingly planted 35.2 million hectares of forests, India 4.5 million, Vietnam 2.5 million and Thailand 1.3 million. Sub-regional and national performance varied significantly. East and South Asia both showed gains in forest cover, while Southeast Asia and the Pacific continued to lose forest cover (Table 1). In Indonesia, the rate of forest lost has greatly declined, but annual forest loss remained still high (100,000 hectares of primary forests and 30,000 hectares of planted forests).

Besides helping to reverse the loss of forest cover, planted forests are an important and efficient source of wood and non-wood products. In 2000, forest plantations accounted for approximately 5% of global forest cover, with industrial forest plantations accounting for only 3% but supplying 35% of global roundwood (FAO 2000). By reducing production pressure, planted forests may have a tempering effect on the rate of natural forest loss.

Established for diverse reasons, tree plantations have generally limited species diversity and they have frequently been monocultures of exotic species. Such systems are much inferior to natural forests in supporting most of the main ecosystem services: biodiversity and habitat conservation, genetic conservation, ecological resilience, water and soil conservation, and carbon storage. Additionally, in many cases, forest plantations are a main cause of natural forest conversion and loss, thus being a direct cause of natural forest growth, biodiversity and carbon stock loss. Hence, forest plantations are a paradox: they are an important and efficient source of wood and non-wood products but are also a main cause of the forest conversion and of the loss of environmental services provided by these natural systems.

Population growth, economic development and demand for forest products

While the forest base will decrease, human populations and economic development will grow, increasing the demand for, and the consumption of, forest and wood products throughout Asia and the

Table 1 - Forest areas in Asia and the Pacific, 1990-2010 (FAO 2011).

Sub-region[1]	Area (,000 ha)			Annual change (,000 ha)		Annual change %	
	1990	2000	2010	1990-2000	2000-2010	1990-2000	2000-2010
East Asia	209,108	226,815	254,626	1762	2781	0.81	1.16
South Asia	78,163	78,098	80,039	-7	221	-0.01	0.28
SE Asia	247,260	223,045	214,063	-2422	-898	-1.03	-0.41
Pacific	198,744	198,381	191,384	-36	-700	-0.02	-0.36
Asia-Pacific	733,364	726,339	740,383	-703	1404	-0.10	0.19
World	4,168,399	4,085,063	4,032,905	-8334	-5216	-0.20	-0.13

[1] **East Asia:** China, North Korea, Japan, Mongolia, South Korea
South Asia: Bangladesh, Bhutan, India, Maldives, Nepal, Pakistan, Sri Lanka
SE (Southeast) Asia: Brunei, Cambodia, Indonesia, Laos, Malaysia, Myanmar, Philippines, Singapore, Thailand, Timor-Leste, Vietnam
Pacific: American Samoa, Australia, Cook Islands, Micronesia, Fiji, French Polynesia, Guam, Kiribati, Marshall Islands, Nauru,, New Caledonia, New Zealand, Niue, Norfolk Island, Northern Marina Islands, Palau, Papua New Guinea, Pitcairn, Samoa, Solomon Islands, Tokelau, Tonga, Tuvalu, Vanuatu, Wallis and Futuna Islands.

rest of the world. In 1995, South and Southeast Asia were home to, respectively, 1,109 million (23% of the world's population) and 437 million (9%) (ADB 2004). By 2010, the human populations of the regions had grown to 1,598 million and 593 million, respectively, with their proportional share of the global population remaining steady (United Nations 2010).

Current annual population growth rates for individual countries in South and Southeast Asia ranged between 0.7% in Thailand and Myanmar; 1.8% in Pakistan, Nepal, Bhutan and Brunei; 2.1% in Timor-Leste; and 3.5% in Singapore (United Nations 2010). Gross national income (GNI) per capita in the regions in 2005 varied from US $270 in Nepal through US $430 in Cambodia and Laos to US $2,720 in Thailand and US $4,970 in Malaysia. Increases in GNI per capita between 2000 and 2005 varied from 17% in Nepal and Bhutan to 35% in Thailand, 62% in India and Vietnam, and 125% in Indonesia (ADB 2006). The gross domestic products (GDP) of most South and Southeast Asian countries have grown at annual rates of about 5-8% between 2010 and 2012 (CIA 2012). At such growth rates, the size of the middle classes in Asia's developing economies (excluding Japan) would double or triple in the first decade of the new millennium, numbering 0.8 to 1 billion people and forming a middle-class market equal to or surpassing that of the US and Europe combined (Chipeta et al. 1998). Population growth and expansion of middle classes with greater disposable incomes will increase the consumption of forest products, which, in turn, will be reflected in the expanded global trade of these products.

The demand for all forest products is significant and is projected to increase, from the regional to the international level. For instance, the global demand for industrial roundwood is expected to increase to 2,165 and 2,436 million m^3 by 2020 and 2030, respectively, these representing increases of 29% and 49% over global production in 2005. During the same time, Asia-Pacific demand for industrial roundwood will increase by 58% and 78%, respectively, but production will not keep pace with demand (FAO 2009). This projection emphasizes the urgent need to expand the regional forest base, a process that should include afforestation, reforestation, and smallholder agroforestry systems (Roshetko et al. 2008).

Smallholder tree-farming systems: origin, multiple goods and environmental benefits in Southeast Asia

In this paper, the term 'smallholder tree-farming systems' is interchangeable with 'smallholder agroforestry systems'. Depending on local needs or opportunities, smallholder systems may focus on tree crops, agricultural crops, livestock or a combination of the three. These various systems will differ greatly in size, species components, tree density, tree longevity, and management intensity.

Smallholder tree-farming systems may originate from natural forests that have been altered in composition or structure by local people, tree-based systems established on agricultural or fallowed land, or a combination of both. There are examples of forest degradation being deflected by the establishment of smallholder tree-farming systems which avoid the more serious stages of environmental degradation. In these situations, good markets for non-wood products, such as fruits, resins and latex, have allowed a transition of substantial areas of Southeast Asian forests into 'agroforests', i.e. a land use that combines 'planted trees' with forest flora and fauna, with either retained or naturally regenerated vegetation (de Jong et al. 2001, Michon and de Foresta 1990, 1995). Similarly, by the production of wood and non-wood products on farms, smallholder agroforestry systems have been identified as a means of reducing pressure on and conserving natural forests (de Foresta et al. 2003, Scherr and McNeely 2008, Strandby-Andersen et al. 2008). Farmers in Sumatra cultivating agroforestry systems relied less on wood supplies harvested from natural forests than those without agroforestry systems (Murniati et al. 2001). Dawson et al. (2013) comment on agroforestry's role in the conservation of tropical tree diversity through use.

In deforested regions, a shortage of local forest resources is often the catalyst of spontaneous expansion of smallholder agroforestry systems (Santos-Martín et al. 2012). This type of farmer-led, spontaneous, smallholder tree-farm development has been documented in Sri Lanka (Gunasena 1999), Philippines (Garrity and Agustin 1995, Schuren and Snelder 2008), Kenya (Scherr 1995) and Indonesia (Michon and Bompard 1987). In addition, proximity to urban centers creates high demand for timber, fruit and other forest products and stimulates spontaneous smallholder agroforestry. This is especially true for areas far from the extractive forest frontier and/or with farms large enough to support tree crops, in addition to seasonal cash crops. In other situations, the temporary migration of young people to cities results in the extensification of land use with tree farming evolving as a lower labour input alternative to annual crops (Bertomeu 2006). Thus, smallholder tree planting has led to land rehabilitation (Pulhin et al. 2006) and regional forest transition by restoring tree cover (Mather and Needle 1998).

Smallholder farmer tree-planting systems are generally successful on their own terms. Smallholders commonly have limited time, labour and financial resources. Planting trees represent a conscious investment chosen among other available options. Farmers generally restrict plantings to the number of trees that can be maintained and integrate tree-growing with agricultural crops and animal husbandry activities. The management practices undertaken to ensure good food crop yields, cultivation, weed control and fertilization, and tree pruning, also benefit trees (Bertomeu et al. 2011). The available land, labor and other resources are allocated according to the farmer's objectives. Because landholdings are small, farmers can select the farm niches most appropriate for tree production. The combination of limited resources, small individual plantings, and intimate familiarity with the planting site result in high species diversity, tree survival and good growth rates. Smallholder tree-growing activities benefit from intensive management over limited areas and vested self-interest: the desire of the farmer to profit from their investment of time and resources (Roshetko et al. 2008).

As opposed to forest plantations and other public-planted forests, smallholder tree-farming systems provide an array of tree and forest products and environmental benefits, including support of local livelihoods. Smallholder timber systems are frequently used to grow assets for emergencies or specific cash needs (Roshetko and Westley 1994, Bertomeu 2004, Perdana et al. 2012). Under these conditions, smallholder farmers look at tree farming as a means of diversifying their production, reducing risk, and building assets to enhance family incomes and security (Roshetko et al. 2007b, Schuren and Snelder 2008). Besides supporting family livelihoods, smallholder agroforestry systems also make a significant contribution to national economies and global trade. In the Philippines and Indonesia smallholder-produced timber has become an important source of raw material for the local industry and the international markets (Bertomeu 2004, Tukan et al. 2004, Bertomeu, 2008, Roshetko et al. 2013). Products from smallholder systems in Indonesia include rattan, forest honey, sandalwood, gaharu, damar, benzoin, cinnamon, cloves, nutmeg, candlenut, rubber, cacao, coffee, oil palm and tea (Dove 2004, de Foresta et al. 2003, Garcia Fernandez 2004, Rohadi et al. 2003, Sunderlin et al. 2000, DGEC 2012). The five major global tree commodities are oil palm, coffee, rubber, cacao and tea, with an export value of roughly US $80 billion in 2009 (Dawson et al. 2014). Indonesia is a major producer of all the five commodities. In 2011, smallholders in Indonesia produced most of the coffee and cacao, 80% of the rubber, 39% of the oil palm, and 26% of the tea (Table 2).

Smallholders' agroforestry also provide many environmental benefits, including soil fertility replenishment, water catchment protection, biodiversity conservation, genetic conservation, reforestation, carbon storage, besides the reduction of pressure on natural forests (Garrity 2004, Roshetko et al. 2007a, Idol et al. 2011, Dawson et al. 2013). In societies where the majority of people live in urban areas, concerns over the accelerating loss of open and green space are becoming prominent. This is a quality-of-life issue to many and increases the recognition of agroforestry systems value to provide ecological functions also.

The main purpose of diversified productions and the complementary achievement of private and public environmental benefits, attribute dynamic, productive, risk-averse values to agroforestry systems. Additionally, but less recognized, agroforestry systems demonstrate a valuable potential in both

Table 2 - Smallholder production of oil palm, coffee, rubber, cacao and tea in Indonesia, 2011 (DGEC 2012).

	Smallholder area (,000 ha)	% of total area	Smallholder production	% of total production (,000 ton)
Oil palm	3,315	42	7,774	39
Coffee	1,255	96	679	96
Rubber	2,935	85	2,104	80
Cacao	1,641	94	828	92
Tea	56	46	40	26

mitigation and adaptation to climate change. On a per area basis, tree-rich, smallholding systems store a significant amount of carbon. Multi-storey agroforests and tree gardens 60 years-old or more can store up to 350 Mg ha^{-1}; home-gardens and smallholders' timber systems around 40 years-old can store up to 280-300 Mg ha^{-1}. The amount of carbon stored by specific systems will depend on biophysical conditions and tree density. However, the amounts of carbon stored in smallholders' agroforestry systems are similar to those in some secondary forests over similar time periods and greatly exceed the carbon stored in the low-biomass systems - such as fallow agricultural land and *Imperata* grasslands - that smallholders' systems often replace (Roshetko et al. 2002).

Other factors bearing on carbon stock are the species composition and management practices, specifically, the time trees are being mantained in a system. For carbon storage it may be beneficial to limit the number of low-biomass species - such as coconuts (*Cocos nucifera*) and bananas (*Musa x paradisica*) - and the amount of timber harvested. These options have to be balanced with farmers' goals for their own livelihood and land management. To farmers, carbon is even less tangible than other environmental benefits, such as watershed protection or biodiversity conservation. First, farmers' agroforestry systems must be socially and economically viable because then they are less likely to be converted to other, lower carbon, land uses. Carbon payments to farmers would promote transparency and increase farmers' understanding of the services the agroforestry systems provide. Any income received for carbon should be treated as an additional return for services already provided by such systems. This approach would help protect smallholders from project or market failure (Roshetko et al. 2007a).

Smallholders can also play a key role in protecting, through use, plant and tree diversity. Although smallholder tree-based systems are less diverse than native forest, agroforestry landscapes of the tropics usually contain dozens or even hundreds of tree species (Idol et al. 2011, Dawson et al. 2013). In complex agroforests, farmers may retain remnants from natural forests because of the products they provide or for religious, cultural or aesthetic reasons. The high tree species richness found in these agroforests suggests a strong role for smallholders to conserve plants and trees in farmland (i.e., circa situm conservation) (Dawson et al. 2013).

In deforested landscapes, smallholder tree planting can increase tree diversity and density, using indigenous or exotic planting materials produced either on-farm or in commercial tree nurseries (Ordoñez et al. 2014). Farmer-led reforestation may initiate forest transition, accelerating the restoration of degraded lands and resulting in higher biodiversity value (van Weerd and Snelder 2008, Idol et al. 2011). Compared to large-scale forest plantations, smallholder systems contain a much greater number of plant and animal species (Michon and de Foresta 1995, Murdiyarso et al. 2002). This diversity can provide ecological resilience and contribute to the recovery and maintenance of beneficial ecological functions. Smallholder tree-based systems, similar to plantations, are 'working forests' and they can help relieve some of the pressure to harvest native forests (although their presence as such is not a sufficient condition for the protection of old-growth forests (Angelsen and Kaimowitz, 2001, Tomich et al. 2002). For instance, in watersheds, linked systems of upland and riparian tree-based buffer systems, designed with regard to other landscape practices and features, can optimize soil and water conservation (van Noordwijk et al. 1998), along with other economic and social services.

Conclusions

Smallholders with diverse, risk-averse farms that include a significant tree component, have been producing a range of tree-commodities and could be efficient producers of other tree-commodities in the future. As described above, their tree-farming systems have high potential to yield both wood and non-wood products and play a key role in the recovery of degraded lands. Smallholder tree-farming systems have the potential to be one component of a general poverty alleviation strategy for agrarian-based, poor rural communities (Roshetko et al. 2007a, Snelder 2008). Although the social potential of tree-farming systems has not been fully exploited and the extent to which these systems can alleviate poverty and enhance food security is poorly documented, the importance and potential of the systems will continue to rise, especially with the continued development of market economies and rural infrastructure (Roshetko et al 2002). The importance of smallholder systems first as a source of forest and tree products and secondly as provider of environmental benefits will only increase as global forest resources continue to shrink and human populations expand (Roshetko 2013).

Acknowledgements

This work was partially supported by the CGIAR Consortium Research Program No. 6: Forests, Trees and Agroforestry: Livelihoods, Landscapes and Governance (CRP6).

The authors want to thank the anonymous reviewers for their helpful suggestions.

References

[ADB] Asian Development Bank 2004 - *Key indicators 2004: poverty in Asia: measurements, estimates, and prospects.* Manila: Asian Development Bank.

[ADB] Asian Development Bank 2006 - *Key indicators 2006: measuring policy effectiveness in health and education.* Manila: Asian Development Bank.

Angelsen A., Kaimowitz D. (eds). 2001 - *Agricultural Technologies and Tropical Deforestation.* Wallingford (UK): CABI Publishing.

Bertomeu M. 2004 - *Smallholder Timber Production on Sloping Lands in the Philippines: A Systems Approach. World Agroforestry Centre.* Southeast Asia Regional Research Programme, 257 p. I.S.B.N.: 971-93153-0-X

Bertomeu M. 2006 - *Financial Evaluation of Smallholder Timber-based Agroforestry Systems in Claveria, Northern Mindanao, the Philippines.* Small-scale Forest Economics, Management and Policy 5 (1): 57-82.

Bertomeu M. 2008 - *Can Smallholder Tree Farmers Help Revive the Timber Industry in Deforested Tropical Countries? A Case Study from Southern Philippines.* In: D.J. Snelder and R.D. Lasco (eds.) Smallholder Tree Growing for Rural Development and Environmental Services. Lessons from Asia. Springer: 177-191.

Bertomeu M., Roshetko J.M., Rahayu S. 2011 - *Optimum pruning strategies for reducing crop suppression in a gmelina-maize smallholder agroforestry system in Claveria, Philippines.* Agroforestry Systems 83: 167–180.

Chipeta M.E., Whiteman A., Brooks D.J. 1998 - *Review of social and economic developments in the Asia-Pacific region with projections to 2010.* Asia-Pacific Forestry Sector Outlook Study Working Paper Series No: APFSOS/WP/49. Bangkok: Food and Agriculture Organization of the United Nations, Forestry Policy and Planning Division, Regional Office for Asia and the Pacific.

[CIA] Central Intelligence Agency 2012 - *The World Factbook.* Washington DC: Central Intelligence Agency. [Online]. Available: https://www.cia.gov/library/publications/the-world-factbook/fields/2003.html [2015]

Dawson I.K., Guariguata M.R., Loo J., Weber J.C., Lengkeek A., Bush D., Cornelius J., Guarino L., Kindt R., Orwa C., Russell J., Jamnadass R. 2013 - *What is the relevance of smallholders' agroforestry systems for conserving tropical tree species and genetic diversity in circa situm, in situ and ex situ settings? A review.* Biodiversity Conservation 22: 301-324.

Dawson I.K., Leakey R., Clement C.R., Weber J.C., Cornelius J.P., Roshetko J.M., Vinceti B., Kalinganire A., Tchoundjeu Z., Masters E., Jamnadass R. 2014 - *The management of tree genetic resources and the livelihoods of rural communities in the tropics: Non-timber forest products, smallholder agroforestry practices and tree commodity crops.* Forest Ecology and Management 333 (1): 9-21. http://dx.doi.org/10.1016/j.foreco.2014.01.021

De Jong W., van Noordwijk M., Sirait M., Liswanti N., Suyanto S. 2001 - *Farming secondary forests in Indonesia.* Journal of Tropical Forest Science 13: 705–726.

de Foresta H., Michon G., Kusworo A., Levang P. 2003 - *Damar agroforests in Sumatra, Indonesia.* In: "Forest products, livelihoods and conservation. Case studies of non-timber forest product systems. Vol.1: Asia. Bogor, Indonesia." Kustersand K., Belchers B., eds. Center for International Forestry Research.

Dove M. 2004 - *So far from power, so near the forest: a structural analysis of gain and blame in tropical forest development.* In: "Borneo in transition: people, forests, conservation and development. Second edition." Padoch C., Peluso N., eds. Oxford UK, Oxford University Press.

[DGEC] Director General of Estate Crops 2012 - *Commodities data.* Director General of Estate Crops, Department of Agriculture, Government of Indonesia. [Online]. Available: http://ditjenbun.deptan.go.id/cigraph/index.php/viewstat/komoditiutama. [2015]

[FAO] Food and Agriculture Organization of the United Nations 2000 - *Global forest resource assessment 2000.* Rome: Food and Agriculture Organization of the United Nations.

[FAO] Food and Agriculture Organization of the United Nations 2009 - *State of the world's forests 2009.* Rome: Food and Agriculture Organization of the United Nations.

[FAO] Food and Agriculture Organization of the United Nations 2010 - Global forest resources assessment 2010. Rome: Food and Agriculture Organization of the United Nations.

[FAO] Food and Agriculture Organization of the United Nations 2011 - *State of the world's forests 2011.* Rome: Food and Agriculture Organization of the United Nations.

Garcia Fernandez C. 2004 - *Benzoin resin: scent from the Far East.* In: "Riches of the forest: food, spices, crafts and resins of Asia. Bogor, Indonesia". Lopez C., Shanley P., eds. Center for International Forestry Research; London: Department for International Development.

Garrity D.P., Agustin P.C. 1995 - *Historical land use evolution in a tropical acid upland agroecosystem.* Agriculture, Ecosystems and Environment 53: 83-95.

Garrity D.P. 2004 - *Agroforestry and the achievement of the Millennium Development Goals.* Agroforestry Systems 61: 5–17.

Gunasena H.P.M. 1999 - *Domestication of agroforestry trees in Sri Lanka.* In: "Domestication of agroforestry trees in Southeast Asia. Forest, Farm, and Community Tree Research Reports", Roshetko J.M., Evans D.O. Special issue 1999. Morrilton USA: Winrock International: 49–53.

Idol T., Haggar J., Cox L. 2011 - *Ecosystem services from smallholder forestry and agroforestry in the tropics.* In: "Integrating agriculture, conservation and ecotourism: examples from the field", Campbell W.B., Lopes Ortiz S., New York: Springer. DOI 10.1007/978-94-007-1309-3_5

Leakey R.R.B., Weber J.C., Page T., Cornelius J.P., Akinnifesi F.K., Roshetko J.M., Tchoundjeu Z., Jamnadass R. 2012. *Tree domestication in agroforestry: progress in the second decade.* 145-173 p. In: "The future of agroforestry", Nair P.K.R., Garrity D.P.,. New York: Springer, 541 p.

Mather A., Needle C. 1998 - *The forest transition: a theoretical basis.* Area 30 (2): 117-124.

Mead D.J. 2004 - *Agroforestry.* In: "Forests and forest plants. Encyclopedia of life science systems". Vol 1. Oxford UK: EOLSS Publishers: 324-355.

Michon G., Bompard J.M. 1987 - *Indonesian agroforestry: peasants' contributions to the conservation of natural forests and their resources.* Revue of Ecology 42 (1): 3–34.

Michon G., de Foresta H. 1990 - *Complex agroforestry systems and conservation of biological diversity.* Agroforestry in Indonesia: a link between two worlds. The Malayan Nature Journal. Golden Jubilee 45: 457-473.

Michon G., de Foresta H. 1995 - *The Indonesian agro-forest model.* In: "Conserving biodiversity outside protected areas. The role of traditional ecosystems", Halladay P., Gilmour D.A., Gland, Switzerland: International Union for the Conservation of Nature: 90–106.

Murdiyarso D., van Noordwijk M., Wasrin U.R., Tomich T.P., Gillison A.N. 2002 - *Environmental benefits and sustainable land-use options in the Jambi transect, Sumatra, Indonesia.* Journal of Vegetation Science 13: 429–438.

Murniati, Garrity D.P., Gintings A.N. 2001 - *The contribution of agroforestry systems to reducing farmers' dependence on the resources of adjacent national parks: a case study from Sumatra, Indonesia.* Agroforestry Systems 52: 171–184.

Ordoñez J.C., Luedeling E., Kindt R., Tata H.L., Harja D., Ramni J., van Noordwijk M. 2014. *Constraints and opportunities for tree diversity management along the forest transition curve to achieve multifunctional agriculture.* Current Opinion in Environmental Sustainability 6: 54–60.

Perdana A., Roshetko J.M., Kurniawan I. 2012 - *Forces of competition: smallholding teak producers in Indonesia.* International Forestry Review 14 (2): 238-248.

Pulhin M.J., Chokkalingam U., Peras J.R.J., Acosta T.R., Carandang P.A., Natividad Q.M., Lasco D.R., Razal A.R. 2006 - Chapter II. *Historical Review.* In: "One century of forest rehabilitation in the Philippines: Approaches, outcomes and lessons", Chokkalingam U., Carandang A.P., Pulhin M.J., Lasco R.D., Peras R.J.J., Toma T. Center for International Forestry Research (CIFOR): 6-41.

Rohadi D., Retno M., Widyana M., Azhar I. 2003 - *Case study of the production to consumption system of sandalwood in South Central Timor, Indonesia.* In: "Forest products, livelihoods and conservation: case studies of non-timber forest product systems." Kusters K., Belcher B., Vol. 1: Asia. Bogor, Indonesia. Center for International Forestry Research; London: Department for International Development.

Roshetko J.M., Westley S.B. 1994 - *Dalbergia sissoo production and use: a field manual.* Winrock International, Morrilton, Arkansas, USA, 48 p.

Roshetko J.M., Delaney M., Hairiah K., Purnomosidhi P. 2002 - *Carbon stocks in Indonesian homegarden systems: Can smallholder systems be targeted for increased carbon storage?* American Journal of Alternative Agriculture 17: 138–148.

Roshetko J.M., Lasco R.D., Delos Angeles M.D. 2007a - *Smallholder agroforestry systems for carbon storage.* Mitigation and Adaptation Strategies for Global Change 12: 219–242.

Roshetko J.M., Nugraha E., Tukan J.C.M., Manurung G., Fay C., van Noordwijk M. 2007b - *Agroforestry for livelihood enhancement and enterprise development.* "Integrated rural development in East Nusa Tenggara, Indonesia", Djoeroemana S., Myers B., Russell-Smith J., Blyth M., Salean I.E.T. In: Proceedings of a workshop to Identify Sustainable Rural Livelihoods, Kupang, Indonesia, 5–7 April 2006. ACIAR Proceedings No.126. Canberra: Australian Centre for International Agricultural Research.

Roshetko J.M., Snelder D.J., Lasco R.D., van Noordwijk M. 2008 - *Future challenge: a paradigm shift in the forestry sector.* In: "Smallholder tree growing for rural development and environmental services", Snelder D.J, Lasco R, Amsterdam. Springer Science + Business Media: 453–485.

Roshetko J.M., Rohadi D., Perdana A., Sabastian G., Nuryartono N., Pramono A.A., Widyani N., Manalu P., Fauzi M.A., Sumardamto P., Kusumowardhani N. 2013 - *Teak agroforestry systems for livelihood enhancement, industrial timber production, and environmental rehabilitation.* Forests, Trees, and Livelihoods 22 (4): 251-256. DOI: 10.1080/14728028.2013.855150

Roshetko J.M. 2013 - *Smallholder tree farming systems for livelihood enhancement and carbon storage.* IGN PhD Thesis, Department of Geosciences and Natural Resource Management, University of Copenhagen, Frederiksberg, 205 p.

Santos-Martín F., Bertomeu M., van Noordwijk M., Navarro R. 2012 - *Understanding Forest Transition in the Philippines: main farm-level factors influencing smallholders' capacity and incentive to plant native timber trees.* Small-scale Forestry 11: 47-60. DOI 10.1007/s11842-011-9166-y

Scherr S.J. 1995 - *Economic factors in farmer adoption of agroforestry: patterns observed in western Kenya.* World Development 23 (5): 787–804.

Scherr S.J., McNeely J.A. 2008 - *Biodiversity conservation and agricultural sustainability: towards a new paradigm of 'ecoagriculture' landscapes.* Philosophical Transactions of the Royal Society B 363: 477–494.

Schuren S.H.G., Snelder D.J. 2008 - *Tree growing on farms in Northeast Luzon (The Philippines): smallholders' motivations and other determinants for adopting agroforestry systems.* In: "Smallholder tree growing for rural development and environmental services", Snelder DJ, Lasco R, eds. Amsterdam: Springer Science + Business Media: 75–97.

Snelder D.J. 2008 - *Smallholder tree growing in Philippine back yards: homegarden characteristics in different environmental settings.* In: "Smallholder tree growing for rural development and environmental services", Snelder DJ, Lasco R, eds. Amsterdam. Springer Science + Business Media: 37–74.

Strandby-Andersen U., Prado Cordova J.P., Nielsen U.B., Smith-Olsen C., Nielsen C., Sørensen M., Kollmann J. 2008 - *Conservation through utilization: a case study of the vulnerable Abies guatemalensis in Guatemala.* Oryx 42: 206–213.

Sunderlin W.D., Resosudarmo I.A.P., Rianto E., Angelsen A. 2000 -*The effect of Indonesia's economic crisis on small farmers and natural forest cover in the outer islands.* Occasional Paper 28 (E). Bogor, Indonesia: Center for International Forestry Research.

Tomich T.P., de Foresta H., Dennis R., Ketterings Q., Murdiyarso D., Palm C., Stolle F., van Noordwijk M. 2002 - *Carbon offsets for conservation and development in Indonesia?* American Journal of Alternative Agriculture 17: 125–137.

Tukan J.C.M., Yulianti, Roshetko J.M., Darusman D. 2004 - *Pemasaran Kayu dari Lahan Petani di Provinsi Lampung (Marketing Timber from Farmers' Land in Lampung Province).* Agrivita 26: 131-140.

United Nations 2010 - *World Population Prospects, 2010 Edition.* Department of Economic and Social Affairs, Population Division, Population Estimates and Projections Section, United Nations. [Online]. Available: http://esa.un.org/unpd/wpp/Sorting-Tables/tab-sorting_population.htm. [2015]

United Nations 2012 - *United Nations Millennium Development Goals.* [Online]. Available: http://www.un.org/millenniumgoals/bkgd.shtml. [2015]

United Nations 2015 - *World population prospects: the 2015 revision. Key findings and advance tables.* Working Paper No. ESA/P/WP.241. Department of Economic and Social Affairs, Population Division, United Nations, New York, USA.

van Noordwijk M., Van Roode M., McCallie E.L., Lusiana B. 1998 - *Erosion and sedimentation as multiscale, fractal processes: implications for models, experiments and the real world.* In: "Soil Erosion at Multiple Scales, Principles and Methods for Assessing Causes and Impacts." F. Penning de Vries, F. Agus and J. Kerr (Eds.). CAB International, Wallingford: 223-253.

van Noordwijk M., Roshetko J.M., Murniati, Delos Angeles M., Suyanto, Fay C., Tomich T.P. 2008 - *Farmer tree planting barriers to sustainable forest management.* In: "Smallholder tree growing for rural development and environmental services", Snelder D.J., Lasco R., Amsterdam: Springer Science + Business Media: 429–451.

van Weerd M., Snelder D.J. 2008 - *Human-altered tree-based habitats and their value in conserving bird and bat diversity in northern Luzon, The Philippines.* In: "Smallholder tree growing for rural development and environmental services", Snelder D.J., Lasco R. Amsterdam. Springer Science + Business Media: 347–377.

6

Forest-food nexus: a tropical opportunity for human well-being and silviculture

P. Corona[1], A. Cutini[1], Ugo Chiavetta[1], Elena Paoletti[2*]

Abstract - As population will reach over 9 billion by 2050, interest in the forest-food nexus is rising. Forests play an important role in food production and nutrition. Forests can provide nutritionally-balanced diets, woodfuel for cooking and a broad set of ecosystem services. A large body of evidence recommends multi-functional and integrated landscape approaches to reimagine forestry and agriculture systems. Here, after an in-depth commented discussion of the literature produced in the last decade about the role for forests with respect to the food security global emergency, we summarize the state of the art in Italy as a country-case-study. This commentary aims to increase awareness about the potential of silviculture in Italy for combining ecological resilience with economic resilience, and for reasonably increasing non-wood products supply by means of a sustainable intensification of forest management at national level. Chain-supply fragmentation, landowner inertia, and lack of governance and cooperation may hamper an effective exploitation of non-wood products. The strategies to guarantee an effective supply of non-wood products require appropriate business skills and the presence of a structured business service. A transparent market is also essential; therefore, the introduction of standards (e.g. grading rules and forest certification schemes) is important since they can add value to products and services, and emphasize the importance and complexity of the forest sector. However, the implementation of sustainable forest management for an effective supply of non-wood products is affected by the availability of appropriate planning tools, and the public officers need a new mindset to stimulate and support the business capacity of forest owners.

Keywords - Forest, Food, Silviculture, Security, Safety

Introduction

Up to 805 million people are undernourished worldwide (FAO, 2014) and malnutrition affects nearly every country on a global scale (IFPRI, 2014). As the world population was 7.2 billion in 2013 and is projected to reach over 9 billion by 2050 (Roberts, 2011), the demand for food, feed, fibre and energy will increase, while per-capita land availability will decline. Therefore, the issues of food security and nutrition are now strategical in policy debates. In 2012, the UN Secretary General proposed to eliminate global hunger by 2025 – the so-called "Zero Hunger Challenge". In parallel, interest in the role of forests and tree-based systems in complementing agricultural production has been rising (Vira et al. 2015).

Forests provide food for one billion people, e.g. by providing ≈20% of proteins in the diet in 62 countries (FAO, 2013). However, the forest-food nexus is complex with many and strong connections. Forests produce carbohydrates, proteins, fats, vitamins, fuels, medicinals, wood for construction, fencing and furniture, as well as essential ecosystem services such as water control and protection of biodiversity, soil, and quality of water and air. The intensity by which forests are managed affects forest structure (Vilén et al., 2012), soils (Jandl et al., 2007), biogeochemical cycles (Luyssaert et al., 2007), biodiversity (Paillet et al., 2010), and other ecosystem services provisioning (Gamfeldt et al., 2013). Growing demand for food, energy and land is increasing the pressure over forests. Loss and degradation of forests worsen food insecurity both directly – by affecting the availability of fruits, wildlife, mushrooms and other products of use in the food industry (tannins, cork, truffles, aromatic herbs, honey, etc.) – and indirectly – by modifying the factors that are important for crop and livestock production (van Noordwijk et al., 2014).

A recent Global Assessment Report prepared by the International Union of Forest Research Organizations (Vira et al. 2015) highlights that the complex processes linking tree products and services to food security and nutrition are currently not adequately incorporated into global and national strategies. Although the focus is mostly on those parts of the world that are characterized by extensive hunger

[1] Consiglio per la ricerca in agricoltura e l'analisi dell'economia agraria, Forestry Research Centre (CREA-SEL, Arezzo, Italy)
[2] Consiglio Nazionale delle Ricerche, Institute for Sustainable Plant Protection (CNR-IPSP, Sesto Fiorentino, Italy)
*elena.paoletti@cnr.it

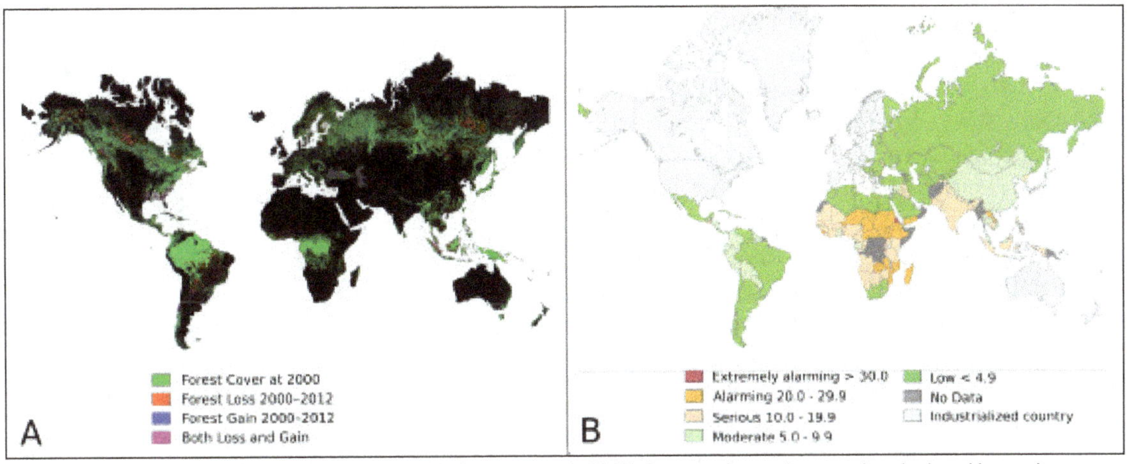

Figure 1 - A, Global forest cover change from 2000 to 2012 (Hansen et al., 2013). Green marks, no change; red marks, loss; blue marks, new forests; purple marks, areas with both losses and gains. B, Global Hunger Index 2014 (Von Grebner et al., 2014).

and malnutrition, primarily in poorer nations and in the tropics (Figure 1), also the most industrialized countries can contribute to a sustainable use of their own forests for improving global food security.

After an in-depth commented discussion of recent scientific literature about forest contribution to food production and the main drivers of forest systems for food nutrition, this commentary addresses the state of the art in Italy as a country-case-study, that is representative of the situation in developed countries. The aim is to increase awareness about the potential of silviculture in Italy by means of a sustainable intensification of forest management for combining ecological resilience with economic resilience.

How forests contribute to food production and nutrition

Non-wood and non-timber products (NWFP and NTFP) are defined as products of biological origin other than wood derived from forests, other wooded land and trees outside forests" (FAO 1999) and as "all biological materials other than timber which are extracted from forest for human use" (De Beer and McDermott 1989), respectively. Therefore NWFPs include animal products (bush meat, trophy, skin, fish, insects), soil (litter, clay, chalk, sand), fungi (mushroom, truffle, spawn), and plants (trees, shrubs, herbs, grasses), which are further subdivided into flowers and fruits (food, oil, spices, honey), leaves (forage, fodder), stem and bark (latex, gum, resin, fibre, dye, sap, cork, bark pieces), while NTFPs include also wood in forms of fuelwood, poles, derivatives (Vidale et al., 2015). All these products may have either a direct or an indirect use in the food industry.

Natural forests, agroforestry systems, single-species tree crop systems and orchards support food production and contribute to dietary diversity and quality. They are a vital source of food to millions of people on the planet, although this service is not well recognized yet. Around one out of every six persons in the world directly depends on forests, with food being one essential aspect of this dependence (Agrawal et al., 2013; Vira et al. 2015).

Much attention is nowadays on agroforestry systems that involve the cultivation and management of trees and/or shrubs for food and/or non-food values (such as soil conservation or providing shelter for crops), generally in combination with agricultural crops. A geospatial analysis by Zomer et al. (2014) estimated extent and recent changes in agroforestry practices at a global scale, based on remote sensing-derived global datasets of land use, tree cover and population: agroforestry systems (defined in this study as agricultural lands with > 10 % tree cover) were 43 % (over 1 billion ha) of global agricultural land in 2010. Globally, the amount of tree cover on agricultural land increased substantially between 2000 and 2010, with the agroforestry area increasing by 3 % (+82.8 million ha). The proportions of agroforestry lands and of people living in these landscapes in Europe were 45 and 46 %, respectively, that basically correspond to the averages at world level.

All forest-based systems represent a steady supply of fruits, vegetables, seeds, nuts, oils, roots, fungi, herbs and animal protein. For instance, around 50 % of the fruit consumed globally comes from trees (Powell et al., 2013): most of these fruits are from fully-domesticated, cultivated sources, but native forests are important genetic resources for the improvement of planted stock (Dawson et al., 2014). A limited number of plant species (20-30) is nowadays used in conventional agriculture all over the world (Ducci et al., 2015), while natural forests and agroforestry systems often harbour high biodiversity and can deliver a wide array of tree foods. As an example, Mediterranean forests include 25,000 plant species (Myers et al., 2000).

Wild meat, fish and insects are other important food sources from forest systems. In Europe, wild ungulate populations of roe deer, red deer, wild boar and alpine chamois have been expanding in recent years (Ramanzin et al., 2010). At present, there are 20 ungulate species in Europe, with an estimated total number of 18 millions heads and a total biomass of about 770 000 tons (Apollonio et al., 2010). The growth of ungulates in many areas has turned into overabundance, originating conflicts with human activities and biodiversity. Marketing of meat from hunted wild ungulates is already a practice in various European countries (Winkelmayer and Paulsen, 2008), and has been proposed as a way of counteracting overabundance (Thogmartin, 2006). Game meat production as alone was estimated over 23,000 tons in EU-28, corresponding to a total value of above 321 M € (FOREST EUROPE 2015). Safety requirements of game meats have been addressed by Regulations (EC) No. 853/2004, 854/2004 and 178/2002. The value of fish as a nutritious food is well established (Kawarazuka and Béné, 2011). In many tropical forests, wild fish represents the main source of animal protein in the diet (daSilva and Begossi, 2009). The importance of insects as a source of food and livestock feed has recently gained momentum (FAO, 2013). Insects are a cheap, available source of proteins, fats, and, to a lesser degree, carbohydrates. Some species are also considered good sources of vitamins and minerals (FAO, 2013; Schabel, 2010).

Trees also provide fodder, green fertiliser and fuel that are essential to food production. Animal fodder enables communities to keep livestock that provides them with nutritionally important products, such as milk and meat. Trees also provide green manure that replenishes soil fertility and supports crop production (Jamnadass et al., 2013). Many forest products are also used in ethnoveterinary treatments that support animal health and hence human food production (Dharani et al., 2014).

In developing countries, 2.4 billion households still use conventional biofuels for cooking and heating. Firewood is the most important rural domestic biofuel in the world, and is expected to further increase (IEA, 2006).

Forest products are also an important source of revenue, which can contribute to food supply. A multitude of NTFPs harvested from natural and cultivated forests and woodlands provide a range of resources that are used directly, or are sold for income that can be used to purchase a variety of products, including food. As NWFP consumption is rarely reported by the national statistical agencies, an estimation of their economic value is complex.

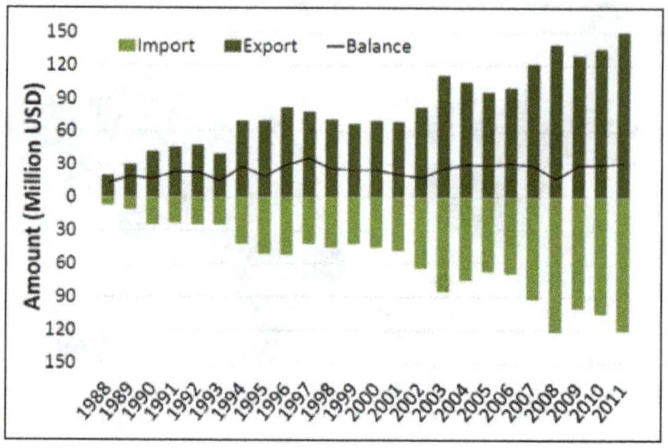

Figure 2 - EU-World trade balance for non-wood forest products (Vidale et al., 2015).

According to UN (2000) and FAO (2010), however, the market in Europe is rising, as it totaled 1.10 billion € in 1995 and 4.53 billion € in 2005. Both import and export of European NWFPs have been considerably increasing in the last 25 years, with a net balance of more than 30 million USD in 2011 (Figure 2). When there is availability but relatively low NTFP food use in areas of dietary need, reasons can include high labour costs, low yields, high phenotypic variability (with large proportions of non-preferred products), and lack of knowledge on appropriate tree management (Jamnadass et al., 2011).

Apart from these direct roles, forests provide ecosystem services which underpin the agricultural production and support the diversification of livelihoods. Forests, agroforests and, within certain conditions, plantations provide important ecosystem services, including water provision, soil protection, nutrient cycling, climate regulation, clean air and water, biodiversity conservation, and pollination, all of which are essential for crop production and ultimately affect food and nutritional security (Figure 3). Here below, we summarize the major links between food security and these forest ecosystem services.

Forests, woodlands and trees play a vital role in controlling water flows and in supplying farmers with water (Malmer et al., 2010). If rainfall does not provide sufficient water supply, households depend on sources of groundwater that are often found in or near the forest. Moreover, forests play a basic role in the quality of groundwater since they act as filters and remove pollution from water and air, with benefits for human and crop health. Trees also prevent soil erosion and nutrient leaching, both of which are critical functions for food production systems. At the same time, green manures and forest litter maintain and enhance soil fertility, supporting crop yields when external fertiliser inputs are not available (Garrity et al., 2010). Nitrogen-fixing trees have received considerable attention for their ability

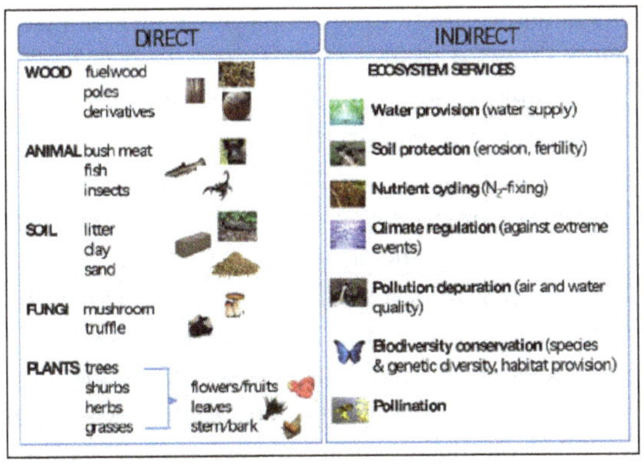

Figure 3 - Effects of forest-based systems to support agricultural production.

to cycle atmospheric nitrogen in cropping systems (Sileshi et al., 2012). Climate regulation by trees can promote more resilient and productive food-cropping systems, such as through the provision of a canopy that protects crops from direct exposure to the sun, extreme rainfall events and high temperatures (Pramova et al., 2012). Forests are centres of plant and animal biodiversity, protecting species and their genetic variation, which may be essential for human food security (Dawson et al., 2014). Pollination is one of the most studied ecosystem services (Klein et al. 2007). A diversity of trees can support populations of pollinator species such as insects and birds (Garibaldi et al., 2013). In addition, forests provide important habitat for a range of other fauna that include the natural predators of crop pests, although forests may also host the crop pests themselves.

Drivers of forests and tree-based systems for food security and nutrition

Interconnected environmental, social, economic and governance factors affecting forests and tree-based systems for food security and nutrition have been classified into the following major drivers: population growth, urbanisation, governance shifts, climate change, commercialization of agriculture, industrialisation of forest resources, gender imbalances, conflicts, formalisation of tenure rights, rising food prices and increasing per capita income (Kleinschmit et al. 2015).

The shift from forests and tree-based systems towards agriculture is among the many inter-related factors that continue to drive deforestation and forest degradation. Deforestation and forest degradation interact with food security and nutrition by affecting both the direct and indirect provision of goods and services. During the past decade, deforestation rates have decreased globally, while some countries are showing increasing rates of reforestation (Meyfroidt and Lambin, 2011). However, deforestation continues unabated in many parts of the world, and is in large part the result of agricultural expansion, cattle ranching (FAO, 2010), urbanization, and globalization of agricultural trade (De Fries et al., 2010). Recent trends show that agriculture is the biggest driver accounting for 73 % of deforestation worldwide, while mining accounts for 7 %, infrastructure for 10 % and urban expansion for 10 % (Hosonuma et al., 2012). Agri-businesses such as cattle ranching, soybean farming and oil palm plantations are now the most important drivers of forest loss globally (Boucher et al., 2011).

Further, an increasing proportion of the world forests have been degraded both structurally and functionally. Forest degradation is the long-term decline in forest ecosystem function and productivity caused by disturbances from which land cannot recover without human intervention. Land degradation currently affects hundreds of millions of hectares of agricultural lands and forests, and an estimated 1.5 billion people who live in these landscapes (Zomer et al., 2009). Land degradation is the long-term result primarily of poor agricultural management, associated with the expansion of extensive and intensive agricultural production practices into lands that are only marginally suitable for such activities. Without adequate organic or fossil fuel-derived fertilisers or other agricultural inputs (e.g. irrigation, pesticides, etc.), agricultural productivity typically declines in such areas. The drivers of forest degradation include unsustainable forest management for timber, fuelwood, wildlife and other NTFPs, air pollution, and human-induced fires, exacerbated in many regions by a number of factors, including climate change (Chazdon, 2014) and changing rural demographics (Uriarte et al., 2012).

As already stressed, deforestation and forest degradation interact with food security and nutrition. For instance, they affect forest carbon stocks and have implications for the governance and local use of forests (Phelps et al., 2010). Studies have shown that there is a direct relationship between tree cover, tree species diversity and food security especially of vulnerable groups (van Noordwijk et al., 2014). Changes in the extent and type of forests have implications for food provisioning, and for food security and nutrition of local and distant human populations (Sunderland et al., 2015). Habitat loss, largely driven by agricultural expansion, has been identified as the single largest threat to biodiversity worldwide (Newbold et al., 2014). Agricultural activities are intensifying, particularly in the tropics (Shackelford et al., 2015). The tropics host the majority of biodiversity-rich areas on the planet.

Consequently tropical land is increasingly subject to competing claims (Giller et al., 2008). A range of concepts and frameworks for implementation are now being discussed which aim to consider land-use change in forested landscapes in such a way that competing demands for food, commodities and forest services may be, hopefully, reconciled (Pirard and Treyer, 2010).

In a world characterized by increasing resource and land scarcity, the traditional conflicts between farming and foresting are aggravated by the increased demands for land to allow for the expansion of urban settlements, industrial development and resource extraction. Under such increasing pressures, hard choices have to be made about land and forest management. Sustainable multi-functional integrated landscape approaches aim at balancing livelihood security and nutritional needs of people with other land management goals (Vira et al. 2015). The contribution of forests to these approaches is of high significance for the implementation of existing international commitments. Forests and tree-based systems are embedded within a mosaic of food production systems and other land uses. An integrated governance is thus needed for securing these multi-functional landscapes.

Present pressures on forests, including climate change, population growth, urbanisation, deforestation, are often interrelated. Thus, designing appropriate responses requires multiple, nested-scales approaches. Managing resilient and climate-smart landscapes on a multi-functional basis that combines food production, biodiversity conservation, other land uses and the overall maintenance of ecosystem services should be at the forefront of efforts to achieve global food security (Vira et al. 2015). Applying an integrated landscape approach provides a unique opportunity for forestry and agriculture to coordinate efforts. Not all tree commodities are, however, amenable to production in diversified systems; for example, oil palm is not well suited (Donald, 2004).

Greater attention from the scientific and policy communities is required for reimagining forests for food security. In particular, a supportive policy framework needs to be developed that considers both the forestry and agriculture sectors in tandem. A better quantification of the benefits received by rural communities from different tree production categories is required (de Foresta et al., 2013): in many tropical countries, laws for timber extraction were largely designed around large-scale export-oriented forestry operations rather than to sustain healthy small-scale domestic markets (Cerutti et al., 2013).

Non-wood forest products in Italy

FOREST EUROPE (2015) estimated that a marketed value of around 2.3 billion €/year is provided by plant (73%) and animal (27%) products from European forests, but the statistics may be incomplete. With respect to the total, 1.7 billion €/year is from plant products (73%), with the main part represented by decorative and ornamental plants (47%), while the value of animal products is around 0.6 billion €/year, mainly due to wild meat (51%) and wild honey (45%). Overall, NWFPs trade is increasing in Europe (Figure 4), where raw NWFPs account for ~20% of timber trade (Vidale et al., 2015). Italy is first in Europe as ratio of annual NWFP value to annual value of industrial roundwood (Figure 4).

Recent results from the COST Action StarTree (Vidale et al., 2015) show that Italy is among the four top European exporters of cork stoppers, is one of the three top countries for chestnut seed processing, and is among the leading exporters of wild mushroom, while it is the only European country among the top five global importers of tannins.

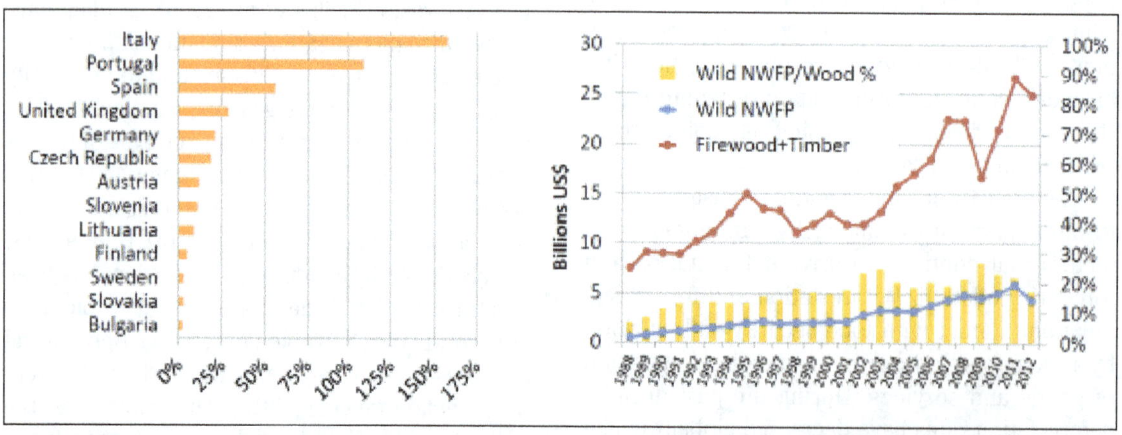

Figure 4 - Ratio of annual non-wood forest product (NWFP) production to industrial roundwood (left), and trade of NWFP and wood in Europe (right) (Vidale et al., 2015; FOREST EUROPE 2011).

The annual value of marketed NWFP in Italy is estimated around 100 M €, but the statistics may be largely incomplete (FOREST EUROPE, 2015). Among NWFPs, food products are also relatively relevant. For instance, Italy is the second largest European chestnut (*Castanea sativa*) producer for fresh and dry products and flour; walnut (*Juglans regia*) production is 10,500 tons per year (Ducci et al. 2015). The market for pine (*Pinus pinea* and *Pinus cembra*) fruits in shell represents over 208,000 tons per year, 80 % absorbed by the industry. Collecting mushrooms and truffles has considerable importance in the economy of rural mountainous and hilly areas: reliable statistics are not available for mushrooms while Italy is the 3rd European producer of truffles, with a turnover of over 19 M €/year (Ducci et al., 2015).

Game meat market trend observed at European level is similar even in Italy, where the increase of total forest coverage and a cautious approach in forest harvesting have enhanced the expansion of ungulates. This trend involved mainly roe deer (*Capreolus capreolus* L.), wild boar (*Sus scrofa* L.) and red deer (*Cervus elaphus* L.), whose populations are estimated to be over 400,000, 1,000,000 and 65,000 heads, respectively, with increasing pressure on agricultural crops and forests in many areas (Chianucci et al., 2013). Roe deer, wild boar and red deer represents over 80% of total ungulates biomass and contribute to the market of wild meat with important economic revenues, which are estimated around 25 M €/year just for Tuscany, a region in Central Italy (Cutini et al., 2015).

The Italian trade of honey is estimated in 38 M €: transhumance of hives to the woods affects honey quality and organoleptic traits determined by the forest species that provide pollen and nectar. Another important example of high value production at local level is that of *manna*, a natural product, at high content of mannithol harvested by the incision of the bark from two species of Ash (*Fraxinus* spp) trees: Italy is the first world producer of manna, with 3200 kg per year (Ducci et al., 2015).

Grounds for intensifying silviculture and food products from forests in Italy

Albeit trade-offs between wood and NWFPs cannot be excluded as it is often the case in developing countries (Chakravarty et al., 2015), in Italy an effective joint impulse for exploitation of wood and NWFPs may come from a sustainable intensification of forest management, with a reasonable increase of the marketed NWFPs too: currently these products are, in many cases, excluded from the market and fostering payments for them would encourage landowners to sustainably manage their forests on the whole (Prokofieva et al., 2012).

Chain-supply fragmentation, ownership fragmentation (Paletto et al., 2013), landowner inertia, and lack of governance and cooperation may hamper an effective exploitation of food products from Italian forests. However, these issues impact all the product chains from forestry in Italy. Generally, the increase of forest service demand and the gradual abandonment of mountainous land have caused a decrease of forestry and significant changes in land management. Only in the case of coppice, wood production has remained relatively high. The wood harvesting rate of Italian forests is ~14 Mm^3/yr, i.e. 1.5 m^3/ha yr (Gasparini and Tabacchi, 2011) and is among the lowest rates in Europe. As a consequence, also the mean value of marketed roundwood (74 €/ha) is much lower than in the neighbouring countries (FOREST EUROPE, 2015). In contrast, the current increment of wood volume is around 36 Mm^3/yr (Gasparini and Tabacchi, 2011), and thus the harvesting rate (≈40 %) is largely lower than in the EU-28 and Europe (71 % and 66 %, respectively, FOREST EUROPE, 2015).

To develop Italy's forest sustainability and resilience and favour forest bioeconomy, an intensification of forest management is the possible solution to the conundrum that increasing demand for conservation areas and increasing pressure for good production have created, similarly to what is happening in Europe (Carnus et al., 2012) and other world areas (e.g. Canada, Mathey et al., 2008). An improved awareness of policy makers and the general public may translate these unexploited Italian forest assets into employment (e.g. a gradual and sustainable increase of the wood harvesting rate up to a sustainable threshold of 20-21 Mm^3/y would translate into ~35,000 new jobs) and gross domestic product.

Developing measures targeted at increasing wood and non-wood supply from forests requires policy decisions and expert knowledge. A forest management map of European forests has been recently developed (Hengeveld et al., 2012): approaches of this kind may greatly help in selecting the areas suitable for intensification. Moreover, the implementation of sustainable forest management for an effective supply of wood and non-wood products is conditioned by an appropriate use of planning tools, and the public officers need to develop a new mindset for stimulating and supporting the business capacity of forest owners.

As concerns distinctively the NWFPs, it should be stressed that they can be effectively exploited under the broader perspective of territorial marketing (Pettenella and Secco, 2006): well known success cases are those of the Road of Porcino mushroom (http://www.stradadelfungo.it) and the Road of Truffle and Chestnut (http://www.tartufoe-castagna.it). Under such a perspective, the strategies to guarantee an effective supply of NWFPs require appropriate business skills and the presence of structured business services. A transparent market is also essential: the introduction of standards (such as grading rules and forest certification schemes) is important since they can also add value to products and services, and emphasize the importance and complexity of the forest sector.

Conclusions

Policy processes towards a bio-based economy should seek to produce decisions that are evidence-based (Corona, 2014). Contextually, the use of scientific knowledge to support evidence-based decisions requires suitable communication of figures and key findings: this paper has been targeted to contribute to this end.

The adoption of large-scale industrial agriculture has resulted in negative impacts on the environment (Cassman, 2012), public health (Bandara et al., 2010) and even nutrition (Ellis et al., 2015), suggesting the paradigm itself needs to be challenged (Tilman and Clark, 2014). This approach was appropriate to the context of the 1960s and 1970s, when water and nutrients were abundant, energy was cheap, and ecosystems were able to detoxify pollutants. The global context today is very different with growing scarcity of cheap energy (Day et al., 2009), water (Wallace, 2000) and nutrients (e.g. phosphorus, Cordell et al., 2009).

The development of crop agriculture and animal husbandry over the past few centuries, and particularly since the early 20th century, has diminished dependence on forests for food security and nutrition in many societies. Nonetheless, forests continue to play a very important role, often complementing other food production systems, and, on a global level, can contribute to the "Zero Hunger Challenge" (Vira et al. 2015). While forests are not a solution for global hunger in themselves, in many circumstances they play a vital supplementary role, especially during periods of unpredictability (such as long drought spells), as they complement conventional staple diets derived from agricultural production systems. To do this efficiently, an improved knowledge of the most effective management of landscapes and the role of forests in the provision of nutritious diets is required.

Evolving strategies to respond to the "Zero Hunger Challenge" primarily focus on achieving a sustainable intensification, by improving the productivity of agricultural and forest systems without causing ecological harm or compromising biodiversity and other ecosystem services (FAO, 2011; Garnett et al., 2013). Paradigms for forest and tree management have evolved considerably in the last 50 years, away from a state-controlled, production-centric approach to more collaborative systems which prioritise the needs of local people, and value the provision of ecosystem services (Mace, 2014). Landscapes are now managed for a much more diverse (often non-local) set of purposes (Ribot et al., 2006). It is time to develop a vision where economic resilience is joint with ecological resilience towards actual sustainability.

Managing landscapes on a multi-functional basis that combines local and global scales, food production, forest conservation and the maintenance of ecosystem services will help to achieve food security (Godfray, 2011). This provides a unique opportunity for silviculture and agriculture to coordinate efforts at the conceptual and implementation levels and achieve more sustainable systems.

Italian forests are well suited for a sustainable intensification of forest management, i.e. for suitably increasing the intensity of forest harvesting while maximizing the provision of forest ecosystem services and products. Ultimately, national production of wood and non-wood goods, including food products, may reduce the pressure on global forests, in particular in the areas at higher risk of deforestation and hunger.

Acknowledgments

We would like to thank Davide Pettenella (University of Padova, Italy) for his very constructive comments on an earlier version of the paper.

References

Agrawal A., Cashore B., Hardin R., Shepherd G., Benson C., Miller D., 2013 - *Economic Contributions of Forests.* Background Paper to UNFF Tenth Session, Istanbul, 8-19 April 2013.

Apollonio M., Andersen R., Putman R., 2010 - *European ungulates and their management in the 21th century.* Cambridge University Press, Cambridge.

Bandara J.M.R.S., Wijewardena H.V.P., Liyanege J., Upul M.A., Bandara J.M.U.A., 2010 - *Chronic renal failure in Sri Lanka caused by elevated dietary cadmium: trojan horse of the green revolution.* Toxicology Letters 198: 33-39.

Boucher D.H., Elias P., Lininger K., May-Tobin C., Roquemore S., Saxon E., 2011 - *The Root of the Problem: What's Driving Tropical Deforestation Today?* Cambridge, MA: Union of Concerned Scientists.

Carnus J.M., Hengeveld G.M., Mason B., 2012 - *Sustainability impact assessment of forest management alternatives in Europe: an introductory background and framework.* Ecology and Society 17 (4): 49. http://dx.doi.org/10.5751/

ES-04838-170449

Cassman K.G., 2012 - *What do we need to know about global food security?* Global Food Security 1: 81-82.

Cerutti P.O., Tacconi L., Lescuyer G., Nasi R., 2013 - *Cameroon's hidden harvest: commercial chainsaw logging, corruption and livelihoods.* Society and Natural Resources 26: 539-553.

Chakravarty, S., Subba, M., Dey, T., Rai, P., Shukla, G., Pala, N.A., Puri A., 2015 - *Value addition of non-timber forest products: prospects, constraints, and mitigation.* In: Value Addition of Horticultural Crops: Recent Trends and Future Directions (Springer, India, ISBN 978-813222262-0;978-813222261-3), 213-244.

Chappell M.J., LaValle L.A., 2011 - *Food security and biodiversity: can we have both? An agroecological analysis.* Agriculture and Human Values 28: 3-26.

Chazdon R. L., 2014 - *Second growth: The promise of tropical forest regeneration in an age of deforestation.* Chicago: University of Chicago Press.

Chianucci F., Mattioli L., Amorini E., Giannini T., Marcon A., Chirichella R., Apollonio M., Cutini A., 2015 - *Early and long-term impacts of browsing by roe deer in oak coppiced woods along a gradient of population density.* Annals of Silvicultural Research 39(1): 32-36. DOI: 10.12899/ASR-945.

Cordell D., Drangert J.-O., White S., 2009 - *The story of phosphorus: global food security and food for thought.* Global Environmental Change 19: 292-305.

Corona P., 2014 - *Forestry research to support the transition towards a bio-based economy.* Annals of Silvicultural Research 38: 37-38.

Cutini A., Chianucci F., Apollonio M., 2015 - *Wild ungulates and forests in Europe: insights from long term studies in central Italy.* In Ciancio O. (ed), Proceedings of II International Congress of Silviculture, Designing the future of the forestry sector, Florence, 26-29 November 2014, Italian Academy of Forest Sciences: 509-517.

daSilva A., Begossi A., 2009 - *Biodiversity: food consumption and ecological niche dimension: A study case of the riverine populations from the Rio Negro, Amazonia, Brazil.* Environment Development and Sustainability 11:489-507.

Dawson I.K., Leakey R., Clement C.R., Weber J., Cornelius J.P., Roshetko J.M., Vinceti B., Kalinganire A., Tchoundjeu Z., Masters E., Jamnadass R., 2014 - *The management of tree genetic resources and the livelihoods of rural communities in the tropics: non-timber forest products, smallholder agroforestry practices and tree commodity crops.* Forest Ecology and Management 333: 9-21.

Day J.W. Jr., Hall C.A., Yáñez-Arancibia A., Pimentel D., 2009 - *Ecology in times of scarcity.* BioScience 59: 321-331.

de Beer JH, McDermott M 1989 - *The Economic Value of Non-Timber Forest Products in South- East Asia.* Amsterdam, The Netherlands Committee for IUCN

de Foresta H., Somarriba E., Temu A., Boulanger D., Feuilly H., Gauthier M., 2013 - *Towards the assessment of trees outside forests.* FAO Resources Assessment Working Paper No. 183. Rome: FAO.

DeFries R.S., Rudel. T., Uriarte M., Hansen M., 2010 - *Deforestation driven by urban population growth and agricultural trade in the twenty-first century.* Nature Geoscience 3, 178-181. doi:10.1038/ngeo756

Dharani N., Yenesew A., Ermais B., Tuei B., Jamnadass R., 2014 - *Traditional Ethnoveterinary Medicine in East Africa: a manual on the use of medicinal plants,* edited by I.K. Dawson. Nairobi: ICRAF.

Donald P.F., 2004 - *Biodiversity impacts of some agricultural commodity production systems.* Conservation Biology 18: 17–38.

Ducci F., Cantiani P., Cutini, A., Dettori, S., 2015 - *Forests and non-wooden products in Italy.* In Ciancio O. (ed), Proceedings of II International Congress of Silviculture, Designing the future of the forestry sector, Florence, 26-29 November 2014, Italian Academy of Forest Sciences: 650-657.

Ellis A., Myers S., Ricketts. T., 2015 - *Do pollinators contribute to nutritional health?* PLoS One 10(1): e114805. doi:10.1371/journal.pone.0114805.

FAO 1999 - *FAO Forestry – Towards a Harmonised Definition of Non-Wood Forest Products. Unasylva*– No. 198, Vol. 50.

FAO 2010 - *Global forest resources assessment.* Main report 163, FAO, Rome, pp. 340 Available at http://www.fao.org/forestry/fra/80298/en/ [Accessed on 15 February 2015]

FAO 2011 - *Save and grow: a policy makers guide to the sustainable intensification of crop production.* Rome: FAO. Available at: http://www.fao.org/docrep/014/i2215e/i2215e.pdf [Accessed on 15 February 2015].

FAO 2013 - *The State of Food and Agriculture. Better food systems for better nutrition.* Rome: FAO. Available at: http://www.fao.org/docrep/018/i3300e/i3300e00.htm [Accessed on 15 February 2015].

FAO 2014 - *Food and Nutrition in Numbers.* Rome: FAO. Available at http://www.fao.org/3/a-i4175e.pdf [Accessed on 15 February 2015]

FOREST EUROPE 2015 - *State of Europe's Forests 2015.* Forest Europe Liason Unit, Grøset Trykk AS.

FOREST EUROPE 2011 - *State of Europe's Forests 2011. Status and Trends in Sustainable Forest Management in Europe.* Forest Europe Liason Unit, Grøset Trykk AS

Gamfeldt L., Snall, T., Bagchi R., Jonsson M., Gustafsson L., Kjellander P., Ruiz-Jaen M.C., Froberg M., Stendahl J., Philipson C.D., Mikusinski G., Andersson E., Westerlund B., Andren H., Moberg F., Moen J., Bengtsson J., 2013 - *Higher levels of multiple ecosystem services are found in forests with more tree species.* Nat. Commun. 4, 1340.

Garibaldi L.A., Steffan-Dewenter I., Winfree R., Aizen M.A., Bommarco R., Cunningham S.A., Kremen C., Carvalheiro L.G., Harder L.D., Afik O., Bartomeus I., Benjamin F., Boreux V., Cariveau D., Chacoff N.P., Dudenhöffer J.H., Freitas B.M., Ghazoul J., Greenleaf S., Hipólito J., Holzschuh A., Howlett B., Isaacs R., Javorek S.K., Kennedy C.M., Krewenka K.M., Krishnan S., Mandelik Y., Mayfield M.M., Motzke I., Munyuli T., Nault B.A., Otieno M., Petersen J., Pisanty G., Potts S.G., Rader R., Ricketts T.H., Rundlöf M., Seymour CL., Schüepp C., Szentgyörgyi H., Taki H., Tscharntke T., Vergara C.H., Viana BF., Wanger T.C., Westphal C., Williams N., Klein A.M., 2013 - *Wild pollinators enhance fruit set of crops regardless of honeybee abundance.* Science 339(6127): 1608-1611. doi: 10.1126/science.1230200.

Garnett T., Appleby M., Balmford A., Bateman I., Benton T., Bloomer P. Burlingame B., Dawkins M., Dolan L., Fraser D., Herroro M., Hoffman I. Smith P., Thornton P., Toulmin C., Vermeulen S., Godfray C., 2013 - *Sustainable intensification in agriculture: Premises and policies.* Science 341: 33-34.

Garrity D., Akinnifesi F., Ajayi O., Weldesemayat S., Mowo J., Kalinganire A., Bayala J., 2010 - *Evergreen Agriculture: a robust approach to sustainable food security in Africa.* Food Security 2: 197-214.

Gasparini P., Tabacchi G. (eds) 2011 - *L'Inventario Nazionale delle Foreste e dei serbatoi forestali di Carbonio INFC 2005. Secondo inventario forestale nazionale italiano. Metodi e risultati. [National Forest and Carbon Sink Inventory INFC 2005. Italian second national forest inventory. Methods and results]*. Ministero delle Politiche Agricole, Alimentari e Forestali. Corpo Forestale dello Stato. Consiglio per la Ricerca e la Sperimentazione in Agricoltura, Unità di ricerca per il Monitoraggio e la Pianificazione Forestale. Edagricole-Il Sole 24 ore, Bologna. 653 pp

Giller K.E., Leeuwis C., Andersson J.A., Andriesse W., Brouwer A., Frost P., Hebinck P., Heitkönig I., Van Ittersum M. K., Koning N., 2008 - *Competing claims on natural resources: what role for science.* Ecology and Society 13(2): 34.

Godfray C.J., 2011 - *Food and biodiversity.* Science 333: 1231-2132.

Hansen M.C., Potapov P.V., Moore R., Hancher M., Turubanova S.A., Tyukavina A., Thau D., S.V. Stehman, Goetz S.J., Loveland T.R., Kommareddy A., Egorov A., Chini L., Justice C.O., Townshend J.R.G., 2013 - *High-Resolution Global Maps of 21st-Century Forest Cover Change.* Science 342(6160): 850-853. [DOI:10.1126/science.1244693]

Hengeveld G.M., Nabuurs G.J., Didion M., Van den Wyngaert I., Clerkx A.P.P.M., Schelhaas M.J, 2012 - *A forest management map of European forests.* Ecology and Society 17(4): 53. http://dx.doi.org/10.5751/ES-05149-170453

Hosonuma N., Herold M., De Sy V., DeFries R.S., Brockhaus M., Verchot L., Angelsen A., Romijn, E., 2012 - *An assessment of deforestation and forest degradation drivers in developing countries.* Environmental Research Letters 7(4): 044009.

IEA, 2006 - *World Energy Outlook.* IEA/OECD, Paris.

IFPRI, 2014 - *Global Nutrition Report. Actions and Accountability to Accelerate the World's Progress on Nutrition.* IFPRI, Washington DC.

Jamnadass R., Place F., Torquebiau E., Malézieux E., Iiyama M., Sileshi G.W. Kehlenbeck K., Masters E., McMullin S., Dawson I.K., 2013 - *Agroforestry for food and nutritional security.* Unasylva 241(64): 2.

Jamnadass R.H., Dawson I.K., Franzel S., Leakey R.R.B., Mithöfer D., Akinnifesi F.K., Tchoundjeu Z., 2011 - *Improving livelihoods and nutrition in sub-Saharan Africa through the promotion of indigenous and exotic fruit production in smallholders' agroforestry systems: A review.* International Forestry Review 13: 338–354.

Jandl R., Lindner M., Vesterdal L., Bauwens B., Baritz R., Hagedorn F., Johnson D.W., Minkkinen K., Byrne K.A., 2007 - *How strongly can forest management influence soil carbon sequestration?* Geoderma 137, 253–268.

Kawarazuka N., Béné C., 2011 - *The potential role of small fish species in improving micronutrient deficiencies in developing countries: building evidence.* Public Health Nutrition 14:1927-1938.

Klein A.M., Vaissiere B.E., Cane J.H., Steffan-Dewenter I., Cunningham S.A., Kremen C., Tscharntke T., 2007 - *Importance of pollinators in changing landscapes for world crops.* Proceedings of the Royal Society B: Biological Sciences 274 (1608): 303-313.

Kleinschmit D., Basnett B.S., Martin A., Rai N.D., Smith-Hall C., Dawson N.M., Hickey G., Neufeldt H., Ojha H.R., Walelign S.Z. 2015 - *Chapter 4 Drivers of Forests and Tree-based Systems for Food Security and Nutrition.* In: Vira B, Wildburger C, Mansourian S (Eds) "Forests, Trees and Landscapes for Food Security and Nutrition A Global Assessment Report". IUFRO World Series Vol 33: 87-110

Luyssaert S., Abril G., Andres R., Bastviken D., Bellassen V., Bergamaschi P., Bousquet P., Chevallier F., Ciais P., Corazza, M., Dechow R., Erb K.H., Etiope G., Fortems-Cheiney Grassi G., Hartmann J., Jung M., Lathière J., Lohila A., Mayorga, E., Moosdorf N., Njakou D.S. Otto, J., Papale D., Peters W., Peylin P., Raymond P., Rödenbeck C., Saarnio S., Schulze E.D., Szopa S., Thompson R., Verkerk P.J., Vuichard N., Wang R., Wattenbach M., Zaehle S., 2007 - *The European land and inland water, CO2, CO, CH4 and N2O balance between 2001 and 2005.* Biogeosciences 9, 3357–3380.

Mace G., 2014 - *Whose Conservation?* Science 345 (6204): 1558-1560.

Malmer A., Ardö J., Scott D., Vignola R., Xu J., 2010 - *Forest water and global water governance.* In: G. Mery P. Katila G. Galloway, I. Alfaro M. Kanninen, M. Lobovikov, J. Varjo. (eds) "Forests and Society – Responding to Global Drivers of Change": 75-93.

Mathey AH, Krcmar E, Innes J,, Vertinsky I 2008 - *Opportunities and costs of intensification and clustering of forest management activities.* Can. J. For. Res. 38: 711-720

Meyfroidt P., Lambin E.F., 2011 - *Global forest transition: prospects for an end to deforestation.* Annual Review of Environment and Resources 36: 343–371. doi:10.1146/annurevenviron-090710-143732.

Myers N., Mittlemeier R.A., Mittlemeier C.G., Da Fonseca G.A.B., Kent J., 2000 - *Biodiversity hotspots for conservation priorities.* Nature, 403: 853–858.

Newbold T., Huson L., Phillips H., Hill S., Contu S., Lysenko I., Blandon A., Butchart S., Booth H., Day J., de Palma A., Harrison M., Kirkpatrick L., Pynegar E., Robinson A., Simpson J., Mace G., Scharlemann J., Purvis A., 2014 - *A global model of the response of tropical and sub-tropical biodiversity to anthropogenic pressures.* Proceedings of the Royal Society B: Biological Sciences 281 (1792): 20141371.

Paillet Y., Berges L., Hjalten J., Odor P., Avon C., Bernhardt-Roemermann M., Bijlsma R.J., De Bruyn L., Fuhr M., Grandin U., Kanka R., Lundin L., Luque S., Magura T., Matesanz S., Meszaros I., Teresa Sebastia M., Schmidt W., Standovar T., Tothmeresz B., Uotila A., Valladares F., Vellak K., Virtanen R., 2010. *Biodiversity differences between managed and unmanaged forests: metaanalysis of species richness in Europe.* Conserv. Biol. 24: 101-112.

Paletto A., De Meo I, Ferretti F. 2013 - *The influence of common property rights on forest management policy: A case study in Sardinia region, Italy.* Forestry Studies 56, 16-26

Pettenella D., Secco L., 2006 - *Small-scale forestry in the Italian Alps: from mass market to territorial marketing.* In: Wall, S. (Ed.), Small-scale forestry and rural development. The intersection of ecosystems, economics and society, Proceedings of IUFRO 3.08 Conference, Galway, Galway-Mayo Institute of Technology. Galway-Mayo Institute of Technology, Galway, pp. 398–408.

Phelps J., Webb E.L., Agrawal A., 2010 - Does REDD+ Threaten to Recentralize Forest Governance? Science, 328: 312-313.

Pirard R., Treyer S., 2010 - *Agriculture and deforestation: What role should REDD+ and public support policies play?* Working Papers N°10/2010. IDDRI. Available at: http://www.iddri.org/publications/agriculture-and-deforestation-what-roleshould-redd+-and-public-support-policies-play [Accessed on 22 February 2015].

Powell B., Ickowitz A., McMullin S., Jamnadass R., Padoch C., Pinedo-Vasquez M., Sunderland T., 2013 - *The role of forests, tress and wild biodiversity for nutrition-sensitive food systems and landscapes.* Expert background paper for the International Conference on Nutrition 2. FAO, Rome.

Pramova E., Locatelli B., Djoudi H., Somorin O.A., 2012 - *Forests and trees for social adaptation to climate variability and change. Wiley Interdisciplinary Reviews: Climate Change* 3(6): 581-596.

Prokofieva I., Wunder S., Vidale E., ... - *Payments for Environmental Services: A Way Forward for Mediterranean Forests?* European Forest Institute Policy Brief 7: 16p.

Ramanzin M., Amici A., Casoli C., Esposito L., Lupi P., Marsico G., Mattiello S., Olivieri O., Ponzetta M.P., Russo C., Trabalza Marinucci M. 2010 - *Meat from wild ungulates: ensuring quality and hygiene of an increasing resource.* Italian Journal of Animal Science 9: 61.

Ribot J.C., Agrawal A., Larson A.M., 2006 - *Recentralizing while decentralizing: How national governments reappropriate forest resources.* World Development 34(11): 1864-1886.

Roberts L., 2011 - *9 Billion?* Science 333(6042): 540-543.

Schabel H.G., 2010 - *Forests insects as food: A global review.* In Durst P.B., Johnson D.V., Leslie R.N., Shono K. (eds) "Forest Insects as Food: Humans Bite Back": 37-64.

Shackelford G., Steward P., German R., Sait S., Benton T., 2015 - *Conservation planning in agricultural landscapes: hotspots of conflict between agriculture and nature.* Diversity and Distributions [early on-line publication]

Sileshi G.W., Debusho L.K., Akinnifesi F.K., 2012 - *Can integration of legume trees increase yield stability in rainfed maize cropping systems in Southern Africa?* Agronomy Journal 104: 1392-1398.

Sunderland T., Baudron F., Ickowitz A., Padoch C., Ros-Tonen M., Sandbrook C., Vira B., Chambers J., Deakin E., Foli S., Jeary K., Parrotta J.A., Powell B., Reed J., Ogalleh S.A., Neufeldt H., Serban A. 2015 - *Chapter 5 Response Options Across the Landscape.* In Vira B, Wildburger C, Mansourian S (eds) "Forests, Trees and Landscapes for Food Security and Nutrition A Global Assessment Report": 113-151

Thogmartin W., 2006 - *Why Not Consider the Commercialization of Deer Harvests?* Bioscience 56:957.

Tilman D., Clark M., 2014 - *Global diets link environmental sustainability and human health.* Nature 515: 518-522.

UN, 2000 - *Forest Resources of Europe, CIS, North America, Australia, Japan and New Zealand (industrialized temperate/boreal countries). UN-ECE/FAO Contribution to the Global Forest Resources Assessment 2000.* New York and Geneva. http://www.unece.org/fileadmin/DAM/timber/fra/pdf/fullrep.pdf [Accessed on 29 January 2015].

Uriarte M., Pinedo-Vasquez M., DeFries R.S., Fernandes K., Gutierrez-Velez V., Baethgen W. E., Padoch C., 2012 - *Depopulation of rural landscapes exacerbates fire activity in the western Amazon.* Proceedings of the National Academy of Sciences 109(52): 21546-21550.

Van Noordwijk M., Bizard V., Wangpakapattanawong P., Tata H.L., Villamor G.B., Leimona B., 2014 - *Tree cover transitions and food security in Southeast Asia.* Global Food Security 3 (3-4): 200-208.

Vidale E., Da Re R., Loric M., Pettenella D. 2015 - *The Non-Wood Forest Product Market: any opportunity for the Italian forest sector?* COST Action FP1203, 3rd Workshop, Zagreb, 18th - 20th February 2015. Available at http://www.nwfps.eu/wp-content/uploads/2012/07/2015-02-18-Vidale-et-al.pdf [Accessed on 22 January 2015].

Vilén T., Gunia K., Verkerk P.J., Seidl R., Schelhaas M.J., Lindner M., Bellassen V., 2012 - *Reconstructed forest age structure in Europe 1950–2010.* For. Ecol. Manage. 286, 203-218.

Vira B., Wildburger C., Mansourian S. (eds.), 2015 - *Forests, Trees and Landscapes for Food Security and Nutrition. A Global Assessment Report.* IUFRO World Series Volume 33. International Union of Forest Research Organizations (IUFRO) Vienna, 172 p.

Von Grebmer K., Saltzman A., Birol E., Wiesmann D., Prasai N. Yin S., Yohannes Y., Menon P., Thompson J., Sonntag A., 2014 - *2014 Gobal Hunger Index: The Challenge of Hidden Hunger.* Welthungerhilfe, International Food Policy Research Institute, and Concern Worldwide, Bonn, Washington DC and Dublin: http://www.ifpri.org/publication/2014-globalhunger-index [Accessed on 15 February 2015]

Wallace J.S., 2000 - *Increasing agricultural water use efficiency to meet future food production.* Agriculture, Ecosystems and Environment 82: 105-119.

Winkelmayer R., Paulsen P., 2008 - *Direct marketing of meat from wild game in Austria: A guide to good practice according to Regulations (EEC) 852 and 853/2004.* Fleischwirtschaft 88:122-125.

Zomer R.J., Trabucco A., Coe R., Place F., 2009. *Trees on Farm: Analysis of Global Extent and Geographic Pattern of Agroforestry.* ICRAF Working Paper no. 89. World Agroforestry Centre, Nairobi.

Zomer R.J., Trabucco A., Coe R. Place F., van Noordwijk M., Xu J.C., 2014 - *Trees on farms: an update and reanalysis of agroforestry's global extent and socio-ecological characteristics.* Working Paper 179. World Agroforestry Centre (ICRAF) Southeast Asia Regional Program, Bogor, Indonesia. DOI: 10.5716/ WP14064.PDF

7

Factors affecting branch wound occlusion and associated decay following pruning - a case study with wild cherry (*Prunus avium* L.)

Jonathan Sheppard[1*], Matthias Urmes[1], Christopher Morhart[1], Heinrich Spiecker[1]

Abstract - Pruning wild cherry (*Prunus avium* L.) is a common silvicultural practice carried out to produce valuable timber at a veneer wood quality. Sub-optimal pruning treatments can permit un-occluded pruning wounds to develop devaluing decay. The aim of this study is to determine relevant branch, tree and pruning characteristics affecting the occlusion process of pruning wounds. Important factors influencing occlusion time for an optimised pruning treatment for valuable timber production utilising wild cherry are derived. 85 artificially pruned branches originating from ten wild cherry trees were retrospectively analysed. Branch stub length, branch diameter and radial stem increment during occlusion were found to be significant predictors for occlusion time. From the results it could be concluded that for the long term success of artificial pruning of wild cherry it is crucial to (*i*) keep branch stubs short (while avoiding damage to the branch collar), (*ii*) to enable the tree to maintain significant radial growth after pruning, (*iii*) to avoid large pruning wounds (>2.5 cm) by removing steeply angled and fast growing branches at an early stage.

Keywords - high value timber production, wound occlusion, stub occlusion

Introduction

Modern forestry management should simultaneously satisfy ecological, economical and social demands of both the public and of forest owners. Valuable broadleaf tree species can be especially regarded as an option to fulfil the multidimensional requirements of today's forestry sector, where an economic return is crucial. The inclusion of valuable broadleaves can also simultaneously increase biodiversity and aesthetics within forests and agricultural land (Rapey 1994, Dupraz 1994, Bell 2009).

High value logs destined for veneer must meet a number of quality parameters: large diameter, cylindrical shaft forms, high volume of knotless wood, a uniform colour and an absence of decay (Mahler 1988, Spiecker 2003, Kupka 2007, Springmann et al. 2011a). Large dimensioned high quality logs with a small knotty core reach the highest prices. The reduction of the width of the occlusion zone to a minimum in order to harvest greater amounts of branch free timber is one important goal for silvicultural management. This goal requires timely and repeated interventions especially in juvenile stages. Of particular importance is artificial pruning in order to reach the production target of branch free valuable timber (Balandier 1997, Oosterban et al. 2009). This is especially true for open grown trees (Balandier and Dupraz 1999) where natural pruning occurs to a lesser extent. The same is true for tree species that do not self prune well (Röhrig et al. 2006). Both points are relevant for wild cherry (*Prunus avium* L.) cultivated in widely spaced systems.

Wild cherry is often considered to be a valuable broadleaved species yielding high prices under relatively short rotations (Bulfin and Radford 1998, Balandier and Dupraz 1999, Morhart et al. 2014). In order to obtain a high quality product, artificial pruning is of upmost necessity (Otter 1954, Balandier and Dupraz 1999, Thies et al. 2009, Springmann et al. 2011b). This results in a small dimensioned knotty core surrounded by branch free wood. The longer it takes for a pruning wound to occlude the larger the knotty core becomes. Therefore, in order to maximise clear wood, the time taken for complete branch stub occlusion must be minimised. The influencing factors for occlusion time and their effects play an important role for silvicultural management. Pruning operations undertaken in conifer stands have been the subject of much research (Møller 1960, Långström and Hellqvist 1991, O'Hara and Buckland 1996) while studies centred around hardwood species are less frequent (Kerr and Morgan 2006, Hein and Spiecker 2007, Dănescu et al. 2015). Previous research conducted on ash (*Fraxinus excelsior* L.) and sycamore (*Acer pseudoplatanus* L.) by Hein and Spiecker (2007) and Dănescu et al.

[1] Chair of Forest Growth and Dendroecology, Albert-Ludwigs-Universität, Freiburg, Germany
[*]jonathan.sheppard@iww.uni-freiburg.de

(2015) could identify significant effects presented by branch diameter and radial increment during the time of wound occlusion. Current literature advises that branch diameters of wild cherry at the time of pruning should not exceed 3.0 cm at the branch collar (Pryor 1988, Spiecker and Spiecker 1988, Seifert et al. 2010, Springmann et al. 2011b). Additionally, the quality of the pruning cut, as expressed by the roughness of the cut surface and the length of the remaining stub is believed to have an effect on occlusion time (Dujesiefken et al. 1998, Springmann et al. 2011b).

Pruning methods such as selective pruning (Spiecker 2010, Springmann et al. 2011b) can improve the production of veneer quality timber from wild cherry due to the early removal of the most vital branches. This results in smaller branch diameter of the pruned branches, and hence, smaller pruning wounds. An employment of this method means that occlusion time should be reduced, which in turn implies a lesser risk of infection of the pruning wound (Spiecker 2010) by wood decaying fungi such as *Phellinus tuberculosus* Baumg. and *Trametes* spp. (Seifert et al. 2010). Hence, the first objective of this study is to evaluate the influence of different parameters such as branch-stub length, branch diameter, radial growth, insertion angle of the branch and cardinal direction on the occlusion time of pruned branches. The second objective was to reveal how branch diameter and occlusion time influence the development of wood decay.

Materials & Methods

The 2.5 ha experimental site is located on the floodplain of the river Rhine in south western Germany close to the town of Breisach (48.071N; 7.589E, 182 m a.s.l.) situated on former agricultural land. The soils are not considered deficient in essential minerals and nutrients. A detailed description of the site, soils and climate conditions can be found in Morhart et al. (2016). The site and climatic conditions can be considered suitable for the growth of wild cherry. The site index for pruned cherry is described as class II: a tree height of 26.7 m at an age of 60 years is to be expected according to applicable yield tables (Spiecker 1994).

The research site was planted in 1997 with 1+1 wild cherry stock derived from the Liliental seed orchard, established within a randomised block design. The initial spacing of all trees on the research plot was 1.5 m x 7.5 m and 1.5 m x 15.0 m in a mixture with other broadleaves species including European ash, pedunculate oak (*Quercus. robur* L.), sycamore, small-leaved lime (*Tilia cordata* Mill.) and European hornbeam (*Carpinus betulus* L.). This mixture relates to the site's aim of investigating the growth of valuable broadleaved tree species within a widely spaced planting design. Thinning was carried out to favour future crop trees, i.e. those displaying a high radial increment and suitable form.

In 2015 a total of 10 wild cherry classified as dominant and co-dominant trees (Kraft 1884) were sampled. These trees were artificially pruned in three year intervals (2007, 2010 and 2013), thus, reflecting representative crop tree management practices. After felling the trees, each whorl was carefully examined for partially or fully occluded branches and the respective whorls were sampled. This yielded 63 occluded and 22 partially occluded artificially pruned branches for further analyses. In addition stem discs were taken for retrospective radial increment analysis at 1.3 m, 2.5 m, 5.0 m and 7.5 m from the ground.

Samples were air-dried under ventilated conditions for three weeks before being prepared for examination. Branch whorls were cut parallel to the stem axis so that the included branches could be divided centrally (see Fig.1 left).

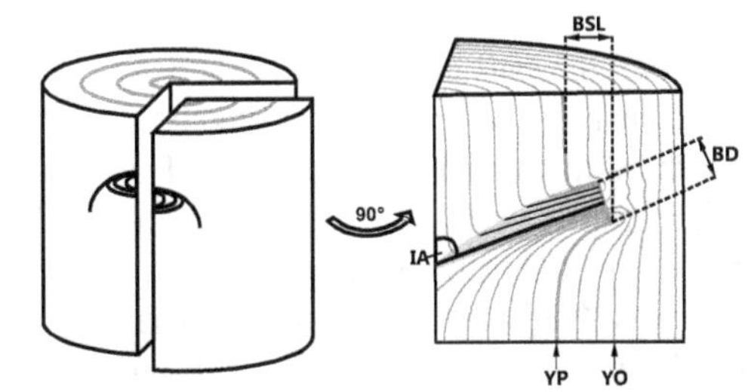

IA = Insertion angle
BSL = Branch stub length
BD = Branch diameter
YP = Year of pruning
YO = Year of complete occlusion

Figure 1 - Measured parameters within sampled whorl sections.

All cut surfaces, plus the stem discs, were sanded with 240 grit sandpaper to prepare them for further analyses. Cardinal direction (CD) of the occluded branches was measured in respect to a north datum that was marked in the field. Tree ring width, as a proxy for radial increment during pruning wound occlusion, was measured on all stem discs, using the software WinDENDRO TM (Regent Instruments Inc.). The yearly quadratic mean radial increment during occlusion was calculated from four orthogonal radii.

For determination of occlusion time (*OT*), the year of pruning (*YP*) and the year of complete occlusion (*YO*) were established (*OT* = *YO* – *YP*). *YP* was defined as the last year in which a defined tree ring originates in the stem and converges within the branch, while *YO* was identified as the first defect free tree ring entirely covering the pruning wound.

The branch cross sectional diameter (*BD*) at the site of the cut face, and the insertion angle (*IA*) of the pruned branch were additionally measured. To obtain the best possible measurement of IA, 50% of BD was utilised as a point of measurement with the vertex of the angle located where the branch pith met the stem pith. Branch stub length (*BSL*) was measured on the underside of the branch using the distance between *YP* and the lower lip of the pruning cut. All measurement sites are illustrated within Fig. 1. To determine the radial increment during the time of wound occlusion (*irO*) we used the yearly quadratic mean radial increment of the stem disc that was located closest to the pruned branch.

Investigation of stem decay as a result of artificial pruning is based on data derived from a total of 85 branches, 63 of which were completely occluded and 22 that remained un occluded, In order to retrospectively investigate stem decay associated with fungal infection, branches were first visually sampled for signs of decay. Three levels of decay were defined: No visible wood decay, wood decay limited to the pruned branch (Level I) and wood decay spreading through the pruned branch into the stem wood (Level II).

Statistical analysis and modelling

Data analysis was carried out using SPSS for Windows 22.0 software (IBM Corp. 2013). The level of significance was set at $p = 0.05$ for all analyses. Normality was tested with a Shapiro-Wilk test. Stepwise ordinary least squares regression was performed to model branch occlusion time as a consequence of *BD*, *irO*, *BSL*, *IA* and *CD*. The independent predictor variables were included within the model in a stepwise method ensuring significance. Residuals were assessed for homoscedasticity by visual analysis of the studentised residuals plotted against predicted values and a normal distribution was ensured using normal Q-Q plots. A Durbin-Watson test was carried out to assess whether there was autocorrelation between residuals and the Variance Inflation Factor (VIF) evaluated in order to assess multicolliniarity within the model. The Akaike Information Criterion (AIC) was utilised to reinforce the choice of model (values can be seen in Tab. 2). The derived model followed the form of Eq. 1 where *OT* is pruning wound occlusion time (years) and X the respective independent predictor variables with associated beta values (β) as derived from the regression analysis.

$$OT = \beta_0 + \beta_1 X_1 + \beta_2 X_2 ... \beta_p X_p \qquad \text{Eq. 1}$$

Results

The following data were attained from 63 completely occluded and 22 non occluded pruned branches.

Tab. 1 displays descriptive statistics for the parameters measured for all fully occluded branches within the sample (n=63). Occlusion time ranged between two and seven years, branch diameter was measured between 7.0 mm and 40.0 mm, branch stub length presented a maximum length of 25.0 mm within the sample while branch radial increment growth ranged between 1.60 mm/year to 4.59 mm/year for fast growing branches. *BSL*, *BD* and *irO* all proved to significantly contribute to the model and hence, model III (F(3,59) = 14.4, p=<0.001) was chosen providing the best model fit and lowest AIC value (see Tab. 2).

Table 1 - Descriptive statistics for occlusion time (*OT*), branch diameter (*BD*), radial increment during occlusion (*irO*), branch stub length (*BSL*) and insertion angle (*IA*) for completely occluded branches (n=63).

Variable	Min.	Max.	Mean.	S.D.
OT (years)	2.0	7.0	4.3	1.2
BD (cm)	0.7	4.0	1.9	0.7
irO (mm)	1.6	4.6	3.0	0.8
BSL (cm)	0.0	2.5	0.6	0.7
IA (°)	31.0	86.0	53.7	12.2

Table 2 - Model parameters for the prediction of wound occlusion time (*OT*) utilising branch stub length (*BSL*), branch diameter (BD) and radial increment during occlusion (*irO*) as predictor variables (n=63).

Model	Included Variables	F	df	R	r^2_{adj}	p	AIC
I	*BSL*	27.5	1,61	0.557	0.299	<0.001	-0.978
II	*BSL, BD*	17.7	2,60	0.609	0.350	<0.001	-4.789
III	*BSL, BD, irO*	14.4	3,59	0.650	0.394	<0.001	-8.223

Table 3 - Regression parameters of model III for the prediction of wound occlusion time (*OT*) utilising branch stub length (*BSL*), branch diameter (*BD*) and radial increment during occlusion (*irO*) as predictor variables (n=63), where β is the beta coefficient and Se the standard error.

Var.	β	Se	*p*-Value	Partial r²	Structure Coefficients
Constant	4.206	0.595	<0.001	---	---
BSL	0.818	0.174	<0.001	0.22	0.857
BD	0.352	0.149	0.0214	0.05	0.315
irO	-0.350	0.152	0.0247	0.05	-0.632

The model was statistically significant and accounts for approximately 40% of the variance of pruned branch occlusion time ($r^2_{adj} = 0.394$). Partial r^2 values suggest that BSL provides the backbone of the model (22%) with *BD* and *irO* supporting with 5% respectively. This is also supported by the stated structure coefficients also given in Tab. 3. Both *IA* (*p*=0.239) and *CD* (*p*= 0.277) did not contribute significantly to the model and were, therefore, excluded.

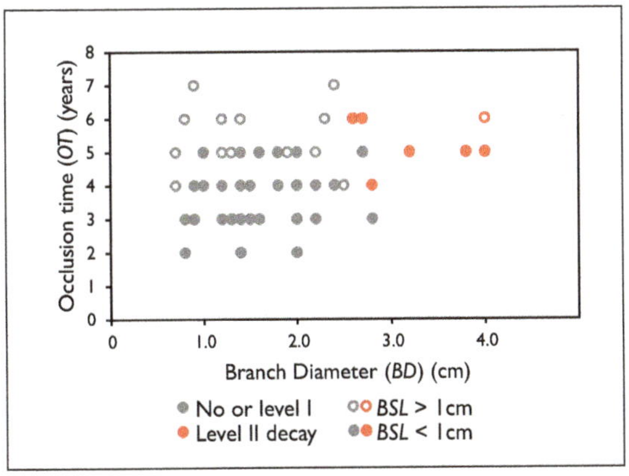

Figure 2 - Occlusion time (*OT*) as a function of branch diameter (*BD*) with marker types indicating branch stub length (*BSL*). Level II decay is indicated by red markers (n=63).

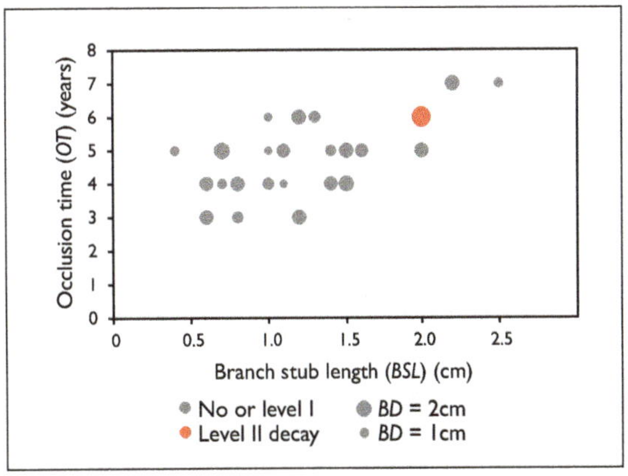

Figure 3 - Occlusion time (*OT*) as a function of branch stub length (*BSL*) (n=30). Size of the markers corresponds to branch diameter (*BD*), size of data points are proportional to branch diameter. Level II decay is indicated in red.

To visually demonstrate the influence of *BD*, *BSL* and *irO* as important contributing factors affecting *OT* Fig. 2, Fig. 3 and Fig. 4 are presented. Fig. 2 shows occlusion time against branches of different diameters. A significant increase (p=0.021) in occlusion time related to increasing branch diameter can be observed. The pruning of larger branches, thus, results in an extension of the time of wound occlusion. It can be seen that even small branches of less than 1.0 cm in diameter can take up to seven years to be completely occluded. Such long occlusion times for small diameter branches can be attributed to other factors such as the *BSL* (Fig. 2 *BSL* > 1.0 cm shown with unfilled points) and the *irO*. *OT* for pruning wounds with a *BSL* of more than 1.0 cm in length is significantly longer (p=0.001) than that for branch stubs shorter than 1.0 cm. Fig. 3 illustrates the correlation between occlusion time and branch stub length with longer branch stubs resulting in a significant (p<0.001) prolongation of occlusion time, irrespective of branch diameter. Fig. 4 exhibits the relationship between occlusion time and radial increment during the period of wound occlusion, where a significant negative correlation (p=0.025) can be observed.

From the sampled branch stubs 89% of fully occluded branches show no decay or decay classed as level I, while 50% of surveyed branches which have not fully occluded after eight or more years show a decay of level II. 18% of un-occluded branches showed no symptoms of decay at the time of analysis. A Kruskal-Wallis H test showed that the distribution of occlusion time as a function of *BD*, *BSL* and *irO* between decay levels is not the same ($\chi^2(2)$ = 22.248, p = <0.001). mean rank scores of 31.03, 43.29 and 64.39 were calculated for no decay, decay level I and decay level II respectively. Fig. 5 illustrates the correlation between the occurrence of decay in artificially pruned branches and the duration of wound occlusion. Pruning wounds that are completely occluded within two to three years exhibit negligible amounts of level I decay and none of the 17 branches in that category could be diagnosed with severe stem decay (level II). If complete occlusion takes up to four years, a large increase (55% of studied branches) of decay within the pruned branches can be observed. Occlusion times of four years and beyond show the first occurrences of severe stem decay, spreading through the pruning wound in an axial direction into the stem wood. Analysis of Fig. 5 suggests that if the pruning wound takes up to five years to be completely occluded almost 20% of the branches will exhibit decay at level II. While at six years and above there are 19% of branches that are completely unaffected by wood decay while 45% are severely infected by level II decay.

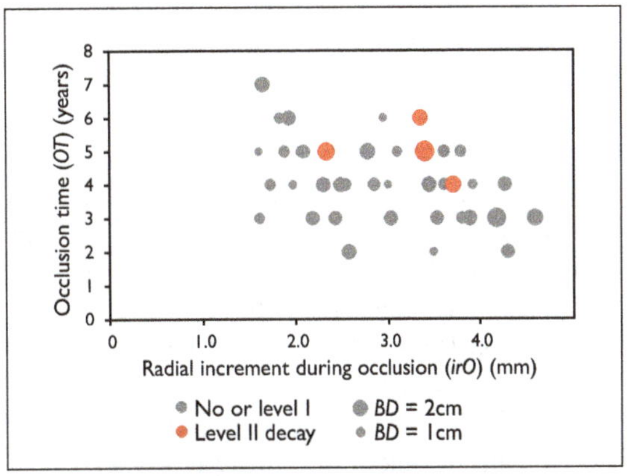

Figure 4 - Occlusion time (*OT*) as a function of radial increment during occlusion (*irO*) (n=63). Size of the markers corresponds to branch diameter (*BD*), size of data points are proportional to branch diameter. Level II decay is indicated in red.

To examine the correlation between the formation of decay and branch diameter we investigated those branches that were pruned with a *BSL* of less than 0.5 cm. This way we could include branches that had been pruned with a correct cut peripheral

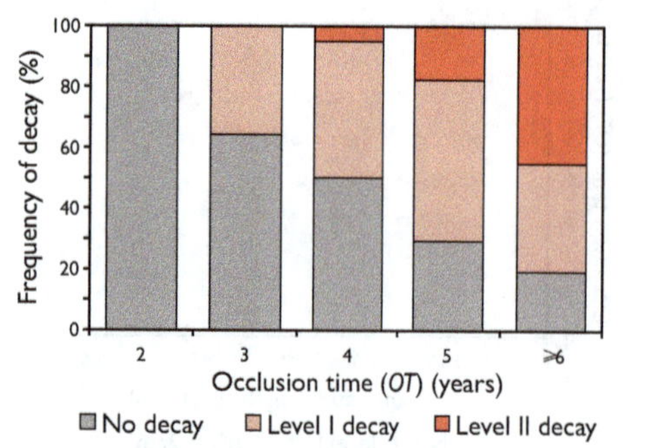

Figure 5 - Percentage frequency of the three levels of decay following artificial pruning (n=85).

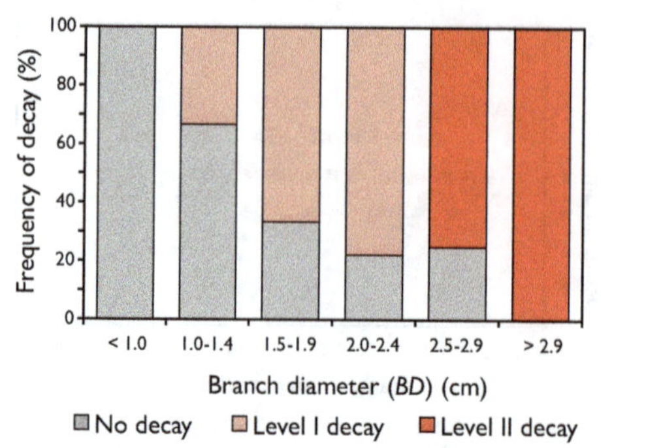

Figure 6 - Percentage frequency of the three levels of decay following artificial pruning for different branch diameters with BSL <0.5 cm (n=34).

to the branch collar. Stubs resulting from the existence of a branch collar never exceeded 0.5 cm in all samples. By setting this threshold we eliminated the large influence of *BSL* on occlusion time which results from poor pruning cuts to be able to assess the occurrence of decay for arising from pruning cuts that follow best practice. This limited the sample size to 34 branches. The correlation between pruned branch diameter and the occurrence of wood decay is illustrated in Fig. 6. Level I decay can be observed with increasing branch diameters starting in the diameter range from 1.0 to 1.5 cm. About 40% of the 22 sampled branches in the diameter range from 1.0 to 2.5 cm exhibit level I decay. Severe stem decay (i.e. level II) can be found commencing at pruned branch diameters of 2.5 cm with 85% of pruned branches showing axial spread of decay into the stem wood.

Discussion

This article retrospectively investigates the occlusion of pruning wounds on wild cherry trees. The impact of different stub parameters such as branch diameter, stub length or branch insertion angle in relation to the bole on the occlusion zone was analysed. It was shown that increased radial increment within the stem results in a reduction in occlusion time, i.e. fast growing trees occlude pruning wounds faster, a logical conclusion. Occlusion time is delayed by an increase in the diameter of pruned branches and by the increased length of the remaining branch stub.

Modelled results suggest that branch stub length, branch diameter and radial increment of the stem during the time of occlusion are all significant predictors of pruned branch occlusion time with a standard error of the estimate of 0.9 years.

The largest contributory factor towards a fast occlusion time is branch stub length, a consequence of the accuracy of pruning operations. Within the model *BSL* contributed 22% of the total for the explanation of variance. This variable has a larger effect on the duration of wound occlusion than the diameter of the branch itself or the radial increment of the stem during the occlusion period. Multiple studies found similar relations in various conifer and broadleaved tree species (Anderson 1951, Finnis 1953, O'Hara and Buckland 1996, Metzler 1997, Dănescu et al. 2015). It could be shown that the majority of fully occluded branches presented a pruned branch stub <1.0 cm. Such pruning wounds were occluded within two to five years dependent on the rate of radial increment within the stem. A recommendation suggesting that branch stubs are a maximum of 1 cm in length constitutes clear easy to follow advice for the practitioner. Nevertheless, an optimal pruning cut should be made close to the bole, as longer branch stubs will significantly extend occlusion time, and therefore, increase the risk of severe stem decay. However, care must be taken not to cut into the branch collar if present (Springmann et al. 2011a). Correct pruning practice must be carried out by making the cut adjacent to the outer perimeter of a branch collar.

These finding concur with current literature: Hein (2003) explicitly suggests the elimination of long branch stubs as one of the key advantages of artificial pruning compared to natural pruning in ash and sycamore, since the wound occlusion time is significantly shortened. Petruncio et al. (1997), utilising a stepwise regression method in order to model branch occlusion in coastal Douglas fir (*Pseudotsuga menziesii* var. menziesii [Mirb.] Franco), concluded that total occlusion time increases with large diameter stem sections (whorl diameter), with long branch stubs and if branches are dead at the time of pruning.

Results of the regression suggest that branch diameter contributes 5% towards the explanation of wound occlusion arising from pruning treatments. Wounds resulting from the artificial pruning of smaller branches occlude significantly faster than those of larger ones. This correlation has been confirmed by various studies on a broad range of tree species including valuable broadleaved species such as red oak (*Q. rubra* L.), ash or sycamore (Petruncio et al. 1997, Joyce et al. 1998, Hein and Spiecker 2007, Seifert et al. 2010, Nicolescu et al. 2013, Dănescu et al. 2015). The diameter of pruned branches is essential when considering wound occlusion time. Recommendations that branches of wild cherry should be pruned before they reach a maximum threshold of 2.5 cm to 3.0 cm can be reinforced (Pryor 1988, Spiecker and Spiecker 1988).

The results of our study confirm a significant reduction of occlusion time for increasing *irO*. Moreover, it could be proven that *irO* outweighs the influence of *BD* as far as occlusion time is concerned. This corresponds with the findings of Finnis (1953) relating to *P. menziesii*. The correlation between *irO* and *OT* suggest that thresholds for ideal pruning diameters cannot be generalised but must be site related and that artificial pruning should be focused on the most vital individuals within a stand. Similar conclusions have been drawn by Žumer (1966) and Dănescu et al. (2015). The sampled wild cherry within this study were derived from a site displaying an average annual ring width increment of approximately 3 mm during the time of occlusion. Since growth rates of 4 to 5 mm have also been measured for trees on this site, the application of silvicultural treatments to promote fast growth can also be suggested. This may include a reduction of competitive pressure amongst trees with lower annual radial increment applied after pruning treatments in order to support rapid wound closure.

The results of this study show a correlation between pruning wound occlusion time and the occurrence of decay. With increasing occlusion time the frequency and severity of decay also increases. This concurs with results presented by both Pretzsch et al. (2010) and Seifert et al. (2010). We suggest that if a pruning wound is occluded within two years, there is limited risk of decay. This study showed that coupled with an occlusion time of four years the first instances of stem decay are observed. Any further prolongation of occlusion time increases the frequency and intensity of infections and consequent decay. As a result, it is of vital importance for silvicultural management to aim for full occlusion within a maximum of three years. Location and site quality are important parameters for radial increment increases, and therefore also play a major role in the occlusion process of pruning wounds. Seifert et al. (2010) developed a model whereby dependent on annual radial increases, branch diameter thresholds to avoid decay arising from pruning wounds can be suggested. They also found that wild cherry located on poor sites invested greater energy in the defence of fungal infection of pruning wounds, and thus presenting a lower decay rate for the same occlusion time as trees located on higher quality sites. The sampled wild cherry within this study were derived from only one site, displaying an average ring width increment of approximately 3 mm during the time of occlusion. According to the model by Seifert et al. (2010), on a site comparable to the one we investigated, branches in the diameter range of up to 2.5 cm can be pruned without or with modest risk of severe stem decay. This is in agreement with results presented within this study.

All three variables manipulating wound occlusion can be influenced by forest managers: *BD* can be minimised by pruning both early and repeatedly in the rotation, *BSL* can also be minimised by improved pruning technique and accuracy, this can be reinforced by training forest workers in best practice techniques. Finally, *irO* can be maximised by utilising targeted silvicultural measures that encourage diameter growth coupled with applied pruning treatments. The selective pruning methodology where large diameter and steeply angled branches are removed within the entire length of the planned branch free bole vs. a classical whorl-wise pruning methodology allows the forest manager to effectively control factors such as *BD* while retaining a high radial increment for a decreased wound occlusion time resulting in a reduced risk of decay.

Conclusion

We conclude that if valuable timber is to be produced the artificial pruning of wild cherry must aim for an accurate pruning cut, leaving the remaining stubs short. We suggest that branches to be pruned should not exceed a diameter of 2.5 cm on order to promote a rapid wound occlusion which can limit the potential for decay. Finally, it can be concluded that the promotion of large radial increments during wound occlusion is of major importance to ensure swift occlusion of the pruning wound.

References

Anderson E.A. 1951 - *Healing time for pruned Douglas-fir.* USDA Forest Service Technical Report No. R1907.

Balandier P. 1997 - *A method to evaluate needs and efficiency of formative pruning of fast-growing broad-leaved trees and results of an annual pruning.* Canadian Journal of Forest Research 27 (6): 809-816.

Balandier P., Dupraz C. 1999 - *Growth of widely spaced trees. A case study from young agroforestry plantations in France.* Agroforestry Systems 43 (1): 151-167.

Bell S. 2009 - *Valuable broadleaved trees in the landscape.* In: "Valuable broadleaved forests in Europe". Spiecker H., Hein S., Makkonen-Spiecker K., Thies M. Eds., Brill: 171-199.

Bulfin M., Radford T. 1998 - *Effect of early formative shaping on newly planted broadleaves-Part 1: Quality.* Irish Forestry 55 (2): 35-51.

Dănescu A., Ehring A., Bauhus J., Albrecht A., Hein S. 2015 - *Modelling discoloration and duration of branch occlusion following green pruning in Acer pseudoplatanus and Fraxinus excelsior.* Forest Ecology and Management 335: 87-98.

Dujesiefken D., Stobbe H., Eckstein D. 1998 - *Langzeitwirkungen von Astungen im Holz von Linde und Roßkastanie.* Forstwissenschaftliches Centralblatt vereinigt mit Tharandter forstliches Jahrbuch. 117 (1-6): 305-315.

Dupraz C. 1994 - *Prospects for easing land tenure conflicts with agroforestry in Mediterranean France: a research approach for intercropped timber orchards.* Agroforestry Systems 25 (3): 181-192.

Finnis J.M. 1953 - *Experimental Pruning of Douglas Fir in British Columbia.* British Columbia Forest Service, Research Division.

Hein S. 2003 - *Zur Steuerung von Astreinigung und Dickenwachstum bei Esche (Fraxinus excelsior L.) und Bergahorn (Acer pseudoplatanus L.).* Dissertation, Freiburg, Germany.

Hein S., Spiecker H. 2007 - *Comparative analysis of occluded branch characteristics for Fraxinus excelsior and Acer pseudoplatanus with natural and artificial pruning.* Canadian Journal of Forest Research 37 (8): 1414-1426.

IBM Corp. 2013 - IBM SPSS Statistics for Windows. IBM Corp., Armonk, NY.

Joyce P.M., Huss J., McCarthy R., Pfeifer A., Hendrick E. 1998 - *Growing broadleaves. Silvicultural guidelines for ash, sycamore, wild cherry, beech and oak in Ireland.* COFORD, Dublin, 144p.

Kerr G., Morgan G. 2006 - *Does formative pruning improve the form of broadleaved trees?* Canadian Journal of Forest Research 36 (1): 132-141.

Kraft G., 1884 - *Beiträge zur lehre von den durchforstungen, schlagstellungen und lichtungshieben.* Klindeorth, 147p.

Kupka I., 2007 - *Growth reaction of young wild cherry (Prunus avium L.) trees to pruning.* Journal of Forest Science 53 (12): 555-560.

Långström B., Hellqvist C. 1991 - *Effects of different pruning regimes on growth and sapwood area of Scots pine.* Forest Ecology and Management 44 (2–4): 239-254.

Mahler G. 1988 - *Erfahrungen bei der Verwertung des Kirschenholzes in Süddeutschland.* AFZ, 20: 559-561.

Metzler B. 1997 - *Infektionsrisiko und Überwallungszeit bei grüngeästeten Fichten.* AFZ 52, 149-151.

Møller C.M. 196o - *The Influence of Pruning on the Growth of Conifers.* Forestry 33 (1), 37-53.

Morhart C., Sheppard J.P., Schuler J.K., Spiecker H. 2016 - *Aboveground woody biomass allocation and within tree carbon and nutrient distribution of wild cherry (Prunus avium L.) – a case study.* Forest Ecosystems 3 (4).

Morhart C.D., Douglas G.C., Dupraz C., Graves A.R., Nahm M., Paris P., Sauter U.H., Sheppard J., Spiecker H. 2014 - *Alley coppice - a new system with ancient roots.* Annals of Forest Science 71 (5): 527-542.

Nicolescu V.-N., Sandi M., Păun M. 2013 - *Occlusion of pruning wounds on northern red oak (Quercus rubra) trees in Romania.* Scandinavian Journal of Forest Research 28 (4): 340-345.

O'Hara K.L., Buckland P.A. 1996 - *Prediction of Pruning Wound Occlusion and Defect Core Size in Ponderosa Pine.* Western Journal of Applied Forestry 11 (2): 40-43.

Oosterbaan A, Hochbichler E, Nicolescu V.-N., Spiecker, H. 2009 - *Silvicultural principles, goals and measures in growing valuable broadleaved tree species.* Die Bodenkultur, 60 (3): 45-51.

Otter M. 1954 - *Le cerisier, son importance sylviculturale et son traitement.* Schweizerische Zeitschrift für Forstwesen 105 (12): 697-711.

Petruncio M., Briggs D., Barbour R.J. 1997 - *Predicting pruned branch stub occlusion in young, coastal Douglas-fir.* Canadian Journal of Forest Research 27 (7): 1074-1082.

Pretzsch H., Nickel M., Dietz E. 2010 - *Wachstum und waldbauliche Behandlung der Kirsche in Abhängigkeit von den Standortsbedingungen.* In: "Beiträge zur Vogelkirsche" Bayerische Landesanstalt für Wald und Forstwirtschaft (LWF) Ed., Freising: 13-23.

Pryor S.N., 1988 - *The silviculture and yield of wild cherry.* Her Majesty's Stationery Office London, 23p.

Rapey H., 1994 - *High value tree orchards in grazing pastures: objectives, principles and references.* Revue Forestière Française 46 (Numéro spécial): 61-71.

Röhrig E., Bartsch N., v. Lüpke B. 2006 - *Waldbau auf ökologischer Grundlage.* UTB Stuttgart, 479p.

Seifert T., Nickel M., Pretzsch H. 2010 - *Analysing the long-term effects of artificial pruning of wild cherry by computer tomography.* Trees - Structure and Function 24 (5): 797-808.

Spiecker H., 2003 - *Laubholzerziehung und Wertleistungsgrundsätze.* Österreichische Forstzeitung 114: 10-11.

Spiecker H., 2010 - *Valuable wood production - An option for the future?* Forestry Ideas 16 (1): 11-18.

Spiecker M., 1994 - *Wachstum und Erziehung wertvoller Waldkirschen.* Forstliche Versuchs- und Forschungsanst Baden-Württemberg, Freiburg, 92p.

Spiecker M., Spiecker H. 1988 - *Erziehung von Kirschenwertholz.* AFZ (20): 562-565.

Springmann S., Morhart C., Spiecker H. 2011a - *Astung von Edellaubbaumarten zur Wertholzproduktion.* AFZ (6): 4-7.

Springmann S., Rogers R., Spiecker H. 2011b - *Impact of artificial pruning on growth and secondary shoot development of wild cherry (Prunus avium L.).* Forest Ecology and Management 261 (3): 764-769.

Thies M., Hein S., Spiecker H. 2009 - *Results of a Questionnaire on Management of Valuable Broadleaved Forests in Europe.* In: "Valuable broadleaved forests in Europe". Spiecker H., Hein S., Makkonen-Spiecker K., Thies M. Eds., Brill: 27-42.

Žumer M., 1966 - *Astungsversuche an Föhre, Fichte, Birke, Aspe, Esche und Eiche.* Det Norske Skogforsøksvesen (20): 405-581.

8

Use of innovative groundcovers in Mediterranean afforestations: aerial and belowground effects in hybrid walnut

Angelo Vitone[1*], Jaime Coello[2], Miriam Piqué[2], Pere Rovira[2]

Abstract - Forest restoration in the Mediterranean area is particularly limited by water scarcity in summer and by weed competition, especially within the first years after establishment. The negative impact of these factors can be mitigated through environmentally friendly and cost-effective techniques which favour root development. This study describes the results of innovative weeding techniques in a reforestation carried out in a former agricultural field in Solsona, NE Spain, under Continental Mediterranean Sub-humid climate conditions. The tested weeding techniques included both novel groundcovers (based on prototypes built on a new biodegradable biopolymer, jute treated with resin and recycled rubber) and reference techniques, i.e. herbicide application, polyethylene and commercial biofilm groundcovers. We studied the response of hybrid walnut (*Juglans x intermedia*) to the application of these techniques during the first vegetative period in terms of survival, aerial growth and aboveground and belowground biomass allocation. The innovative groundcovers produced generally similar outcomes as the reference techniques with regard to tree survival and growth and resulted better in the case of belowground and, to a lesser extent, total tree biomass. Although preliminary, our results suggest that the tested novel groundcovers, notably the model based on treated jute, may represent a promising alternative to plastic mulching and herbicide application in afforestation of agricultural lands under Mediterranean-continental conditions. Besides these promising productive results, the novel groundcovers bring together relevant technical and environmental benefits, related to their use (not requiring removal or being reusable) and composition, based on biodegradable or recycled materials.

Keywords - Biomass allocation, Eco-innovation, Forest restoration, *Juglans,* Mulching, Reforestation

Introduction

Mediterranean climate is characterized by summer drought and high air temperatures (IPCC 2007, Senatore et al. 2011). The annual potential evapotranspiration (PET) often doubles the rainfall, causing significant water stress to plants (Flexas et al. 2006). It is expected that this condition becomes worse in the future because of climate change, as a result of the projected temperature rise and the more intense drought periods (Giorgi and Lionello 2007). Indeed, the Mediterranean region has faced wide climate shifts in the past (Luterbacher et al. 2006), and it has been identified as one of the most prominent 'hotspots' in future climate change scenarios (Giorgi 2006, Ulbrich et al. 2006).

The most critical factors limiting the successful establishment of a new plantation in the Mediterranean context are the high incoming radiation, the air temperature and the low summer rainfall (Verdú and Garcìa Fayos 1996). The heavy radiation may lead to photo-damage (Methy 1996) and cause significant soil water losses due to evaporation (Rey Benayas 1998). This factor, together with low summer rainfall, results in severe water stress particularly critical for young plants with an underdeveloped root system (Coll et al. 2003). Another major factor leading to an enhanced water stress in young trees is the weed vegetation (Rey Benayas et al. 2003). In the Mediterranean area, weeds affect negatively young trees especially with regard to water (Picon-Cochard et al. 2001) and, then, to soil nutrients (Nambiar and Sands 1993) and light availability.

Together with an adequate soil preparation (deep soil ripping, micro-catchments), and the use of adapted seedlings (in terms of species, provenance and root/shoot ratio), there are many techniques that can be applied to facilitate young trees in the first years since the plantation (i.e the installation phase). These can reduce summer water shortage, exacerbated by weed competition, even though their effectiveness and feasibility depend upon the characteristics of the sites of concern. In the case of competing vegetation, the most common techniques applied in temperate areas are the mechanical and chemical weeding (Willoughby et al. 2009). Mechanical weeding has the disadvantage of requiring a not negligible use of resources and may damage plants

[1] Università degli Studi del Molise, Campobasso, Italy
[2] Forest Sciences Centre of Catalonia, Solsona, Spain
*angelo.vitone@studenti.unimol.it

(George and Brennan 2002). Chemical weeding is an adequate cost-effectiveness solution in many circumstances (Thiffault and Roy 2011). This technique requires however recurrent application and raises significant social/environmental opposition (Bond and Grundy 2001). Its utilization in forest ecosystems is thus increasingly regulated or even banned (Willoughby et al. 2009). Mulching soil with groundcovers is being increasingly considered as a suitable technique alternative to recurrent weeding, especially in the framework of minimizing the number of interventions (Scott Green et al. 2003). Mulching has proven to reduce the vegetative competition in the root zone (Adams 1997) and to mitigate soil water evaporation (McDonald et al. 1994), thus increasing soil water availability. Mulching also increases the availability of nutrients for trees (Wilson and Jefferies 1996). Some mulching models (e.g., films) raise the soil temperature favouring the nutrient cycle and therefore root growth (Ghosh et al. 2006, Kasirajan and Ngouajio 2012).

The application of mulching often results in growth gain in juvenile phase, especially under condition of high vegetative competition (Scott Green et al. 2003). The most common products for film mulching are polyethylene and polypropylene. The main advantage of these materials concerns affordability, easiness of application, long-lasting duration and effectiveness against weeds (Arentoft et al. 2013). Their main drawback is that these products derive from unsustainable raw material (petroleum) and their removal is very expensive (McCraw and Motes 1991). In the last years, bio-based film mulches (i.e. biodegradable and obtained from renewable sources) are becoming available in the market. This allows keeping the advantages of plastic mulches while avoiding the need for removal (Kasirajan & Ngouajio 2012). The main factor limiting the use of biofilm mulches in comparison with plastic-based ones is their higher cost. While a 100 x 100 cm polyethylene sheet can cost 0.4-0.5 €, a similar piece of biofilm may be as expensive as 2-2.5 €.

Goal of the study is to evaluate the effectiveness of new groundcover types in controlling weeds and stimulating seedling growth, both above-ground and below-ground, within the first years in a hybrid walnut plantation in a Mediterranean-continental area. These techniques were developed with the aim of improving forest restoration projects in Mediterranean and temperate conditions from an environmental, technical and economic viewpoint. Our hypothesis is that the new groundcovers should increase tree growth compared to the unweeded trees, similarly to the reference weeding techniques (i.e. plastic mulching and herbicide application).

Materials and methods

Study area and plantation description

The study was carried out in Solsona, Lleida, NE Spain, at coordinates 42°00'09.71"N 1°31'46.09"E, elevation of 672 m a.s.l.. The climate is Mediterranean-continental sub-humid (Martín-Vide 1992), with an average annual temperature of 12.0°C (average temperature at the warmest and coldest month is 21.4°C and 3.7°C, respectively). Mean annual precipitation is 683 mm, 165 mm during summer period (Ninyerola et al. 2005). The plantation was established in spring 2014 in a flat field, previously used for cereal production (wheat and barley).

Soil texture is loamy-clay (32% clay, 40% silt and 28% sand). During the plantation the soil was prepared by crossed sub-soiling with ripper (45 cm depth). Plants were installed manually, in pits sized 40 x 40 x 40 cm. The plantation scheme was regular, 3 x 3 m. The vegetative material was hybrid walnut (*Juglans* x *intermedia*) MJ-209xRa, bare rooted, 40/60 cm high. Aim of the plantation was the production of valuable timber for veneer industry.

Experimental design and treatments

The seven experimental treatments, described in Table.1, include three innovative groundcovers, three reference techniques against competing vegetation and a control (unweeded trees). These treatments were compared following a randomized block design with 6 blocks, each of them containing 10 trees per treatment (60 trees per treatment, 420 experimental trees).

Table 1 - Description of the 7 experimental treatments. Each individual treatment refers to 80x80 cm area.

Treatment type	Description	Treatment code
Control	*No weeding treatment	Unweeded
Reference weeding techniques	* Herbicide application (glyphosate, 14.4 cm3 tree-1 at 1.25%) applied in May via backpack sprayer	Comm_HER
	*Commercial black polyethylene film, anti-UV treated, 80 μ thick	Comm_PE
	* Commercial green biodegradable woven biofilm	Comm_BF
Innovative groundcovers	* Recycled rubber based mulch, anti-UV treated, long-lasting and not requiring fixation (1.5 mm thick)	New_RUB
	* Woven jute cloth treated with bio-based resin for increased lifetime, 100% biodegradable	New_JUTE
	* Black new biopolymer-based frame 100% biodegradable, fused to a black commercially available biodegradable film	New_BF

Data collection and measurements

Survival and vegetative status

Tree mortality and vegetative status were assessed visually in October 2014. A tree was considered to reveal vegetative problems when the apical shoot was dead or when showing basal sprouting.

Stem growth

Basal diameter (measured with digital caliper 5 cm over the ground level on a painted mark) and total height (by measuring tape) were collected both at planting (April) and at the end of the vegetative period (October). Annual diameter and height increment were calculated as the difference between first and second measurement.

Biomass allocation study

Two living trees per treatment and block (84 in total) were randomly chosen in November 2014. These trees were pulled out carefully by a small bobcat-type backhoe excavator, keeping a full rootball with all fine roots intact. Uprooted trees were immediately placed in labelled paper bags and stored at 4°C until being processed at the lab within the subsequent 7 days. Following Schall et al. (2012) and Nanayakkara et al. (2013), the roots were put in a bucket with water to soften them and dilute soil and then rinsed gently with tap water, without damaging the roots and recovering all broken roots from each tree. Then, the trees were divided into three components: stem, coarse roots and fine roots (thickness >2 and <2mm, respectively), by using scissors for cutting and digital caliper for measuring thickness. Each component was located into a labelled aluminium tray, where they were oven-dried at 70°C for 72 hours. Finally, dry components were weighed. The variables resulting from this process are the absolute (g) and relative (% with respect to the total of the tree) biomass for each component.

Weather in 2014

The study took place within the 2014 growing season, characterized by an anomalous high rainfall amount in the July-September months (313 mm), value that almost doubled the average historical reference value of 165 mm. There was no relevant dry period during the year according to the Bagnouls & Gaussen diagram from January to October 2014 (Fig. 1).

Data analysis

Data analysis consisted of evaluating the effect of the different weeding techniques. The survival and vegetative status were analyzed by means of the Pearson's Chi-square (χ^2) test. Tree growth and biomass allocation were analyzed through a one-way analysis of variance (ANOVA; p-level < 0.05). When significant differences between treatments resulted from ANOVA, these were evaluated by

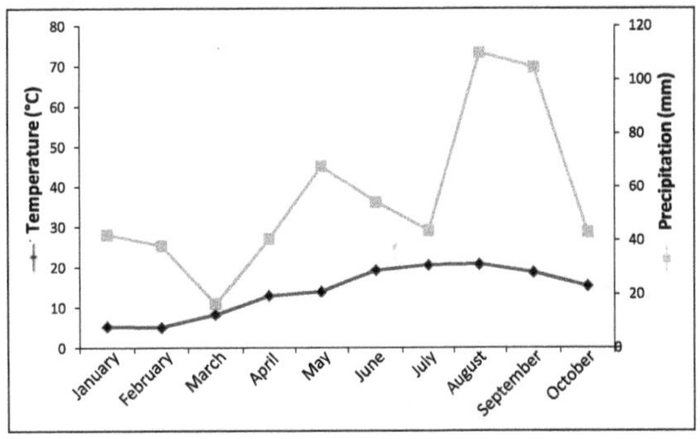

Figure 1 - Bagnouls & Gaussen diagram for the vegetative period 2014 in the study area.

means of post-hoc Tukey's HSD test. The analysis was performed with the software Statistica 7.1, 2005 (StatSoft Inc. USA).

Results

Survival and vegetative status

During the first vegetative period overall mortality rate was very low (five trees or 1.2%) with the higher mortality rate in the unweeded trees (four trees out of five). As for the vegetative status, 16.7% of trees showed some kind of vegetative problem (Fig. 2). The Pearson's χ^2 test did not explain any relationship between treatment and incidence of mortality or vegetative problems.

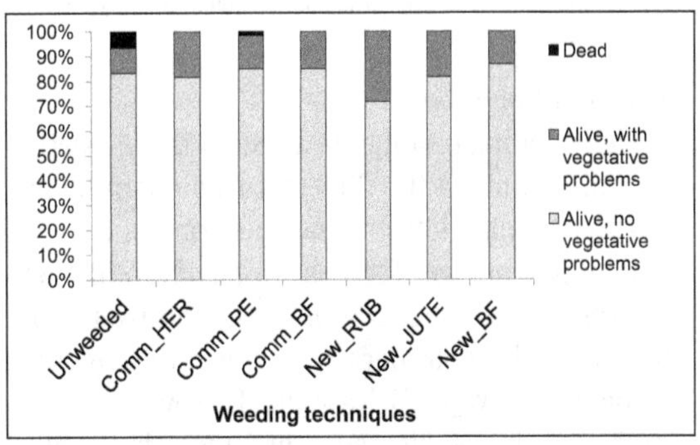

Figure 2 - Percentage of trees under each category of survival and vegetative status.

Aerial tree growth: diameter and height

Aerial growth (both in diameter and height) was affected by the different weeding treatments applied. Diameter growth was significantly improved by all the weeding treatments compared to the Unweeded-control, with diameter values which were doubled or tripled. However, no significant difference was found between the 6 weeding techniques tested (Fig. 3).

Height growth showed the same trend, with Unweeded-control providing the lowest values, although only New_JUTE and New_BF provided significantly higher results (Fig. 4).

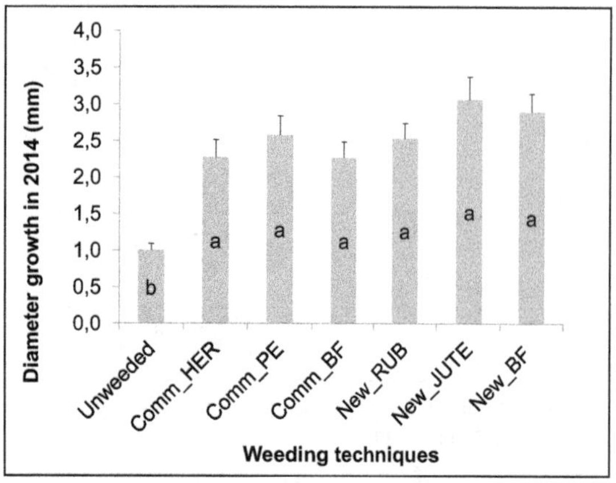

Figure 3 - Mean walnut diameter growth in 2014, subject to different weeding techniques; whiskers indicate standard error of the mean. Different letters indicate significant differences at the p<0.05 level, grouping according to Tukey's HSD post hoc test.

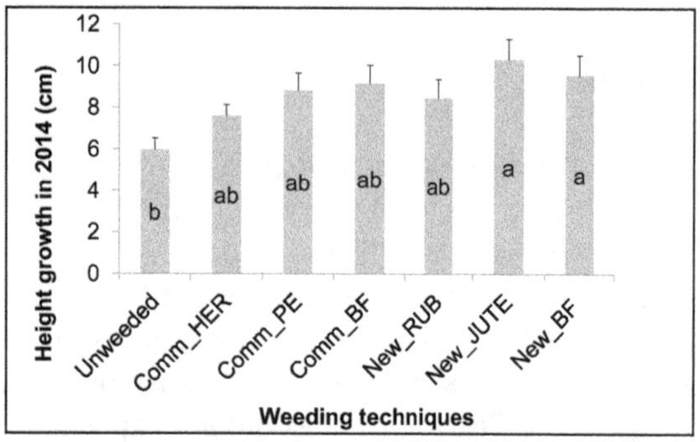

Figure 4 - Mean walnut height growth in 2014, subject to different weeding techniques; whiskers indicate standard error of the mean. Different letters indicate significant differences at the p<0.05 level, grouping according to Tukey's HSD post hoc test.

Biomass allocation

The weeding treatments affected the allocation of biomass among the different tree components, both for absolute (total biomass per component) and relative (percentage of biomass for each component with respect to total tree biomass and root/shoot ratio) variables considered. All innovative weeding techniques (New_RUB, New_JUTE, New_BF) led to significantly higher biomass than Unweeded-control for all tree components, while Comm_BF improved biomass production in coarse roots, total roots and total biomass in comparison to Unweeded-control (Figure 5). The comparison between the different weeding treatments highlights a consistent trend: New_JUTE and New_BF resulted in higher biomass allocation than all three reference techniques for most compartments, while New_RUB only improved biomass production with respect to Comm_PE and Comm_HER at most cases.

In the case of the relative distribution of biomass among the various compartments, Unweeded and Comm_PE provided a lower share of biomass in coarse roots than most of other treatments, and a higher proportion of fine roots and aboveground biomass (Fig.6).

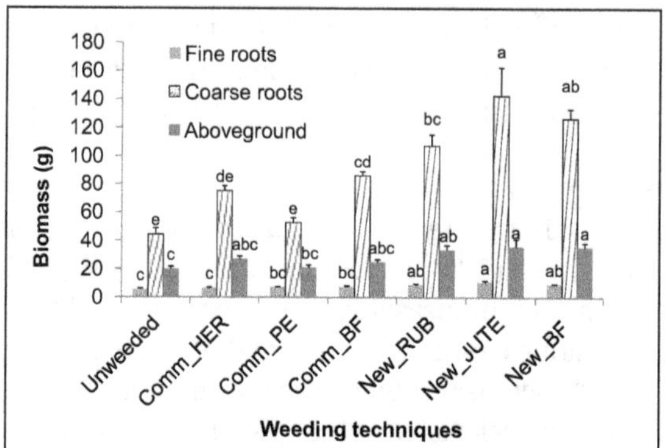

Belowground	49.8 (e)	81.6 (de)	59.8 (e)	93.3 (cd)	116.1 (bc)	153.4 (a)	135.8 (ab)
Total biomass	69.8 (d)	108.7 (bcd)	80.7 (cd)	118.2 (bc)	149.5 (ab)	189.4 (a)	171.2 (a)

Figure 5 - Mean biomass allocated to each tree compartment (g) as a response to the different treatments. Whiskers indicate standard error of the mean. Different letters indicate significant differences at the p<0.05 level, grouped according to Tukey's HSD test. Below the graph are given the mean total root biomass (belowground) and total seedling biomass, with the letters in brackets showing the grouping according to Tukey's HSD test.

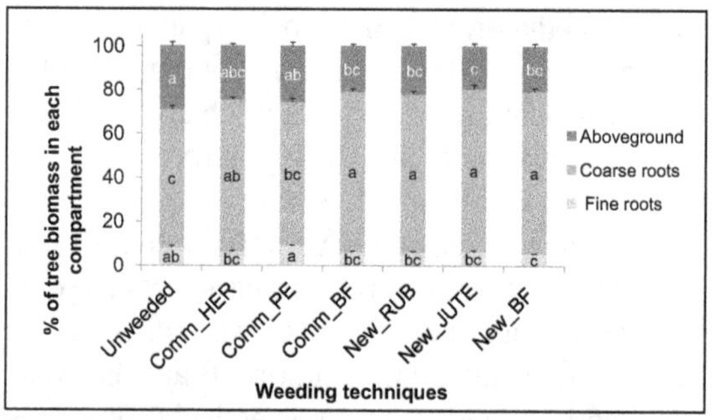

Figure 6 - Mean relative allocation of biomass in each compartment, with respect to total tree biomass (%); whiskers indicate standard error of the mean. Different letters indicate significant differences (p<0.05), grouped by Tukey's HSD test.

Weeding techniques had a significant effect on root/shoot ratio. Also for this parameter New_JUTE showed significant higher values than Comm_PE, while both New_JUTE, New_BF and Comm_BF provided significant higher values than Unweeded (Fig.7).

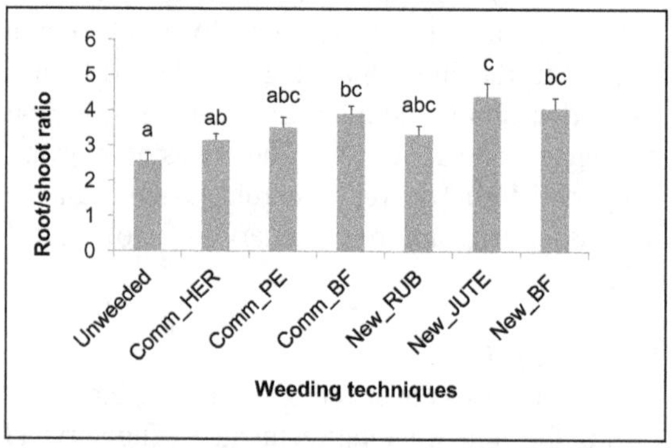

Figure 7 - Mean root/shoot ratio for the different weeding treatments; whiskers indicate mean standard error. Different letters indicate significant differences (p<0.05), grouped by Tukey's HSD test.

Discussion

The analysis highlighted the effect of different

weeding techniques on survival, vegetative status, aerial growth and biomass allocation of hybrid walnut seedlings, during the first growing season. We found a positive effect of all weeding treatments in most variables, compared to the unweeded trees.

Seedling survival and vegetative status

The survival of seedlings after one growing season is high (98.8%). Among the five dead plants of the study, four corresponded to the Unweeded treatment, suggesting the negative effect of competition (Chaar et al. 2008). The vegetative status was generally good, 82.1% of seedlings being healthy. No linkage was checked between the different treatments applied and the occurrence of vegetative problems.

Tree aerial growth

Weeding techniques had a remarkable positive effect compared to unweeded trees (Haywood 2000, Athy et al. 2006). Diameter growth of all weeding treatments resulted to be superior to unweeded trees as already observed in other research trials (Bendfeldt et al. 2001, Paris et al. 2005). Tree height growth, in New_JUTE and New_BF showed higher values than unweeded trees, as observed by previous studies on walnut (Haywood 1999, Smith et al. 2000). Significant tree growth differences between weeded and unweeded trees but not significant differences among different weeding techniques was also found in other analyses (Abouziena et al. 2008, Maggard et al. 2012).

Absolute biomass allocation

The analysis of biomass allocation among the different tree compartments (fine roots, coarse roots, aerial biomass) showed a clear effect of the different treatments applied. All innovative mulching techniques, and Comm_BF in most cases, increased the biomass of all tree compartments compared to unweeded trees, as found in previous studies assessing mulching techniques (Scott Green et al. 2003). By excluding weed competition, mulching increases water, nutrients and light available for the trees (Wilson and Jefferies 1996, Liu et al. 2003), and in turn facilitates an increase in biomass productivity (Haywood 1999). The case of root biomass is especially relevant in the study of young tree plantations, since it allows the exploration of larger soil volume and increases its resistance to prolonged water stresses in the subsequent years (Vallejo et al. 2012). The innovative mulches improved seedling rooting during the installation phase compared to Unweeded and reference weeding techniques.

Only one mulching technique (Comm_PE) did not increase significantly tree biomass of any compartment compared to unweeded trees. The reason could be related to the particular features of polyethylene mulch, i.e. a continuous plastic layer fixed to the ground, thus limiting water and air circulation and considerably reducing the soil water evaporation (Sharma et al. 1998). In general, this feature is an advantage when water availability is the main limitation (McCraw and Motes 1991) linked to drought and/or poor soil water retention. However, in the case of concern, the trial took place on a heavy textured soil and under a very wet year, and the same feature may have resulted in excessive water accumulation in the soil, which is in turn an undesirable condition for walnut (Becquey 1997). Indeed, the best results were obtained by New_JUTE mulch, a woven-structure mat with rather high permeability to water, allowing infiltration during a rainfall event and evaporation afterwards thus enhancing tree growth (Debnath 2014). The effectiveness of jute mulching has also been demonstrated in other studies, especially in agriculture (Tomar et al. 2005) not only for its weed control potentialities, but also for its mechanical and hydraulic characteristics, even concerning the regulation of soil temperature (Sanjoy 2014).

Comm_BF, leading to slightly better results than the other reference techniques, is also a woven mat with a similar structure as New_JUTE (55 threads cm^{-2} in front of 58 threads cm^{-2}, respectively) but with a notably different density (110 g m^{-2} in front of 460 g m^{-2}) and hygroscopic properties: while Comm_BF does not absorb water, New_JUTE consists of vegetal fibres which need to be moistened before becoming permeable. As a result, Comm_BF may allow a slight rainfall to penetrate in the soil but its capacity to prevent its evaporation will not stand for long time. On the other hand, New_JUTE is expected to require a heavier rainfall to allow water to reach the soil, but it would be therefore retained for longer time. In our study, with a rather high supply of water, New_JUTE performed better than Comm_BF in terms of belowground biomass, probably linked to a more efficient prevention of soil water evaporation during the warmest periods of summer, while it still allows sufficient soil aeration, opposed to Comm_PE.

Regarding New_BF and New_RUB, they are both black film mulches, although contrary to Comm_PE their surface is not continuous but consists on assembled pieces which leave some air exchange through the film, although closed enough as to impede weed growth through the mulch.

Finally, Comm_HER did not improve the results of unweeded trees regarding biomass production of any tree component. In contrast, the use of the same herbicide with the same concentration and time of application yielded during four years (2011-2014) created much more favourable results than the use of four different mulching models in a trial located two kilometers away from the field of this study, with similar soil features and the same tree species (Coello et al. *submitted*). In the mentioned study, the area per tree treated with herbicide was 100 x 100 cm, while in our study the treated area was similar to mulch size (80 x 80 cm), thus one third smaller. Moreover, in our study the wet conditions in 2014 might have boosted weed proliferation and its competitiveness during summer, reducing the efficiency of a one-time, spring herbicide application.

Relative biomass allocation

The survival of seedlings during summer is closely related to the development of the root system, rather than to the aerial organs (Lloret et al. 1999, Villar-Salvador et al. 2012). A more developed root system can more easily absorb deep (Canadell and Zedler 1995, Pemán et al. 2006) and surface water (Hilbert and Canadell 1995), and allows a more efficient extraction of nutrients (Wein et al. 1993, Lambert et al. 1994). The root/shoot ratio is a trait describing how the tree distributes the available resources (Lamhamedi et al. 1998), it being very important under water-limited condition, like Mediterranean climates (Lloret et al. 1999). It is generally accepted that a high root/shoot ratio indicates a better chance of survival under Mediterranean conditions (Navarro et al. 2006, Jacobs et al. 2009). Plants with a high root/shoot ratio are considered as better performers in water-limited sites (Royo et al. 1997), as they consume less water than plants with opposite traits (Leiva and Fernàndez-Ales 1998). In our study, the root/shoot ratio was found to be especially high for the treatments resulting in highest biomass production of compartments altogether (New_JUTE and, to a lesser extent, New_BF and Comm_BF).

Fine roots and coarse roots are the components of belowground biomass, their dry weight being proportional to the volume of soil that they can explore (Tufekcioglu et al. 1999). Fine roots represent a dynamic portion of belowground biomass, i.e. the main component dedicated to nutrients uptake, and representing a significant part of net primary production (Buyanovsky et al. 1987). On the other hand, coarse roots production is closely linked to resource availability (Albaugh et al. 1998) and especially involved in carbohydrate and nutrient storage (Comas et al. 2013) making the tree more resistant to stresses. In our study, mulching technology, particularly the new versions, enhanced a higher production of root biomass, especially among the coarse component, thus being an indicator of successful plantation.

Conclusions

Our preliminary study demonstrated the importance of weeding for increasing both aerial and belowground early tree growth in productive Mediterranean-continental conditions.

Mulching was especially effective on reducing competition by herbaceous species for water and nutrients. Among the mulching techniques tested, a novel model based on woven jute proved to be particularly successful, with enhanced results for all the traits studied, probably associated with its adequate permeability rate, which was beneficial in a wet year as 2014. Other innovative film mulches based on new biopolymers and on recycled rubber, and to a lesser extent a commercial biofilm, also led to superior results in comparison to unweeded trees, and to trees subjected to herbicide application or to polyethylene mulching.

The evaluated novel mulches demonstrated their technical adequacy, in addition to the environmental advantages compared to reference techniques; both the new biopolymer formulation and the treated jute films are 100% biodegradable, this avoiding the need for their removal and being a notable advantage compared to plastic groundcovers. The new mulch based on recycled rubber is not biodegradable, but it is made of recycled waste and its long durability (estimated in up to 15 years in outdoor conditions) makes it especially suitable for long-term applications where removal costs should be minimized: urban forestry, afforestation of easily accessible land, etc.

These results correspond to a preliminary phase of study, and must be confirmed with data from further years and further study areas in order to adequately assess the potential of these techniques for forest restoration under Mediterranean conditions with regard to their effect on tree survival and growth.

Acknowledgements

The research leading to these results has received funding from the European Union Seventh Framework Programme managed by REA-Research Executive Agency http:// ec.europa.eu/research/rea (FP7/2007-2013) under grant agreement n° 606554 - SUSTAFFOR project 'Bridging effectiveness and sustainability in afforestation / reforestation in a climate change context: new technologies for improving soil features and plant performance'. We are grateful to Sergio Martínez for his support during field and lab work and to Alejandro Borque, Sílvia Busquet, Ina Krahl, Guillem Martí and Aida Sala for their essential support during the establishment of the field trial.

References

Abouziena H.F., Hafez O.M., El-Metwally I.M., Sharma S.D., Singh M. 2008 - *Comparison of weed suppression and mandarin fruit yield and quality obtained with organic mulches, synthetic mulches, cultivation, and glyphosate.* HortScience 43: 795-799.

Adams J.C. 1997 - *Mulching improves early growth of four oak species in plantation establishment.* South Journal Applied Forestry 21 (1): 44-46.

Albaugh T.J., Allen H.L., Dougherty P.M., Kress L.W., King J.S. 1998 - *Leaf-area and above- and belowground growth responses of loblolly pine to nutrient and water additions.* Forest Science 44: 317-328.

Arentoft B.W., Ali A., Streibig J.C., Andreasen C. 2013 - *A new method to evaluate the weed-suppressing effect of mulches: a comparison between spruce bark and cocoa husk mulches.* Weed Research 53 (3): 169-175. doi: 10.1111/wre.12011

Athy R.E., Keiffer H.C., Stevens M.H. 2006 - *Effects of mulch on seedlings and soil on a closed landfill.* Restoration Ecology 14: 233-241. doi: 10.1111/j.1526-100X. 2006.00125.x

Becquey J. 1997 - *Les noyers à bois. Institut pour le développement forestier.* Paris, 143 p.

Bendfeldt E.S., Feldhake C.M., Burger J.A. 2001 - *Establishing trees in an Appalachian silvopasture: response to shelters, grass control, mulch, and fertilization.* Agroforestry Systems 53: 291-295. doi: 10.1023/A:1013367224860

Bond B., Grundy A.C. 2001 - *Non-chemical weed management in organic farming systems.* Weed Research 41: 383-405. doi: 10.1046/j.1365-3180.2001.00246.x

Buyanovsky G.A., Kucera C.L., Wagner G.H. 1987 - *Comparative analyses of carbon dynamics in native and cultivated ecosystems.* Ecology 68 (6): 2023-2031.

Canadell J., Zedler P.H. 1995 - *Underground structures of woody plants in Mediterranean ecosystems of Australia, California and Chile.* In: Arroyo MTK. et al. (ed.) Ecology and Biogeography of Mediterranean Ecosystems in Chile, California and Australia. Springer Verlag, New York: 177-210.

Chaar H., Mechergui T., Khouaja A., Abid H. 2008 - *Effects of tree-shelters and polyethylene mulch sheets on survival and growth of cork oak* (Quercus suber L.) *seedlings planted in north-western Tunisia.* Forest Ecology and Management 256 (4): 722-731. doi: 10. 1016/j.foreco.2008.05.027

Coello J., Coll L., Piqué M. - *Can bioplastic or woodchips groundcovers be an alternative to herbicide or plastic mulching on valuable broadleaved plantations in Mediterranean areas? (submitted)*

Coll L., Balandier P., Picon-Cochard C., Prévosto B., Curt T. 2003 - *Competition for water between beech seedlings and surrounding vegetation in different light and vegetation composition conditions.* Annals of Forest Science 60 (7): 593-600. doi: 10.1051/forest:2003051

Comas L.H., Becker S.R., Cruz V.M.V., Byrne P.F., Dierig D.A. 2013 - *Root traits contributing to plant productivity under drought.* Frontier in Plant Science 4: 442. doi: 10.3389/fpls.2013.00442

Flexas J., Bota J., Galmés J., Medrano H., Ribas-Carbó M. 2006 - *Keeping a positive carbon balance under adverse conditions: responses of photosynthesis and respiration to water stress.* Physiologia Plantarum 127: 343-352. doi: 10.1111 /j. 13993054.2006.00621.x

Ghosh P.K., Dayal D., Bandyopadhyay K.K., Mohanty M. 2006 - *Evaluation of straw and polythene mulch for enhancing productivity of irrigated summer groundnut.* Field Crops Research 99: 76-86. doi: 10.1016/j.fcr.2006.03.004

Giorgi F. 2006 - *Climate change Hot-spots.* Geophysical Research Letters 33 (8): L08707. doi: 10.1029/2006GL025734

Giorgi F., Lionello P. 2007 - *Climate change projections for the Mediterranean region.* Global and Planetary Change 63: 90-104. doi:10.1016/j.gloplacha.2007.09.005

Haywood J.D. 1999 - *Durability of selected mulches, their ability to control weeds and influence growth of loblolly pine seedlings.* New Forests 18: 263-275. doi: 10.1023/A:1006699910149

Haywood J.D. 2000 - *Mulch and hexazinone herbicide shorten the time longleaf pine seedlings are in the grass stage and increase height growth.* New Forests 19: 279-290. doi: 10.1023/A:1006673509218

Hilbert D.W., Canadell J. 1995 - *Biomass partitioning and resource allocation of plants from Mediterranean-type ecosystems: possible responses to elevated atmospheric CO2.* In: Moreno JM. Oechel WC. (eds.) Global Change and Mediterranean-Type Ecosystems Springer-Verlag, New York 76-101. doi: 10.1007/978-1-4612-4186-7_4

IPCC 2007 - *Working Group I Assessment Report,* Ch. 3: Observations: Surface and Atmospheric Climate Change.

Jacobs D.F., Salifu K.F., Davis A.S. 2009 - *Drought susceptibility and recovery of transplanted* Quercus rubra *seedlings in relation to root system morphology.* Annals of Forest Science 66 (5): 235-251. doi: 10.1051/forest/2009029

Kasirajan S., Ngouajio M. 2012 - *Polyethylene and biodegradable mulches for agricultural applications: a review.* Agronomy for Sustainable Development 32: 501-529. doi: 10.1007/s13593-s011-0068-3

Lamhamedi M.S., Bernier P.Y., Hébert C., Jobidon R. 1998 - *Physiological and growth responses of three sizes of containerized* Picea mariana *seedlings outplanted with and without vegetation control.* Forest Ecology and Management 110: 13-23. doi: 10.1016/S0378-1127(98)00267-9

Leiva M.J., Fernàndez-Alés R. 1998 - *Variability in seedling water status during drought within a* Quercus ilex *subsp. ballota population, and its relation to seedling morphology.* Forest Ecology and Management 111: 147-156. doi: 10. 1016/S0378-1127(98)00320-X

Liu X.J., Wang J.C., Lu S.H., Zhang F.S., Zeng X.Z., Ai Y.W., Peng B.S., Christie P. 2003 - *Effects of non-flooded mulching cultivation on crop yield, nutrient uptake and nutrient balance in rice-wheat cropping systems.* Field Crops Research 83 (3): 297-311. doi: 10.1016/S0378-4290(03)00079-0

Lloret F., Casanova C., Peñuelas J. 1999 - *Seedling survival of Mediterranean shrubland species in relation to root: shoot ratio, seed size and water and nitrogen use.* Function Ecology 13 (2): 210-216. doi: 10.1046/j.1365-2435.1999 .00309.x

Luterbacher J., Xoplaki E., Casty C., Wanner H., Pauling A., Kuttel M., Rutushauser T., et al. 2006 - *Mediterranean climate variability over the last centuries.* A review. In : Lionello P., Malanotte-Rizzoli P., Boscolo P. (eds.) Mediterranean Climate Variability. Elsevier, Amsterdam: 27-148.

Maggard A.O., Will R.E., Hennessey T.C., McKinley C.R., Cole J.C. 2012 - *Tree-based Mulches Influence Soil Properties and Plant Growth.* HortTechnology 22 (3): 353-361.

Martìn-Vide J. 1992 - *El clima.* Geografia general dels Països Catalans. Volume 1, Barcelona, Enciclopèdia Catalana: 1-110.

McCraw D., Motes JE. 1991 - *Use of plastic mulch and row covers in vegetable productions.* The Oklahoma Cooperative Extension Service. Oklahoma State University. OSU Extension Facts F-6034.

McDonald P.M., Fiddler G.O., Harrison H.R. 1994 - *Mulching to regenerate a harsh site: effects on Douglas-fir seedlings, forbs, grasses and ferns.* In: McDonald PM. (ed) USDA For. Srv. Gen. Tech. Rep., PSW-222, 10 p.

Nambiar E.K.S., Sands R. 1993 - *Competition for water and nutrients in forests.* Canadian Journal of Forest Research 23 (10): 1955-1968.

Nanayakkara B., Lagane F., Hodgkiss P., Dibley M., Smaill S., Riddell M., Harrington J., Cown D. 2014 - *Effects of induced drought and tilting on biomass allocation, wood properties, compression wood formation and chemical composition of young Pinus radiata genotypes (clones).* Holzforschung 68 (4): 455-465. doi: 10.1515/hf-2013-0053

Navarro R.M., Villar Salvador P., Del Campo A. 2006 - *Morfologìa y establecimiento de plantones.* In: Cortina J., Penuelas JL., Puértolas J., Savé R., Vilagrosa A. (eds.) Calidad de Planta Forestal Para la Restauraciòn en Ambientes Mediterràneos. Estado Actual de Conoscimientos. Ministerio de Medio Ambiente, Serie Forestal, Madrid: 67-88.

Ninyerola M., Roure J.M., Fernàndez X.P. 2005 - *Atlas climático digital de la Peninsula Ibérica.* Bellaterra, Centre de Recerca Ecològica y Aplicacions Forestals.

Paris P., Pisanelli A., Todaro L., Olimpieri G., Cannata F. 2005 - *Growth and water relations of walnut trees (*Juglans regia *L.) on a mesic site in central Italy: effects of understorey and polyethylene mulching.* Agroforestry System 65: 113-121. doi: 10.1007/s10457-004-6719-5

Pemán J., Voltas J., Gil-Pelegrin E. 2006 - *Morphological and functional variability in the root system of* Quercus ilex *L. subject to confinement: consequences for afforestation.* Annals of Forest Science 63: 425-430. doi: 10.1051/forest:2006 022

Picon-Cochard C., Nsourou-Obame A., Collet C., Guel J.M., Ferhi A. 2001 - *Competition for water between walnut seedlings (*Juglans regia*) and rye grass (*Lolium perenne*) assessed by carbon isotope discrimination and δ18O enrichment.* Tree Physiology 21 (2-3): 183-191. doi: 10.1093/treephys/21.2-3. 183

Rey Benayas J.M. 1998 - *Growth and mortality in* Quercus ilex *L. seedlings after irrigation and artificial shading in Mediterranean set-aside agricultural lands.* Annals of Forest Science 55 (7): 801-807.

Rey Benayas J.M., Espigares T., Castro-Dìez P. 2003 - *Simulated effects of herb competition on plant* Quercus faginea *seedlings in Mediterranean abandoned cropland.* Applied Vegetation Science 6: 213-222. doi: 10.1111/j.1654-109X. 2003.tb00582.x

Royo A., Fernandez M., Gil L., Gonzales E., Puelles A., Ruano R., Pardos J.A. 1997 - *La calidad de la planta de vivero de* Pinus halepensis *Mill. Destinada a repoblaciòn forestall. Tres años de resultados en la Comunidad Valenciana.* Montes 50: 29-39.

Sanjoy D. 2014 - *Jute-Based sustainable agro textiles, their properties and case studies.* In: Subramanian SM. (ed.) Roadmap to Sustainable Textiles and clothing, Eco-friendly Raw Materials, Technologies, and Processing Methods. Springer, Singapore: 327-355. doi: 10.1007/978-981-287-065-0_10

Senatore A., Mendicino G., Smiatek G., Kunstmann H. 2011 - *Regional climate change projections and hydrological impact analysis for a Mediterranean basin in southern Italy.* Journal of Hydrology 399 (1-2): 70-92. doi: 10.1016/j.jhydrol. 2010.12.035

Schall P., Lödige C., Beck M., Ammer C. 2012 - *Biomass allocation to roots and shoots is more sensitive to shade and drought in European beech in Norway spruce seedlings.* Forest Ecology and Management 266: 246-253. doi: 10.1016/j.foreco .2011.11.017

Scott Green D., Kruger E.L., Stanotcz G.R. 2003 - *Effects of polyethylene mulch in a short-rotation poplar plantation vary with weed-control strategies, site quality and clone.* Forest Ecology and Management 173: 251-260. doi: 10.1016/S0378-1127(02)00003-8

Sharma N.K., Singh P.N., Tyagi P.C., Mohan S.C. 1998 - *Effect of leucaena mulch on soil-water use and wheat yield.* Agriculture Water Management 35 (3): 183-192. doi: 10.1016/S0378-3774(97)00047-4

Smith M.W., Carroll B.L., Cheary B.S. 2000 - *Mulch improves pecan tree growth during orchard establishment.* HortScience 35 (2): 192-195.

Thiffault N., Roy V. 2011 - *Living without herbicides in Québec (Canada): historical context, current strategy, research and challenges in forest vegetation management.* European Journal of Forest Research 130 (1): 117-133. doi: 10. 1007/s10342-s010-0373-4

Tomar R.K., Gupta V.K., Garg R.N., Dwivedi B.S., Sahoo R.N., Chakrabarty D. 2005 - *An overview of mulching in agricultural and horticultural crops.* Intensive Agric. March-April 3 p.

Tufekcioglu A., Raich J.W., Isenhart T.M., Schultz R.C. 1999 - *Fine root dynamics, coarse root biomass, root distribution, and soil respiration in a multispecies riparian buffer in Central Iowa, USA.* Agroforestry Systems 44 (2-3):163-174.

Ulbrich U., May W., Li L., Lionello P., Pinto J.G., Somot S. 2006 - *The Mediterranean climate change under global warming.* In : Lionello P., Malanotte-Rizzoli P., Boscolo P. (eds.) Mediterranean Climate Variability. Elsevier, Amsterdam: 398-415.

Vallejo R., Smanis A., Chirino E., Fuentes D., Valdecantos A., Vilagrosa A. 2012 - *Perspectives in dry land restoration: approaches for climate change adaptation.* New Forests 43 (5): 561-579. doi: 10.1007/s11056-012-9325-9

Verdú M., García Fayos P. 1996 - *Nucleation process in a Mediterranean bird-dispersed plant.* Functional Ecology 10: 275-280.

Villar-Salvador P., Planelles R., Enrìquez E., Penuelas Rubira J. 2004 - *Nursery cultivation regimes, plant functional attributes, and field performance relationships in Mediterranean oak Quercus ilex L.* Forest Ecology and Management 196 (2-3): 257-266. doi: 10.1016/j.foreco.2004.02.061

Wein H.C., Minotti P.L., Grubinger V.P. 1993 - *Polyethylene mulch stimulates early root growth and nutrient uptake of transplanted tomatoes.* HortScience 118: 207-211.

Willoughby I., Balandier P., Bentsen N.S., McCarthy N., Claridge J. 2009 - *Forest vegetation management in Europe: current practice and future requirements.* COST Office, Brussels. 156 p.

Wilson D.J., Jefferies R.L. 1996 - *Nitrogen mineralization, plant growth and goose herbivory in an arctic coastal ecosystem.* Journal Ecosystem 84 (6): 841-851. doi:10.2307/2960556

9

Assessing the bibliometric productivity of forest scientists in Italy

Francesca Giannetti[1], Gherardo Chirici[1*], Elena Paoletti[2], Marco Borghetti[3], Piermaria Corona[4], Gabriele Bucci[5]

Abstract - Since 2010, the Italian Ministry of University and Research issued new evaluation protocols to select candidates for University professorships and assess the bibliometric productivity of Universities and Research Institutes based on bibliometric indicators, i.e. scientific paper and citation numbers and the h-index. Under this framework, the objective of this study was to quantify the bibliometric productivity of the Italian forest research community during the 2002-2012 period. We examined the following issues: (i) the bibliometric productivity under the Forestry subject category at the global level; (ii) compared the aggregated bibliometric productivity of Italian forest scientists with scientists from other countries; (iii) analyzed publication and citation temporal trends of Italian forest scientists and their international collaborations; and (iv) characterized productivity distribution among Italian forest scientists at different career levels. Results indicated that: (i) UK is the most efficient country based on the ratio between Gross Domestic Spending on Research and Development and bibliometric productivity under the Forestry subject category, followed by Italy; (ii) Italian forest scientist productivity has a significant positive time trend, but is characterized by high inequality across authors; (iii) one-half of the Italian forest scientist publications are written in collaboration with foreign scientists; (iv) a strong relationship exists between bibliometric indicators calculated by WOS and SCOPUS, suggesting that these two databases have the same potential to evaluate the forestry research community; and (v) self-citations do not significantly affect the rank of Italian forest scientists.

Keywords - Scientometrics, Forestry, Web of Science, SCOPUS, SCImago

Introduction

In the last few decades, increased attention has been paid to the scientific productivity of researchers and research institutions (Abramo & D'Angelo 2014; Adams 1990; Griliches 1998). Science policy increasingly includes productivity as a key factor in determining the financial budgets for research projects and scientists' careers (Bouyssou and Marchant 2010; Buela-Casal et al. 2010).

Chirici (2012) reported two main approaches applied to evaluate scientific productivity: (i) peer-review, where panels of appointed experts perform a qualitative evaluation; and (ii) bibliometrics, where a quantitative analysis of publications and citations is performed. In the last two decades, evaluation of researchers' work and careers has increasingly transitioned from peer-review to bibliometric evaluation (e.g. Seglen 1997b; Rogers 2002; Cameron 2005). Several studies were conducted that confirmed the use of bibliometric indicators as a suitable evaluation method (e.g. Falagas et al. 2006; Kumari 2006, Li & Zhao 2015). A measure of the publication and citation numbers provides an assessment of the respective quantity and quality of the research within a given field of science. For example, Vergidis et al. (2005) generated an analysis of microbiology researcher productivity; Falgas et al. (2006) examined global trends of research productivity in tropical medicine; Kumari (2006) compared the trends in different countries regarding synthetic organic chemistry research; Chirici (2012) analyzed Italian research productivity in forestry; and Li and Zao (2015) published a bibliometric assessment of global environmental research.

Measuring research strength is considered essential for a modern country's ongoing innovative and competitive capacity at the global level. A country's success in science, technology, and research determines its ability to compete for increasingly mobile resources and investment capital and to participate in global knowledge-sharing networks (The Council of Canadian Academies 2006). Monitoring research achievements in a specific field is crucial to measure a country's vitality in a specific research sector. The number of research publications in a certain scientific field reflects a country's commitment to science and is a reasonable indicator of its research

[1] Department of Agricultural, Food and Forestry Systems, University of Florence, Firenze, Italy
[2] Institute for Sustainable Plant Protection, National Research Council (CNR), Firenze, Italy
[3] Scuola di Scienze Agrarie, Forestali, Alimentari e dell'Ambiente, Università della Basilicata, Potenza, Italy
[4] Consiglio per la ricerca in agricoltura e l'analisi dell'economia agraria, Forestry Research Centre (CREA SEL), Arezzo, Italy
[5] Institute of Bioscience and Bioresources, National Research Council (CNR), Firenze, Italy
[*] gherardo.chirici@unifi.it

and development (R&D) efforts in that field (Li & Zhao 2015; Falgas et al. 2006; Rajendram et al. 2006). Moreover, Hagen (2015) reported that participation in top-level international research is an indicator of national competitive ability and academic achievement. Multi-national teams has generated more than one-quarter of all publications in the world (Royal Society UK 2011). Collaboration has many benefits and is considered essential for groundbreaking research (Bidault & Hildebrand 2014; Sonnenwald, 2007).

The scientific output of a country is evaluated by assessing institutions or individual scientists. Two important parameters are examined, including overall production and impact of scientific publications (Bornmann 2011; Cronin 1984; Franceschini et al. 2007). The following three approaches are applied to evaluate these parameters using bibliometric indicators: i) counting the publication number; ii) counting the citation number; or iii) combining the first two counts to create hybrid indicators. Publication and citation counts are traditionally employed to indicate the influence or impact an author has within the research community (Adams 1990; Abramo & D'Angelo 2011, Wildegaard 2015). Hybrid indicators, such as the Hirsh index (h-index), provide a productivity measure and its impact using a single numerical value (Hirsch 2005; Jacso 2009; Alonso et al. 2009). An approach that offers an alternative to the combination of absolute output count and citation weight is to adjust citation measures directly for a range of factors, most commonly research age. For instance, the age-weighted citation rate (AWCR) adjusts citations by a given publication age (Jin 2007; Fedderke 2013).

A direct relationship exists between research and the overall development of a country (UNESCO, 2010). A viable approach to provide evidence of research productivity is to compare bibliometric indicators with Gross Domestic Spending (GDS) on research and experimental development (R&D) of a country (Meo et al. 2013; Leydesdorffa et al. 2009, Matthew et al. 2006). For example, the perfomance of a country in a specific field can be expressed as the number of scientific papers published or the number of citations received per 1 million USD investment in R&D (Clarke et al. 2007, Tarkowski 2007).

Another important performance aspect to analyze in a country or institution within specific scientific fields is productivity distributions among authors. Inequality indicators are applied to understand if productivity rates in a specific area are due to the efforts of a few or many authors (Cole & Eales 1917; Fuyuki et al. 2003; Bornmann et al. 2008).

Glänzel and Thjis (2004) has stressed the influence of self-citations in calculating bibliometric indicators. In fact, where citations are used as a proxy to evaluate impact on the scientific community, self-citations are problematic, as they do not represent the influence of the work on other researchers, and therefore might distort citation rates (Asknes 2003; Glänzel et al. 2006).

In the present study, these issues of individual and institutional productivity were examined with reference to Italy, and specifically to the Forestry subject category. Italy is a suitable case study: since 2010, the Ministry of University and Research introduced new evaluation protocols to select candidates under the National Scientific Habilitation (ASN – Abilitazione Scientifica Nazionale) and University and Research Institute productivity is assessed under the Evaluation of Research Quality (VQR – Valutazione della Qualità della Ricerca). Both evaluations are based on bibliometric indicators, i.e. number of scientific papers, citations and h-index (MIUR 2012). The assessments are also used to determine fund allocations to Universities.

Citation databases are employed to calculate bibliometric indicators. Comparisons of existing citation databases have been performed to assess scientific productivity of authors or organizations using Thomson Reuters Web of Science (WOS), Elsevier SciVerse Scopus (SCOPUS), and Google Scholar (Chirici 2012; Abrizah et al. 2013; Bartol et al. 2014). Franchescet (2010) completed a detailed literature review and demonstrated a moderate to high correlation between h-indexes produced by WOS and SCOPUS. In Italy, the National Agency for the Evaluation of Italian Universities and Research Institutes (ANVUR, Agenzia Nazionale per la Valutazione del Sistema Universitario e della Ricerca) recommended calculating bibliometric indicators on the basis of either WOS or SCOPUS.

The aim of the present study was to conduct a quantitative assessment of the bibliometric productivity of the Italian forest research community for the 2002-2012 publication period. Specific objectives were targeted to: (i) assess the global aggregated bibliometric productivity of Italian forest scientists using SCOPUS data available from the SCImago Journal & Country Rank (SCImago) systems; (ii) compare aggregated bibliometric productivity of Italian forest scientists with the most productive countries in Forestry on a global level (USA, UK, China, Germany, and France) on the basis of GDS on R&D; (iii) show publication and citation temporal trends by Italian forest scientists; (iv) analyze international collaborations by Italian forest scientists; (v) investigate inequality of bibliometric productivity among Italian forest scientists; (vi) show main subject categories of publications by Italian forest scientists; and (vii) compare productivity of Italian forest scientists at different career levels.

2. Materials and methods

2.1 Time frame and indicators

The 2002-2012 time period was analyzed, the same officially adopted by the Italian Ministry of University and Research for the last ASN evaluation.

The following bibliometric indicators were obtained from WOS and SCOPUS databases: i) number of publications (NP); ii) number of citations, including self-citations (NC) and without self-citations (NCws); and iii) h-index. NP is the number of scientific papers published by a given author; authorship sequence and journal ranking were not factored into the analysis. NC is the number of times papers written by an author were cited by other papers; the journal ranking where the citation was referenced was not considered; self-citations, defined as citations from papers authored or co-authored by the individual were either included or excluded. The last indicator was the well known Hirsch or h-index (Hirsch 2005). The h-index is defined by how many h of a researcher's publications have at least h citations each. The h-index requires the following: the total number of papers published by an author (NP) and the total citation number (here NC and NCws).

We also evaluated the following two additional bibliometric indicators useful in analyzing author efficiency: (i) mean citation number per paper, i.e. CPP (with self-citations) or CPPws (without self-citations); and (ii) age-weighted citation rate (AWCR), which enhances contributions from early stage researchers (Jin 2007; Fedderke 2013).

2.2 Global level analysis

Global comparisons of aggregated bibliometric productivity under the Forestry subject category were conducted on the basis of the SCImago database (SCImago 2007). For each year, NP, NCws, and AWCR were queried to determine the most productive countries of those included in the analysis, i.e. USA, France, Germany, China, UK, and Italy. The perfomance of a country was evaluated by calculating the GDS Index-NP as the ratio between GDS on R&D (GDS, in millions $USD); and NP and the GDS Index-NC as the ratio between GDS and NC. GDS was defined as the total expenditure (current and capital) on R&D conducted by all resident companies, research institutes, universities, and government laboratories in a country; it included R&D funded entities from abroad, but excluded domestic funds for R&D spent outside the domestic economy; this indicator was measured in millions USD and as a percentage of Gross Domestic Product (OECD 2015). The data on GDS were acquired from OECD (2015) as an mean over the 2002-2012 period.

The aggregated bibliometric productivity in Forestry of Western Europe was also compared with the USA and Italy based on NP, NC, and CPP. We integrated the definition adopted in SCIMAGO (SCImago, 2007) for Western Europe.

The international collaboration rate of a country was calculated as the percentage of publications whose affiliations of the authors include other countries on the total publications of the considered country.

2.3 Italian level analysis

An Italian forest scientist database including individuals with permanent positions was created from an official list of professors and researchers at Italian Universities and the Agricultural Research Council (CRA). The database included two different subject subcategories: forest management and silviculture (coded AGR05 for VQR and ASN), and wood technology and forest operations (coded AGR06 for VQR and ASN). As concerns the researchers at the National Research Council (CNR), who were not included in the list, we selected individuals officially affiliated with the Italian Society of Silviculture and Forest Ecology. The final database resulted in a total of 144 authors.

For each author, the following indicators were obtained: NP, NC, NCws, and h-index derived from WOS and SCOPUS, and h-index, excluding self-citations (NC), from SCOPUS only. We extracted author publications from 2002-2012 and citations attributed to those publications until the end of 2014. Differences in indicator means were tested for statistical significance using Wilcoxon's signed ranks test (Wilcoxon 1945). Statistical association among indicators derived from the two databases was calculated using the methodology of González-Pereira et al. (2010) and further developed by Chirici (2012).

We also analyzed NP, NC, and CPP temporal trends of the 144 authors during 2002-2012, the authors' international collaborations, and the specific subject categories where the papers were published. The international collaboration rate was calculated as the percentage of publications with international co-authors based on the total publication number (Morel et al. 2009). Co-author country affiliations were used to analyze the international co-authorship network (ICN) (Leydesdorff et al., 2014).

The Lorenz curve displays statistical distributions and the dimension of production unevenness or inequality. The approach was applied to conduct an in-depth investigation of differences between NP and NC among authors. The information in the Lorenz curve was also examined using the Gini coef-

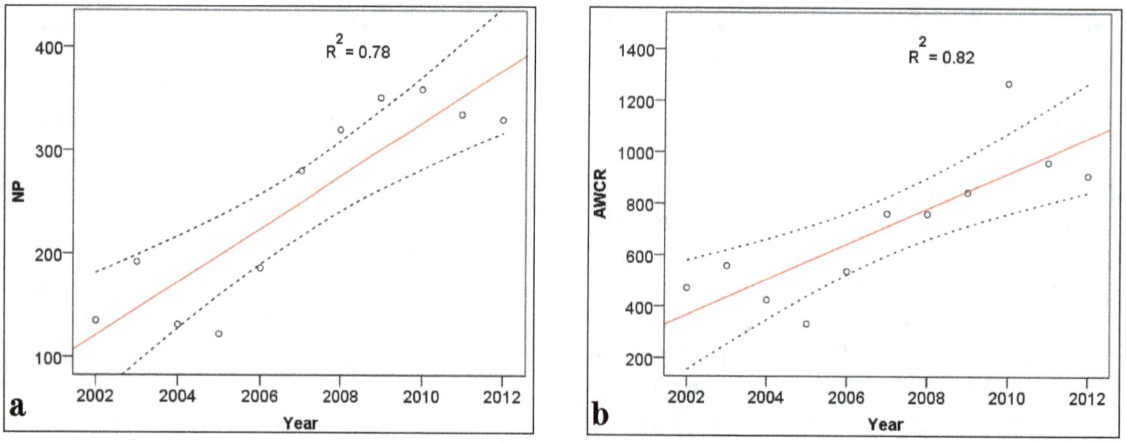

Figure 1 - Time trend of number of publications (a) and AWCR (age-weighted citation rate) (b) in Italy under the subject category Forestry in the period 2002-2012. Dotted lines are the 95% confidence interval of the linear regression. Data source: SCImago database.

ficient, a measure of statistical dispersion (Allison & Stewart 1974), similarly to Dundar and Lewis (1998) and Hagen (2015).

Self-citation relevance was analyzed to determine the rank position of individual authors in terms of citations and h-index. We chose the SCOPUS database and calculated Spearman's correlation coefficient (Spearman 1904) between each indicator determined with and without self-citations.

In Italy, scientists of Universities (UNI), the National Research Council (CNR), and the Agricultural Research Council (CRA) achieve three career levels: A-level (UNI: full professor; CNR: executive researcher; CRA: executive researcher); B-level (UNI: associate professor; CNR: first researcher; CRA: first researcher); C-level (UNI, CNR, CRA: researcher). Forest scientist productivity at different career levels (A-level, B-level, C-level) was analyzed using Wilcoxon signed-rank test; the mean, median, variance, maximum and minimum values, and standard deviation of each indicator were generated. We also compared CPP temporal trends per author per year among scientists at different career levels. The Gini coefficient was employed to quantify productivity inequality among authors at the same career level.

3. Results

3.1 Global results

Throughout 2002-2012, the cumulative bibliometric productivity of forest scientists at the global level was 0.60% of the total productivity of scientists for NP (118,561 vs. 20,117,441) and NC (1,503,622 vs. 249,752,677). The cumulative bibliometric productivity of Italian forest scientists (NP = 2824; NC = 49,214) was 0.013% NP and 0.015% NC in global bibliometric productivity and 2.3% NP and 2.6% NC in Forestry global bibliometric productivity. On a national level, the cumulative bibliometric productivity of Italian forest scientists was 0.44% of the total number of Italian scientific publications and 0.27% of the total citations received by Italian scientists.

Globally, the four scientific subject areas with the highest bibliometric productivity were Medicine (30% NP and 35% NC); Biochemistry, Genetics, and Molecular Biology (12% NP and 24% NC); and Engineering (19% NP and 9% NC). Agriculture and Biological Sciences, which include Forestry, represented 6.8% NP and 7.8% NC. The results from Italy were similar. The highest bibliometric productivity was represented by Medicine (35% NC and 41% NC); Biochemistry, Genetics, and Molecular Biology were consistent with global results (15% NP and 21% NC); and Engineering (15% NP and 9% NC). Results indicated Agriculture and Biological Sciences produced 6.6% of total NP and 6.9% of total NC in Italy.

In the Forestry subject category, the USA was the most productive country, with 32,032 total publications (35% of the total at the global level) and 71,808 citations (40% of the total), resulting in an h-index = 241. In 2012, Italy was ranked 9th based on its h-index (97); and 13th and 10th respectively from NP (2782), and NC and NCws (2722); and 8th from average citations per publication (CPP). While NP and AWCR increased from 2002-2012 (Fig. 1), Italy's h-index and NC rank remained stable over the examined period (Fig. 2).

On a global level, France, Germany, and the United Kingdom, which traditionally publish the largest number of European forestry papers, were respectively 2nd (142), 3rd (136), and 4th (133) in h-index results; and respectively 6th (5124), 4th (5931), and 5th (5280) in NP; and 5th (16015), 3rd (5732), and 2nd (17233) in NC, respectively.

The country demonstrating the highest improvement during the 2002-2012 period was the P.R. China, with results showing increased NP (from 7th position in 2002 to 2nd in 2006) and NC (from the 18th position in 2002 to 5th in 2012).

Among the countries examined, China had the

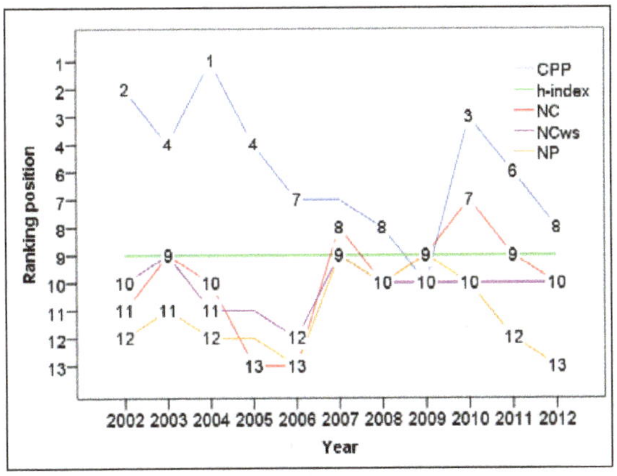

Figure 2 - Trend of number of publications (NP), number of citations with self-citations (NC) and without self-citations (NCws), mean citations per publication (CPP) and h-index of Italian authors under the subject category Forestry in the period 2002-2012. Bold numbers mark the position of Italy in the annual international ranking. Data source: SCImago database.

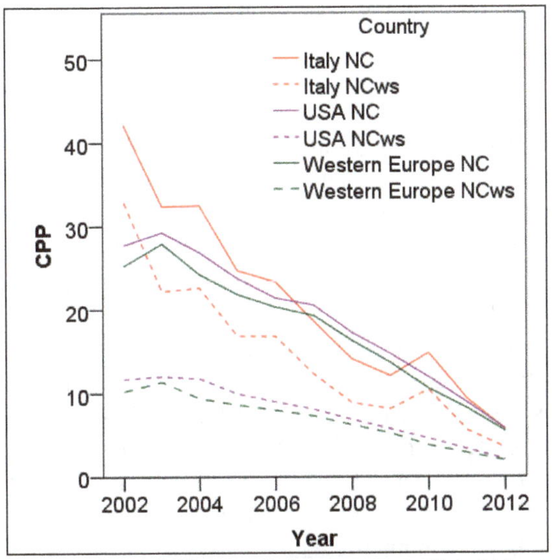

Figure 3 - Trend of CPP (mean citations per paper) with (solid lines) and without self-citations (dotted lines) for USA, Western Europe and Italy under the subject category Forestry. Data source: SCImago.

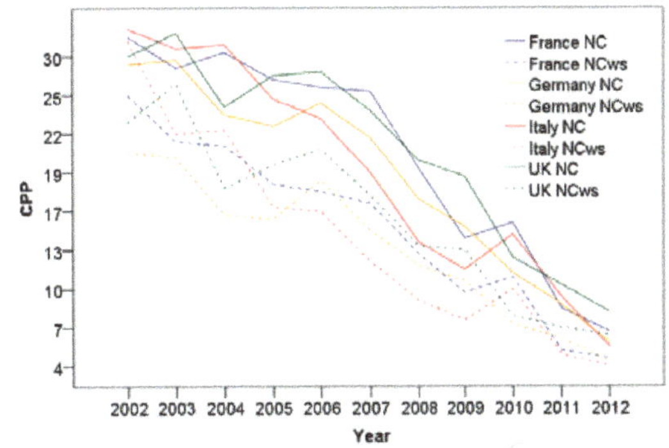

Figure 4 - Trend of CPP (mean citations per paper), with (solid lines) and without (dotted lines) self-citations, as concerns scientific papers from France, UK, Germany and Italy under the subject category Forestry.

lowest pro-capita GDS (106 USD), followed by Italy (392 USD) (Table 1). The UK was the most efficient in terms of total expenditure per article and per citation, followed by Italy. On average, Italy spent 32% less than China, 17% less than the USA, 23% less than Germany, 5% less than France, and 9% more than the UK to publish a paper. Italy spent 72% less than China, 22% less than Germany, 15% less than the USA, the same as France, and 13% more than the UK to generate a citation.

From 2002-2006 and 2010-2012, Italy demonstrated higher CPP values than the USA and Western Europe. Furthermore, Italy showed higher CPPws values than the USA and Western Europe over the entire period (2002-2012) (Fig. 3).

Italy's CCP was comparable to the three European countries with the most productive h-indices (France, UK, Germany). Italy ranked first in 2002 and 2004 and second in 2010 and 2011 (Fig. 4).

Italy demonstrated active international collaboration, evidenced by at least one co-author from a different country, always for at least 42% of the papers (minimum value in 2008) and the highest result was observed in 2002 with 64% of the papers (Fig. 5). Italy was more active than the USA with its international co-authorship.

Table 1 - Comparison of the efficiency of bibliometric productivity with respect to the gross domestic spending on research and experimental development (GDS). Data source: GDS: OECD (2015); NP and NC: SCImago.

	OECD data			Productivity		Productivity in Forestry				GDS Indices in Forestry	
Country	GDS (Million USD)	Population (Millions)	GDS pro capita (USD)	Total number of publications	Total number of citations	Number of publications (NP)	Number of citations (NC)	% of total number of publications	% of total number of citations	GDS Index-NP (Million USD per publication)	GDS Index-NC (Million USD per citation)
United Kingdom	36632	63.70	575	1526627	44011201	5344	102236	0.35	0.23	6.86	0.36
France	48185	65.63	734	984010	24700140	5248	103566	0.53	0.42	9.18	0.47
United States	381343	314.11	1214	5494335	177434935	32452	594488	0.59	0.34	11.75	0.64
Germany	80159	80.42	997	1141980	35721869	5977	107474	0.52	0.30	13.41	0.75
Italy	23316	59.53	392	648963	18019464	2824	49214	0.46	0.27	8.26	0.47
PR China	143672	1350.69	106	2482078	19110353	8882	48494	0.36	0.25	16.18	2.96

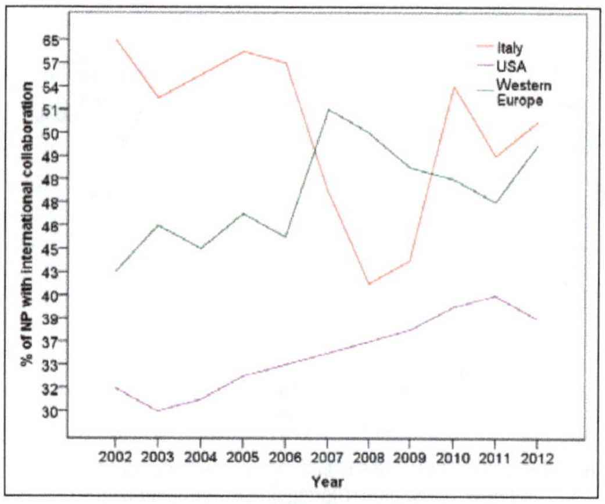

Figure 5 - Trend of the percentage of papers with international coauthorship under the subject category Forestry as concerns USA, Western Europe and Italy. Data source: SCImago

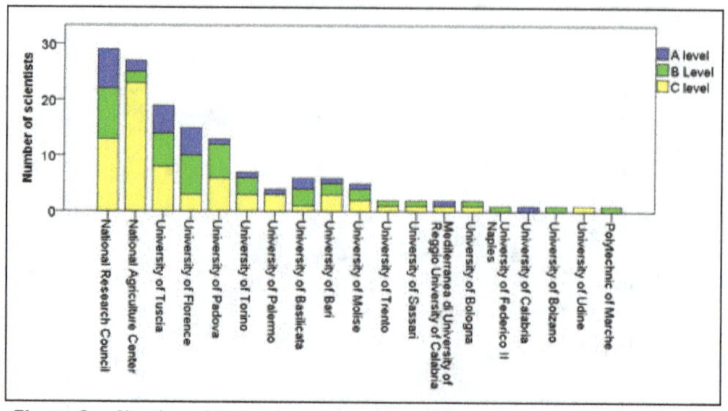

Figure 6 - Number of Italian forest scientists with a permanent position in the period 2002-2012, reported by Institution (National Research Council – CNR, Agricultural Research Council – CRA, and several universities - UNI) and career level.

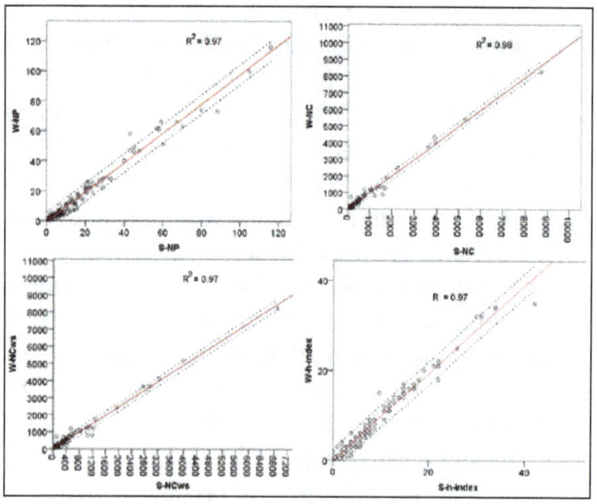

Figure 7 - Linear regressions between the WOS and SCOPUS values concerning: (a) NP - number of publications; (b) NC - number of citations with self-citations; (c) NCws - number of citations without self-citations; (d) h-index. Dotted lines are the 95% confidence interval of the linear regression. Data source: SCOPUS and WOS.

3.2 Italian forest scientist results

One hundred forty-four Italian forest scientists with permanent positions at 19 Italian research institutions (17 Universities [UNI]; the National Research Council [NRC]; and the Agricultural Research Council [CRA]) were analyzed. The forest scientists were classified as 28 A-level scientists, 46 B-level scientists, and 70 C-level scientists (Fig. 6).

NP per author ranged from 0 to 116 using both databases (WOS and SCOPUS) (Table 2). WOS results indicated NC per author was between 0 and 8323 (8227 excluding self-citations) and SCOPUS queries resulted in 0 to 8697 NP (6903 excluding self-citations). The h-index ranged between 0 and 42 using SCOPUS and 0 and 35 using WOS.

The Wilcoxon signed ranks test found no significant differences among mean values for the three indicators following WOS and SCOPUS bibliometric queries (NP: $Z = 0.274$, $P = 0.073$; NC: $Z = 0.323$, $P = 0.342$; NCws: $Z = 0.267$, $P = 0.0789$; h-index: $Z = 0.765$, $P = 0.393$). Correlation analyses showed strong atatistical association ($P < 0.001$) between WOS and SCOPUS for all the three indicators: $R = 0.98$ for NP; $R = 0.99$ for NC and NCws; $R = 0.98$ for h-index (see also Fig. 7). These results confirm that SCOPUS and WOS produce comparable and closely related bibliometric data.

The notable differences between mean and median values for NP and NC (Table 2) were due to variability in productivity among scientists. The Gini coefficient for NP (Gini = 0.84 SCOPUS; Gini = 0.85 WOS) and NC (Gini = 0.81 SCOPUS; Gini = 0.80 WOS) provided support for these observations. Among Italian forest scientists, we found the absence of publications for 9.1% of them in WOS and

Table 2 - Bibliometric indicators of Italian forest scientists over the period 2002-2012. Values refer to individual scientists. Data source: SCOPUS and WOS.

	Number of publications		Number of citations		Number of citations without self-citations		h-index		h-index without self-citations
	SCOPUS	WOS	SCOPUS	WOS	SCOPUS	WOS	SCOPUS	WOS	SCOPUS
Average	15	14	421	423	320	402	7	7	6
Maximum	116	116	8697	8323	6903	8227	42	35	37
Median	8	6	75	70	52	61	5	4	4
Minimum	0	0	0	0	0	0	0	0	0
Standard Deviation	20	20	1040	1039	813	1005	7	7	6

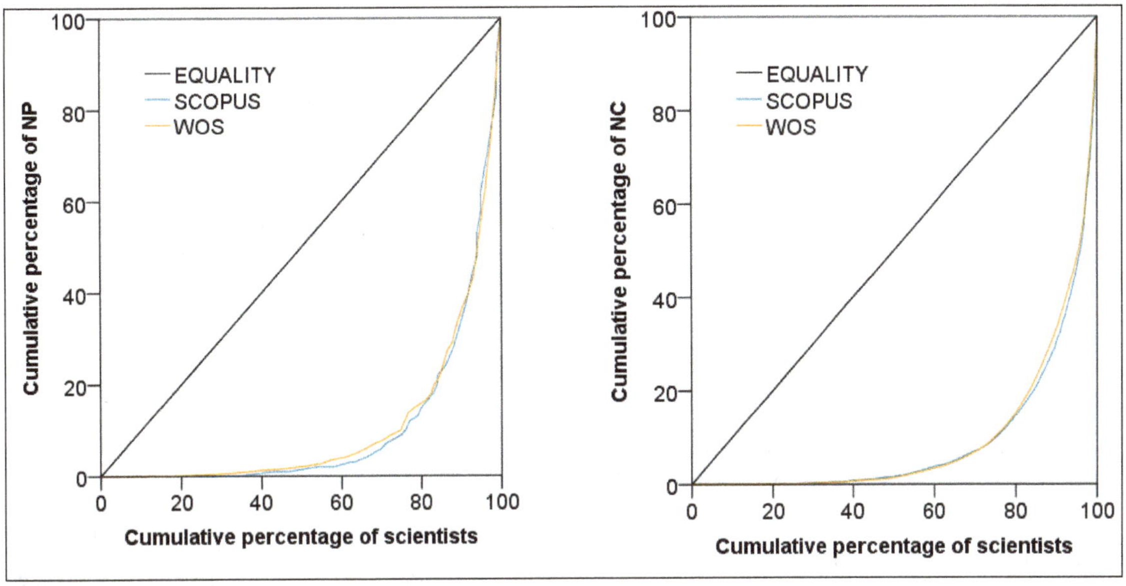

Figure 8 - The Lorenz line plots the cumulative percentage of authors vs. (a) the cumulative percentage of number of publications (NP) and (b) the cumulative percentage of number of citations. The Gini coefficient represents the area between the equality line (dotted) and the Lorenz curves: the larger the area, the higher the inequality indicator.

Table 3 - Classification by subject areas of the publications by the Italian forest scientists in the period 2002-2012. Data source: SCOPUS.

Subject Area	Number of publications
Agricultural and Biological Sciences	995
Environmental Science	619
Biochemistry Genetics and Molecular Biology	234
Earth and Planetary Sciences	231
Social Sciences	64
Medicine	61
Engineering	53
Mathematics	34
Materials Science	30
Immunology and Microbiology	27
Energy	26
Physics and Astronomy	25
Pharmacology Toxicology and Pharmaceutics	22
Chemistry	19
Computer Science	17
Arts and Humanities	16
Multidisciplinary	14
Business Management and Accounting	13
Chemical Engineering	11
Economics Econometrics and Finance	11
Decision Sciences	7
Neuroscience	3
Veterinary	2
Health Professions	1

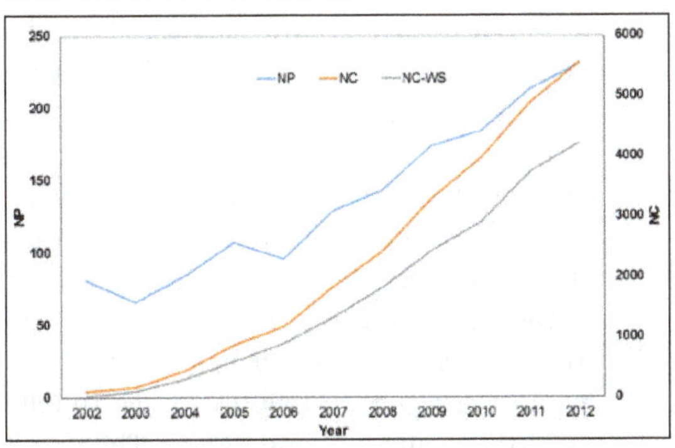

Figure 9 - Trend of number of publications (NP), number of citations with self-citations (NC) and number of citations without self-citations (NCws) of the Italian forest scientists. Data source: SCOPUS.

SCOPUS, while 35% of them represented over 90% of the total NP and NC (Fig. 8).

Following analysis of SCOPUS data over the 2002-2012 period, papers published by the Italian scientists totaled 1508, with 38723 citations (29318 NCws) and h-index = 91 (h-index ws = 80). The papers were classified under a wide range of subject areas in SCOPUS and some were classified in more than one subject area. Agricultural and Biological Science (66%) was the most common subject area, which include the subject category Forestry. However, a large number of publications were also included in Environmental Science (41%), Biochemistry, Genetics, and Molecular Biology (15%), Earth and Planetary Science (15%), and other subject areas (20%) (Table 3).

The annual figures for total NP, NC, and NCws for the Italian forest scientists strongly increased from 2002-2012. In terms of publications, 81 were found in 2002 and 231 in 2012 (Fig. 9).

More than 42% of the publications had one or more international co-authors. The level of international collaboration remained stable over the considered time period (Fig. 10).

INC, calculated on the basis of SCOPUS data in the period 2002-2012, reported Italian forest scientists co-authored publications with 64 different countries, including the USA (co-authorship number = 459), France (380), Germany (236), and the UK (182) (Fig.11). Interestingly, these countries are those with the highest h-indices in Forestry during the analyzed time period.

Author rank was not influenced by self-citations; in fact, author position with and without self-citation

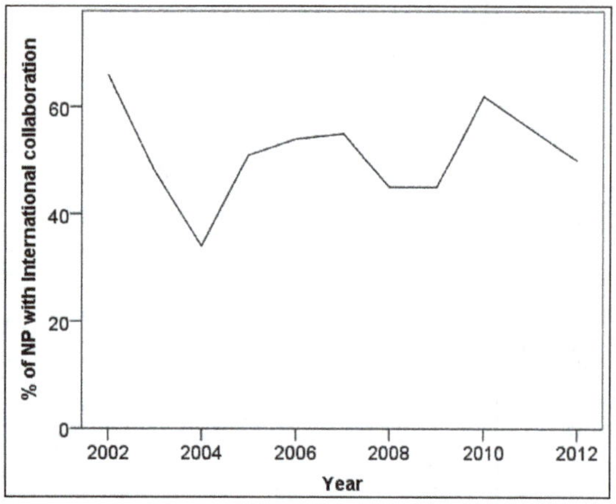

Figure10 - Temporal trend of percentage of papers written by Italian forest scientists in collaboration with foreign scientists. Data source: SCOPUS.

showed a high linear relationship based on NC and h-index (Fig. 12).

The comparison of scientists at different career levels (A, B, C) showed that A-level scientists exhibited higher mean values than B- and C-levels for NP, NC, and h-index (Table 4). However, A-level authors showed the highest variability relative to the other two groups for all three indicators. In fact, within the A-level group, 70% of the total NP were authored by only 20% of the A-group scientists (Gini SCOPUS = 0.62; Gini WOS = 0.63), 10% of the A-group had not published any paper (NP = 0), and 90% of the total NC, including NCws (Gini SCOPUS and WOS=0.79), were represented by 10% of the authors.

Results showed 40% of the B-level group authors published 80% of the total NP (Gini SCOPUS = 0.53; Gini WOS = 0.57); and 4% of authors did not have any publications (NP = 0). Eighty percent of NC was attributed to 20% of the B-level scientists (Gini SCOPUS = 0.72; Gini WOS = 0.74). Analysis results indicated 90% of the C-level group publications were authored by 40% of the scientists (Gini SCOPUS = 0.57; Gini WOS = 0.61), publications were not detected in the databases for 13% of the authors (NP = 0), and 80% of NC were attributed to 80% of C-level authors (Gini SCOPUS = 0.77; Gini WOS=0.78).

Analyzing the mean CPP per author, we found C-level scientists exhibited the lowest values during the 2002-2012 analysis period (Fig. 13).

Discussion and conclusions

At the global level, the Forestry subject category represented 0.6% of the total number of scientific publications and citations, and in Italy the subject category was detected in 0.4% of NP and 0.3% of NC. Italy published fewer scientific papers in Forestry compared with the USA, China, France, Germany and UK, which were the most productive countries in terms of NP during the analysis period (2002-

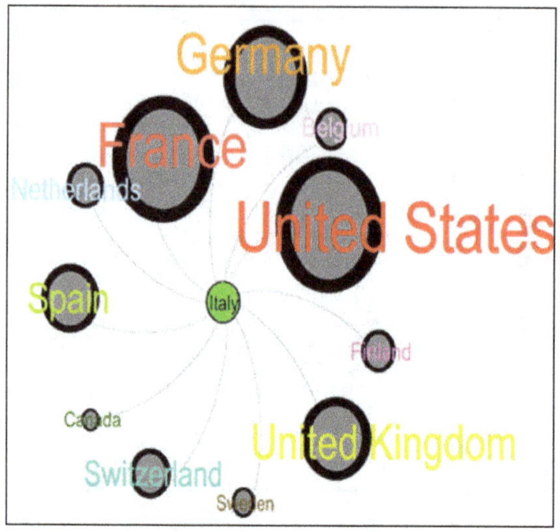

Figure 11 - International collaboration network showing the top 11 countries linked with Italy under the Forestry subject category. The circle size is proportional to the number of collaborative papers.

2012). However, if the economic investments in research (on the basis of GDS in R&D) are considered, then Italy becomes the most productive country following UK. These results are consistent with global research efficiency analysis conducted by the Royal Society of UK (2011), reported by Nature (2013). China and the USA, the most productive countries per NP were last in terms of CPP (mean citation per paper), emphasizing these two countries produce a high number of publications with fewer citations compared to Italy, UK, France, and Germany. Based on aggregated bibliometric productivity under the Forestry subject category, results showed Western Europe exceeded the USA in terms of NP. Comparable results are reported for other scientific fields, including Parasitology (Falagas et al. 2006) and Microbiology (Vergidis et al. 2005).

Overall, our study identified the following essential bibliographic results to assess scientific performance of forest scientists in Italy.

(i) Bibliometric indicators (number of publications; number of citations; h-index) shows a strong relationship between WOS and SCOPUS, suggesting the two databases have the same potential to evaluate the Italian forestry research community.

(ii) Self-citations do not significantly affect author rank under the Forestry subject category, therefore evaluation of individual productivity can be conducted using indicators with or without self-citations.

(iii) Bibliometric productivity under the Forestry subject category in Italy increased rapidly over the evaluated time period. This trend was also observed for other subject categories in Italy (Aspen Report 2012; Dario & Moed 2011).

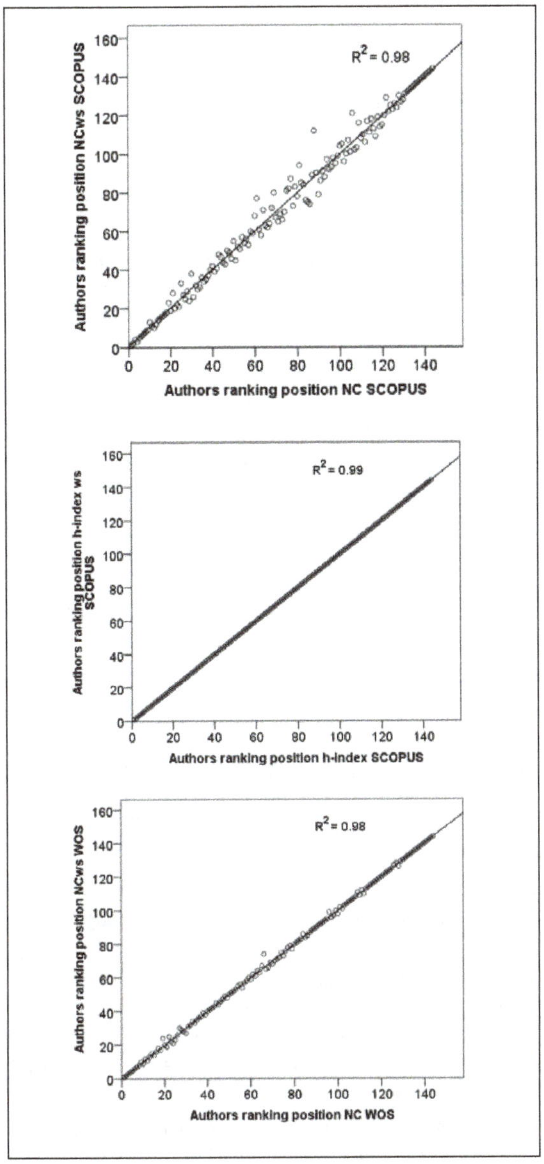

Figure 12 - Correlation between the ranks of Italian forest scientists calculated with and without self-citations: (a) SCOPUS citations, (b) WOS citations, (c) SCOPUS h-index.

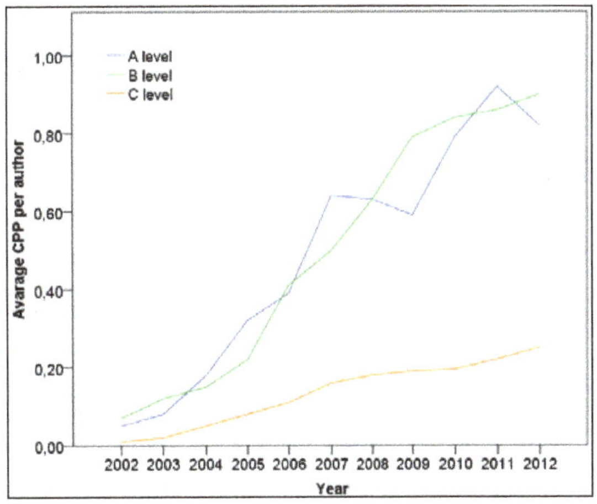

Figure 13 - Trend of mean CPP (NC/NP) of the Italian forest scientists by career level.

(iv) The productivity of Italian forest scientists is not equitable; a small number of active researchers produces the largest number of scientific publications, while a small number of forest scientists are inactive (with no publications registered on WOS or SCOPUS during the 2002-2012 period). This variability is even higher for scientists at top career levels (A-level). These results are consistent with Paulina and Francesconi (2007) for other subject categories in Agricultural and Biological Sciences in Italy.

(v) A high number of publications by the Italian forest research community (50% of the total) is written in collaboration with one or more foreign scientists. This result reflects the global internationalization trend of Italian research emphasized by Glänzel and Schlemmer (2007). Elsevier (2013) reported on a global level the rate of co-authorship among different countries increased from 14% in 2003 to 17% in 2011. The countries exhibiting more co-authorship with the Italian forest

Table 4 - Bibliometric indicators of the Italian forest scientists by career level. Means are calculated per author.

Career level		Number of publications		Number of citations		Number of citations without self-citations		h-index		h-index without self-citations
		SCOPUS	WOS	SCOPUS	WOS	SCOPUS	WOS	SCOPUS	WOS	SCOPUS
A	Average	24	23	891	900	698	864	10	10	9
	Maximum	116	116	8697	8323	6903	8227	42	35	37
	Median	7	7	55	80	38	75	6	5	5
	Minimum	0	0	0	0	0	0	0	0	0
	Standard Deviation	32	32	1848	1822	1468	1771	11	11	9
B	Average	17	16	423	408	322	387	8	7	7
	Maximum	80	74	3620	3709	2749	3665	22	22	19
	Median	11	8	109	95	81	92	7	5	6
	Minimum	0	0	0	0	0	0	0	0	0
	Standard Deviation	18	18	707	711	548	689	6	6	5
C	Average	11	10	232	242	169	228	6	5	5
	Maximum	67	66	5317	5410	3966	5190	31	32	28
	Median	7	5	58	48	39	46	4	3	3
	Minimum	0	0	0	0	0	0	0	0	0
	Standard Deviation	14	14	675	695	501	662	6	6	5

research community (USA, France, Germany, UK) are also the most productive on a global level under the Forestry subject category.

References

Abramo G., D'Angelo C.A., 2014 - *Assessing national strengths and weaknesses in research fields.* Journal of Informetrics 8: 766–775. http://dx.doi.org/10.1016/j.joi.2014.07.002 1751-1577

Abramo G., D'Angelo C, 2011a - *Evaluating research: from informed peer review to bibliometrics.* Scientometrics 87 (3): 499-514.

Abramo G., D'Angelo C. A., 2011 - *National-scale research performance assessment at the individual level.* Scientometrics 86(2): 347–364.

Abrizah A., Zainab A. N., Kiran K., & Raj R. G., 2013 - *LIS journals scientific impact and subject categorization: A comparison between Web of Science and Scopus.* Scientometrics 94(2): 721–740.

Adams J. ,1990 - *Fundamental stocks of knowledge and productivity growth.* Journal of Political Economy 98(4): 673–702.

Allison P.D., Stewart J.A., 1974 - *Productivity Differences Among Scientist: Evidence for Accumulative Advantage.* American Sociological Review 39: 596-606.

Alonso S., Cabrerizo F.J., Herrera-Vidma E, Herrera F., 2009 - *h-index: A Review focused in tis Variants Computation and Standardization for Diferent Sicentific Field.* Journal of Informetrics 3(4): 273-289.

Archambault E., Gagné E.V., 2004 - *The Use of Bibliometrics in the Social Sciences and Humanities.* Science-Metrix 514.495.6505 4572 De Lorimier Avenue Montreal Quebec Canada.

Aksnes DW, 2003 - *A macro study of self-citation.* Scientometrics 56: 235–246.

Bar-Ilan J. , 2008 - *Which h-index? A comparison of WoS Scopus and Google Scholar.* Scientometrics 74(2): 257–271.

Bartol T., Budimir G., Dekleva-Smrekar D., Pusnik M., Juznic P., 2014 - *Assessment of research fields in Scopus and Web of Science in the view of national research evaluation in Slovenia.* Scientometrics 98(2): 1491-150.

Bidault F., T. Hildebrand T., 2014 - *The distribution of partnership returns: Evidence from co-authorships in economics journals.* Research Policy 43: 1002–1013. DOI: http://dx.doi.org/10.1016/j.respol.2014.01.008

Blagus R., Leskošek B.L., Stare J., 2015 - *Comparison of bibliometric measures for assessing relative importance of researchers.* Scientometrics 105: 1743-1762. DOI: http://dx.doi.org/10.1007/s11192-015-1622-6

Bornmann L., Mutz R., Neuhaus C., Daniel H., 2008 - *Citation counts for research evaluation: tandards of good practice for analyzing bibliometric data and presenting and interpreting results.* Ethics in Science and Environmental Politics. 8: 93-102.

Bouyssou D., Marchant T. - 2010 - *Consistent bibliometric rankings of authors and of journals.* Journal of Informetrics 4: 365-378.

Buela-Casal G., 2010 - *Scientific journal impact indexes and indicators for measuring researchers' performance.* Journal of Psychodidactics 15: 3-19.

Cameron B.D., 2005 - *Trends in the usage of ISI bibliometric data: uses abuses and implications.* Portal - Libraries and the Academy 5(1): 105-125. DOI: http://dx.doi.org/10.1353/pla.2005.0003

Chirici G., 2012 - *Assessing the scientific productivity of Italian forest researchers using the Web of Science SCOPUS and SCIMAGO databases.* IForest 5: 101-107. doi: http://dx.doi.org/10.3832/ifor0613-005

ChartsBin statistics collector team, 2011 - *Number of Researchers per million inhabitants by Country (*ChartsBin.com viewed 26th November 2015) <http://chartsbin.com/view/1124>.

Clarke A., Gatineau M., Grimaud O., Royer-Devaux S., Wyn-Roberts N., Le Bis I, Lewison G., 2007 - *A bibliometric overview of public health research in Europe.* European Journal of Public Health 17 (SUPPL.1): 43-49.

Cole F.J., Eales N.B., 1917 - *The history of comparative anatomy Part I: A statistical analysis of the literature.* Science Progress in the Twentieth Century 11: 578-596.

Cronin B., 1984 - *The Citation Process: The Role and Significance of Citations in Scientific Communication.* Taylor Graham, London.

Dundar H., Lewis D.R., 1998 - *Determinants of Research Productivity in Higher Education.* Research in Higher Education 39: 607-630.

Elsevier, 2013 - *Comparative Benchmarking of European and US Research Collaborations and Researcher Mobility.* A report prepared in collaboration between Science Europe and Elsevier's SciVal Analytics September 2013.

Falagas M.E., Karavasiou A., Bliziotis A, 2006 - *Bibliometric analysis of global trends of research productivity in tropical medicine.* Acta Trop 99(2–3):155–9.

Falagas M. E., Papastamataki P. A., Bliziotis L.A., 2006 - *A bibliometric analysis of research productivity in Parasitology by different world regions during a 9-year period (1995–2003).* BMC Infectious Diseases 6 (56). DOI: http://dx.doi.org/10.1186/1471-2334-6-56

Franceschet M., 2010 - *A comparison of bibliometric indicators for computer science scholars and journals of science and Google Scholar.* Scientometrics 83(1): 243-258.

Franceschini F., Maisano D., 2011 - *Proposal for evaluatin the regularity of scientist's research output.* Scientometrics (2011) 88: 279-295. DOI: http://dx.doi.org/10.1007/s11192-011-0371-4.

Franceschini F., Galetto M., Maisano, D., 2007 - *Management by measurement: Designing key indicators and performance measurement systems.* Berlin, Springer.

Fedderke J. W., 2013 - *The objectivity of national research foundation peer review in South Africa assessed against bibliometric indexes.* Scientometrics 97:177–206. DOI: http://dx.doi.org/10.1007/s11192-013-0981-0

Fuyuki Y., Kyo K., Keita T., 2003 - *A method for the comparative analysis of concentration of author productivity giving consideration to the effect of sample size dependency of statistical measures.* Journal of the American Society for Information Science and TechnologyVolume 54 (6): 521–528. DOI: http://dx.doi.org/10.1002/asi.10239.

Glänzel W., Schlemmer B., 2007 - *National research profiles in a changing Europe (1983-2003) an exploratory study of sectoral characteristics in the Triple Helix.* Scientometrics 702: 267-275.

Glänzel W., Debackere K., Thijs B., Schubert, A., 2006 - *A concise review on the role of author self-citations in information science bibliometrics and science policy.* Scientometrics 67: 263–277.

Glänzel W., Thjis B., 2004 - *The influence of author self-citations on bibliometric macro indicators.* Scientometrics 59: 3281–310.

González-Pereira B., Guerrero-Bote V.P., Moya-Anegón F., 2010 - *A new approach to the metric of journals' scientific prestige: the SJR indicator.* Journal of Informetrics 4: 379-391.

Griliches Z., 1998 - *R&D and productivity.* Chicago: Chicago University Press. ISBN 9780226308869

Hagen N.T., 2015 - *Contributory inequality alters assessment of academic output gap between comparable countries.* Journal of Informetrics 9 (3): 629-641. DOI: http://dx.doi.org/10.1016/j.joi.2015.06.002

Hirsch J., 2005 - *An index to quantify an individual's scientific research output.* Proceedings of the National Academy of Sciences of the United States of America (PNAS) 102(46): 16569–16572. http://arxiv.org/abs/physics/0508025

Hirsch J. E. , 2005B - *An index to quantify an individual's scientific research output.* PNAS 102 (46): 16569–16572.

Jacso P., 2009f - *The h-index for countries in Web of Science and Scopus.* Online Information Review 33(4): 831–837.

Jin B., 2007 - *The AR-index: Complementing the h-index.* ISSI Newsletter, 3(1): 6.

Kumari L., 2006 - *Trends in synthetic organic chemistry research.* Cross-country comparison of Activity Index. Scientometrics 67(3): 467–76.

Li W., Zhao Y., 2015 - *Bibliometric analysis of global environmental assessment research in a 20-year period.* Environmental Impact Assessment Review 50: 158–166.

Leydesdorff L., Park H. W., Wagner C., 2014 - *International coauthorship relations in the Social Sciences Citation Index: Is internationalization leading the Network?* Journal of the Association for Information Science and Technology 65: 2111–2126. DOI: http://dx.doi.org/10.1002/asi.23102

Leydesdorff L., Wagnerb C., 2009 - *Macro-level indicators of the relations between research funding and research output.* Journal of Informetrics 3 (4): 353–362. DOI: http://dx.doi.org/10.1016/j.joi.2009.05.005

Meo S.A., Al Masri A.A., Usmani A.M., Halepoto D.M., 2013 - *Impact of GDP, spending on R&D, the number of universities and scientific journals on research publications in environmental sciences in the Middle East.* International Journal of Occupational Medicine and Environmental Health 26 (5): 702-709. doi: 10.1371/journal.pone.0066449

Matthew E. F., Argyris S. M., Ioannis A. B., Elpidoforos S. S., 2006 - *A bibliometric analysis by geographic area of published research in several biomedical fields, 1995–2003.* CMAJ 175 (11): DOI: http://dx.doi.org/10.1503/cmaj.060361

MIUR, 2012 - *Regolamento recante criteri e parametri per la valutazione dei candidati ai fini dell'attribuzione dell'abilitazione scientifica nazionale per l'accesso alla prima e alla seconda fascia dei professori universitari nonche' le modalita' di accertamento della qualificazione dei Commissari ai sensi dell'articolo 16 comma 3 lettere a) b) e c) della legge 30 dicembre 2010 n. 240 e degli articoli 4 e 6 commi 4 e 5 del decreto del Presidente della Repubblica 14 settembre 2011 n. 222. Roma.*

Morel C.M., Serruya S.J., Penna G.O., Guimarães R, 2009 - *Co-authorship Network Analysis: A Powerful Tool for Strategic Planning of Research Development and Capacity Building Programs on Neglected Diseases.* PLoS Negl Trop Dis 3(8): e501. DOI: http://dx.doi.org/10.1371/journal.pntd.0000501

Nature, 2013 - 504, 192–193 (12 December 2013) DOI: http://dx.doi.org/10.1038/504192a

OECD, 1997 - *Bibliometric indicators and analysis of research systems: methods and examples.* Paris: OECD-STI working papers 1997/1.

OECD, 2015 - *Gross domestic spending on R&D (indicator).* DOI: http://dx.doi.org/10.1787/d8b068b4-en (Accessed on 02 December 2015).

Pulina G., Francesconi A.H.D., 2007 - *Some bibliometric indexes for members of the Scientific Association of Animal Production (ASPA).* Italian Journal of Animal Science 6: 83-103.

Rajendram R., Lewison G., Preedy V., 2006 - *Worldwide alcohol-related research and the disease burden.* Alcohol Alcohol 41 (1): 99–106.

Rogers L.F. , 2002 - *Impact Factor: The numbers game.* American Journal of Roentgenology178 (3): 541 –542. DOI: http://dx.doi.org/10.2214/ajr.178.3.1780541

Royal Society UK, 2011 - *Knowledge networks and nations: Global scientific collaboration in the 21st century.* Report Royal Society UK (2011). http://royalsociety.org/policy/projects/knowledge-networks-nations/report/

SCImago, 2007 - *SJR — SCImago Journal & Country Rank.* Retrieved September 24 2015 from http://www.SCImagojr.com

Seglen P.O. , 1997b - *Why the impact factor of journals should not be used for evaluating research.* British Medical Journal 314(7079): 498–502. DOI: http://dx.doi.org/10.1136/bmj.314.7079.497

So C.Y.K., 1998 - *Citation ranking versus expert judgment in evaluating communication scholars: Effects of research specialty size and individual prominence.* Scientometrics 41 (3): 325 – 333.

Sonnenwald D.H., 2007 - *Scientific collaboration.* Annual Review of Information Science and Technology 41: 643–681. DOI: http://dx.doi.org/10.1002/aris.2007.1440410121

Spearman C., 1904 - *The Proof and Measurement of Association between Two Things.* The American Journal of Psychology 15 (1): 72-101. DOI: 10.2307/1412159

Tarkowski S.M., 2007 - *Environmental health research in Europe – bibliometric analysis.* The European Journal of Public Health 17 SUPPL.1, 2007: 14-18. DOI: http://dx.doi.org/10.1093/eurpub/ckm065.

Thomson Reuters, 2008 - *Using bibliometrics: a guide to evaluating research performance with citation data.* http://scientific.thomsonreuters.com/news/newsletter/2008-07/8465001/

UNESCO, 2010 - *Measuring R&D: Challenges Faced by Developing Countries.* UNESCO Institute for Statistics P.O. Box 6128, Succursale Centre-Ville Montreal, Quebec H3C 3J7, Canada.

Van Raan A. F. J., 2000 - *The Pandora's box of citation analysis: measuring scientific excellence the last evil?* In B. Cronin & H. B. Atkins (Eds.) The Web of knowledge: a Festschrift in honor of Eugene Garfield 301–319. Medford NJ: ASIS Monograph Series Information Today Inc.

Vergidis P.I., Karavasiou A.I., Paraschakis K., Bliziotis I.A., Falagas M.E., 2005 - *Bibliometric analysis of global trends for research productivity in microbiology.* Eur J Clin Microbiol Infect Dis 24 (5): 342–5.

The Council of Canadian Academies, 2006 - *The state of science & technology in Canada / the Committee on the State of Science & Technology in Canada.* ISBN 0-9781778-2-7. Link (19.10.2015) http://www.scienceadvice.ca/uploads/eng/assessments%20and%20publications%20and%20news%20releases/sandt/complete%20report.pdf

Weingart P., 2005 - *Impact of bibliometrics upon the science system: Inadvertent consequences?* Scientometrics 62 (1): 117 - 131.

Wilcoxon F., 1945 - *Individual Comparisons by Ranking Methods.* Biometrics Bulletin 1 (6): 80-83.

Wildgaard L., 2015 - *A comparison of 17 author-level bibliometric indicators for researchers in Astronomy Environmental Science Philosophy and Public Health in Web of Science and Google Scholar.* Scientometrics 104: 873-906. DOI: http://dx.doi.org/10.1007/s11192-015-1608-4

10

Assessing the mechanical stability of trees in artificial plantations of *Pinus nigra* J. F. Arnold using the LWN tool under different site indexes

Maurizio Marchi[1*], Ugo Chiavetta[1], Paolo Cantiani[1]

Abstract - In young black pine plantations, the most valuable and interesting thinning scheme is mainly based on the positive selection of dominant and well-shaped trees to be candidates for carbon sequestration, timber production and natural regeneration. The mechanical stability of candidate trees is here a fundamental skill that must be taken into account and the slenderness ratio (HD) is one of the main indicators. HD has been recently proved to be correlated to the living whorl number (LWN) by Cantiani & Chiavetta (2015). In this study, the statistical model was re-calibrated in order to study the influence of soil fertility on the HD - Living whorls number (LWN) relationship. The fertility-balanced models estimated a different LWN threshold. The model for the highest fertility class (Site index 24) estimated 12 LWN (RMSE of 20%). Similarly, a lower value were detected for the other two fertility classes, SI20 and SI16, where 10 LWN were considered enough with an associated RMSE of 16% and 17% respectively. Compared to the general model provided by Cantiani & Chiavetta (11 LWN with 18% of RMSE) the site index approach improved the accuracy and reliability.

Keywords - *Pinus nigra*; living whorl number; tree stability; slenderness ratio; site index; silviculture

Introduction

Artificial tree plantations are among the most fragile and dynamic forest systems. In such forests a variable horizontal and vertical structure may help the development of an adequate mechanical and biological stability. Indeed, the regular planting scheme and the reduced space for roots on the soil (especially for protective forests on thin soils), are among the most important factors of mechanical instability (Coutts 1983). Nowadays, forest management of protective plantations is often carried out following the natural dynamics, modelling gaps and species abundance (Brang et al. 2014, Muscolo et al. 2014). On one hand, thinning is used in young forests to regulate the competition between trees and the amount of solar radiation on the soil. In early stages of young stands (20-40 years), the removal of dominated and deformed trees can increase growth trends of dominants ones and accelerate biological dynamics into the soil (Cantiani et al. 2015). A too high density may cause an excessive competition for natural resources (i.e. drought stresses and reduced growth) and an unbalanced vertical increment, leading trees to be more susceptible to wind storms (Chirici et al. 2015, Dupont et al. 2015).

Trees' mechanical stability is probably the most important skill to be evaluated in any silvicultural practices (e.g. thinning or timber harvesting). Tree density plays a fundamental role in biodiversity conservation and also during the regeneration processes. In this framework, a correct forest management is mandatory to speed up ecological processes and to improve resilience and biodiversity of artificial plantations (Anfodillo et al. 2013, Kerr 1996, Shaw 2006, Tomaiuolo 2010, Valbuena et al. 2008, Yen 2015). It is well known that the health and the mechanical stability of trees is highly related to tree density. While many indicators has been developed to assess biodiversity in forest stands (Becagli et al. 2013, Chiavetta et al. 2016, Latham et al. 1998, Neumann and Starlinger 2001, Pretzsch et al. 2016), just the slenderness ratio (Height/diameter at breast height) has been demonstrated to be a fast and good proxy for mechanical stability of trees (Wang et al. 1998), especially for conifers. Anyway, an innovative method for assessing trees stability in black pine plantations was recently provided by Cantiani & Chiavetta (2015), where the relationships between Living Whorl Number (LWN) and the slenderness ratio (HD) was modelled. The aim was to assess the mechanical stability of pines in artificial pinewoods

[1] Consiglio per la ricerca in agricoltura e l'analisi dell'economia agraria, Research Centre for Forestry and Wood, Arezzo (Italy)
[*]maurizio.marchi@crea.gov.it

simply by counting the LWN. A total of 1098 trees were randomly extracted from the experimental plots of the Italian Research Centre for Forestry and Wood" (CREA FL) and analysed using a non-parametric approach, the Local Weighted Scatter-plot Smoothing (LOWESS). Due to the modelling procedure and the sampling scheme, the final result was considered to be good for almost all the artificial pinewoods along the Italian Apennines chain. The LWN demonstrated to be a simple and informative indicator to estimate the mechanical stability with an average error of 18%. In few words: a black pine with more than 11 living branches (16 taking into account the error of the model) was considered to be very likely to have a HD ratio below 90, that is considered as the threshold for stable black pines.

In this paper an update of the research on HD~LWN relationship is proposed, using the same modelling approach to evaluate the performace of this indicator under different fertility conditions. The novelty of this research consists in a deeper analysis of the data, combining the information derived from the database with the soil's fertility: the Site Index class (SI). To update the same dataset analysed by Cantiani & Chiavetta (2015) with soil characteristics, an overlay procedure in a GIS environment was performed with the management plans of the properties. Afterwards, trees were aggregated following this new approach and a possible influence of soil fertility among different site indexes was investigated. The hypothesis is that with a different vertical growth rate, which is influenced by the SI, the threshold to be used could change.

Materials and Methods

The European black pine (*Pinus nigra* Arnold spp.) is one of the most common species for reforestation programmes on bare or poor soil and in ecologically-demanding situations (Enescu et al. 2016). Despite the differences among provenances, reflected by a different gene pool and consequence of local adaptation in a fragmented natural distribution (Marchi et al. 2016, Nikolic and Tucic 1983, Vidakovic 1974), more subspecies were used during the nursery activities as seed sources (Isajev et al. 2004). Unfortunately, after plantation, the scheduled treatments were rarely applied (e.g. thinning at age 30) (Emer et al. 2011, Lehtimäki and Nurmi 2011) and the management of these plantations is nowadays very challenging. Recently, some black pine plantations in Central Italy were treated with innovative silvicultural treatments (e.g. selective thinning) demonstrating the species to be highly reactive to late-thinning (Cantiani et al. 2010).

The methodology of this paper partially reflects the one used by Cantiani & Chiavetta (2015). The same tree-level dataset, composed by 1098 trees randomly extracted from the CREA FL database, was used and the same modelling approach (LOWESS) was implemented. This database was compiled between 2000 and 2016 and includes 10 stands distributed across the Apennines Range, between Tuscany and Umbria. The analysed forests are mainly pure black pine stands in public properties around age of 55 years. According to Cantiani and Chiavetta (2015), trees were selected randomly from each stand. A total of 329 dominant trees, 371 co-dominant trees, and 398 sub-dominant trees were used for the statistical analysis. Further data and the spatial distribution of the 10 experimental plot are reported in Cantiani and Chiavetta (2015).

The SI for each of the 10 experimental plot was assessed following the most used and widely applied yield tables for *Pinus nigra* spp. in Italy (Bernetti et al. 1969). Such tables were built considering 87 research plots distributed across the whole Apennines chain and is the reference model for artificial plantations of laricio pine (*Pinus nigra* spp. *laricio*) and Austrian pine (*Pinus nigra* spp. *nigra*) in Italy. In these tables three fertility classes were delineated and named "Fertility I", "Fertility II" and "Fertility III", without any information about the regressive model used for the calculation neither a reference value of the dominant height at a defined age (the classical Site Index approach). From this starting point and assuming the SI as the value of the dominant height at age 50 (Nyland 2011), the value for the *Pinus nigra* species in Italy was calculated. Due to the unavailability of the regressive model used by Bernetti et al., the goodness of fit of four widely-used models was tested (linear, logarithmic, power and polynomial). Each model was fitted on tables' data and the one with the highest r^2 was used to calculate the dominant height at age 50 for each research plot, similarly to other researches (Carmean et al. 1989, Hann and Scrivani 1987, Means and Helm 1985, Means and Sabin 1989). Thus, the LOWESS models for each SI were studied, compared among them and to the "complete model" (i.e. with the 1098 trees together). The errors of prediction were calculated to check the quality of the study as well as to test the biased distribution of errors. A leave-one-out cross validation procedure was implemented to calculate BIAS, relative BIAS (RBIAS), root mean square error (RMSE) and root mean square percentage error (RMSPE). Afterwards, a visual comparison was used to asses differences between models and data structure plotting each fertility-balanced LOWESS fit versus the one obtained by Cantiani & Chiavetta (2015), afterwards called "reference model". Finally a non parametric ANOVA was performed to assess differ-

Table 1 - Fitting results on the data of the yield table (Bernetti et al. 1969).

Fertility	Explained proportion of variance (r^2) Linear	Logarithmic	Power	Polynomial (2ndorder)	Site Index
I class	0.9450	0.9956	0.9769	1.0000	24
II class	0.9383	0.9945	0.9694	1.0000	20
III class	0.9405	0.9949	0.9676	1.0000	16

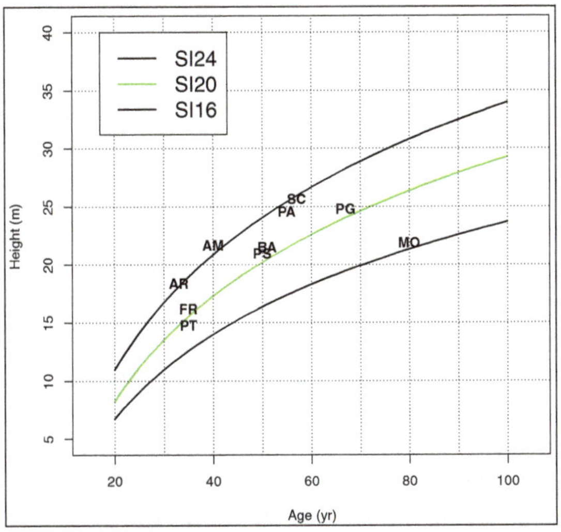

Figure 1 - Site Index classes for Pinus nigra spp. in Italy (Bernetti et al. 1969) and study sites.

ences among the average value of LWN for stable trees (HD<=90). Considering the different number of observations for each SI across the database as well as the lack of normality of the distribution, the Wilcoxon Rank sum test (Wilcoxon 1946) was used.

Results

As expected, the second order polynomial curve was detected as the best fitting model for height (Tab.1) with a perfect fit (r^2=1) and representing the

Table 3 - Correlation between LWN and HD values.

Dataset	Correlation value (Spearman)	p-value	N. of observations	Av. LWN
Complete	-0.67	<0.0001	1098	18.23
SI24 only	-0.47	<0.0001	411	19.61
SI20 only	-0.70	<0.0001	550	17.86
SI16 only	-0.73	<0.0001	137	15.30

Table 4 - Results of the models. BIAS; RBIAS=Relative BIAS; MAE=Mean absolute error; MAPE=Mean absolute percentage error; RMSE=Root mean square error; RMSPE=Root mean square percentage error; Av.LWN=Average LWN of a stable tree.

Dataset	BIAS	RBIAS	MAE	MAPE	RMSE	RMSPE	Threshold
Refer. model	**-0.3232**	**-0.68%**	**10.89**	**14.70%**	**14.01**	**18.25%**	**11**
SI24	-0.6677	-4.54%	11.94	15.66%	15.28	19.90%	12
SI20	-0.0609	-2.65%	9.33	13.03%	12.12	15.79%	10
SI16	0.1652	-2.18%	9.36	11.86%	13.09	17.05%	10

same height model (with the same coefficients) used by Bernetti et al. (1969). With this model, the dominant height at age 50 was estimated for each fertility class. For the first class we calculated a site index of 24 meters (SI24), 20 meters (SI20) were calculated for the second class while a dominant height of 16 meters (SI16) was calculated for the third class. In this framework, among the ten research plots, five (411 trees in total) were located into SI24, four in SI20, (550 trees) and only one (137 trees) belonged to the third class, SI16 (Fig.1). The descriptive statistics and the sample plot codes of the updated database are reported in Tab.2. Correlation between LWN and HD were calculated for each strata using the Spearman non-parametric method (Spearman 1987). Compared to the correlation coefficient of the complete dataset, (-0.67), only the first class showed a lower value (-0.47) while the other two were higher (-0.70 for SI20, -0.73 for SI16). However, all the correlations were statistically significant (Tab.3).

Table 2 - Research plots extracted from the CREA FL database (Cantiani & Chiavetta 2015) with the Site Index.

Site	Code	Elevation (m)	Aspect	Age (yr)	Dom.Height (m)	Density (tr.ha^{-1})	Basal area (m^2)	Quadratic Mean DBH (cm)	Mean Heigth (m)	Sample trees	Fertility (Site Index)
Monte Amiata	AM	930	NE	40	21.7	998	48.30	24.3	22.3	70	I (SI24)
Fonte Archese	AR	750	SW	33	18.4	1342	66.75	22.7	15.65	85	I (SI24)
La Baita	BA	880	SW	51	21.5	1216	88.60	30.5	20.4	226	II (SI20)
Fonte dei Frassini	FR	1020	W	35	16.2	1286	61.20	24.75	15.55	59	II (SI20)
Monte Modina	MO	1050	SE	80	21.9	956.5	60.00	26.7	20.6	137	III (SI16)
Palmoline	PA	1000	NE	55	24.6	946.5	61.80	29.45	21.6	135	II (SI20)
Pigelleto	PG	765	NW	67	24.8	510	35.50	34.8	24.3	20	I (SI24)
Fonte del Pesce	PS	1210	W	50	21.0	896.5	73.35	33	19.8	102	I (SI24)
Monte Pettenaio	PT	1000	NW	35	14.8	1377	35.35	18.75	13.05	130	II (SI20)
Lo Scoiattolo	SC	900	W	57	25.7	1048	61.34	27	23.7	134	I (SI24)

The LOWESS models were not biased (RBIAS was lower than ±5% in all cases) and the MAPE ranged between 15.66% of SI24 and 11.86% of SI16, very similarly to the complete model which was 14.70% (Tab.4). The curves' shapes were very similar for SI24 and SI20 and not much different from the reference model. On the contrary, a different form was recognized for SI16. Despite that, the thresholds for stable trees did not differed too much each other and to the reference curve. For SI24 a threshold of 12 LWN (15 considering the error of prediction) was found, and 10 LWN (13 considering the error of prediction) for SI20 and SI16. These latter values differred from the threshold of the reference model, 11 (+5 taking into account the error of the model). A graphical comparison between the four models is provided in Fig.2.

The ANOVA analysis demonstrated that the average value of LWN for stable trees was statistically different for all classes. Considering the whole dataset and without differentiation in SI classes, the average value of LWN for a stable tree (HD<=90) was 18.23. Clustering the dataset into SI classes, SI24 had an average LWN of 19.61, while for the SI20 and SI16 this value was 17.86 and 15.30 respectively.

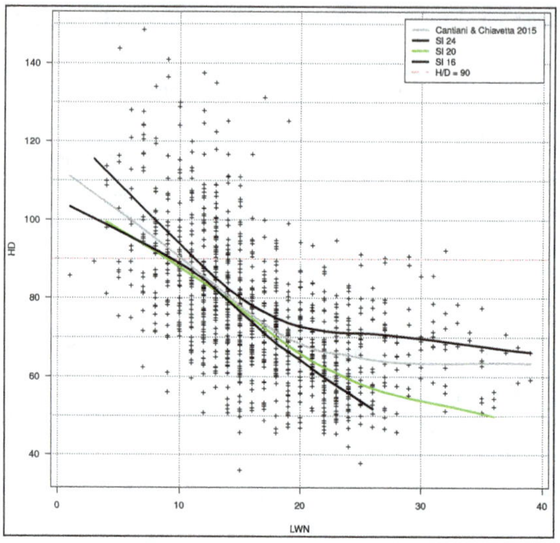

Figure 2 - HD~LWN relationships based on the Site index

Discussion

The analysis demonstrated that even if the general method can be a fair choice for all the conditions, the updated version of the LOWESS model can be more adequate and reliable when a major precision is required and site fertility information are available. In fact the use of the additional information provided by the site index class made the LOWESS model much more accurate for local conditions. The threshold of 11 LWN (+5 in case of inclusion of the error) provided by the reference model demonstrated to be higher for the first class (SI24, with LWN of 12) and more precautionary in case of the second and third classes (SI20 and SI16, with LWN of 10) where a lower value of LWN can assure a reasonable degree of single-tree stability.

Concerning the average number of LWN of stable trees (HD<=90) statistical differences were detected among fertility classes. Anyway, this aspect has a different meaning from the minimum number to be used as threshold. In fact, the average value does not take into account the "minimum LWN" and is not informative for stability. Indeed, this value is connected more to the shape of the line than to a practical use. In fact, this parameter well describes the last part of the curves (Fig.2) where the values of LWN are higher and the HD is lower. While after 20 LWN the LOWESS curve for SI24 becomes asymptotic to horizontal, this aspect is not present in case of SI 20 and SI16. The meaning of such shape is that when the site fertility is low, a stable tree with a very high LWN is very rare. For the same reason, the lower height growth rate in lower fertility stands can explain the lower LWN required for stable trees. In addition, this asymptotic trend of the curves is strictly connected to reduced height increment in lower SI classes and has an important biological meaning: the higher the fertility is, the deeper the crown is expected to be. Finally it is also reasonable that shorter trees have a higher stability also with the same HD of taller trees: this is due to the reduced torque for a shorter tree during the application of the same force.

Natural regeneration processes are often desired during the conversion of artificial plantations, such as conifer stands, into more compositionally and structurally complex stands. This aspect currently represents one of the main silvicultural objective in many European countries (Aerts and Honnay 2011, Ciccarese et al. 2012, Sayer et al. 2004, Stanturf et al. 2014). The greater structural and functional stability of trees and forests is here one of the main goals to recover degraded or simplified ecosystems due to direct or indirect human actions (Davis and Slobodkin 2004, Young et al. 2005). Under a practical point of view, this improved model could be proficiently used both in young plantations, when thinning is the main silvicultural treatment, as well as in mature stands, when the aim is to naturally regenerate them or to drive pure stands towards mixed stands (Bravo-Oviedo et al. 2014, Klopcic and Boncina 2012). The selection of stable and mature trees, for example, is fundamental to model the microclimatic variability and gaps dynamics on forests (Muscolo et al. 2014), to obtain the maximum results in terms of wood production, carbon storage and non-wood productions (Cantiani et al. 2015).

In both cases the estimation of stable trees plays a fundamental role. In addition, the LWN indicator can be seen as a double-faced tool: on one side, it is a simple method for selecting trees; on the other side it can be a quick way to control the harvesting activities, i.e. to check if released trees are likely to be stable or not.

The use of the LWN instrument into the Italian context could be easily connected to the Italian regulation. Artificial pinewoods are currently regulated with a very short rotation period (between 40 and 90 years, e.g. Tuscany Regulation) depending on the regional administrations and in relation to the Regional policies (i.e. with the aim to transform them into mixed forests shortly). In this view, the LWN indicator could be successfully used in both cases: selective thinning or selection of candidates trees for natural regeneration. The increased interest in bioenergy from pinewoods (Emer et al. 2011, Lehtimäki and Nurmi 2011) as well as the role of potential substitute for indigenous coniferous species in Central Europe under future climate scenarios (Enescu et al. 2016), currently made the European black pine one of the most interesting species to be used on degraded lands thanks to its ability to tolerate drought stresses especially in case of southern provenances and Mediterranean marginal forest populations (Amodei et al. 2012, Marchi et al. 2015). Additionally, the use of shelterwood system to naturally regenerated this species could become the main field of application of the LWN tool. Similarly the implementation of such modelling approach to other species could increase its potential use. However, a more comprehensive evaluation of trees is always mandatory to take into account further instability factors (e.g. asymmetric crown, deformed stem, etc.).

Conclusions

The general LOWESS model provided by Cantiani & Chiavetta (2015) demonstrated to be a good and stable method also under different soil fertilities. Anyway, in case of lower fertility classes the effect of site index was evident. Specifically site index can contribute to evaluate as stable also trees that with the reference model were classified instable. For this reason, when the Site Index is known and low, the updated thresholds is a more adequate and safer choice when planning and controlling in field activities for thinning or selection of candidate trees.

References

Aerts R., Honnay O. 2011 - *Forest restoration, biodiversity and ecosystem functioning*. BMC Ecology 11(1): 29. doi:10.1186/1472-6785-11-29

Amodei T., Guibal F., Fady B. 2012 - *Relationships between climate and radial growth in black pine (Pinus nigra Arnold ssp. salzmannii (Dunal) Franco) from the south of France)*. Annals of Forest Science 70(1): 41–47. doi:10.1007/s13595-012-0237-9

Anfodillo T., Carrer M., Simini F., Popa I., Banavar J.R., Maritan A. 2013 - *An allometry-based approach for understanding forest structure, predicting tree-size distribution and assessing the degree of disturbance*. Proceedings. Biological sciences / The Royal Society 280(1751): 20122375. doi:10.1098/rspb.2012.2375

Becagli C., Puletti N., Chiavetta U., Cantiani P., Salvati L., Fabbio G. 2013 - *Early impact of alternative thinning approaches on structure diversity and complexity at stand level in two beech forests in Italy*. Annals of Silvicultural Research 37(1): 55–63. doi:10.12899/asr-802

Bernetti G., Cantiani M., Hellrigl B. 1969 - *Ricerche alsometriche e dendrometriche sulle pinete di pino nero e laricio in Toscana*. L'Italia Forestale e Montana 1: 10–40.

Brang P., Spathelf P., Larsen JB., Bauhus J., Boncina A., Chauvin C., Drossler L., Garcia-Guemes C., Heiri C., Kerr G., Lexer MJ., Mason B., Mohren F., Muhlethaler U., Nocentini S., Svoboda M. 2014 - *Suitability of close-to-nature silviculture for adapting temperate European forests to climate change*. Forestry 87(4): 492–503. doi:10.1093/forestry/cpu018

Bravo-Oviedo A., Pretzsch H., Ammer C., Andenmatten E., Barbati A., Barreiro S., Brang P., Bravo F., Coll L., Corona P., Den Ouden J., Ducey M. J., Forrester D. I., Giergiczny M., Jacobsen J. B., Lesinski J., Löf M., Mason B., Matovic B. et al. 2014 - *European mixed forests: Definition and research perspectives*. Forest Systems 23(3): 518–33. doi:10.5424/fs/2014233-06256.

Cantiani P., Chiavetta U. 2015 - *Estimating the mechanical stability of Pinus nigra Arn. using an alternative approach across several plantations in central Italy*. iForest-Biogeosciences and Forestry 8(6): 846-852. doi:10.3832/ifor1300-007.

Cantiani P., De Meo I., Becagli C., Bianchetto E., Cazau C., Mocali S., Salerni E. 2015 - *EffEcts of thinnings on plants and fungi biodiversity in a Pinus nigra plantation: a case study in central Italy*. Forestry ideas 21(2): 149–62.

Cantiani P., Plutino M., Amorini E. 2010 - *Effects of silvicultural treatment on the stability of black pine plantations*. Annals of Silvicultural Research 36: 49-58. doi:10.12899/asr-819

Carmean WH., Hahn JT., Jacobs RD. 1989 - *Site Index Curves for Forest Tree Species in the Eastern United States*. North Central General Technical NC: 128 142.

Chiavetta U., Skudnik M., Becagli C., Bertini G., Ferretti F., Cantiani P., Di Salvatore U., Fabbio G. 2016 - *Diversity of structure through silviculture Diversity of structure through silviculture*. Italian Journal of Agronomy 11(April): 18–22.

Chirici G., Bottai L., Bottalico F., Bronzi A., Chiostri C., Ciancio O., Fioravanti M., Germani M., Giannetti F., Gozzini B., Gravano E., Melara AM., Nocentini S., Travaglini D. 2015 - *Attività di monitoraggio dei danni da vento ai comprensori forestali della Regione Toscana a seguito dell'evento del 5 marzo 2015*. In SiSEF (Ed.), SISEF National Congress - Abstract Book. Florence

Ciccarese L., Mattsson A., Pettenella D. 2012 - *Ecosystem services from forest restoration: Thinking ahead*. New Forests 43(5–6): 543–60. doi:10.1007/s11056-012-9350-8

Coutts M. 1983 - *Root architecture and tree stability*. Plant Soil. 71(1): 171-188. doi:10.1007/BF02182653

Davis MA, Slobodkin LB. 2004 - *The science and values of restoration ecology*. Restoration Ecology 12(1): 1–3. doi:10.1111/j.1061-2971.2004.0351.x

Dupont S., Pivato D., Brunet Y. 2015 - *Wind damage propagation in forests*. Agricultural and Forest Meteorology 214–215: 243–51. doi:10.1016/j.agrformet.2015.07.010

Emer B., Grigolato S., Lubello D., Cavalli R. 2011 - *Comparison of biomass feedstock supply and demand in Northeast Italy*. Biomass and Bioenergy 35(8): 3309–17. doi:10.1016/j.biombioe.2010.09.005

Enescu C., de Rigo D., Caudullo G., Mauri A., Houston Durrant T. 2016 - *Pinus nigra in Europe: distribution, habitat, usage and threats*. In: San-Miguel-Ayanz, J., de Rigo, D., Caudullo, G., Houston Durrant, T., Mauri, A. (Eds.), European Atlas of Forest Tree Species.

Hann DW., Scrivani J. A 1987 - *Dominant-height growth and site-index equations for Douglas-fir and ponderosa pine in southwest Oregon*. Research Bulletin 59.

Isajev V., Fady B., Semerci H., Andonovski V. 2004 - *EUFORGEN Technical Guidelines for genetic conservation and use for European black pine (Pinus nigra)*. International Plant Genetic Resources Institute: 6p.

Kerr G. 1996 - *The effect of heavy or "free growth" thinning on oak (Quercus petraea and Q-robur)*. Forestry 69(4): 303–17. doi:10.1093/forestry/69.4.303

Klopcic M., Boncina A. 2012 - *Recruitment of tree species in mixed selection and irregular shelterwood forest stands*. Annals of Forest Science 69 (8): 915–25. doi:10.1007/s13595-012-0224-1

Latham P., Zuuring H., Coble D. 1998 - *A Method For Quantifying Vertical Forest Structure*. Forest Ecology & Management 104(1–3): 157–70.

Lehtimäki J., Nurmi J. 2011 - *Energy wood harvesting productivity of three harvesting methods in first thinning of scots pine (Pinus sylvestris L.)*. Biomass and Bioenergy 35(8): 3383–88. doi:10.1016/j.biombioe.2010.09.012

Marchi M., Castaldi C., Merlini P., Nocentini S., Ducci F. 2015 - *Stand structure and influence of climate on growth trends of a Marginal forest population of*. Annals of silvicultural research 39(2): 100–10. doi:10.12899/ASR-1066

Marchi M., Nocentini S., Ducci F. 2016 - *Future scenarios and conservation strategies for a rear-edge marginal population of Pinus nigra Arnold in Italian central Apennines*. Forest Systems 25(3): e072. doi:10.5424/ fs/2016253-09476

Means JE., Helm ME. 1985 - *Height Growth and Site Index Curves for Douglas-Fir on Dry Sites in the Willamette National Forest*. Forestry sciences (July): 17.

Means JE., Sabin TE. 1989 - *Height Growth and Site Index Curves for Douglas-Fir in the Siuslaw National Forest, Oregon*. Western Journal of Applied Forestry 4(4): 136–42. http://andrewsforest.oregonstate.edu/pubs/pdf/pub1115.pdf

Muscolo A., Bagnato S., Sidari M., Mercurio R. 2014 - *A review of the roles of forest canopy gaps*. Journal of Forestry Research 25(4):725-736. doi:10.1007/s11676-014-0521-7

Neumann M., Starlinger F. 2001 - *The significance of different indices for stand structure and diversity in forests*. Forest Ecology and Management 145(1–2): 91–106. doi:10.1016/S0378-1127(00)00577-6

Nikolic D., Tucic N. 1983 - *Isoenzime variation within and among populations of european black pine (Pinus nigra Arnold)*. Sylvae Genetica 32(3/4): 80-89.

Nyland RD. 2011 - *Silviculture: concepts and applications*. Plant Biosystems 145(2): 411-418. doi:10.1080/11263504.2011.558705

Pretzsch H., del Río M., Schütze G., Ammer C., Annighöfer P., Avdagic A., Barbeito I., Bielak K., Brazaitis G., Coll L., Drössler L., Fabrika M., Forrester DI., Kurylyak V., Löf M., Lombardi F., Matović B., Mohren F., Motta R. 2016 - *Mixing of Scots pine (Pinus sylvestris L.) and European beech (Fagus sylvatica L.) enhances structural heterogeneity, and the effect increases with water availability*. Forest Ecology and Management 373: 149–66. doi:10.1016/j.foreco.2016.04.043

Sayer J., Chokkalingam U., Poulsen J. 2004 - *The restoration of forest biodiversity and ecological values*. Forest Ecology and Management doi:10.1016/j.foreco.2004.06.008

Shaw JD. 2006 - *Reineke's Stand Density Index: Where Are We and Where Do We Go From Here?* Society of American Foresters 2005 National Convention. http://www.fs.fed.us/rm/pubs_other/rmrs_2006_shaw_j006.pdf

Spearman C. 1987 - *The proof and measurement of association between two things. By C. Spearman, 1904*. The American journal of psychology 100(3–4): 441–71. doi:10.1037/h0065390

Stanturf JA., Palik B. J., Williams MI., Dumroese RK., Madsen P. 2014 - *Forest Restoration Paradigms*. Journal of Sustainable Forestry 33(1): 161–94. doi:10.1080/10549811.2014.884004

Tomaiuolo M. 2010 - *Analisi del fenomeno dell' autodiradamento in piantagioni di pino laricio (Pinus laricio Poiret) della Calabria*. 36: 5–14.

Valbuena P., Peso C., Bravo F. 2008 - *Stand Density Management Diagrams for two Mediterranean pine species in Eastern Spain*. Investigación Agraria: Sistemas y Recursos Forestales 17(2): 97–104. doi:10.5424/srf/2008172-01026

Vidakovic M. 1974 - *Genetics of European Black Pine (Pinus nigra Arn.)*. Academia Sci. et Art. Slavorum Meridionalum. Zagreb. Annales Forestales, Anali za Šumarstvo 6(3): 57–86.

Wang Y., Titus SJ., LeMay VM. 1998 - *Relationships between tree slenderness coefficients and tree or stand characteristics for major species in boreal mixedwood forests*. Canadian Journal of Forest Research 28(1995): 1171–83. doi:10.1139/x98-092

Yen TM. 2015 - *Relationships of Chamaecyparis formosensis crown shape and parameters with thinning intensity and age*. Annals of Forest Research 58(2): 323–32. doi:10.15287/afr.2015.408

Young TP., Petersen DA., Clary JJ. 2005 - *The ecology of restoration: Historical links, emerging issues and unexplored realms*. Ecology Letters 8(6): 662–673. doi:10.1111/j.1461-0248.2005.00764.x

11

Effects of potting mixture on the growth of seedlings of *Entandrophragma cylindricum* (SPRAGUE) and *Tamarindus indica* L

Enoma Aigbedion[1], William W. Modugu[1], Friday Nwabueze Ogana[2*]

Abstract - The study describes the effects of biochar and NPK fertilizer on *Entandrophragma cylindricum* and *Tamarindus indica* seedlings in the nursery of the Department of Forestry and Wildlife, University of Benin, with a view to evaluating the seedling growth response to the quantity of potting mixture. The potting mixtures used were soil only which is the control, mixture of soil and biochar, mixture of soil, biochar and fertilizer, mixture of soil and fertilizer. The seeds of *Entandrophragma cylindricum* and *Tamarindus indica* were sourced for and planted. Transplanting was carried out for germinated seeds until the required weeks number of seedlings needed for the experiment was obtained. Measurements were taken at one week interval for ten weeks plant height, stem height, collar diameter and leaf number. The experiment was a completely randomized design. Data were analyzed using ANOVA. Result indicates that there was no significant difference in the growth variables except for stem height of *Entandrophragma cylindricum* at 5% level. Control i.e. forest top soil had the highest height of stem of 5.90 cm and 8.49 cm for *Entandrophragma cylindricum* and T*amarindus indica*, respectively. Marginal differences in the growth variables were observed. Other silvicultural practices that would promote the growth of the species should be considered.

Keywords - Biochar; seedlings, fertilizer; *Entadrophragma cylindricum*; *Tamarindus indica*

Introduction

Most tropical trees have natural durability and fruit edibility; they are preferred for use as materials for housing, construction, building erection, food and medicine. The majority of these timber and non timber trees possess extractives that are of biological and economic values. The peculiar nature of tropical climate and soils predisposes the soils to heavy leaching due to their acidic nature (Isikhuemen 1995). The need to buffer and augment the soil nutrients supply by artificial means becomes important. Aluko (1982) reported that in developed countries, forest fertilization is an accepted tool for increasing wood production.

The use of biochar as soil amendment is modeled on the carbon rich anthropogenic soils known as "Indian black earth" (Glaser 2007). Recently, efforts to duplicate the "Indian black earth" techniques using biochars produced from modern pyrolysis prove that charcoal additions can have an ameliorating effect on excessive weathered, infertile tropical soils by increasing cation exchange capacity (CEC) and plant nutrient supply, reducing soil acidity and aluminum toxicity, and improving fertilizer efficiency due to reduced nutrient leaching (Glaser et al. 2007; Lehmann et al 2003). Studies have reported that plant growth responses are largest when charcoal and fertilizers are combined (Lehmann et al. 2003; Steiner et al 2007; Chan et al. 2008; Mete et al. 2015).

Today, the need to carry out research on the protection, conservation as well as biology of indigenous tree species is increasing in Nigeria because of their importance (Isikhuemen 1995). *Entandrophragma cylindricum* and *Tamarindus indica* are two important species in Nigeria. *Entandrophragma cylindricum* and *Tamarindus indica* belong to the family of Meliaceae and Caesalpinioideae, respectively. *Entandrophragma cylindricum*, also called Sapelli, is a long-living timber tree species of high commercial value found from Sierra Leone East to the Democratic Republic of Congo and to Uganda, in tropical evergreen and semi deciduous forests (Lehmann et al 2003). *Tamarindus indica*, commonly known as Tamarind tree is one of the most important multipurpose tropical fruit tree species in the Indian subcontinent. It is a large evergreen tree up to 30 m tall, bole usually 1-2 m, up to 2 m diameter, crown is dense, widely spreading, rounded; bark rough, fissured, grayish- brown (Santosh et al. 2011). *Tamarindus indica* has a hard and heavy wood for general carpentry, boat build-

[1] University of Benin, Department of Forestry and Wildlife, Benin City, Edo State (Nigeria)
[2] University of Ibadan, Department of Forest Reaources Management Ibadan, Oyo State (Nigeria)
*ogana_fry@yahoo.com

ing, firewood, charcoal, the tree is used for shade, amenity, bee forage and windbreak.

To effectively harness the potentials of these two species, emphasis must be made for improving their demand as plantation trees, and also their early growth, and speed up their full development rate. In this regard, the use of biochar application and fertilization into forest soil management practice becomes essential. Miller (1981) reported that during the early stages of growth, trees are very well dependent on soil nutrient supply. At this point, the forest manager, by judicious application of biochar and or fertilizer, can influence the subsequent development of stand, perhaps saving many years in rotation length.

Materials and methods

Experimental site

The study was carried out in the nursery of the Department of Forestry and Wildlife, Faculty of Agriculture, University of Benin, Benin City, Nigeria. The University is located in the southern part of Edo State and lies on Latitude 5°34'N and Longitude 6°24'E with an elevation of 134 m above sea level. The mean annual temperature is about 27°C and receives an annual rainfall of 1500mm to 2000mm. The mean relative humidity ranges from about 75% at 12:00 a.m. to over 95% at 6:00 a.m. The topography of the study area is gentle slope (University of Benin Master Plan, 1993)

Seed procurement and sowing

150 Seeds of *Entandrophragma cylindricum* were purchased from Okomu National Park, Udo, Edo State, on 6th February, 2016; and kept in a cool dry place. While 495 seeds of *Tamarindus indica* were purchased from Forestry Research Institute, Otagba, Benin City, Edo State, on 15th March, 2016. The seeds were soaked overnight. This was done for faster and easier germination of the seeds. The seeds of *E. cylindricum* and *T. indica* were sown in germination bed and germination trays, respectively. The germination tray was filled with top soil to three-quarter its size to prevent water run-off during watering. Watering was carried out regularly every other day.

Soil collection and preparation

Top soil was collected from the forest floor in the nursery and was properly mixed (See Fig. 1). The polythene pots were filled with top soil and each given a uniform weight of 6 kg for the pots without biochar. The pots containing biochar were filled in a ratio four to half, that is, four head pans of top soil to half head pan of biochar for every eighteen poly pots (See Fig. 2). The treatments were:

1. Top soil only i.e. control (T1)
2. Soil + Biochar (T2)
3. Soil + Biochar + Fertilizer (T3)
4. Soil + Fertilizer (T4)

Figure 1 - Top soil from the nursery.

Figure 2 - Biochar before pounding.

Seedling transplant

112 seedlings of *Entandrophragma cylindricum* were transplanted to the prepared poly pots, one seedling per poly pot. 80 seedlings of *Tamarindus indica*, were transplanted to 6 kg poly pots. Fertilizer was applied six days after the transplant of all the species in linear form and 3 cm deep for T3 and T4 i.e. treatments involving fertilizer as aforementioned. The level of fertilizer used was 400kg/ha. That is, 1.2 g of NPK per 6 kg of poly pot.

Experimental design

The experiment was laid in a Completely Randomized Design (CRD) with seven replications for both species independently. Data collection started ten days after transplanting with the following parameters measured at one week interval per plant, per treatment, for ten weeks. The parameters were height, leaf number, collar diameter. Data were analyzed using one-way analysis of variance (ANOVA); least significant difference (LSD) was used to separate means that were significant at 5%.

Results

Characterisation of the Soil Mixture

We carried out laboratory analysis on the soil samples prior to the experiment so as to ascertain the nutrient status of the samples. The preceding biochar, top soil and soil mixtures analysis presented in Table 1 show their nutrient status and pH. The nutrient status of top soil alone appears to have the lowest Nitrogen (0.82g/kg), while the mixture of soil+ biochar+ fertilizer has the highest amount of Nitrogen (3.80g/kg), Phosphorus (22.06mg/kg) and Potassium (5.35cmol/kg). The pH level for biochar is 7.8, fairly neutral and that of the mixture of soil+ fertilizer is 4.8, fairly acidic.

Table 1 - Physio-chemical properties of the treatments.

Parameters	Biochar	Soil	Soil+ biochar	Soil+ biochar+ fertilizer	Soil+ fertilizer
pH	7.8	5.8	6.8	6.9	4.8
Org. C. (g/kg)	31.92	17.2	33.73	35.44	25.40
Total N (g/kg)	1.60	0.82	1.69	3.80	3.27
P. (mg/kg)	10.22	6.54	12.36	22.06	20.40
K (cmol/kg)	3.65	0.35	3.87	5.35	5.11
Ca (cmol/kg)	3.00	0.67	2.15	2.25	0.52
Mg (cmol/kg)	1.21	0.28	0.63	0.92	0.21
Na (cmol/kg)	0.64	0.21	0.33	0.36	0.18
H (cmol/kg)	0.20	0.40	0.22	0.21	1.6
Al^3 (cmol/kg)	0.04	0.04	0.05	0.05	0.56
Sand (g/kg)	0	886.4	868	866	864
Silt (g/kg)	0	56.2	100	93	86
Clay (g/kg)	0	57.4	32	41	50

Growth parameters

Plant height

The results of the effects of the four treatments i.e. Soil + Fertilizer, Soil + Biochar + Fertilizer, Soil + Biochar, and Control (forest top soil) on plant heights are presented in Table 2. We defined plant height as the vertical distance from seedling collar to the tip of the apical bud. The mixture of soil and fertilizer had the highest mean plant height growth of 14.975 cm for *Entandrophragma cylindricum*. Control (forest top soil) had 14.767 cm. While the mixture of soil and biochar had the lowest mean total height growth of 14.096 cm for same species. However, there is no statistical difference among the treatments at 5% level with respect to plant height growth for *Entandrophragma cylindricum*. In the case of *Tamarindus indica*, the mixture of Soil + Biochar + Fertilizer had the highest mean height growth of 15.00 cm. This was followed by forest top soil (control) with 14.75 cm. The mixture of Soil + Fertilizer had the lowest value of 13.944 cm. No significant difference was observed among the treatment at 5% level when the data were analyzed regardless of marginal increase.

Table 2 - Separation of means for plant height.

Species	Treatments	Mean
E. cylindricum	Soil + Fertilizer	14.975[a]
	Control	14.767[a]
	Soil + Biochar + Fertilizer	14.504[a]
	Soil + Biochar	14.096[a]
T. indica	Soil + Biochar + Fertilizer	15.000[a]
	Control	14.750[a]
	Soil + Biochar	14.397[a]
	Soil + Fertilizer	13.944[a]

Mean with the same superscript are not significantly different at 5%

Stem height

The result for stem height is presented in Table 3. Stem height is defined as the vertical distance between the oldest/lowest leaf of the crown and the collar of the seedling. The result showed that *Entandrophragma cylindricum* raised on forest top soil (control) only had the highest average stem height of 5.902 cm. This was followed by mixture of forest soil + fertilizer average stem height of 5.496 cm. The mixture of Soil + Biochar and the mixture of Soil + Biochar + Fertilizer had the lowest average stem height of 3.992 cm and 3.946 cm, respectively for *Entandrophragma*. The analysis of variance revealed a high level of significance at 5% ($P<0.05$). Control i.e. forest top soil was significantly different from other treatment. While the mixture of Soil + Biochar and the mixture of Soil + Biochar + Fertilizer were not statistically different. The result for *Tamarindus indica* showed no significant difference at 5% ($P>0.05$) among the treatment for stem height. However, there were numerical variations; with Control having the heighest average stem height of 8.488 cm. The mixture of Soil + Biochar + Fertilizer had the second height average stem height of 8.375 cm. The mixture of Soil + Fertilizer had the lowest mean stem height (7.309 cm).

Table 3 - Separation of means for stem height.

Species	Treatments	Mean
E. cylindricum	Control	5.902[a]
	Soil + Fertilizer	5.496[b]
	Soil + Biochar	3.992[c]
	Soil + Biochar + Fertilizer	3.964[c]
T. indica	Control	8.488[a]
	Soil + Biochar + Fertilizer	8.375[a]
	Soil + Biochar	7.841[a]
	Soil + Fertilizer	7.309[a]

Mean with the same superscript are not significantly different at 5%

Collar diameter

The result of the effect of the treatment on seedling collar diameter is presented in Table 4. From the Table it can be observed that the four treatments i.e. Control, Soil + Fertilizer, Soil + Biochar and Soil + Biochar + Fertilizer did not significantly affect the collar diameter growth for both *Entandrophragma cylindricum* and *Tamarindus indica*. However, there were numerical variations among the treatments. The mixture of Soil + Fertilizer had the largest collar diameter of 3.177 cm for *Entandrophragma cylindricum*, while the mixture of Soil + Biochar had the largest mean collar diameter of 2.249 cm for *Tamarindus indica*.

Table 4 - Separation of means for collar diameter.

Species	Treatments	Mean
E. cylindricum	Soil + Fertilizer	3.177[a]
	Soil + Biochar	3.064[a]
	Soil + Biochar + Fertilizer	3.052[a]
	Control	3.029[a]
T. indica	Soil + Biochar	2.249[a]
	Soil + Biochar + Fertilizer	2.247[a]
	Soil + Fertilizer	2.238[a]
	Control	2.194[a]

Mean with the same superscript are not significantly different at 5%

Mean leaf number

Similarly, the leaf number of *Entandrophragma cylindricum* and *Tamarindus indica* were counted and analyzed (see Table 5). No significant difference was observed at 5% ($P>0.05$) among the four treatment for the two species. Little or no variation could be seen from the Table. However, the mixture of Soil + Biochar had the highest average leaf number of 6.271 and 13.84 for *Entandrophragma cylindricum* and *Tamarindus indica*, respectively.

Table 5 - Separation of means for leaf number.

Species	Treatments	Mean
E. cylindricum	Soil + Biochar	6.271[a]
	Soil + Biochar + Fertilizer	6.164[a]
	Soil + Fertilizer	5.971[a]
	Control	5.707[a]
T. indica	Soil + Biochar	13.840[a]
	Soil + Fertilizer	13.314[a]
	Soil + Biochar + Fertilizer	13.107[a]
	Control	12.764[a]

Mean with the same superscript are not significantly different at 5%

Discussion

This study has demonstrated the potential of biochar to promote growth in *Entandrophragma cylindricum* and *Tamarindus indica*. The major morphological criteria used to describe seedling quality are shoot height, collar diameter, leaf number, etc. These are the bases for grading seedlings in the nursery. The result of this study reveals that the total height, collar diameter and leaf number were not statistically significant but the stem height was significant for only *Entandrophragma* seedlings. The stem height of *Entandrophragma* was statistically different within the treatments, but *Tamarindus* appeared marginally better than *Entandrophragma* seedlings. The stem height appears to do better in control and soil + fertilizer for *Entandrophragma* and control and soil + biochar + fertilizer for *Tamarindus*. The stem height of *Tamarindus* was consistently better than *Entandrophragma* seedlings all through the experiment for all the treatments.

This study is not in line with Mete et al. (2015) who studied the synergistic effects of biochar and NPK on soybean yield in an alkaline soil. The authors reported a significant increase in soybean yield when biochar and NPK was combined compared to control. The low yield in the control was attributed to a high soil pH of 8.8. The pH of control in the current study was 5.8 which may have favoured the growth of the seedlings. Similar observation was reported by Liang et al. (2014), Oram et al. (2014), Smider and Singh (2014) and Tammeorg et al. (2014).

The total height of *Tamarindus* was marginally better than that of the *Entandrophragma* seedlings throughout the experiment. Though there was no statistical significant difference among the treatments, but it appears that *Tamarindus* seedlings did better in all the treatments than *Entandrophragma* seedlings. Although all the means of collar diameters were not statistically significant, the collar diameter of *Entandrophragma* appears to be better than that of *Tamarindus* marginally. *Tamarindus* seedlings appear marginally better than *Entandrophragma* seedlings among the treatments for leaf number. Also, we observed pest attack on the leaf of *Entandrophragma cylindricum* which could have affected the growth of this plant. As the pest was sighted feeding on the leaf of the plant; thereby reducing the leaf surface area available for photosynthesis.

The results obtained from pot trials cannot be estimated for the field due to differences in soil volume, presence of impeding horizons or moisture stress Onuwaje and Uzu (1982). The pot trials should rather serve as a guide for field application or further field studies into the effect of biochar and NPK fertilizer on seedlings growth.

Conclusions

The result of the study showed that application of biochar and NPK fertilizer at four ratio half head pans and 400kg/ha respectively did not significantly affect most of the growth parameters measured; although there was improvement and marginal increase. Therefore, the use of biochar and NPK fertilizer in raising *Entandrophragma* and *Tamarindus* seedlings may not be necessary. We recommend that other silvicultural practices that would promote the growth of *Entandrophragma* and *Tamarindus* should be considered. And whenever an experiment like this is being carried out in the nursery, arrangements should be made to eradicate pest and insect attack.

References

Aluko AP. 1982 - *Mineral Nutrition,* Terminalia ivorensis *and* Terminalia superb *seedlings grown in a typical rain forest soil of South West Nigeria,* M. Phill thesis University of Ife, Nigeria: 156-159.

Chan KY., Van Zwieten L., Meszaros I., Downie A., Joseph S. 2008 - *Using poultry litter Biochars as soil amendments.* Australian J. Soil Res. 46: 437-444.

Glaser B. 2007 - *Prehistorically modified soils of central Amazonia: a model for sustainable Agriculture in the twenty-first century.* Philosophical transactions of the Royal Society. 362: 187-196.

Isikhuemen EM. 1995 - *The Effect of NPKMg Fertilizer on Growth and Dry Matter Accumulation of* Terminalia ivorensis *(A.Chev.) and* Khaya Grandifoliola *(C.D.C.) Seedlings,* A project report.

Lehmann J., Gaunt J., Rondon M. 2003 - *Biochar sequestration in terrestrial ecosystems-a review. Mitigation Adaptation Strategies.* Global Change, 11: 395-419.

Liang F., Li GT., Lin QM., Zhao, XR. 2014 - *Crop yield and soil properties in the first 3 years after biochar application to a calcareous soil.* J. Integr Biol, 14: 975-979.

Mete FZ., Mia S., Dijkstra FA., Abuyusuf M., Hossain ASMI. 2015 - *synergistic effects of biochar and NPK on soybean yield in an alkaline soil.* Pedosphere, 25(5): 713-719.

Miller HG. 1981 - *Forest fertilization: Guiding concepts.* Journal of the Institute of Foresters 54(2): 157-167.

Onuwaje O., Uzu F. 1982 - *Growth response of rubber seedlings to N, P and K fertilizer in Nigeria.* Fert. Res.: 169-175.

Oram NJ., van de Voorde TFJ., Ouwehand GJ., Martijn-Bazemer TM., Mommer L., Jeffery S., van Groenigen JW. 2014 - *Soil amendment with biochar increases the competitive ability of legumes via increased potassium availability.* Agr Ecosyst Environ. 191: 92-98.

Santosh SB., Aditya G., Jitendra N., Gopal R., Alok PJ. 2011 - *Extent of explored potential.* Pharmacognosy Review, 5(9): 73-81 DOI: 10.4103/0973-7847.79102.

Smith JHG., Kozak A., Sziklai O., Walters J. 1996 - *Relative importance of seedbed fertilization. Morphological grade, site, provenance and parentage to juvenile growth and survival of Douglas-fir.*: 83-86.

Steiner C., Teixeira WG., de Macêdo JLV., Lehmann J., Blum WEH. 2007 - *Long term effects of manure, charcoal and mineral fertilization on crop production and fertility on a highly weathered Central Amazonian upland soil.* Plant Soil: DOI 10.1007/s11104-007-9193-9.

Tammeorg P., Simojoki A., Mäkelä P., Stoddard FL., Alakukku L., Helenius J. 2014 - *Biochar application to a fertile sandy clay loam in boreal conditions: effects on soil properties and yield formation of wheat, turnip rape and faba bean.* Plant Soil, 374: 89-107.

University of Benin. 1993 - *Master Plan.* University of Benin Printing Press, 360p.

Wilde SA., Wittenkamp R., Stone EL., Galloway H. 1940 - *Effect of high fertilizer treatments of nursery stock upon its survival and growth in the field.*: 806-809.

12

Tree root system imaging using Ground Penetrating Radar

Carlotta Ferrara[1*], Luca Salvati[1], Pier Matteo Barone[2]

Abstract - Evaluating tree roots systems without compromising their environment with destructive and laborious methods is of crucial importance for preserving plant resources. Ground Penetrating Radar (GPR) technology applied to root and forest ecology studies is a key active remote sensing technology, based on the use of electromagnetic waves, providing unique, non-invasive resources to sample root biomass and spatial distribution. This paper reviews the use of GPR techniques, as near surface sensor radar scanning technology using a ground-coupled single-offset antenna configuration, to yield accurate tree roots system information. Based on the analysis of both geometric and general characteristics of underground soil layers, the GPR can be used as a rapid and high-spatial resolution tool for the analysis of roots distribution, morphology, orientation, and the occupied soil volume. Finally, this paper will give a brief description of a remote technique supporting silvicultural research in terms of application of radar technology to tree root systems.

Keywords - Ground Penetrating Radar; tree root systems; forest ecology

Introduction

Thanks to the latest advancements in geophysical research applied to forestry and agricultural issues, the accurate characterization of subsurface can be obtained using new spatial and temporal monitoring techniques. In particular, the application of Ground Penetrating Radar (GPR) has gradually increased in the last few years, mainly due to high-frequency antenna development (Allred et al. 2008).

In general, GPR can be used to determine physical parameters of both large-scale (km) and small-scale areas (cm), in rapid and non-destructive ways. The use of GPR to study plants, specifically tree root distribution, size, morphology and their biomass, has progressively received more attention for their valuable capability of providing *in situ* non-destructive high-precision measurements (Wu et al. 2014; Ferrara et al. 2014a; Butnor et al. 2008). Moreover, the usefulness of high-frequency GPR for monitoring spatial-temporal variation of tree root systems has become very valuable, given the importance of understanding carbon cycle balance between sources and sinks due to the increasing atmospheric carbon dioxide (CO2) levels and climate global changes (Danyagri and Dang 2013; Wielopolski et al. 2002; Stover et al. 2007).

Assessing tree health plays a key role in urban and peri-urban environments (Salvati 2014), tree roots are vital for plant healthiness and longevity as they play a fundamental role in water and nutrient intake and plant anchorage and their decline anticipates that of the whole plant (Barone et al. 2016). Tree roots are concentrated close to the trunk base and they are generally included into the canopy projection on the ground, although in some situations they reach a greater extension. In fact, their growth is not only based on the species characteristics, but also on the site-specific pedo-climatic conditions. The root system size can vary to ensure an adequate and constant energetic supply even in non-optimal soil conditions (Merlini and Corona 2007; Zucconi 2003).

GPR has a key role in this detecting, in real time, both vertical and horizontal dielectric anomalies associated with natural or manmade subsurface variability (Barone 2016; Perez-Gracia et al. 2010). Even if, in some cases, the successful application of GPR in root detection is site-specific and some factors can interfere with the root resolution (Hirano et al. 2009), the success of the recent GPR investigations of the tree root architecture is relevant and uncontested (Guo et al. 2013). The goals of this study were to test the feasibility of using GPR to describe root architecture, root orientation, and soil volume utilization in large urban parks and peri-urban/wildland fringe land, by means of a brief technical note on a procedure supporting silvicultural research in

[1] CREA Research Centre of Forestry and Wood, Arezzo (Italy)
[2] The American University of Rome, Roma (Italy)
[*]carlotta.ferrara@crea.gov.it

the northern Mediterranean region (Barbati et al. 2013; Ferrara et al. 2014b; Salvati 2015). In general, tree roots are difficult to assess and delineate without laborious and destructive excavations that are impossible in these places, for this reason, processing speed, repeatability and non-invasiveness of GPR measurements, main characteristics of these geophysical surveys, are valuable for urban forestry applications.

Materials and methods

The GPR technique is based on analysing the Two-Way Traveltime (TWT) of an electromag-netic pulse of a short duration (1-10 ns) and high frequency (MHz-GHz) sent into the ground, and its subsequent reflection. Injecting an electromagnetic (em) field in a non-homogeneous field, there will occur reflection and refraction phenomena, well-known in optics. Due to a contrast of dielectric properties between different materials, by exploiting electromagnetic (em) propagation principles, for each variation of the refractive index, associated with the wavelength of the radiation (1):

$$n(\lambda) = \frac{c}{v(\lambda)} \qquad (1)$$

there is a partial reflection of the em waves, which can be intercepted and analyzed, identifying the position on the time scale (x-axis) and intensity (y-axis) of the reflected or refracted pulses (Annan 2004). The position on the time scale of a peak determines the distance between GPR antennas and the object that produces the spread. Assuming that the investigated medium is homogeneous, the propagation velocity v can be considered constant, so that the depth D of the target is calculated as (2):

$$D = \frac{t \cdot v}{2} \qquad (2)$$

where the em waves velocity is (3):

$$v = \frac{1}{\sqrt{\varepsilon\mu}} \approx \frac{c}{\sqrt{\varepsilon_r}} \qquad (3)$$

which varies depending on soil physical properties. The right expression, in which there are c, the speed of light, and ε_r, the dielectric constant of the material, is valid in the case of materials which are good dielectric, the conductivity, σ, is negligible (Jol 2009) and the magnetic properties are close to those of vacuum ($\mu = \mu_0$), or propagation occurs essentially without dispersion (Annan 2004). The reflection intensity depends both on the wave attenuation and the reflection coefficient, where the latter mainly depends on the contrast between the different dielectric properties of the medium in which the waves propagate and the investigated object (Barone 2016). Because of this em properties contrast, the GPR technique can determine the physical characteristics, and in particular the water content since the relative dielectric permittivity of water is approximately 80, an order of magnitude greater than that characterizing the soils (Ferrara et al. 2013; Ferrara et al. 2014a).

To measure the root morphology and distribution and to assess tree root concentrations GPR measurements have were carried out at in an urban park of Rome (Italy) dominated by sparse old *Pinus pinea* plants. The root systems of two pine trees were selected for GPR survey, where the GPR antennas are moved along the surface of the medium to be investigated, over straight lines, called profiles. The average dimensions characteristics of the considered pines are the diameter at breast height (DBH) that is 72.5 cm, the average tree height (H) that approaches 21.5 m and the average canopy radius that is about 8.5 m.

Finally, the data were collected using a bistatic 500 MHz antennas (FINDAR - Sensors & Software, Inc.) and an odometer. The basic Dewow time filter and Automatic Gain Control were applied to all radargrams. Then, to convert the TWT time into depth, a velocity analysis using the hyperbola calibration technique (Annan, 2004) was performed on all radar sections in which the hyperbolic events were well detectable, providing an average velocity of 0.10 m/ns.

Results and discussion

Through the acquired measures it is possible to built the so-called radargram, i.e. a two-dimensional reconstruction of the medium as if it were seen at different radiofrequencies, resulting from the envelope of the individual tracks acquired along the current profile. On each track the reflections that undergo the signal will be detectable, which may be due to both the different stratification in the soil or to well-defined objects.

During the GPR measurements, it is possible to notice a variation of em parameters in both trees, but the reflections produced are different. In the first case, a sub-horizontal reflector appears in the radargram, while the second displays a reflection hyperbola. This is due to the fact that as the radar is moved horizontally, the signal two-way travel time diffused by the same reflector has initially a descending trend, until the top, when the antennas are above

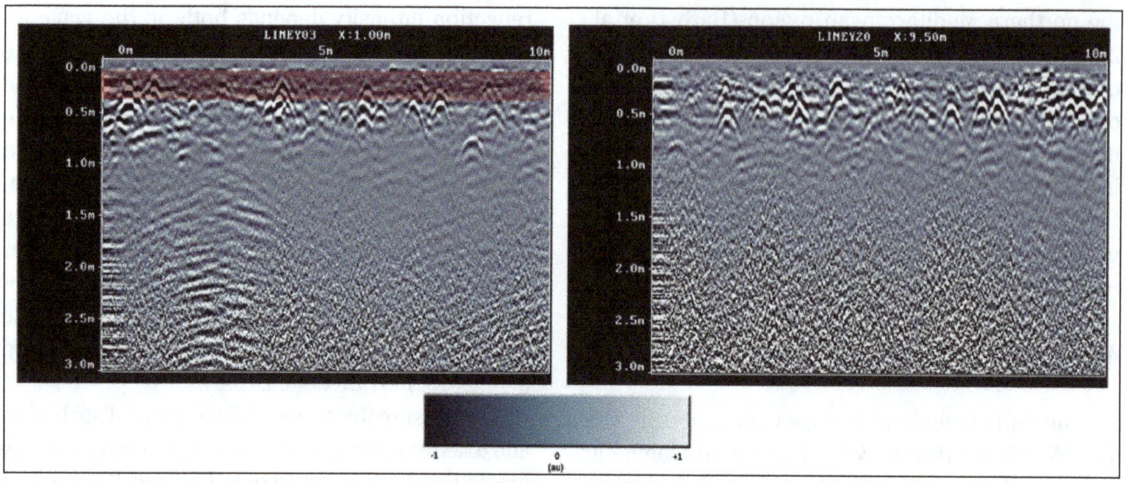

Figure 1 - GPR sections, where the vertical scale is depth (m) and the horizontal scale is the profile length (m). The tree roots of the two different pines are located in correspondence of the hyperbolas events between 0-0.45 m depth (highlighted in red). Note that the scale of values is related to the average envelope amplitude (arbitrary unit).

the target, and then increasing (Annan 2004; Jol 2009; Barone 2016). The radargrams reported in Fig. 1 show the GPR ability of detecting and precisely locating the tree roots, from the above-mentioned hyperbolas.

Furthermore, if the acquisitions have provided parallel profiles within a x-y grid, maps at different depths (depth-slices) can be obtained, displaying not only geometries of buried objects but also their size. Fig. 2 shows two different maps obtained using an envelope algorithm, known as average envelope amplitude (Annan 2004; Ferrara et al. 2013; Barone 2016), displaying the root system distribution.

In particular, the pines roots extend out from the trunk occupying to the fullest extent an area of approximately 100 m^2, where the roots orientation reproduces a peculiar star architecture. From radargram and depth slice analysis the maximum depth of the pine root system can be defined, which is 0.45 m, giving an occupied soil volume of 45 m^3.

The ratio between the canopy projection on the ground and the radial root area is an important indicator of soil fertility (Zucconi 2003), as tree roots tend to extend to compensate low resources availability.

The aim of the data collection was to obtain rapidly and in real time both a qualitative and brief quantitative on-site evaluation of the root architecture. Using different GPR systems, with different frequencies and different geophysical methods, it is possible to obtain more detailed information in a larger time-lapse (Rodríguez-Robles et al. 2017, and literature therein).

Conclusion

The spatial distribution of tree root architecture is of great importance for understanding its role in resource acquisition, storage and structural support. Unfortunately, tree roots are not easy to sample, and typical root detection methods, such as excavations and uprooting, require a large amount of labour and a destructive interference. This creates a limit to quantitative and repeated monitoring in long-term research.

Overall, the application of ground radar technol-

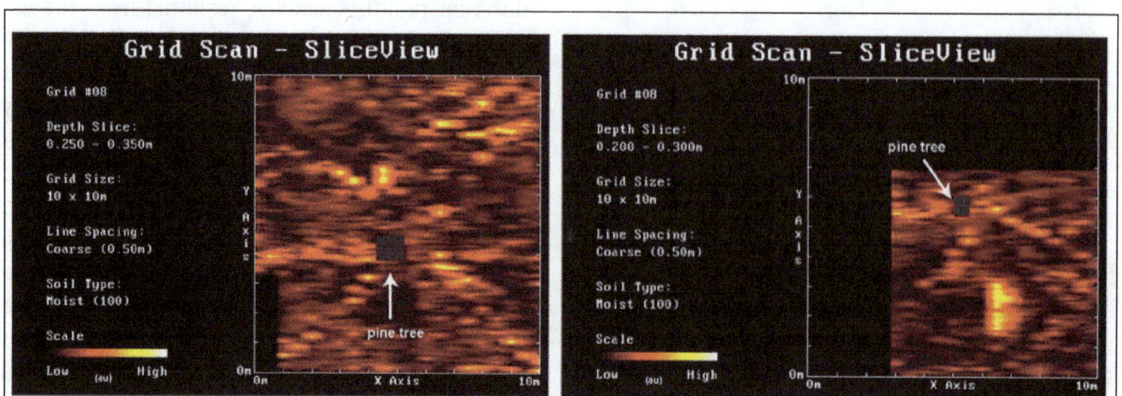

Figure 2 - GPR depth-slices. The tree roots star architectures of the two different pines are precisely mapped at a depth of about 0.30 m. Note that the scale of values is related to the average envelope amplitude (arbitrary unit).

ogy to root and urban forest ecology studies provides unique, non-destructive means to sample root spatial distribution. In this study, the GPR technique can be used as a non-invasive tool to precisely and rapidly quantify root architecture, root orientation and soil volume utilization, of pine trees. Compensatory tree root growth depending on soil mineral resources availability and the canopy/radical ratio is directly proportional to the soil fertility. Based on these information, it is possible to provide valuable support in the urban-forestry sector and in urban-parks management policies to preserve stability and healthiness of urban/forests environments.

References

Allred BJ., Daniels JJ., Ehsani MR. 2008 - *Handbook of Agricultural Geophysics.* CRC Press, Taylor & Francis Group, Boca Raton.

Annan AP. 2004 - *Ground Penetrating Radar: principles, procedures and applications.* Sensors and Software, Elsevier Science, Amsterdam.

Barbati A., Corona P., Salvati L., Gasparella L. 2013 - *Natural forest expansion into suburban countryside: a contrasting perspective about changing forest landscape pattern.* Urban Forestry and Urban Greening 12: 36-43.

Barone PM. 2016 - *Understanding Buried anomalies: A Practical Guide to GPR.* Lambert Academic Publishing, Saarbrucken, Germany.

Barone PM., Ferrara C., Di Maggio R., Salvati L. 2016 - *When the Crime Scene Is the Road: Forensic Geoscience Indicators Applied to Road Infrastructure and Urban Greening.* Geosciences 6, 50. DOI:10.3390/geosciences6040050

Butnor JR., Stover DB., Roth BE., Johnsen KH., Day FP., McInnis, D. 2008 - *Using Ground-Penetrating Radar to Estimate Tree Root Mass: Comparing Results from Two Florida Surveys.* In Allred BJ., Daniels JJ., Ehsani MR. (eds), *Handbook of Agricultural Geophysics.* CRC Press, Taylor & Francis Group, Boca Raton.

Danyagri G., Dang QL. - 2013 - *Effects of Elevated [CO2] and Low Soil Moisture on the Physiological Responses of Mountain Maple (*Acer spicatum *L.) Seedlings to Light,* PLoS ONE 8(10): e76586. DOI:10.1371/journal.pone.0076586

Ferrara C., Barone PM., Steelman C., Pettinelli E.,Endres AL. 2013 - *Monitoring shallow soil water content under natural field conditions using the early-time GPR signal technique.* Vadose Zone Journal 12 (4). DOI:10.2136/vzj2012.0202.

Ferrara C., Barone PM., Pettinelli E., Salvati L. 2014a - *Ground Penetrating Radar as a remote-sensing approach to investigate the root system architecture beneath the soil.* Applied Ecology and Environmental Research 12 (3): 695-702. DOI: 10.15666/aeer/1203_695702

Ferrara A., Salvati L., Sabbi A., Colantoni A. 2014b - *Urbanization, Soil Quality and Rural Areas: Towards a Spatial Mismatch?* Science of the Total Environment 478: 116–122.

Guo L., Chen J., Cui X., Fan B., Lin H. 2013 - *Application of ground penetrating radar for coarse root detection and quantification: a review.* Plant Soil 362: 1–23. DOI 10.1007/s11104-012-1455-5

Hirano Y., Dannoura M., Aono K., Igarashi T., Ishii M., Yamase K., Makita N., Kanazawa Y. 2009 - *Limiting factors in the detection of tree roots using ground-penetrating radar.* Plant Soil 319: 15–24. DOI 10.1007/s11104-008-9845-4.

Jol HM. 2009 - *Ground Penetrating Radar: Theory and Application.* Elsevier Science, Oxford, UK.

Perez-Gracia V., Lorenzo H., Diaz-PeDiaz PN. 2010 - *Forestry applications of ground penetrating radar.* Forest System 19: 5-17, ISSN: 2171-5068.

Rodríguez-Robles U., Arredondo T., Huber-Sannwald E., Ramos-Leal JA., Yépez EA., *Application of geophysical tools for tree root studies in forest ecosystems in complex soils* - Biogeosciences Discussion 91: 1-22. DOI:10.5194/bg-2017-91, 2017

Salvati L. 2014 - *Profiling Forest Fires along the Urban Gradient: a Mediterranean Case Study.* Urban Ecosystems 17 (4): 1175-1189.

Salvati L. 2015 - *Agro-forest Landscape and the 'Fringe' City: a Multivariate Assessment of Land-use Changes in a Sprawling Region and Implications for Planning.* Science of the Total Environment 490: 715-723.

Stover D., Day FP., Butnor JR., Drake BG. 2007 - *Effect of elevated CO2 on coarse-root biomass in Florida scrub detected by ground-penetrating radar.* Ecology 88 (5): 1328–1334.

Wu Y., Guo L., Cui X., Chen J., Cao X., Lin H. 2014 - *Ground-penetrating radar-based automatic reconstruction of three-dimensional coarse root system architecture.* Plant Soil 383, 155–172.

Zucconi F. 2003 - *Declino del suolo e stanchezza del terreno (Soil decline and soil sickness).* Pitagora (ed), Bologna, Italy.

13

Concept to Practices of Geospatial Information Tools to Assist Forest Management and Planning under Precision Forestry Framework

Most Jannatul Fardusi[*,1,2], Francesco Chianucci[3], Anna Barbati[2]

Abstract - Precision forestry is a new direction for better forest management. Precision forestry employs information technology and analytical tools to support economic, environmental and sustainable decision; the use of geospatial information tools enables highly repeatable measurements, actions and processes to manage and harvest forest stands, simultaneously allowing information linkages between production and wood supply chain, including resource managers and environmental community. In this contribution, we reviewed the most recent advances in the use of geospatial information technologies in forestry, and discussed their potential opportunities and challenges towards forest management and planning in the framework of precision forestry.

Keywords - Precision agriculture, Information and Communication Technologies, Decision Support Systems, Site-specific management, Forest planning

Introduction

The concept of precision has emerged in farming since the 1990s and it is based on the recognition that spatial and temporal variability in crop production need to be addressed by means of variable management practices, according to site or soil conditions (Tran and Nguyen 2006). Accordingly, "precision agriculture" means the integration of technology in the collection, interpretation, and analysis of data to support the decision-making system, in order to improve the agricultural processes by precisely managing each step to ensure maximum agricultural production while maintaining the sustainability of natural resources (Martini et al. 2001).

The idea of precision was translated also into the forest sector since the early 2000s in the US (Becker 2001), where the First International Symposium on Precision Forestry attempted to provide the state of the art in the use of information technologies and analytical tools to support site-specific, economic, environmental, and sustainable decision-making for the forest sector (Dyck 2003). Subsequently, precision forestry has rapidly gained ground in various application like forest inventory, management, engineering, and operations community applications (IUFRO 2015), especially in Europe and North America.

Precision forestry is based on similar principles than those of "precision agriculture" (Kovácsová and Antalová 2010). The substantial difference with precision agriculture is the broader spatio-temporal domain covered by precision forestry, ranging from a single tree to global scale, over different temporal frequencies of observations. However, despite its wider potential domain of application, it is remarkable that the term "precision forestry" has received less consideration than the more comprehensive precision agriculture terminology (Corona et al. 2017). This implies the need of a better understanding of precision forestry peculiarities with respect to the agricultural production system.

The precision forestry discipline exploits information technology and analytical tools to support economic, environmental and sustainable decisions (Gallo and Mazzetto 2013, Gallo et al. 2013). The use of such technologies enables highly repeatable measurements, actions, and processes to manage and harvest forest stands, simultaneously allowing information linkages between production and wood supply chain, including resource managers and environmental community (Dyck 2003).

The need for precision forestry is no longer a choice in managing forest and providing forest products. The demand for high-quality geospatial information about forests is supporting technologies

[1] Consiglio per la Ricerca in Agricoltura e l'Analisi dell'Economia Agraria - Research Centre for Forestry and Wood, Arezzo (Italy)
[2] Università degli Studi della Tuscia – Department for the Innovation in Biological, Agro-food and Forest systems, Viterbo (Italy)
[3] Consiglio per la Ricerca in Agricoltura e l'Analisi dell'Economia Agraria - Research Centre for Agriculture and Environment, Rome (Italy)
[*] jfardusi@unitus.it

to converge with the field of applications to provide the measurements, storage, analysis, and decision-making needs of the forest sector. At the same time, multifunctional forest management requires that multiple forest ecosystems services (ES) are simultaneously sustained. This pushes decision-makers towards GIS-based decision support systems to explore trade-offs between forest ES (Bottalico et al. 2016), thereby increasing the interest on reliable spatial estimation approaches to map the supply of ES across multiple spatio-temporal scales (Corona et al. 2014, Ferrara et al. 2017).

Notwithstanding cultural and technological advances, there are currently some difficulties in introducing precision technologies in forestry. In relatively few cases forestry applications have driven research efforts related to precision technologies (Schmoldt and Thomson 2003). More often, once precision technologies have been developed and made available, they often find ready application to research and management of environmental and ecological systems. Recent technologies like biosensing, micro-electromechanical systems (MEMs) and sensor networks are undergoing a rapid expansion, which is continuing modifying the framework of technologies, tools, processes, and materials available for precision forestry. This implies that many precision forestry applications are still at an initial stage of implementation. At another level, the push in technology use in the forestry sector often contrasts with its slow adoption from the end-user community, which is often poorly trained about how technology potential can improve and optimize their work.

In this paper, we discuss the most recent advances in precision forestry technologies and applications with specific reference to geospatial-information tools to assist forest management and planning. Firstly, we provide a definition of precision forestry; then we show the main categories of tools and applications available for precision technologies. Finally, we discuss current challenges and opportunities of precision forestry.

2. Context

As outlined in the 'First international conference on Precision Forestry' in Seattle, USA (Sarre 2014), there are several definitions of precision forestry for instance: "*Precision Forestry uses high technology sensing and analytical tools to support site-specific, economic, environmental, and sustainable decision-making for the forestry sector supporting the forestry value chain from bare land to the customer buying a sheet of paper or board*" (IUFRO 2014, IUFRO 2015). Taylor et al. (2002) proposed to define precision forestry as "*planning and conducting site-specific forest management activities and operations to improve wood product quality and utilization, reduce waste, and increase profits, and maintain the quality of the environment*".

However, the term precision should be understood as a new priority focus of land cultivation, including forestry, where the central concern is to deal with the environment, its variability and heterogeneity (Auernhammer 2001). In this view, precision forestry can be regarded as an "*environment friendly system solution that optimizes product quality and quantity while minimizing cost, human impact and intervention, and the variation caused by unpredictable nature*" (Joint Research Centre of the European Commission 2014). Indeed, other definitions of precision forestry include terms related to risk, environmental effects, and degradation, as key concerns.

Albeit precision forestry has received relatively limited consideration by the European Parliament's Committee on Agricultural and Rural Development, its document "Precision Agriculture: an opportunity for EU farmers potential support with the CAP2014-2020" has outlined the importance of exploiting available and advanced technologies for the smart and sustainable management of natural resources. Further, the Horizon 2020 (block Industrial Leadership) aims to achieve industrial leadership via key enabling technologies such as information and communications technology, nanotechnology, advanced materials, biotechnology, advanced manufacturing and processing, and space technology. On this basis, the "Strategic Research and Innovation Agenda for 2020" of the European Forest Institute and Forest-based Technology Platform intended to address the Societal challenges and improve industrial competitiveness in accordance with the European Strategy. Although not directly mentioned in these documents, precision forestry represents a cornerstone of the technologic development in the forestry sector.

3. Precision forestry applications

Measurements of forest variables are taken at spatial scales from millimeters range (e.g. nitrogen fixation in the soil) to the kilometer range (e.g. stand volume) up to global scale. Such multi-spatial scales also imply that different temporal scales can occur, where application can range from short time periods to much longer-period phenomena.

In this contribution, we focus our analysis on geospatial-information tools and techniques for forest data collection, which can be subdivided into three activity fields:

i) Remote sensing technologies
ii) Field survey technologies
iii) Decision Support Systems

In the following sections, we first briefly describe the tools and techniques available, mainly to outline their strength and weakness. Next, we move to more general considerations about the main challenges and recommendation to make these tools operational in forest management and planning.

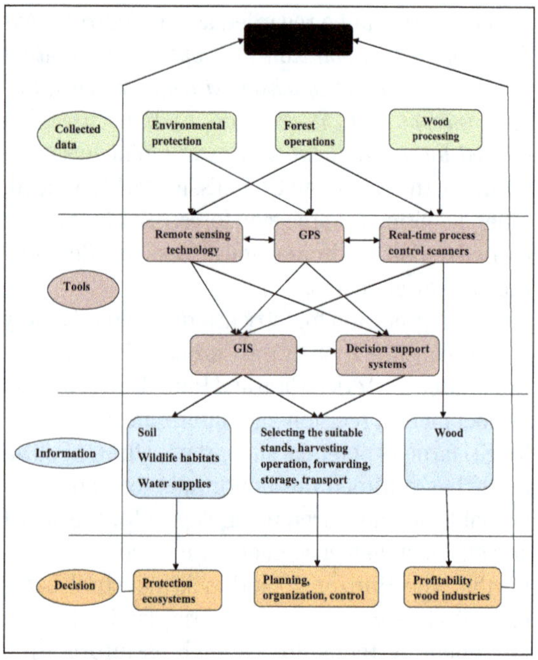

Figure 1 - Diagram of precision forestry (modified from Kovácsová 2016).

3.1 Remote sensing technologies

Airborne and satellite data provides a unique way for mapping forest attributes on spatially extensive areas, from regional to global scale. For this reason, remote sensing products have long been widely used for many forestry application including forest inventory (Corona et al. 2012), biodiversity assessment and monitoring, global change (Trumbore et al. 2015), as well as to support decision-making sustainable forest management (Franklin 2001). Traditionally used remotely sensed products include satellite and/or air-borne passive (multispectral, hyperspectral) and active (radio detection and ranging - RADAR, light detection and ranging - LIDAR) sensors. More recent advancements are bringing unmanned aerial vehicles (UAV) as a promising platform for meeting local objectives (Tang and Shao 2015).

The literature trends showed a constant increase of remote sensing application in forestry over the last few decades. Such trend is mainly supported by the increased availability of airborne hyperspectral and (particularly for the forestry sector) LIDAR data. Both data sources have been frequently applied in forest inventory and forest tree species classification (for a review, see Fassnacht et al. 2016). The significant progress that has been achieved in remote sensing of forests in recent years is related to the four linked drivers: i) technological sophistication in sensor design and deployment; ii) growth in data processing techniques and user-driven tools for image analysis; iii) increasing computational capabilities; iv) development of open source code; v) remote sensing imager archives opened free of charge online; vi) increased understanding of how and why remotely sensed information are important in forestry for information synthesis (Wulder and Franklin 2012) and vii) continuing advance of IMU sensors, particularly low-cost MEMS solutions, which are becoming the primary georeferencing components. In addition, new platforms such as unmanned aerial vehicles are opening an unprecedented opportunity for scale-appropriate measurements of ecological phenomena (Anderson and Gaston 2013), with large potential in precision forestry applications (Tang and Shao 2015).

For a more comprehensive overview of the available platforms and sensors, the reader is referred to review by Toth and Józków (2016). In the next section, we focus mainly on the most consolidated and promising technologies in precision forestry, related to multispectral/hyperspectral satellite sensors, airborne laser scanning, synthetic aperture RADAR, and unmanned aerial vehicles.

3.1.1 Multispectral and Hyperspectral satellite sensors

Spaceborne sensors have been operational for more than 40 years. First Landsat mission started in 1972, followed by SPOT-1 in 1986 and Ikonos in 1999. Based on the number of bands and their spectral resolution, satellite optical sensors can be divided into multispectral and hyperspectral (also called imaging spectroscopy) systems. Whereas most of the multispectral sensor systems typically have 4–8 bands, hyperspectral imagery is acquired in narrow, contiguous bands that can cover the visible (VIS), near-infrared (NIR) and shortwave-infrared (SWIR) portions of the electromagnetic spectrum (0.4–2.5 μm). Both multispectral and hyperspectral sensors have been widely used in forestry application including forest inventory (Corona et al. 2012, Babcock et al. 2016, Ferreira et al. 2016) and tree species classification (Ke et al. 2010, Dalponte et al. 2012, Fassnacht et al. 2016). These systems allow retrieving ecosystem information by measuring the spectral response of vegetation in sensor-specific wavelength regions. The advantage of hyperspectral sensors, compared with multispectral ones, is that the narrow bandwidths characteristic of hyperspectral data permit an in-depth examination of earth surface features which would otherwise be 'lost' within the relatively coarse bandwidths acquired with multispectral data. In this line, the use of narrow-band vegetation indices from hyperspectral sensors allows a variety of ecophysiological processes (e.g. photosynthesis, light use efficiency, chlorophyll content) to be monitored (Marshall et al. 2016, Zarco-Tejada et al. 2016) than is possible with multispectral sensors. However, as some studies (e.g. Galvao et al. 2005, Koch et al. 2005, Govender et al. 2007, Govender et al. 2008) pointed out, the selection of important vegetation wavelength region is more important than the number of available bands from these passive sensors. For example, Govender et al. (2008) concluded that "for describe diverse vegetation classes optimal hyperspectral bands is the best option to use. Optimal hyperspectral bands ought to be fused into multispectral system through sensor adjustments or spectral filters. A definitive objective is to move from unrealistic expansive information volumes and costly sensor advancement towards useful operational low information volumes, high temporal resolution and less expensive sensor improvement". The recent launch of Sentinel-2 mission from the European Science Agency fits well with this strategy; although it is not a hyperspectral sensor, Sentinel-2 is specifically designed for vegetation sensing, due to a combination of fine-tuned spectral capabilities (e.g., three narrow-bands in red edge wavelength), high spatial resolution (10 m), wide swath and quick turnaround (up to 5 days). Because of its relatively recent release (23 June 2015), its capability in forest remote sensing needs to be further verified, but early experience and simulations already indicate a clear potential in a wide range of forest applications at medium to large scale (e.g. Laurin et al. 2016, Immitzer et al. 2016), including forest management and planning.

3.1.2 Airborne Laser Scanning

The height of individual trees or canopy density of the stand can be measured accurately with this active remote sensing technology. Airborne Laser Scanning (ALS) has significant advantages over optical passive remote sensing because it can collect highly detailed 3D data quickly from a large area with varying conditions at repeated time intervals. For example, from airborne LIDAR data, the upper canopy height is correctly measured from the canopy height model (CHM), making this methodology particularly suited to estimate timber volume and biomass in forest stands (Naesset 1997, Barbati et al. 2009, Corona et al. 2012).

From ALS, different data products can be derived, such as digital elevation model grid, contours, raw point data and intensity clouds. Recent advancements in this technology in forestry is related to the increasing sophistication in sensor design and deployment, which currently allows very dense point clouds to be generated at decreasing costs, becoming operational in many regional and national forest inventories (Næsset 2007, Wulder et al. 2012). Currently, ALS represents one of the most promising and effective technology for a wide range of forestry applications, like e.g. above ground biomass estimation (Laurin et al. 2016, Nie et al. 2017), automatic forest area delineation (Alivernini et al. 2016), forest harvesting planning (Heinimann and Breschan 2012). However, the standardization of the point cloud processing, the comparability between different products generated by different resolution ALS data, and the influence of LIDAR accuracy's on topography, land cover categories, canopy density, represent still critical issues for the operational use of ALS data (Scrinzi et al. 2014).

3.1.3 Synthetic Aperture RADAR

Synthetic Aperture RADAR (SAR) is an active remote sensing technology which demonstrated relationships with biomass, and the all-weather data collection capacity (Tsui et al. 2013). According to Wulder et al. (2012), it provides complementary sets of information for estimating forest biomass. Orthorectified Radar Imagery (ORRI) has a wide variety of applications. These products are used in hydrology modelling (Bates et al. 2006), flood risk assessment (Mason et al. 2016), land use and land cover mapping (Yang et al. 2016), earth crust deformation monitoring (Vilardo et al. 2009), riparian studies (Makkeasorn et al. 2009) and forestry mapping (Kovácsová and Antalová 2010, Naidoo et al. 2015). However, SAR is height sensitive, and multiple baseline DEM generation has the potential to resolve phase ambiguities in an area with high vegetation and to solve the problems from foreshortening and layover. In addition, SAR tomography has the potential to resolve image distortion due to layover and foreshortening (Krieger et al. 2010). Compared with other active remote sensing technologies like satellite LIDAR, airborne radar systems are quite expensive and this has limited the number of commercial operators. Spaceborne radar systems have seen major advancement recently, and provided progressively informative data worldwide. In general, the application of RADAR technology has been less frequent in forestry than LIDAR and optical remote sensing. However, the recent Sentinel-1 radar system can represent a potential mission to support improvements in the remote sensing of forest vegetation via SAR sensors.

3.1.4 Unmanned aerial vehicles (UAV)

Also referred to as drones, unmanned aerial vehicles systems have emerged over the past few years as remote sensing platforms. The benefits of UAV include low material and operational costs, flexible control of spatial and temporal resolution, high-intensity data collection, and the absence of risk to crews. The current forestry applications of drone remote sensing are still at an experimental stage, but they are expected to expand very rapidly (Tang and Shao 2015). Recent forest application tested drone remote sensing to measure forest canopy height (Lisein et al. 2013, Tulldahl and Larsson 2014, Wallace et al. 2014, Zarco-Tejada et al. 2014), mapping canopy gap (Getzin et al. 2012), canopy cover and leaf area index (Chianucci et al. 2016), to detect fallen trees (Inoue et al. 2014) or pest infestation (Lehmann et al. 2015) and tracking forest wildfires (Ambrosia et al. 2011, Hinkley and Zajkowski 2011). The sensors currently used on remote sensing UAV are predominantly optical cameras, as they are widely available in lightweight and inexpensive versions. Dense point cloud generation from non-metric cameras (Ostrowski et al. 2014) has matured in recent years, and, as an emerging technology, it is a promising alternative to LIDAR sensors for small-scale inventory of forest structural variables (e.g. Puliti et al. 2015, Wallace et al. 2016). The relatively recent upsurge and the exponential growth in this technology system are rapidly changing the benchmark of UAV applications. Currently, the main challenge in UAV technology is related to the development of sensor technologies, particularly regarding the radiometric and geometric resolution of payloaded cameras. Also, portable LIDAR and hyperspectral sensors are introduced recently, but their accuracy and reliability still need to be verified. Developments are also still required in the navigational capabilities, in the miniaturization of measurement technologies (i.e. smaller, lighter payloads, GPS, and inertial units) and in the navigation accuracy, as particularly laser and Hyperspectral imaging HSI sensor requires accurate georeferencing. However, the convergence process is strongly accelerated by the customer market, as the demand for anywhere and anytime navigation with full visualization of the environment is quickly increasing, supporting a rapid increase in autonomous system and vehicle navigation (Toth and Jóźków 2016).

3.2 Field survey technologies

Field survey technologies can support forest management and planning by reducing the time and cost of ground sampling. Among the available techniques, the most promising and effective ones are related to terrestrial laser scanning and Field Map, which are described in detail in the following sections. Other instruments are also briefly considered.

3.2.1 Terrestrial laser scanning systems (TLS)

TLS has been demonstrated to be a promising mobile-system tool for plot-level advanced field inventories (Liang et al. 2011, Seidel et al. 2011). The main advantages of using TLS to support forest planning and management lie in its capability to document the forest details automatically and at very fine (millimeter) spatial scales. TLS data can theoretically improve the accuracy and efficiency of field measurements, and the capability of locating trees on the plot. In addition, TLS data document the entire plot at a particular time, which enables accurate subsequent measurements and time-series analyses. In the last few years, TLS has been widely used in different forestry applications, such as tree mapping (Pueschel et al. 2013), callipering (Srinivasan et al. 2015), tree height estimation (Olofsson et al. 2014, Srinivasan et al. 2015) and forest biomass calculation (Yao et al. 2011, Seidel et al. 2012, Srinivasan et al. 2014, Greaves et al. 2015). The use of TLS has also been intensively studied, e.g., for the estimation of canopy structural parameters (Huang and Pretzsch 2010, Béland et al. 2011, Van der Zande et al. 2011, Moorthy et al. 2011, Seidel et al. 2012, Zhao et al. 2012, Zheng et al. 2013, Bayer et al. 2013, Hopkinson et al. 2013, van Leeuwen et al. 2013, Cifuentes et al. 2014). Notwithstanding these improvements, the current main limitation of TLS is the cost of the instrument and the complexity of data processing which requires well-trained operators. In addition, TLS data accuracy is still heavily dependent on operator, resolution scan and instrument accuracy, all factors that affect the replicability of TLS products depending on the acquisition setup (point cloud density, scan-time, the field of view) and stand structure (canopy cover, tree density, phenological period). In theory, several scans and high scanning resolution can provide complete information about forest structure. But multiple scans increase measurement times, as well as high scanning resolution, boosts data loading and processing times. In future, best data processing solutions should be found to easily deal with such a huge amount of data, especially in the case of standardized forest inventories. The development of freeware for automated extraction of stand attributes from TLS could improve the exploitation of the method in the nearly future. Another major problem when using TLS in the forest environment is occlusions caused by lower branches, surrounding trees, and understory. This phenomenon leads to lower point density and, therefore, to poor descriptions in the upper part of crowns and partially or totally hidden trees (Van Der Zande et al. 2006).

Recently, handheld TLS has been made available; reducing the cost of instrumentation, but their reliability in forestry remains to be verified. In short, use of TLS is strongly promising, but still far from an operational phase, because the greater 3D characterization available from this system currently does not balance the higher cost, training, processing required by TLS.

3.2.2 Global Navigation Satellite System (GNSS)

Historically, except for space platforms, such as the International Space Station, GPS is the core component of any georeferencing system, providing highly accurate absolute positioning in open areas worldwide (Toth and Jóźków 2016). The GPS/GIS on the GNSS is effective in data collection for predicting important attributes of forest plots including mean tree height, mean diameter at breast height, basal area, stem volume and tree biomass (Liu et al. 2016) and detailed object location in thematic mapping which increase the accuracy and availability of GNSS technology in forest. The application of GNSS on harvesting operations has already permitted to evaluate efficiency of the forest operations (Gallo et al. 2013). But the signal shading from the foliage is one of the main problems of GNSS. When the device follows a path that travels under tree canopy direct signals between receivers and satellites are blocked and weaker reflected signals that are affected by higher noise and multipath are used. Most direct georeferencing solutions are based on integrating GPS and IMU sensor data (Toth and Grejner-Brzezinska 1998, Toth 2002), supplying positioning and attitude data as well as providing better continuity and limited protection against GPS signal anomalies. High-end solutions, typical for airborne remote sensing, use dual-frequency carrier phase GPS data and tactical grade IMU sensors that can provide sub cm level accuracy in post-processing mode. At the other end of the spectrum, a simple code based GPS solution is usually integrated with MEMS IMU data, typically in real-time, providing sufficiently accurate georeferencing for UAV flight control and approximate imaging sensor orientation (Toth and Jóźków, 2016).

3.2.3 FieldMap

Field map is a very useful tool for forest inventorying, which computes field data collected during fieldworks (Figure 2). FieldMap is used for forest mapping and field measurements (tree height, crown projection, and profile, stem profile, estimation of timber volume), assessment of biomass and growth (Kovácsová and Antalová 2010). The disadvantage of the tool is that it is quite expensive, which might limit its application in practical forest inventory.

Figure 2 - FieldMap data collector (Source: http://www.fieldmap.cz/)

3.2.4 Other field instruments

Other commonly employed advanced field instruments in forestry include SilvaScan, and ultrasound decays detectors (UDD), which are used to carry out *in situ* wood quality measurements like density, stiffness (modulus of elasticity), microfibril angle, dimensional stability, and so on (Figure 3). These wood quality measurements can be incorporated into operational inventories to allow the improved characterization of the resources that will allow for raw material be allocated to the most appropriate manufacturing plant (Ritland et al. 2011) (Figure 4). This provides strategic information that can help make economic and environmental management de-

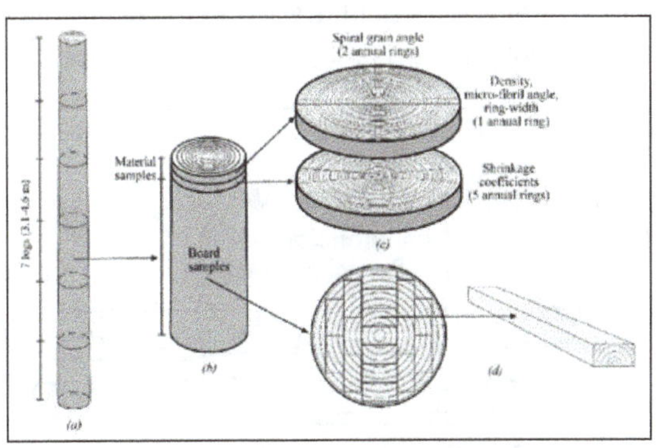

Figure 3 - Wood density and microfibril angle measurement from wood disc by SilvaScan (Source: Ormarsson and Cown, 2005).

cisions on treatments for individual trees and forest stands, improve thinning and harvesting operations, and more efficiently allocate timber resources for optimal utilization. Using acoustic non-destructive evaluation can help to operationalize the collection of wood properties that can support the primary log supply chain. Today, the precision of such technologies has been improved to the point where tree quality and intrinsic wood properties can be predicted and correlated to the performance and economic values of the final products. With continuous advancements and refinements, this technology could assist in managing wood quality, assessing forest value, and improving the timber quality of future plantations (Wang 2012). However, more research is needed to connect wood fibre properties with the performance of huge and growing number of products and processes which will help the forest product industry decision making.

Another ground instrument employed in precision forestry includes Ground Penetrating Radar (GPR), which can be used for estimating soil moisture, root architecture, and wood properties (Ferrara et al. 2013).

3.3 Decision Support System (DSS)

Decision Support Systems refer to an approach that integrates decision maker's own insights with computer's information processing capabilities for improving the quality of decision making. Decision support systems for forest management (FMDSS) have been developed world-wide to account for a broad range of forest ecosystems, management goals, and organizational frameworks. The main features of FMDSS include definition of sustainability of forest management which embraces the ecological, economic and social targets; measures of sustainability regarding the spatial and temporal dimensions; operationalising sustainability in terms of defining strategies to improve forest management. The decision support role of a geographic information system is particularly notable. Geographic information systems (GIS) assist the evaluation of a greater number of alternatives. Examples include the INVEST model for assessing ecosystem services (Tallis et al. 2013), which has been developed in North America and it has been also adopted in Europe (e.g. Bottalico et al. 2016). Many DSS applications are focused on wildfire risk reduction and management (e.g. Bonazountas et al. 2007, Kaloudis et al. 2008) and disaster risks in general (Iliadis et al. 2005). Other application includes planning of forest road (Karlsson et al. 2007), transportation planning (Forsberg et al. 2013), selecting timber harvesting system (Kühmaier and Stampfer 2010) participatory planning (Pastorella et al. 2016) and harvesting plan (Scrinzi and Clementel 2014).

The main current challenge is that there is an increasing number of DSS, which significantly vary in scope and underlined approaches, and which not often allow determining what DSS is more suited to the applied forest reality. To this end, a general conceptual and operational framework is needed, at least as concerns FMDSS. Theoretical developments have moved faster than empirical applications in forest management and planning; FMDSS are less compared to e.g. applications in water resources management. Expanding empirical applications need innovations in several areas; there is need to refine decision criteria to reduce their vagueness, add clarity and limit analysis to a manageable set of attributes. Furthermore, forest management is dynamic and the objectives are evolving towards sustainable/adaptive management. Finally, more efforts are required to develop FMDSS innovatively to capture the changing dynamics of forest management (Ananda and Herath 2009).

4. General challenges and recommendations

Introducing precision tools and technologies into forest management and planning is still challenging. General needs and issues are outlined here below.

4.1 Tools integration

Remote sensing, computing science, and in-

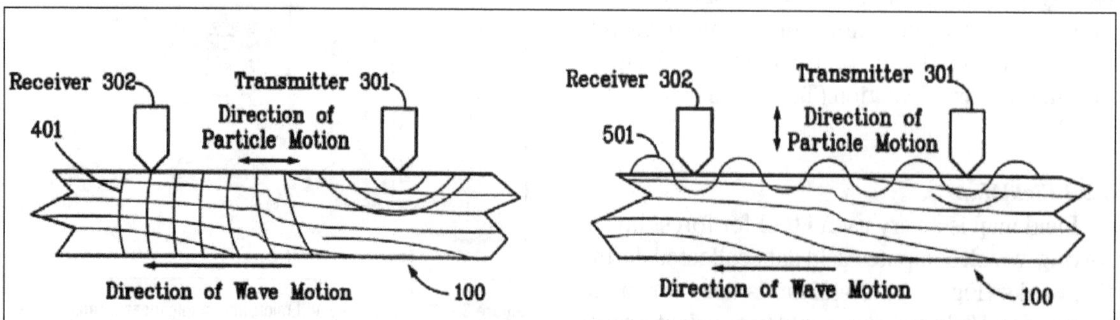

Figure 4 - Schematic diagram of the ultrasonic propagation time tests on the thick wood (Source: https://www.google.si/patents/US6276209).

formation technologies are greatly affected by universal development trends (Toth and Jóźków 2016), which are mainly driven by the fast pace of technological advances in material sciences, optical physics, nanotechnologies, and so on. Consequently, algorithmic research and software are generally lagging the potential offered by emerging hardware. Hence, there is an increasing need for integrating the existing tools, techniques, and sensors in a way that allow the coupling of all the different information available in a multi-scale and multi-source system.

4.2 Open access data

Appropriate access to data and software has both technical and intellectual components. The real question is more about to accessing the sensory data, which directly connects to communication and the general Information Technology infrastructure. In 2012, about 2.5 quintillions bytes of data were acquired worldwide per day (IBM 2012). Exploiting the potential of crowd sensing data leads to new methods, including Data Analytics and Big Data (Najafabadi et al. 2015). Accessibility is also challenging if results are presented inappropriately. For example, data may be aggregated at a fixed scale that may have limited value for many users. In other cases, language and concepts beyond the end user's understanding or vernacular might render a decision support system useless for a large audience segment. The continuous, transparent, and "user-friendly" information access makes interoperability a required attribute of databases, systems, and vocabularies. Because end-users of data are not necessarily local or regional, and because large-scale forest assessments are becoming more important, standards and protocols for forest data are looming on the horizon.

4.3 Skilled operators

Precision forestry technology implementation requires human resources and institutions with the appropriate proneness and knowledge to use the technologies. For example, the technology transfer issue is very worth to be solved to implement REDD+ or other programs in developing countries. Technology transfer must be a prominent consideration in helping solve the challenges we face in managing forests, locally, nationally, and globally.

4.4 Cost-effectiveness

Despite the potential, precision forestry tools have only just begun to be used in forestry communities (especially in the research environment) and are still not considered as standard tools. The main reason lies in the cost of instruments and peripherals (reference targets, software suites, graphic workstations). However, the price of devices has decreased by half during few last years. This trend should continue for some time. When prices become affordable for foresters, precision forestry instruments should become essential tools to provide extensive information about forests with a smaller investment in time and labor than current operations.

5. Conclusion

Managing forests in an increasingly environmental- and profit- concerned market have become a complex challenge for forest planning. Precision forestry could generate future forestry operations at a macro and micro level in a more economically and ecologically sustainable way to satisfy public and environmental demands. Modern Information and Communication Technologies can contribute to higher precision in forest management and decision making process.

Wood and food security are now central concepts in most parts of Europe, but there is evidence that increased production has led to significant harmful environmental consequences in terms of water pollution, greenhouse gas emissions and damage to natural ecosystems (Geiger et al. 2010, Kleijn et al. 2011). Confronted with, apparently antagonistic, pressures to conserve the environment and be careful with natural resources (Tilman et al. 2011), the forestry sector must face this main challenge and produce more with less environmental impact. The way to address this is to look at science and technology for possible answers.

Current precision and information technologies portend a future filled with improved capabilities to manage natural resources with greater skills and understanding (Schmodlt and Thomson 2003). The application of information and communication technologies into precision forestry has clear benefits to optimize production efficiency and to increase quality, but also to minimize environmental impact and risk, which include undesirable variability caused by the human operator (Gebbers and Adamchuck 2010). Precision forestry becomes a management practice of increasing interest because it links to key drivers directly related to worldwide issues such as Sustainable Forestry and Wood Security (Joint Research Centre 2014).

Precision forestry presents also some benefits for social and working conditions. The evaluation of precision technologies holds great potential to develop automated monitoring and management applications and thus reduce labor requirements.

A customer oriented production, an integrated forestry wood chain and the global vision of precision forestry are concepts that researchers and people involved in the forest and wood industry must deal with. Fulfilling, the specific requirements for the site, length, biological features and proof of origin that customers demand to wood products entails the existence of advanced methods and tools for understanding and thus controlling the production process along the forestry-wood chain value.

However, precision forestry could generate future forestry operations at macro and micro level more economically possible way to satisfy public and environment demands. Modern technologies i.e. LIDAR, UAV, TLS, hyper spectral remote sensing etc. may contribute for higher precision in forest management and decision making process and. Also it can be applied to increase the efficiency and information basis of existing national forest inventories and operational forest management planning.

Acknowledgments

This research has been supported by the Project "ALForLab"(PON03PE 00024 1) co-funded by the Italian Operational Programme for Research and Competitiveness (PON R&C) 2007-2013, through the European Regional Development Fund (ERDF) and national resource (Revolving Fund - Cohesion Action Plan (CAP)MIUR). It was also supported by the national project URBANFOR3, funded by Lazio Innova (CUP: C82I16000000005). Most Jannatul Fardusi was also supported by the PhD in Forest Ecology (XXXI Cycle) at Department for the Innovation in Biological, Agro-food and Forest systems, Università degli Studi dellaTuscia, Italy.

References

Alivernini A., Barbati A., Fares S., Corona P. 2016 - *Unmasking forest borderlines by an automatic delineation based on airborne laser scanner data.* International Journal of Remote Sensing 37 (16): 3568-3583.Ambrosia V., Wegener S., Zajkowski T., Sullivan D., Buechel S., Enomoto F., Lobitz B., Johan S., Brass J., Hinkley E. 2011 - *The Ikhana unmanned airborne system (UAS) western states fire imaging missions: from concept to reality (2006-2010).* Geocarto. Int. 26 (2): 85-101.

Ananda J., Herath G. 2009 - *A critical review of multi-criteria decision making methods with special reference to forest management and planning.* Ecological economics 68 (10): 2535-2548.

Anderson K. and Gaston KJ., 2013 - *Lightweight unmanned aerial vehicles will revolutionize spatial ecology.* Frontiers in Ecology and the Environment, 11(3):138-146.

Auernhammer H. 2001 - *Precision farming: the environmental challenge.* Computers and Electronics in Agriculture 32: 31-43.

Babcock C., Finley AO., Cook BD., Weiskittel A., Woodall CW. 2016 - *Modeling forest biomass and growth: Coupling long-term inventory and LiDAR data.* Remote Sensing of Environment 182: 1-12.

Barbati A., Chirici G., Corona P., Montaghi A., Travaglini D. 2009 - *Area-based assessment of forest standing volume by field measurements and airborne laser scanner data.* International Journal of Remote Sensing 30 (19): 5177-5194.

Bates PD., Wilson MD., Horritt MS., Mason DC., Holden N., Currie A. 2006 - *Reach scale floodplain inundation dynamics observed using airborne synthetic aperture radar imagery: Data analysis and modelling.* Journal of Hydrology 328(1): 306-318.

Bayer D., Seifert S., Pretzsch H. 2013 - *Structural crown properties of Norway spruce (*Picea abies *[L.] Karst.)and European beech (*Fagus sylvatica *[L.]) in mixed versus pure stands revealed by terrestrial laser scanning.* Trees 27: 1035-1047.

Becker G. 2001 - *Precision Forestry in central Europe – new perspectives for a classical management concept.* In: Proceedings of the First International Precision Forestry Symposium. University of Washington, Seattle, WA., 2001: 17-20.

Béland M., Widlowski JL., Fournier RA., Côté JF., Verstraete MM. 2011 - *Estimating leaf area distribution in savanna trees from terrestrial LIDAR measurements,* Agric. For. Meteorol. 151: 1252-1266.

Bonazountas M., Kallidromitou D., Kassomenos P., Passas N. 2007 - *A decision support system for managing forest fire casualties.* Journal of Environmental Management 84 (4): 412-418.

Bottalico F., Pesola L., Vizzarri M., Antonello L., Barbati A., Chirici G., Corona P., Cullotta S., Garfì V., Giannico V., Lafortezza R. 2016 - *Modeling the influence of alternative forest management scenarios on wood production and carbon storage: A case study in the Mediterranean region.* Environmental Research 144: 72-87.

Chianucci F., Disperati L., Guzzi D., Bianchini D., Nardino V., Lastri C., Rindinella A. Corona P., 2016 - *Estimation of canopy attributes in beech forests using true colour digital images from a small fixed-wing UAV.* International Journal of Applied Earth Observation and Geoinformation 47: 60-68.

Cifuentes R., Van der Zande D., Farifteh J., Salas C., Coppin P. 2014 - *Effects of voxel size and sampling setup on the estimation of forest canopy gap fraction from terrestrial laser scanning data.* Agric. For. Meteorol. 194: 230-240.

Corona P., Chianucci F., Quatrini V., Civitarese V., Clementel F., Costa C., Floris A., Menesatti P., Puletti N., Sperandio G., Verani S., Turco R., Bernardini V., Plutino M., Scrinzi G. 2017 - *Precision forestry: riferimenti concettuali, strumenti e prospettive di diffusione in Italia.* Forest@ 14: 1-12.

Corona P., Cartisano R., Salvati R., Chirici G., Floris A., Di Martino P., Marchetti M., Scrinzi G., Clementel F., Torresan C. 2012 - *Airborne Laser Scanning to support forest resource management under alpine, temperate and Mediterranean environments in Italy.* European Journal of Remote Sensing 45: 27-37.

Corona P., Fattorini L., Franceschi S., Chirici G., Maselli F., Secondi L. 2014 - *Mapping by spatial predictors exploiting remotely sensed and ground data: A comparative design-based perspective.* Remote Sensing of Environment 152: 29-37.

Dalponte M., Bruzzone L., Gianelle D. 2012 - *Tree species classification in the Southern Alps based on the fusion of very high geometrical resolution multispectral/hyperspectral images and LiDAR data.* Remote sensing of environment 123: 258-270.

Dyck B. 2003 - *Precision Forestry - the path to increased profitability.* In: Proceedings of the Second International Precision Forestry Symposium. University of Washington, Seattle, WA, 2003: 3-8.

Fassnacht FE., Latifi H., Stereńczak K., Modzelewska A., Lefsky M., Waser LT., Straub C. Ghosh, A. 2016 - *Review of studies on tree species classification from remotely sensed data.* Remote Sensing of Environment 186: 64-87.

Ferrara C., Barone PM., Steelman C., Pettinelli E., Endres AL. 2013 - *Monitoring shallow soil water content under natural field conditions using the early-time GPR signal technique.* Vadose Zone Journal DOI:10.2136/vzj2012.0202.

Ferrara C., Carlucci M., Grigoriadis E., Corona P., Salvati L. 2017 - *A comprehensive insight into the geography of forest cover in Italy: Exploring the importance of socio-economic local contexts.* Forest Policy and Economics 75: 12-22.

Ferreira MP., Zortea M., Zanotta DC., Shimabukuro YE., de Souza Filho CR. 2016 - *Mapping tree species in tropical seasonal semi-deciduous forests with hyperspectral and multispectral data.* Remote Sensing of Environment 179: 66-78.

Figueiredo EO., d'Oliveira MVN., Braz EM., de Almeida Papa D., Fearnside PM. 2016 - *LIDAR-based estimation of bole biomass for precision management of an Amazonian forest: Comparisons of ground-based and remotely sensed estimates.* Remote Sensing of Environment 187: 281-293.

Forsberg M., Frisk M., Ronnqvisty M. 2013 - *FlowOpt: a decision support tool for strategic and tactical transportation planning in forestry.* International Journal of Forest Engineering 16: 101-114.

Franklin SE. 2001 - *Remote sensing for sustainable forest management.* CRC Press.

Gallo R., Grigolato S., Cavalli R., Mazzetto F. 2013 - *GNSS-based operational monitoring devices for forest logging operation chains.* Journal of Agricultural Engineering XLIV (S2): 27.

Galvao LS., Formaggio AR., Tisot DA. 2005 - *Discrimination of sugarcane varieties in South Eastern Brazil with EO-1 Hyperion data.* Remote Sens. Environ. 94: 523-534.

Govender M., Chetty K., Bulcock H. 2007 - *A review of hyperspectral remote sensing and its application in vegetation and water resource studies.* Water SA 33:145-151.

Gebbers R. Adamchuk VI. 2010 - *Precision agriculture and food security.* Science 327: 828-831.

Geiger F., Bengtsson J., Berendse F., Weisser WW., Emmerson M., Morales MB., Ceryngier P., Liira J., Tscharntke T., Winqvist C., Eggers S. 2010 - *Persistent negative effects of pesticides on biodiversity and biological control potential on European farmland.* Basic and Applied Ecology, 11 (2): 97-105.

Getzin S., Wiegand K., Schöning I. 2012 - *Assessing biodiversity in forests using very high-resolution images and unmanned aerial vehicles.* Methods Ecol. Evol. 3 (2): 397-404.

Govender M., Chetty K., Naiken V., Bulcock H. 2008 - *A comparison of satellite hyperspectral and multispectral remote sensing imagery for improved classification and mapping of vegetation.* Water SA 34 (2): 147-154.

Greaves HE., Vierling LA., Eitel JU., Boelman NT., Magney TS., Prager CM., Griffin KL. 2015 - *Estimating aboveground biomass and leaf area of low-stature Arctic shrubs with terrestrial LIDAR.* Remote Sens. Environ. 164: 26-35.

Heinimann HR., Breschan J. 2012 - *Pre-harvest assessment based on LIDAR data.* Croatian Journal of Forest Engineering 33 (2): 169-180.

Hinkley E., Zajkowski T. 2011 - *USDA forest service-NASA: unmanned aerial systems demonstrations - pushing the leading edge in fire mapping.* GeocartoInt 26: 103-111.

Hopkinson C., Lovell J., Chasmer L., Jupp D., Kljun N., van Gorsel E. 2013 - *Integrating terrestrial and airborne LIDAR to calibrate a 3D canopy model of effective leaf area index.* Remote Sens. Environ. 136: 301-314.

Huang P., Pretzsch H. 2010 - *Using terrestrial laser scanner for estimating leaf areas of individual trees in a conifer forest.* Trees 24: 609-619.

IBM 2012: InfoSpherSpher. [Online]. Available at: https://www01.ibm.com/software /data/infosphere/streams/. [2016, April 19]

Iliadis LS., Spartalis SI. 2005 - *Fundamental fuzzy relation concepts of a DSS for the estimation of natural disasters' risk (The case of a trapezoidal membership function).* Mathematical and computer modeling 42 (7): 747-758.

Immitzer M., Vuolo F., Atzberger C. 2016 - *First experience with Sentinel-2 data for crop and tree species classifications in central Europe.* Remote Sensing 8(3): 166.

Inoue T., Nagai S., Yamashita S., Fadaei H., Ishii R., Okabe K., Taki H., Honda Y., Koji Kajiwara K., Suzuki R. 2014 - *Unmanned Aerial Survey of Fallen Trees in a Deciduous Broadleaved Forest in Eastern Japan.* PLOSOne 9 (10): 109881.

IUFRO 2014 - *Precision Forestry: The anchor of your value chain.* In: First Announcement, Precision Forestry Symposium. Stellenbosch, South Africa.

IUFRO 2015 - *Collection of definition or related elements Precision Forestry.* SILVAVOC Terminology Project. [Online]. [2016, 18 April]

Joint Research Centre 2014 - *Precision agriculture: an opportunity for EU-farmers - potential support with the CAP 2014-2020.* Policy Department B, European Union, 56 p.

Kaloudis S., Costopoulou CI., Lorentzos NA., Sideridis AB., Karteris M. 2008 - *Design of forest management planning DSS for wildfire risk reduction.* Ecological informatics 3 (1): 122-133.

Karlsson J., Ronnquist M., Frisk M. 2007 - *RoadOpt: a decision support system for road upgrading in forestry.* Scandinavian Journal of Forest Research 21 (S7): 5-15.

Ke Y., Quackenbush LJ., Im J. 2010 - *Synergistic use of QuickBird multispectral imagery and LIDAR data for object-based forest species classification.* Remote Sensing of Environment 114(6): 1141-1154.

Kleijn D., Rundlöf M., Scheper J., Smith HG., Tscharntke T., 2011 - *Does conservation on farmland contribute to halting the biodiversity decline?* Trends Ecol. Evol. 26: 474-481.

Koch M., Inzana J., El Baz F. 2005 - *Applications of Hyperion hyperspectral and aster multispectral data in characterizing vegetation for water resource studies in arid lands.* Proc. Geol. Remote Sens. 44-5.

Kovácsová P. 2016 - *Precision Forestry. [Online]. Available:* gis.tuzvo.sk/tiki-download_file.php?fileId=150. [2016, April 12]

Kovácsová P., Antalová M. 2010 - *Precision Forestry - definition and technologies,* Pregledničlanci . Reviews 11-12: 603-611.

Krieger G., Papathanassion KP., Younis M. 2010 - *Inferometric synthetic aperture Radar (SAR) mission employing formatting flying.* Proceedings of the IEEE 98 (5): 816-843.

Kühmaier M., Stampfer KS. 2010 - *Development of a multi-attribute spatial decision support system in selecting timber harvesting systems.* Croatian Journal of Forest Engineering 31: 75-88.

Laurin GV., Puletti N., Hawthorne W., Liesenberg V., Corona P., Papale D., Chen Q., Valentini R. 2016 - *Discrimination of tropical forest types, dominant species, and mapping of functional guilds by hyperspectral and simulated multi-spectral Sentinel-2 data.* Remote Sensing of Environment 176: 163-176.

Laurin GV., Puletti N., Chen Q., Corona P., Papale D., Valentini R. 2016 - *Above ground biomass and tree species richness estimation with airborne lidar in tropical Ghana forests.* International Journal of Applied Earth Observation and Geoinformation 52: 371-379.

Liang X., Litkey P., Hyyppä J., Kaartinen H., Kukko A., Holopainen M. 2011 - *Automatic plot-wise tree location mapping using single-scan terrestrial laser scanning.* Photogramm. J. Fin. 22: 37-48.

Lehmann JRK., Nieberding F., Prinz T., Knoth C. 2015 - *Analysis of Unmanned Aerial System-Based CIR Images in Forestry - A New Perspective to Monitor Pest Infestation Levels.* Forests 6: 594-612.

Lisein J., Pierrot-Deseilligny M., Bonnet S., Lejeune P. 2013 - *A photogrammetric workflow for the creation of a forest canopy height model from small unmanned aerial system imagery.* Forests 4 (4): 922-944.

Liu J., Hyyppä J., Yu X., Jaakkola A., Liang X., Kaartinen H., Kukko A., Zhu L., Wang Y., Hyyppä H. 2016 - *Can global navigation satellite system signals reveal the ecological attributes of forests?* International Journal of Applied Earth Observation and Geoinformation 50: 74–79.

Makkeasorn A., Chang NB., Li J. 2009 - *Seasonal change detection of riparian zones with remote sensing images and genetic programming in a semi-arid watershed.* Journal of Environmental Management 90(2): 1069-1080.

Marshall M., Thenkabail P., Biggs T., Post K. 2016 - *Hyperspectral narrowband and multispectral broadband indices for remote sensing of crop evapotranspiration and its components (transpiration and soil evaporation).* Agricultural and Forest Meteorology 218: 122-134.

Martinić I., Hengl T., Jurišić M., Husnjak S. 2001 - *Satellite navigation (GPS) - Trends and application.* Strojarstvo 43: 49-56.

Moorthy I., Miller JR., Berni JAJ., Zarco-Tejada P., Hu B., Chen J. 2011 - *Field characterization of olive (Olea europaea L.) tree crown architecture using terrestrial laser scanning data.* Agric. For. Meteorol. 151: 204-214.

Najafabadi MM., Villanustre F., Khoshgoftaar TM., Seliya N., Wald R. Muharemagic, E. 2015 - *Deep learning applications and challenges in big data analytics.* Journal of Big Data 2: 1-21.

Naesset E. 1997 - *Estimating timber volume of forest stands using airborne laser scanner data.* Remote Sensing of Environment 61(2): 246-253.

Næsset E. 2007 - *Airborne laser scanning as a method in operational forest inventory: Status of accuracy assessments accomplished in Scandinavia.* Scandinavian Journal of Forest Research 22(5): 433-442.

Naidoo L., Mathieu R., Main R., Kleynhans W., Wessels K., Asner G., Leblon B. 2015 - *Savannah woody structure modelling and mapping using multi-frequency (X-, C-and L-band) Synthetic Aperture Radar data.* ISPRS Journal of Photogrammetry and Remote Sensing 105: 234-250.

Nie S., Wang C., Zeng H., Xi X., Li G. 2017 - *Above-ground biomass estimation using airborne discrete-return and full-waveform LiDAR data in a coniferous forest.* Ecological Indicators 78: 221-228.

Olofsson K., Holmgren J., Olsson H. 2014 - *Tree stem and height measurements using Terrestrial Laser Scanning and the RANSAC algorithm.* Remote Sens. 6 (5): 4323-4344.

Ostrowski S., Jozkow G., Toth C., Vander Jagt B. 2014 - *Analysis of point cloud generation from UAS images.* ISPRS Annals of the Photogrammetry, Remote Sensing and Spatial Information Sciences 2 (1): 45.

Pastorella F., Borges JG., De Meo I. 2016 - *Usefulness and perceived usefulness of Decision Support Systems (DSSs) in participatory forest planning-the final users' point of view.* iForest-Biogeosciences and Forestry 9: 422-429.

Pueschel P., Newnham G., Rock G., Udelhoven T., Werner W., Hill J. 2013 - *The influence of scan mode and circle fitting on tree stem detection-stem diameter and volume extraction from terrestrial laser scans.* ISPRS J. Photogramm. Remote Sens. 77: 44-56.

Puliti S., Ørka HO., Gobakken T., Næsset E. 2015 - *Inventory of Small Forest Areas Using an Unmanned Aerial System.* Remote Sensing 7: 9632-9654.

Gallo R., Mazzetto F. 2013 - *A Methodology Study for the Application of Precision Forestry Approach in Logging Operation Chains.* In: EFITA-WCCA-CIGR Conference "Sustainable Agriculture through ICT Innovation", Turin, Italy.

Ritland K., Krutovsky KV., Tsumura Y., Pelgas B., Isabel N., Bousquet J. 2011 - *Genetic mapping in conifers.* In: Genetics, Genomics and Breeding of Conifers. Science Publishers, an imprint of Edenbridge Ltd CRC press:196-238.

Sarre A. 2014 - *In pursuit of precision.* ITTO - International tropical Timber organiza- tion. [Online]. Available: www.itto.int/direct/.../topics_id=1250000&no =1. [2016, April 25]

Schmoldt DL., Thomson AJ. 2003 - *Precision technologies: data availability past and future.* In: Proceedings of the First International Precision Forestry Symposium. University of Washington, Seattle, WA, 2003: 9-15.

Scrinzi G., Clementel F. 2014 - *Un sistema software di supporto decisionale per la stima sintetica del parametri dendrometrici di popolamento nella NPFAT.* [Online].Available: http://sito.entecra.it/portale/cra_manuali_dettaglio.php?id_manuale=14 554&lingua=IT. [2016, May 25]

Seidel D., Beyer F., Hertel D., Fleck S., Leuschner C. 2011 - 3D-laser scanning - *A non-destructive method for studying above-ground biomass and growth of juvenile trees.* Agr. Forest Meteorol. 151: 1305-1311.

Seidel D., Fleck S., Leuschner C. 2012 - *Analyzing forest canopies with ground-based laser scanning-a comparison with hemispherical photography.* Agric. Forest Meteorol. 154: 1-8.

Srinivasan S., Popescu SC., Eriksson M., Sheridan RD., Ku NW. 2014 - *Multi-temporal terrestrial laser scanning for modeling tree biomass change.* For. Ecol. Manage. 318: 304-317.

Srinivasan S., Popescu SC., Eriksson M., Sheridan RD., Ku NW. 2015 - *Terrestrial laser scanning as an effective tool to retrieve tree level height, crown width, and stem diameter.* Remote Sens. 7 (2): 1877-1896.

Tallis HT., Ricketts T., Guerry AD., Wood SA., Sharp R., Nelson E., Ennaanay D., Wolny S., Olwero N., Vigerstol K., Pennington D., Mendoza G., Aukema J., Foster J., Forrest J., Cameron D., Arkema K., Lonsdorf E., Kennedy C., Verutes G., Kim CK., Guannel G., Papenfus M., Toft J., Marsik M., Bernhardt J., Griffin R., Glowinski K., Chaumont N., Perelman A., Lacayo M., Mandle L., Griffin R., Hamel P., Chaplin-Kramer R. 2013 - *InVEST 2.6.0 User's Guide*. The Natural Capital Project, Stanford University, Stanford.

Tang L., Shao G. 2015 - *Drone remote sensing for forestry research and practices*. Journal of Forestry Research 26 (4): 791-797.

Taylor SE., Veal MW., Grift TE., Mcdonald TP., Corley FW. 2002 - *Precision Forestry-Operational tactics for today and tomorrow*. In: 25th annual Meeting of the council of Forest Engineers. Auburn University, Auburn, Alabama, USA.

Tilman D., Balzer C., Hill J., Befort BL. 2011 - *Global food demand and the sustainable intensification of agriculture*. In: Proceedings of the National Academy of Sciences, 13 December 2011, n 108, vol. 50: 20260-20264.

Toth C. Jóźków G. 2016 - *Remote sensing platforms and sensors: a survey*. ISPRS Journal of Photogrammetry and Remote Sensing 115: 22-36.

Toth C., Grejner-Brzezinska DA., 1998 - *Performance analysis of the airborne integrated mapping system (AIMSTM), ISPRS commission II symposium on data integration: systems and techniques. Int. Arch. Photogramm.* Remote Sens. XXXII (Part 2): 320–326.

Toth CK. 2002 - *Sensor integration in airborne mapping. IEEE Trans, Instrum.* Meas. 51 (6): 1367–1373

Tran DV., Nguyen NV. 2006 - *The concept and implementation of precision farming and rice integrated crop management systems for sustainable production in the twenty-first century*. International Rice Commission Newsletter (FAO) 55: 91-102.

Trumbore S., Brando P., Hartmann H. 2015 - *Forest health and global change*. Science 349 (6250): 814-818.

Tsui OW., Coops NC., Wulder MA., Marshall PL. 2013 - *Integrating airborne LIDAR and space-borne radar via multivariate kriging to estimate above-ground biomass*. Remote Sensing of Environment 139: 340-352.

Tulldahl HM., Larsson H. 2014 - *LIDAR on small UAV for 3D mapping*. In: Proceedings of SPIE sponsored conference on electro-optical remote sensing, photonic technologies, and applications VIII; and military applications in hyperspectral imaging and high spatial resolution sensing II. Amsterdam, Netherlands, 2014.

Van Der Zande D., Hoet W., Jonckheere I., Van Aardt J., Coppin P. 2006 - *Influence of measurement set-up of ground based for derivation of tree structure*. Agric. For. Meteorol. 141: 147-160.

Van der Zande D., Stuckens J., Verstraeten WW., Mereu S., Muys B., Coppin P. 2011 - *3D modeling of light interception in heterogeneous forest canopies using ground-based LIDAR data*. Int. J. Appl. Earth Obs. Geoinf. 13: 792-800.

Van Leeuwen M., Coops NC., Hilker T., Wulder MA., Newnham GJ., Culvenor DS. 2013 - *Automated reconstruction of tree and canopy structure for modelling the internal canopy radiation regime*. Remote Sens. Environ. 136: 286-300.

Vilardo G., Ventura G., Terranova C., Matano F., Nardò S. 2009 - *Ground deformation due to tectonic, hydrothermal, gravity, hydrogeological, and anthropic processes in the Campania Region (Southern Italy) from Permanent Scatterers Synthetic Aperture Radar Interferometry*. Remote Sensing of Environment 113(1): 197-212.

Wallace L., Watson C., Lucieer A. 2014 - *Detecting pruning of individual stems using airborne laser scanning data captured from an unmanned aerial vehicle*. Int. J. Appl. Earth Obs. Geoinf. 30: 76-85.

Wallace L., Lucieer A., Malenovský Z., Turner D., Petr Vopenka P. 2016 - *Assessment of Forest Structure Using Two UAV Techniques: A Comparison of Airborne Laser Scanning and Structure from Motion (SfM) Point Clouds*. Forests 7: 62.

Wang X. 2012 - *Advanced Sorting Technologies for Optimal Wood Products and Woody Biomass Utilization*. In: Proceedings of 2012 International Conference on Biobase Material Science and Engineering (BMSE 2012). Changsha, China. 21-23 October, 2012: 175-179.

Wulder MA., Franklin SE. (Eds.). (2012) - *Remote sensing of forest environments: concepts and case studies*. Springer Science & Business Media, New York.

Wulder MA., White JC., Nelson RF., Næsset E., Ørka HO., Coops NC., Hilker T., Bater CW., Gobakken T. 2012 - *LIDAR sampling for large-area forest characterization: a review*. Remote Sensing of Environment 121: 196-209.

Yang S., Wang M., Long H., Liu Z. 2016 - *Sparse Robust Filters for scene classification of Synthetic Aperture Radar (SAR) images*. Neurocomputing 184: 91-98.

Yao T., Yang X., Zhao F., Wang Z., Zhang Q., Jupp D., Lovell J., Culvenor D., Newnham G., Ni-Meister W. 2011 - *Measuring forest structure and biomass in New England forest stands using Echidna ground-based LIDAR*. Remote Sens. Environ. 115 (11): 2965-2974.

Zarco-Tejada PJ., Diaz-Varela R., Angileri V., Loudjani P. 2014 - *Tree height quantification using very high resolution imagery acquired from an unmanned aerial vehicle (UAV) and automatic 3D photo-reconstruction methods*. Eur. J. Agron. 55: 89-99.

Zarco-Tejada PJ., González-Dugo MV., Fereres E. 2016 - *Seasonal stability of chlorophyll fluorescence quantified from airborne hyperspectral imagery as an indicator of net photosynthesis in the context of precision agriculture*. Remote Sensing of Environment 179: 89-103.

Zhao F., Strahler AH., Schaaf CL., Yao T., Yang X., Wang Z., Schull MA., Román MO., Woodcock CE., Olofsson P. 2012 - *Measuring gap fraction, element clumping index and LAI in Sierra Forest stands using a full-waveform ground based LIDAR*. Remote Sens. Environ. 125: 73-79.

Zheng G., Moskal LM., Kim SH. 2013 - *Retrieval of effective leaf area index in heterogeneous forests with terrestrial laser scanning*. IEEE Trans. Geosci.

14

Proposal of a simplified method for pastoral value assessment inside forest planning

Giovanni Argenti[1], Elisa Bianchetto[2], Fabrizio Ferretti[3*]

Abstract - Carrying capacity is one of the most important variables that should be assessed for a proper evaluation of forage potentiality of pastures. Nevertheless, in forest management plans (silvo-pastoral plans), the need for reducing costs make the involvement of additional specialized technical staff impossible. For this reason different methodologies to simplify data collection have been proposed for pasture planning. In this paper a simplified method to evaluate the pastoral value in forest planning (one of the most common procedures for carrying capacity estimation) is proposed and assessed in real conditions in order to obtain a dependable potential stocking rate of a whole pasture area. The method is based on a previous research that proposed a simplified method of data collection that is performed by means of different functional groups of species or botanical families: palatable grasses, not palatable grasses, legumes, species belonging to other botanical families, spiny and poisonous species, trees and shrubs. Each category is linked to its feeding behavior by an index that summarizes forage potentiality and this allows classifying the resources in different quality classes, each of them characterized by a given potential stocking rate. The proposed methodology seems easy to be performed also by staff without a specific formation in pasture management, and comparison performed with the traditional procedure produced accurate results. Even if the proposed scheme should not to be considered alternative to the original methodology, it can be useful for acquiring information for pastoral resource management, especially in a silvo-pastoral context.

Keywords - botanical composition; forage quality; pastoral value assessment; forest planning; silvo-pastoral

Introduction

Natural pastures are ecosystems highly comparable to forests concerning complex botanical composition and possibility of combining economic utilisation with the conservation of multiple functions and provision of ecosystems services (Alados et al. 2007, Orlandi et al. 2016). Nowadays, in many parts of Europe, forage production of these peculiar ecosystems is less important than other purposes (Conant et al. 2016). For this reason, conservation of pastures is a primary issue for landscape conservation, for the maintenance of recreational spaces and for the conservation of biodiversity and of habitats for different kinds of wild animals (Porqueddu et al. 2003, Peeters 2008, Argenti et al. 2012, Dossche et al. 2016).

In many cases, grasslands are of an anthropogenic origin and they are defined as semi-natural grasslands as they were created in ancient times to provide forage for animal grazing (Nösberger and Staszewski 2002, Michaud et al. 2012). Due to their reduced utilisation, these resources are currently threatened by shrub encroachment and by re-colonisation of woody species that represent the potential vegetation in many areas currently occupied by pastures (Targetti et al. 2013, Pittarello et al. 2016, Török et al. 2016). In this context, relationships between pastoral and forest resources are often characterised by a high level of conflicts (Bernes et al. 2016) and it is necessary to face this issue following a multidisciplinary approach that should take into account the coexistence of diversified resources inside complex silvo-pastoral systems (Balandier et al. 2003, Rossetti et al. 2015).

One of the most important parameters that is evaluated along with the assessment of pastoral resources inside forest planning is the potential stocking rate or carrying capacity. It can be defined as the maximum number of animals (usually expressed as livestock unit, LU) that can graze a given pasture surface for a given time without deterioration of the resource (Allen et al. 2011). Following different approaches, many methodologies have been developed to calculate this quantitative parameter. So, some methods need the measurement of dry matter production of the pasture which is compared with the requirements of one LU (Pardini et al. 2001). The

[1] Università degli Studi di Firenze, Scienze delle Produzioni Agro-alimentari e dell'Ambiente, Firenze (Italy)
[2] CREA Research Centre for Agriculture and Environment, Firenze (Italy)
[3] CREA Research Centre of Forestry and Wood, Arezzo (Italy)
*fabrizio.ferretti@crea.gov.it

French school proposed the pastoral value approach (Daget and Poissonet 1969 and 1971) in which the carrying capacity is considered to be proportional to this parameter that is obtained only on the basis of botanical transects and for this reason this method is simple and fast. The use of only botanical composition data permits also to utilise vegetation analysis previously conducted not specifically for pastoral purposes (Roggero et al. 2002). Moreover, recently the attention was driven to the pasture type approach (Argenti and Lombardi 2012), for the assessment of wide territories occupied by pastures. A remarkable number of studies evaluated appropriate methodologies for classifying pasture vegetation types occurring in a territory to propose management guidelines (Ziliotto et al. 2004, Cavallero et al. 2007, Forbis et al. 2007, Targetti et al. 2010). Utilisation of remote sensing and the implementation of different GIS techniques can produce remarkable and efficient results in this domain (Boschetti et al. 2007, Yu et al. 2010, Argenti et al. 2011), taking into account also multidisciplinary approaches (Primi et al. 2016).

In this context, it is useful to identify the most important and the most informative parameters to define a minimum data set to be collected for a proper evaluation of pasture characteristics for management purposes. The main aim of the present paper is to validate a simplified methodology for pastoral assessment inside forest planning (Argenti et al. 2006), in order to evaluate its reliability and to evaluate how gained information could be useful not only for the classification of botanical composition but also to insert such findings inside pasture planning.

Materials and methods

In a previous paper Argenti et al. (2006) suggested a simplified method for evaluating the botanical composition, making data collection easier, and for estimating a synthetic pastoral value useful for pasture classification with management purposes. In this work this method was utilised inside a real forest management plan to evaluate its efficiency, tested in comparison with the complete pasture analysis as proposed by the previously mentioned pastoral value approach (Daget and Poissonet 1971 and 1972). In this way, the suggested method is useful to classify pastures in different categories of quality on the basis of their botanical composition and, after attribution to each class of a prescribed level of stocking rate, it becomes an efficient tool for the management practices and for the evaluation of the carrying capacity of a given pastoral area.

As mentioned before, the main difference from the original method is the way in which the botanical composition of the pasture is classified. In the simplified method, vegetation is assessed by identification of large functional plant categories and not by means of complete transect analysis. In this way, pasture vegetation can be assessed also by forest staff not expert in this field, with reduction of length and cost of the survey. Moreover, its application at territorial level is possible in territorial forest plans. The six categories proposed were chosen for their importance in affecting quality and productivity of forage availability and, consequently, of the carrying capacity (Table 1).

Table 1 - Botanical categories and Specific Index (SI_c) calculated for each category, according to Argenti et al. (2006).

Botanical category	Acronym	SI_c
Palatable grasses	PG	1.95
Not palatable grasses	NG	0
Legumes	LE	2.99
Species belonging to other botanical families	OT	0.29
Spiny and poisonous species	SP	0
Trees and shrubs	TS	0.03

In the complete and traditional approach, the pastoral value (PV) is calculated as follows:

$$PV = \frac{\sum SC_i \times SI_i}{5}$$

where SC_i is the specific contribution, i.e. the percentage of each species in the total of the vegetation derived from vertical point quadrat transect (Argenti and Lombardi 2012), and SI_i is the specific index, ranging from 0 to 5, which summarizes the forage value of each species in the pasture (Cavallero et al. 2002, Bagella et al. 2013). According to this formula, PV ranges from 0 to 100, and it estimates the forage potentiality of a pasture area, which is directly proportional to its carrying capacity by means of a coefficient of transformation (Cavallero et al. 2007).

In the proposed method, a different SI is attributed to each category of plants, taking into account the average characteristics of each species belonging to that given category, in order to use the same formula as before and to evaluate the estimated pastoral value (PVe), obtained exclusively on the basis of the vegetation classified through the six proposed categories:

$$PVe = \frac{\sum SC_c \times SI_c}{5}$$

where SC_c is the percentage presence of each of the six botanical categories and SI_c is the specific index for each category (Tab. 1), derived from Argenti et al. (2006).

Estimated pastoral value is thus utilised to arrange a qualitative classification of the forage resources that can be reported in cartography and representing the different types of quality class of pastures. Assessment of carrying capacity for the whole area is then performed assigning to each category a prescribed level of stocking rate which is proposed in relation to PV of each category. Proposed categories of pasture quality and their potential stocking rate are reported in Tab. 2.

Table 2 - Classification scheme of pasture quality classes according to VPe values and annual potential stocking rate per class to be used for pasture planning.

Class level	Pasture quality class	VPe	Prescribed stocking rate (LU ha^{-1} y^{-1})
1	Poor pasture	VPe < 15	0.15
2	Medium pasture	15 ≤ VPe ≤25	0.30
3	Good pasture	VPe > 25	0.45

Assessment of the proposed method here presented was conducted in the northern Apennines, between Tuscany and Emilia-Romagna regions, and it interested different pastoral areas, i.e. clearings of limited surface inside forest or open areas. Main forest vegetation is represented by broadleaves species, such as *Quercus cerris*, *Q. pubescens*, *Acer campestre*, *A. pseudoplatanus*, *Fraxinus excelsior* and *Fagus sylvatica* and by some shrubs encroaching open areas, such as *Rosa canina*, *Crataegus monogyna*, *Prunus spinosa*, and *Juniperus communis*.

In the experimental sites, 138 botanical analyses were conducted on the same point in two different periods of the year (69 in spring and 69 in summer) to verify also the influence of the period of data collection on the accuracy of the botanical analysis. Usually, in forest planning, the botanical survey of herbaceous resources is performed simultaneously to the survey of the forest resources, so it is possible that the period is not adequate for perfect identification of the herbaceous species that present a different phenology from forest ones. In each sample area, the synthetic method of botanical analysis was conducted evaluating, by a visual estimation, the ground cover percentage occupied by the six categories of plants and then the estimated pastoral value was calculated (PVe). Afterwards, on the same sample area, the complete transect analysis according to the traditional methodology was performed to obtain the botanical characterisation and the control pastoral value (PVt). In this way, it was possible to compare the pastoral value estimated with the synthetic and the original approach and it was possible to analyse how the period of survey could affect the accuracy of the pastures assessment. Moreover, the potential stocking rate was assessed on the experimental area by means of the two methods to obtain a proper evaluation of the simplified method on a pastoral surface.

Statistical analysis, using RStudio (RStudio Team 2015) and R (R Core Team 2016), was performed to investigate differences between PVe and PVt. To test if the collected data were normally distributed, the Shapiro Wilk normality test was performed (Shapiro and Wilk 1965). Since, in one case, data distribution was not normal ($p<0.05$), nonparametric tests were used to analyse data among the quantitative variables. The Wilcoxon signed-rank test (Siegel 1956) is a non-parametric statistical hypothesis test used to compare repeated measurements on a single sample and to assess whether their population mean ranks differ. In this case we apply the small samples version, where N (≤25) is the number of pairs minus any pairs whose difference is zero and T is the smaller sum of like-signed ranks. When "an observed T is equal to or less then the critical value of T under a particular significance level for the observed value of N, the null hypothesis may then be rejected at that level of significance" (Siegel 1956). The test was applied to verify if there were differences in four cases: a) between spring PVt and PVe, b) between summer PVt and PVe, c) between spring and summer PVt, d) between spring and summer PVe. All statistical tests were assessed at the α =0.05 level.

Results

Applying the proposed method, we can highlight a slight statistical distortion comparing the two distributions (PVt and PVe) for spring and summer botanical surveys (Tab. 3), where PVt distribution is more similar to a normal one, as shown by kurtosis and skewness parameters.

Table 3 - Comparison of descriptive statistics for PVt and PVe for spring and summer surveys.

	spring surveys		summer surveys	
Spring	PVt	PVe	PVt	PVe
Mean	32.75	29.19	31.64	27.64
Standard Error	1.16	0.68	1.10	0.78
Median	32	30	31	28
Mode	29	32	30	30
Standard Deviation	9.63	5.61	9.17	6.51
Sample Variance	92.72	31.45	84.06	42.41
Kurtosis	-0.13	0.71	-0.03	0.15
Skewness	0.00	-0.61	-0.08	-0.52
Range	45	30	43	30
Minimum	10	13	9	10
Maximum	55	43	52	40
Sum	2260	2014	2183	1907
Count	69	69	69	69

In both sampling dates, PVt shows a wider range of values, standard deviation always higher in comparison to PVe and average values higher than those found with PVe. This difference can be affected by sampling errors, but it is certainly partially due to calculation methodology that reduces variability in PVe assessment. Indeed, for each pair PVt-PVe whose calculation is respectively based on SI and SI_C, PVt can be higher of the corresponding PVe, because, for example for legumes, SI_c is 2.99 while SI it could get up to 5.

PVt and PVe data values were used for the classification in three classes of forage quality previously identified (see Table 2). Based on this grouping, we observe slight differences between spring PVt and PVe, summer PVt and PVe, spring and summer results (Fig. 1).

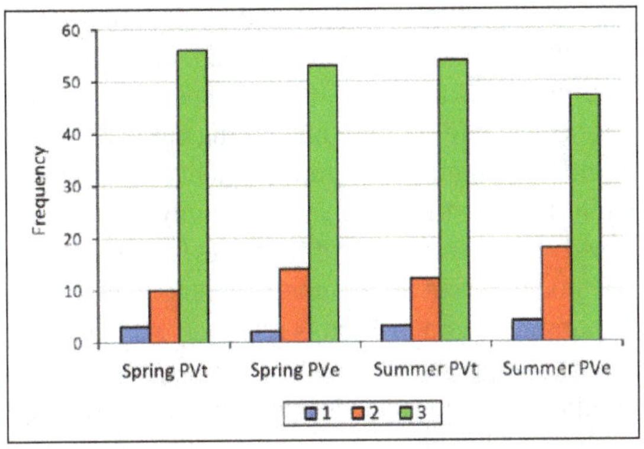

Figure 1 - Distribution of PVt and PVe in each class level for spring and summer surveys.

In both seasons, the two methods used to determine pastoral value were able to identify roughly the same absolute value of quality classes (on the total 69 survey per season), with the same trend. To analyse differences in identifying the same quality of pasture class, in Tab. 4 the differences between PVt and PVe class levels are reported.

Table 4 - Differences between PVt and PVe class levels (absolute and percentage values) for each season and in total.

Difference	Spring	%	Summer	%	Total	%
-2	0	0	0	0.0	0	0.0
-1	4	5.8	7	10.1	11	8.0
0	59	85.5	48	69.7	107	77.5
1	6	8.7	13	18.8	19	13.8
2		0.0	1	1.4	1	0.7
total	69	100	69	100	138	100

The value 0 means that both methods identified the same class level, while positive or negative values mean that PVt overestimated (positive values) or underestimated (negative values) the class level obtained by PVe. We observe no differences in more than 85% and 69% of spring and summer samples and only in one case the difference is of two class levels. An overall assessment of both methods, independently of the survey season, produced a score of 77.5% for identification of the same class. Generally, we observed the summer survey tendency to underestimate the PV, probably related to the major difficulties to estimate in a perfect way ground cover of species during the summer period.

The Wilcoxon signed-rank test, for small samples ($N \leq 25$), was applied to test if there were differences in four cases namely a) between spring PVt and PVe (N=11, T=30), b) between summer PVt and PVe (N=23, T=80.5), c) between spring and summer PVt (N=13, T=24), d) between spring and summer PVe (N=17, T=36). In every case we accepted the null hypothesis (H_0) that there are no differences between the traditional pastoral value approach and the proposed method. Thus, the results suit the expectancy that the simplified approach is an effective way to estimate the PV expressed in three classes of forage quality.

For a practical application of the proposed method and to assess its accuracy to identify a proper carrying capacity, evaluation of potential stocking rate on the whole investigated area, composed by three different pastoral areas (Lombina, S. Paolo Alto and S. Paolo Basso), adopting the two analyzed scheme was performed (Tab. 5).

Carrying capacity was obtained by pastoral value complete methodology using a transformation coefficient as proposed by Cavallero et al. (2007), while data obtained by the simplified methodology were used to classify pastures in quality classes and to obtain the whole carrying capacity by means of the unitary potential stocking rate reported in Tab. 2. Transformation from yearly to seasonal stocking rate was performed according to Cavallero et al. (2002). Results at territorial level confirm what

Table 5 - Carrying capacity for the experimental area obtained by both tested methods in different seasons of data collection.

Pastoral area	Sampling period	Surface (ha)	PVt	Seasonal potential stocking rate from PVt (LU ha^{-1})	Seasonal potential stocking rate from PVe (LU ha^{-1})	Deviation (%)
Lombina	Spring	21	32.0	24.2	21.6	-10.6
S. Paolo Alto	Spring	40	35.1	50.5	39.8	-21.2
S. Paolo Basso	Spring	73	29.4	77.2	67.8	-12.2
	Total Spring	*134*	*31.5*	*151.9*	*129.2*	*-14.0*
Lombina	Summer	21	31.8	24.0	22.7	-5.6
S. Paolo Alto	Summer	40	29.4	42.4	38.1	-10.1
S. Paolo Basso	Summer	73	31.4	82.6	65.0	-21.3
	Total Summer	*134*	*30.9*	*149.0*	*125.8*	*-15.6*

found in the previous comparison. The difference according to date of survey is really reduced, with values of stocking rate for each pastoral district and on the total area remarkably similar. Differences among methods produce higher value of carrying capacity (with an error roughly 15% as an absolute value) when found by PVt than PVe, confirming an underestimation of the simplified procedure in comparison to the control method.

Discussion and conclusions

As reported by Argenti et al. (2006), the proposed method is an effective way to estimate the PV when we express it by a small number of classes of quality, thus we can consider it as a "qualifying approach". It is very different from the traditional method, which produces a large amount of different data suitable for pastoral resources characterization and management, such as a proper vegetation characterization which is useful for many different purposes (Cavallero et al. 2007). Anyway, this synthetic approach for pastoral value evaluation was already used in different contexts and in previous studies with other aims, such as those concerning analysis of functional traits in herbaceous plant communities (Bolzan 2009) or wild animal habitat selection (Argenti et al. 2015). In these studies, pastures characterization is not a major issue and a simplified procedure to obtain the pastoral value can help the assessment of forage potentiality, especially in areas involving silvo-pastoral systems (Bagella et al. 2013). Moreover, the underestimation of stocking rate performed by the simplified method facilitates a conservative assessment of forage potentiality, and in this way it can assure a prudent exploitation for these resources, highly necessary to preserve natural grasslands (Arponen et al. 2013).

The proposed method is based on a simplified survey applicable by forest staff not well experienced in rangeland planning (Bianchetto et al. 2015) and, obviously, it can not replace the traditional approach. The simplified method is a useful tool for forest planning activities in which an extreme accuracy in pasture assessment is not required, as reported by Bianchetto et al. (2009). More specifically, the "qualifying approach" allows to define a range of carrying capacity or to have an approximate estimation of the potential stocking rate without the involvement of other professional operators. In fact, the participation of different specialized staff may lead to increases in costs of the whole procedure, which is a significant aspect especially in forestry planning (Agnoloni et al. 2009). In this context, other methodologies, well-known to be more accurate, such as those based on productivity measurements (Pardini et al. 2001) are expensive and time consuming, with serious difficulties in their application at territorial level (Argenti et al. 2002). Moreover, utilization of the typological approach, which is increasing especially in the Italian Alps (Argenti and Lombardi 2012), could be an important tool for pasture planning at the district level, but existence of pastoral typologies is limited to a reduced number of areas in Alpine regions, therefor this approach is not generally adoptable, as opposed to what happens in the forestry sector (Strano 2010). Subsequently, the proposed simplified procedure is advisable in a multifunctional approach to forest planning (Paletto et al. 2012).

Previous studies analyzed how different periods of sampling can affect measurements of pasture parameters. Dubbs et al. (2003) observed high differences in late summer and spring data collection in different methods of forage quality assessment, while in our study analysis is mainly vegetation-based and for this reason differences found in potential stocking rate are comparable among methods, as botanical composition is relatively steady along seasons, especially in mountain areas, and its variability in this time span is lower than forage quality evolution (Farruggia et al. 2014). This feature confirms the ability of the simplified method to estimate grazing value using also summer surveys, and this aspect extends the available period of surveying and it enlarges the possibilities for forest planners that usually perform data collection during this season (Ferretti et al. 2011).

In conclusion, the proposed method seems effective in its application and easily adoptable in many contexts. It can represent an acceptable compromise among different requirements in silvo-pastoral resource monitoring: cost reduction of data collection, availability of technical staff able to conduct vegetation assessment and obtainable information for management purposes. This preliminary assessment of the methodology is positive, but further investigations in different environmental conditions and vegetation communities are needed to evaluate its real potentiality for pasture planning.

References

Agnoloni S., Bianchi M., Bianchetto E., Cantiani P., De Meo I., Dibari C., Ferretti F. 2009 - *I piani forestali territoriali di indirizzo: una proposta metodologica*. Forest@ 6: 140-147.

Alados CL., El Aich A., Komac B., Pueyo Y., Garcıa-Gonzalez R. 2007 - *Self-organized spatial patterns of vegetation in alpine grasslands*. Ecological Modelling 201: 233-242.

Allen VG., Batello C., Berretta EJ., Hodgson J., Kothmann M., Li X., McIvor J., Milne J., Morris C., Peeters A., Sanderson M. 2011 - *An international terminology for grazing lands and grazing animals*. Grass and Forage Science 66: 2-28.

Argenti G., Bianchetto E., Ferretti F., Staglianò N. 2006 - *Proposta di un metodo semplificato di rilevamento pastorale nei piani di gestione forestale.* Forest@ 3: 275-280.

Argenti G., Bianchetto E., Sabatini S., Staglianò N., Talamucci P. 2002 - *Indicazioni operative per la gestione delle risorse pastorali nei Parchi Nazionali.* In:"Linee guida per la gestione ecosostenibile delle risorse forestali e pastorali nei Parchi Nazionali" Ciancio O., Corona P., Marchetti M., Nocentini S. eds). Accademia Italiana di Scienze Forestali, Firenze, I: 155-203.

Argenti G., Bottai L., Chiesi M., Maselli F., Staglianò N., Targetti S. 2011 - *Analisi e valutazione di pascoli montani attraverso l'integrazione di dati multispettrali e ausiliari.* Rivista Italiana di Telerilevamento/Italian Journal of Remote Sensing 43: 45-57.

Argenti G., Cervasio F., Ponzetta MP. 2012 - *Control of bracken (Pteridium aquilinum) and feeding preferences in pastures grazed by wild ungulates in an area of the Northern Apennines (Italy).* Italian Journal of Animal Science 11: 336-341.

Argenti G., Ciulla A., Nicoloso S., Brugnoli A., Staglianò N. 2015 - *Post-release habitats' selection of Capreolus capreolus italicus (Festa, 1925) in a protected area in southern Italy* Annals of Silvicultural Research 39 (1), 2015: 37-45 DOI: http://dx.doi.org/10.12899/asr-752

Argenti G., Lombardi G. 2012 - *The pasture-type approach for mountain pasture description and management.* Italian Journal of Agronomy 7: 293-299.

Arponen A., Heikkinen RK., Paloniemi R., Pöyry J., Similä J., Kuussaari M. 2013 - *Improving conservation planning for semi-natural grasslands: Integrating connectivity into agri-environment schemes.* Biological Conservation 160: 234-241.

Bagella S., Salis L., Marrosu GM., Rossetti I., Fanni S., Caria MC., Roggero PP. 2013 - *Effects of long-term management practices on grassland plant assemblages in Mediterranean cork oak silvo-pastoral systems.* Plant Ecology 214: 621-631.

Balandier P., Bergez JE., Etienne M. 2003 - *Use of the management-oriented silvopastoral model ALWAYS: Calibration and evaluation.* Agroforestry Systems 57: 159-171.

Bernes C., Jonsson BG., Junninen K., Lõhmus A., Macdonald E., Müller J., Sandström J. 2016 - *What are the impacts of manipulating grazing and browsing by ungulates on plants and invertebrates in temperate and boreal forests? A systematic review protocol.* Environmental Evidence, 5: 17.

Bianchetto E., Argenti G., Ferretti F. 2009 - *Proposta per una metodologia speditiva di stima del carico animale mantenibile da formazioni forestali nell'ambito della pianificazione territoriale.* Forest@ 6: 129-136.

Bianchetto E., Buscemi I., Corona P., Giardina G., La Mantia T., Pasta S. 2015 - *Fitting the stocking rate with pastoral resources to manage and preserve Mediterranean forestlands: a case study.* Sustainability 7: 7232-7244.

Bolzan A. 2009 - *Analisi dei parametri vegetazionali e dei caratteri funzionali di specie guida, come strumenti di studio di comunità prative.* PhD Dissertation, University of Bologna, 163 p.

Boschetti M., Bocchi S., Brivio PA. 2007 - *Assessment of pasture production in the Italian Alps using spectrometric and remote sensing information.* Agriculture, Ecosystems and Environment 118: 267-272.

Cavallero A., Aceto P., Gorlier A., Lombardi G., Lonati M., Martinasso B., Tagliatori C. 2007 - *I tipi pastorali delle Alpi piemontesi.* Alberto Perdisa Editore, Bologna, I, 467 p.

Cavallero A., Rivoira G., Talamucci P. 2002 - *Pascoli.* In: Coltivazioni erbacee. Foraggere e tappeti erbosi (Baldoni R, Giardini L eds). Pàtron Editore, Bologna, I: 239-294.

Conant RT., Cerri CE., Osborne BB., Paustian K. 2016 - *Grassland management impacts on soil carbon stocks: A new synthesis.* Global Environmental Change-Human and Policy Dimensions 23: 240-251.

Daget P., Poissonet J. 1969 - *Analyse phytologique des prairies. Applications agronomiques.* CNRS CEPE, Montpellier, France, doc. 48, 67 p.

Daget P., Poissonet J. 1971 - *Une méthode d'analyse phytologique des prairies. Critères d'application.* Annales Agronomiques 22: 5-41.

Daget P., Poissonet J. 1972 - *Un procède d'estimation de la valeur pastorale des pâturages.* Fourrages 49: 31-40.

Dossche R., Rogge E., van Eetvelde V. 2016 - *Detecting people's and landscape's identity in a changing mountain landscape. An example from the northern Apennines.* Landscape Research 41: 934-949.

Dubbs TM., Vanzant ES., Kitts SE., Bapst RF., Fieser BG., Howlett CM. 2003 - *Characterization of season and sampling method effects on measurement of forage quality in fescue-based pastures1.* Journal of Animal Science 81: 1308-1315.

Farruggia A., Pomiès D., Coppa M., Ferlay A., Verdier-Metz I., Le Morvan A., Bethier A., Pompanon F., Troquier O., Martin B. 2014 - *Animal performances, pasture biodiversity and dairy product quality: how it works in contrasted mountain grazing systems. Agriculture.* Ecosystems and Environment 185: 231-244.

Ferretti F., Dibari C., De Meo I., Cantiani P., Bianchi M. 2011 - *ProgettoBosco: A data-driven decision support system for forest planning* Mathematical and Computational Forestry and Natural-Resource Sciences, 3 (1):. 27-35.

Forbis TA., Provencher L., Turner L., Medlyn G., Thompson J., Jones G. 2007 - *A method for landscape-scale vegetation assessment: application to Great Basin.* Rangeland Ecosystems 60: 209-217.

Michaud A., Plantureux S., Amiaud B., Carrère P., Cruz P., Duru M., Dury B., Farruggia A., Fiorelli JL., Kerneis E., Baumont R. 2012 - *Identification of the environmental factors which drive the botanical and functional composition of permanent grasslands.* The Journal of Agricultural Science 150: 219-236.

Nösberger J., Staszewski Z. 2002 - *Overview of the changes in research on grasslands in Europe.* In: "Proc. of 19th General Meeting of European Grassland Federation" (Durand J., Emile J., Huyghe C., Lemaire G. eds). La Rochelle, F, 27-30 May 2002, P. Oudin Imprimerie: 17-27.

Orlandi S., Probo M., Sitzia M., Trentanovi G., Garbarino M., Lombardi G., Lonati M. 2016 - *Environmental and land use determinants of grassland patch diversity in the western and eastern Alps under agro-pastoral abandonment.* Biodiversity and Conservation 25: 275-293.

Paletto A., Ferretti F., Cantiani P., De Meo I. 2012 - *Multifunctional approach in forest landscape management planning: an application in Southern Italy.* Forest Systems 21: 68-80.

Pardini A., Longhi F., Lombardi P., Argenti G. 2001 - *Animal stocking rate in tropical and subtropical rangelands. II. Comparison of methods for the calculation.* Journal of Agriculture and Environment for International Development 94: 257-274.

Peeters A. 2008 - *Challenges for grasslands, grassland-based systems and their production potential in Europe.* In: Proceedings of the "22th General Meeting of European Grassland Federation" (Hopkins et al. eds). Uppsala, S, 9-12 June 2008. SLU Repro, Sweden: 9-24.

Pittarello M., Probo M., Lonati M., Lombardi G. 2016 - *Restoration of sub-alpine shrub-encroached grasslands through pastoral practices: effects on vegetation structure and botanical composition.* Applied Vegetation Science 19: 381-390.

Porqueddu C., Parente G., Elsasser M. 2003 - *Potential of grasslands.* In: "Proc. of 12th Symposium of European Grassland Federation" (Kirilov A, Todorov N, Katerov I eds). Pleven, B, 26-28 May 2003. Uniscorp: 11-20.

Primi R., Filibeck G., Amici A., Bückle C., Cancellieri L., Di Filippo A., Gentile C., Guglielmino A., Latini R., Mancini LD., Mensing S.A., Rossi CM., Rossini F., Scoppola A., Sulli C., Venanzi R., Ronchi B., Piovesan G. 2016 - *From Landsat to leafhoppers: A multidisciplinary approach for sustainable stocking assessment and ecological monitoring in mountain grasslands.* Agriculture, Ecosystems and Environment 234: 118-133.

R Core Team. 2016 - *R: A language and environment for statistical computing. R Foundation for Statistical Computing.* Vienna, Austria. URL https://www.R-project.org/.

Roggero PP., Bagella S., Farina R. 2002 - *Un archivio dati di Indici specifici per la valutazione integrata del valore pastorale.* Rivista di Agronomia 36: 149-156.

Rossetti I., Bagella S., Cappai C., Caria MC., Lai R., Roggero PP., Martins da Silva P., Sousa JP., Querner P., Seddaiu G. 2015 - *Isolated cork oak trees affect soil properties and biodiversity in a Mediterranean wooded grassland.* Agriculture, Ecosystems and Environment 202: 203-216.

RStudio Team. 2015 - *RStudio: Integrated Development for R.* RStudio, Inc., Boston, MA URL http://www.rstudio.com/.

Shapiro S., Wilk M. 1965 - *An analysis of variance test for normality (complete samples).* Biometrika 52: 591-611.

Siegel S. 1956 - *Nonparametric statistics for the behavioral sciences.* New York, McGraw-Hill, 312 p.

Strano E. 2010 - *L'approccio tipologico nello studio delle risorse pastorali.* Thesis dissertation, Università degli Studi di Firenze, I, 52 p.

Targetti S., Messeri A., Staglianò N., Argenti G. 2013 - *Leaf functional traits for the assessment of succession following management in semi-natural grasslands: a case study in the North Apennines, Italy.* Applied Vegetation Science 16: 325-332.

Targetti S., Staglianò N., Messeri A., Argenti G. 2010 - *A state-and-transition approach to alpine grasslands under abandonment.* iForest-Bigeosciences and Forestry 3: 44-51.

Török P., Hölzel N., van Diggelen R., Tischew S. 2016 - *Grazing in European open landscapes: How to reconcile sustainable land management and biodiversity conservation?* Agriculture, Ecosystems and Environment 234: 1-4.

Yu L., Zhou L., Liu W., Zhou HK. 2010 - *Using remote sensing and GIS technologies to estimate grass yield and livestock carrying capacity of alpine grasslands in Golog prefecture, China.* Pedosphere 20: 342-351.

Ziliotto U., Andrich O., Lasen C., Ramanzin M. 2004 - *Tratti essenziali della tipologia veneta dei pascoli di monte e dintorni.* Regione Veneto, Accademia Italiana di Scienze Forestali, Venezia, 262 p.

15

The 1936 Italian Kingdom Forest Map reviewed: a dataset for landscape and ecological research

Fabrizio Ferretti[1*], Chiara Sboarina[1], Clara Tattoni[2], Alfonso Vitti[2], Paolo Zatelli[2], Francesco Geri[2], Enrico Pompei[3], Marco Ciolli[2]

Abstract - The recovery, digitalization in vector format and a first analysis of the 1936 Italian Kingdom Forest Map (IKFM) (276 sheets 1:100,000) is described. The original document is available in paper format using a datum and a map projection no longer in use, therefore it is not suitable for analysis using current digital tools. Besides it is a unique document that describes the forest extent and species composition for the whole of Italy. This map provides historical, ecological and landscape information and fills a great temporal gap in those portions of Italy where landscape maps are available for some earlier periods. Its importance is extended to parts of Croatia, Slovenia and France. Therefore, the effort to make it available for analysis using current digital tools was done. The technical problems faced in the recovery and transformation of the cartography into a usable format are described and discussed. A first data overview and analysis based on a test study, and comparisons with current national forest inventory data aimed to highlight potential and limits of IKFM are presented. The results demonstrate the validity and usability of the digital version that is available on-line through a WebGIS at the address carta1936.dicam.unitn.it

Keywords - Forest Landscape; Historical Cartography; Milizia Forestale; Vector Conversion; WebGIS

Introduction

The analysis of landscape and ecological changes that occurred through the years using GIS (Geographic Information System) to compare historical geographical data is being increasingly used to reconstruct and understand the past ecological evolution at different scales, taking into account long periods (Bieling et al. 2013, Mikusinska et al. 2013). Similar examples that analyse only some decades can be widely found, in particular in Italy (Geri et al. 2010, Bracchetti et al. 2012, Palombo et al. 2013, Garbarino et al. 2013), other regions of Europe (Sitzia et al. 2010), South and North America (Renó et al. 2011, Zald 2008) and Asia (Zhou et al. 2011, Tang et al. 2012). This type of data and analysis has also been used to model possible future scenarios (Schirpke et al. 2012, Cimini et al. 2013), to design future interventions (Marignani et al. 2008), and to evaluate the hydrological impact of land use changes in water basins (Glavan et al. 2013, Zlinszky & Timár 2013). It has also been suggested that using a combined approach of GIS modelling and historical data analysis could enable the comparison of biomass production in different periods (Sacchelli et al. 2013).

The evidence demonstrates that past land use has a significant influence on the present biodiversity distribution and status (Falcucci et al. 2006, Pezzi et al. 2011, Aggemyr & Cousins 2012, Amici et al. 2012) and on traditional ecological knowledge (Ianni et al. 2015, Tattoni et al 2017), so historical information is crucial for the interpretation of many of the results of present ecological analysis, including those obtained with field sampling (Geri et al 2016).

The sources of information that can be used in this type of study are mostly satellite imagery, aerial photographs and historical maps, which are often the oldest documents. Some Landsat images can date back to the 1970s, while aerial photographs at the national scale are available for the majority of the countries from the 1940s or 1950s. Occasionally, single aerial photographs taken in earlier periods can be found in the archives for small areas, like the photographs taken during the World War II reconnaissance campaigns over Italy in 1944 by the British Royal Air Force (Merler et al. 2005). In this framework, historical maps are particularly useful, but they are difficult to find and often cover limited areas or have not been specifically created to show ecological features or vegetation features (Marchetti et al. 2009).

Thus, reliable historical maps that report detailed vegetation features over a large area in the past are a treasure, full of information that is particularly useful and important to understand the profound changes that occurred over the years in

[1] CREA Research Centre for Forestry and Wood, Arezzo (Italy)
[2] Dipartimento di Ingegneria Civile Ambientale e Meccanica, Università degli Studi di Trento, Trento (Italy)
[3] MIPAAF, Roma (Italy)
*fabrizio.ferretti@crea.gov.it

the landscape and ecology of an area or the whole of a country.

The first representations of forest coverage in some portions of Italian territory date back to the medieval period (Marchetti et al. 2009), but the first documents based on cartographic projections that can be compared with the current cartography date back to the beginning of the 19th century and were carried out in the framework of a local cadastre creation or updating, sometimes with specific attention given to the forest cadastre (Tattoni et al. 2010). Those maps represent limited areas with a single public or private owners, and the information contained is so heterogeneous that it is almost impossible to compare two different maps or check them against current data (Marchetti et al. 2009, Corona et al. 2001).

The Tourist Map produced in 1914 by the Touring Club Italiano (1:250,000) represents a generic "forest" category distribution in the whole Italian territory derived by enquiries to all CAI (Club Alpino Italiano - Italian Alpine Club) members, Alpine expert trekkers and specialists (Vota 1954, Meyer 2012). No distinctions were highlighted among species, forest structure or management schemes (for example, bushes and trees or coppice and high forest).

This work presents a detailed description of the vector transformation of the first homogenous map that represents Italian forests categorized under a recognizable scheme, the "Carta Forestale della Milizia Forestale del Regno d'Italia del 1936", i.e., the 1936 Italian Kingdom Forest Map (IKFM, hereafter), published by the Milizia Forestale del Regno d'Italia in 1936 (Brengola 1939, Marchetti et al. 2009). The digital map is a very unique and important document that describes the forest extent and main species composition in the whole Italian territory in 1936. The IKFM is the first document that records the distribution of forests at the national level and fills a temporal gap often present in studies in the Alps and in Italy, the critical period between the two world wars (Tappeiner et al. 2007, Ciolli et al. 2012). Moreover, its importance is not limited to Italy because it covers some areas that have now become parts of other countries (Croatia, Slovenia and France). Aim of the work is to describe the different technical problems tied to the quality of the original materials that were faced in the creation of the digital vector map. The technical and practical choices taken to solve these problems are also discussed, and the results are presented.

Despite our long and repeated searches in all the available Italian libraries, universities and research institute archives, as well as interviewing living people who later met some of the people involved in the original work in 1936, it was not possible to find reliable sources explaining in detail the taxonomy and the criteria used in the field work classification and in the final map. Therefore, the only information available are the IKFM legend labels and conventions, and this brings us to the second aim of this work, that is, to carry out a preliminary analysis of the data to determine whether the classification is reliable and evaluate what type of analysis can profitably be carried out.

Tests comparing area calculations on the digital version and old paper documents were carried out. The use of IKFM in the existing studies is highlighted. A comparison of the data in the digital IKFM map with two national forest inventories data, the IFNI 1985 Inventario Forestale Nazionale (IFNI, 1985) and the INFC 2005 Inventario Nazionale delle Foreste (INFC, 2007), is provided. Additionally, a first data overview based on landscape metrics and indices (Uuemaa et al. 2009) in 1936 at the national and regional levels is presented.

Materials and methods

The 1936 map material

The cartographic base for the IKFM map consists of 276 sheets (Fig. 1), with a 1:100'000 scale of representation. The original field sampling was initially reported on 1:25'000 official maps of the time, created by the Istituto Geografico Militare Italiano (IGMI), and then transferred to maps in the final 1:100'000 scale, also provided by IGMI.

The material used for conversion from raster to vector consisted of a set of 276 scans of the sheets of the maps, provided by the former Corpo Forestale

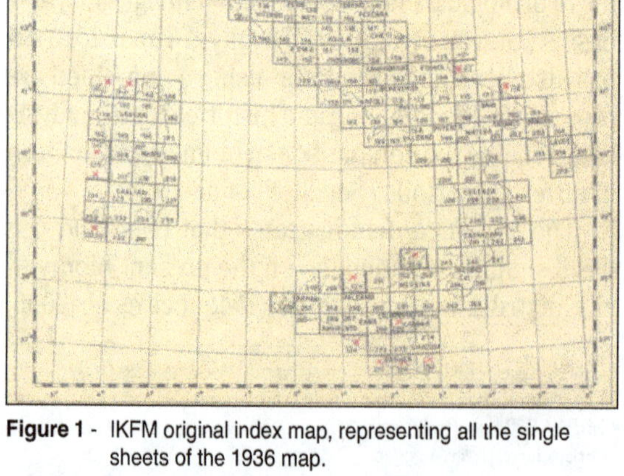

Figure 1 - IKFM original index map, representing all the single sheets of the 1936 map.

dello Stato in the frame of this work. The dimensions of each image was approximately 112 Mb, 6'650 * 5'900 pixels, with a resolution of 400 ppi (pixel per inch), RGB bands, projected in the Italian national Rome40 datum, Gauss-Boaga projection, east or west zone, depending on the location of the map sheet. The division into two different zones was maintained during the entire digitalization process, creating two different GRASS GIS Locations, Fuso Est (East Zone) and Fuso Ovest (West Zone).

Approximately 63'000 areas detected in the raster format were transformed into vector polygon format.

The IKFM forest classification system

The three levels system used in 1936 for the classification of the forest in the IKFM was based on species composition and silvicultural system, highly related to a definite distinction between wood for construction or furniture versus firewood (Marchetti et al. 2009). The 1st World War had ended a couple of decades before, and forests were an irreplaceable source of goods providing income and satisfying many of the primary needs of the Italian population (like firewood, bushmeat, mushrooms, herbs used for food and medicines), not much different from what is happening now in some developing countries, in which the growing population is increasingly impacting on forests and natural resources (Martin et al. 2012). The forest classification system chosen in the map description reflects the practical approach of the period but also contains a large amount of ecological information, like, as an example, the distribution of the main species and the general boundaries of the forests.

The forest cover at national level is divided into 3 macrocategories (conifers, broadleaves and "degraded forests") and 8 different categories, with some sub-categories related to forest system features for broadleaves (Fig. 2 translated in Tab. 1). The conifers sub-categories were classified depending on the main species that can be found in each polygon.

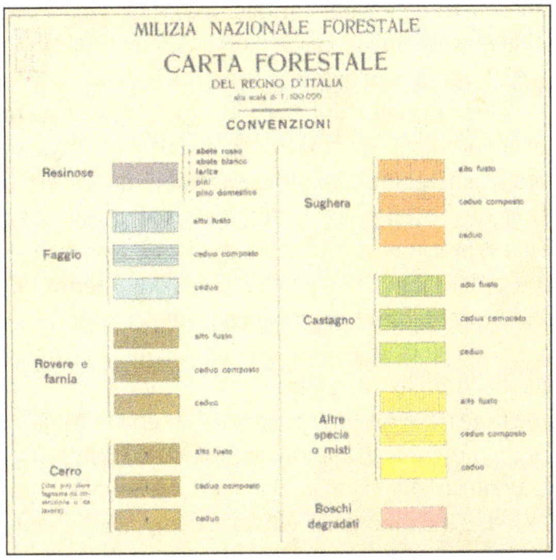

Figure 2 - IKFM original legend showing the different levels of the forest classification system represented with different colours, symbols and hatchings in 1936. The main categories are Resinose (Conifers), Faggio (Beech), Rovere e Farnia (Sessile oak and English oak), Cerro (Turkey oak), Sughera (Cork oak), Castagno (Chestnut), Altre specie o misti (other species or mixed wood), and Boschi degradati (Degraded forest). Detailed English translations of the names can be found in Tab S1.

Table 1 - Original label translated from italian (in brackets the original italian terms)

Macrocategory	Category	Subcategory
Conifer (RESINOSE)	Conifer (RESINOSE)	Norway spruce (ABETE ROSSO) Silver fir (ABETE BIANCO) Larch (LARICE) Stone pine (PINO DOMESTICO) other pines (PINI)
Broadleaves (LATIFOGLIE)	Beech (FAGGIO)	high forest (alto fusto) coppice with standard (ceduo composto) coppice (ceduo)
	Sessile oak and English oak (ROVERE E FARNIA)	high forest coppice with standard coppice
	Turkey oak (CERRO)	high forest coppice with standard coppice
	Cork oak (SUGHERA)	high forest coppice with standard coppice
	Chestnut (CASTAGNO)	high forest coppice with standard coppice
	Other species or mixed wood (ALTRE SPECIE O MISTI)	high forest coppice with standard coppice
Degraded forest (BOSCHI DEGRADATI)		

Map digitalisation

The digitization work was carried out using the open source software GRASS 6.4.3 (Neteler et al. 2012, Preatoni et al. 2012, Zambelli et al. 2013) and QGIS Quantum GIS version 1.4.0-Enceladus.

Because the amount of cartographic material was huge, an approach to limit the use of manual digitalization has been tested, with the use of the supervised image classification procedure available in GRASS to extract the forest data. To facilitate automatic detection, various pre-processing techniques were applied, combining colour separation with filtering and map algebra.

This approach did not give good results, with many problems due to the original data quality and many ambiguous situations in which the intervention of the operator is fundamental. Some map sheets presented defects due to scanning problems, and some imperfections regarding colour representation were due to the original print process.

The use of the automatic extraction technique has been applied to different sample map sheets, and the time necessary to complete the semi-automatic procedure, consisting of the automatic extraction and successive manual correction, has been compared to the time necessary to accomplish the same task using a completely manual procedure, and the latter has been shown to take up to 50% less time.

The semi-automatic procedure was slower because the time needed to correct errors due to the misinterpretation of colours, bad signs on the map or incorrect hatching was much higher than the time the operator needs to make a decision on what to do during the manual digitalization process.

The following rules were observed during the digitization:

a) always assign to each area the corresponding subcategory according to original legend (Fig. 2, Tab S1);
b) always digitize the black borders and not the coloured area extent, unless the black external border is absent. In this case, the coloured area extent is digitized;
c) always follow the black border, even if part of the border covers a lake or the sea surface;
d) in the case of an area without well-defined hatching, common sense has been used;
e) in the case of an area without colour, because it was not possible to assign the area to a specific subcategory, the value 25 in the field VALUE has been inserted, corresponding to the label NON CLASSIFICABILE (not classifiable);
f) regarding the representation of purple (conifers), two cases were possible: the first was that the area was without any species symbol, and in this case we inserted the value 0 in the field VALUE; the second possibility for purple (conifer) areas was that one or more forest subcategories symbols, each indicating the presence of a conifer species, were represented. In this second case, we inserted the value 0 in the field VALUE, but we also counted the number of symbols of different subcategories and inserted it into the field RESINOSE (conifers) according to the following codification:

AR = abete rosso (*Picea abies* Karst, European Spruce)

AB = abete bianco (*Abies Alba* Mill., European silver fir)

L = larice (*Larix decidua* Mill., European Larch)

P = pini (*Pinus sp.* Pines)

PD = pino domestico (*Pinus pinea* L., Stone Pine).

Because the only information available about the abundance of the conifer species is the number of species symbols reported in each conifer area of the map, we decided to preserve the historical information by decoding it into a text label to facilitate data extraction via Structured Query Language (SQL). For example, an area in which we can count symbols for 10 spruces, 4 firs, 3 larches and 1 pine will contain in the RESINOSE (conifer) field the label 10AR4AB3L1P.

Apart from digitalisation errors, the size of the smallest polygon is about 0.14 hectares. Areas of this size are clearly visible in the original map and are compatible with similar maps generated at the same scale in the same period in Europe (Kaim et al 2014).

The Italian boundary was derived from the map (shape file, Gauss-Boaga/Rome40) available on the Italian National Institute of Statistics (ISTAT) website (http://www.istat.it/en).

Projection and reference system transformation errors

The overall geometric uncertainties for the coordinates in the final vector map can be assessed as the sum of three terms: the geometric uncertainties of the original map, errors introduced by the datum transformation and inaccuracy during the vectorialization. The first term is easily estimated from the scale of the original map: a scale of 1:100'000 corresponds to an expected accuracy of 20 m.

Datum transformation errors are not available for the transformation used in this application, but their magnitude for the transformation between these two datums can be reliably assumed to be of the order of 10-20 cm (Radicioni & Stoppini 2009).

Finally, the digitalization error, excluding gross errors in the choice of the correct feature, is less than half the size of a pixel. The pixel sizes vary slightly approximately 6.5 m between raster sheets, so this error can be estimated to be less than 3.5 m. Consequently, the total error on the coordinates is less than 30 m, with the original map accuracy being the main factor.

The transformation from the original datum, using a Bessel ellipsoid oriented around Genoa and the Sanson-Flamsteed equal-area map projection, to the national one, using the Rome40 datum with the Gauss-Boaga conformal projection, has been carried out with a two-step procedure by the former Corpo Forestale dello Stato. This procedure involves the interpolation of the transformation parameters evaluated for the corners of each original map sheet. This transformation inevitably leads to a variation of the surfaces of the areas on the maps: a complete analysis of this issue is under way, but a first assessment has already been carried out. A set of 46 map sheets, selected partly at random and partly by choosing those sheets where the surface variation is expected to be more relevant (i.e., the sheets farthest from the central meridian and on the north and south extremities of the region covered by the map), has been tested by comparing the surface of the original sheets to the reprojected ones.

According to this test, the variation of the surface due to the reprojection procedure has a mean value of 0.2%, with a maximum of 0.6% (std dev 0.2), and it can be considered negligible with respect to the other uncertainties involved.

Errors in the original map

The original map contained some recurring errors, which are unavoidably also present in the raster format and thereby in the vector transformation.

The most frequent errors consist of the area colour filling expanding beyond the black border of the areas, the total or partial absence of the black border with the presence of the colour filling (fig. 3 and fig. 4), the presence of the black border with the total or partial absence of the colour filling and the spreading of the black borders beyond terrain limits into water bodies (fig. 5). Where possible, the black borders have been used to define areas to introduce the minimum modifications to the original paper map.

Another repeated error was the presence of incongruous hatchings that began horizontally and then became vertical or slanting or abruptly disappeared. Some of these errors are probably due to inattention, and they were not corrected on the original map in 1936 because once the drawing of the hatching was begun, it was not possible to erase it and change it without redoing the whole sheet. The classification of these areas was done

Figure 3 - Example of an area without colour filling (near the pink area) and example of an area without a black border (the yellow area at the bottom of the figure) in IKFM original material.

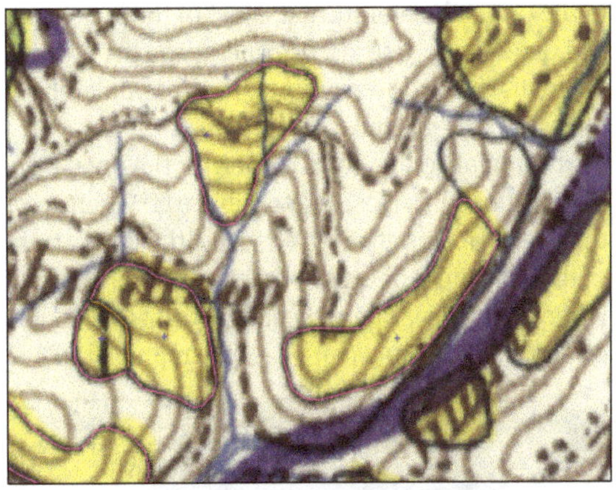

Figure 4 - An example of an area with the black border but without colour filling is visible on the right side of the figure: the forest in this area is not classifiable.

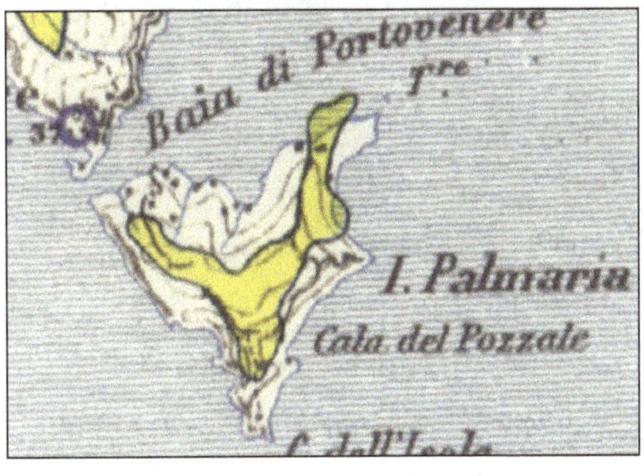

Figure 5 - Example of a forested area that partially overlapping the sea in IKFM original material.

trying to respect the content of the original map as much as possible, but in some cases, the choice of the forest subcategories involves a certain degree of subjectivity.

Errors during digitalization

During the digitalization process, there is the possibility that some polygons may be assigned by the operator to the wrong category. A sample of one hundred areas, chosen randomly, has been checked for misclassification and no errors were found, but it is impossible to guarantee that all 63'000 forested polygons have been correctly classified.

An error that must be taken into account is the error due to the thickness of the lines. This error is present in all paper maps and is due to the width of the lines representing the graphical entities. The smaller the scale, the larger this thickness error is. A one millimetre wide line, if represented in a 1:10'000 map, corresponds to a 10 m thick line on the ground, and it corresponds to a 100 m thick line on the ground if represented in a 1:100'000 map.

To determine how this error can affect the final results, some sampling has been carried out to measure the widths of the boundaries of different areas in the original raster format. Most of the lines are 3 or 4 pixels wide, that is, 20-25 metres on the ground, but some of them are wider, up to approximately 50 metres on the ground.

To minimize this source of error, the lines have been digitized as close as possible to the centre of their section. The effect of choosing the centre rather than the inner or outer border of a boundary line is shown in fig. 6, where the influence of this choice on the resulting area can be easily appreciated.

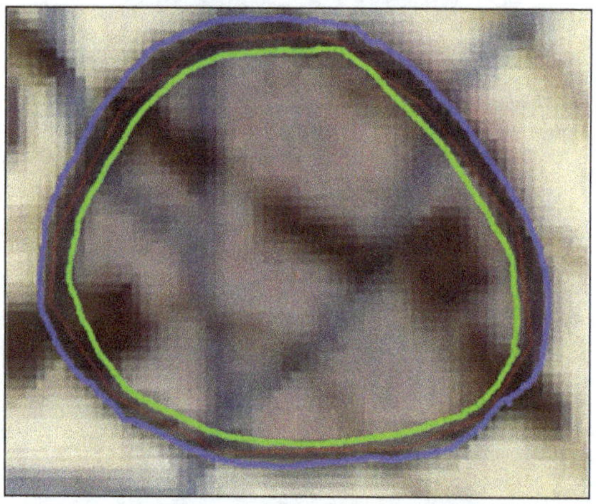

Figure 6 - Variation of an area depending on the location of the digitized boundary with respect to the original line drawing in IKFM, external (blue), intermediate (red) and internal (green) margins of the border are indicated.

The red line (intermediate part of the border) corresponds to an area of 7.13 hectares (assumed as 100%), while the green line (internal margin of the border) corresponds to an area of 6.43 hectares (90.2% of the intermediate) and the blue line (external margin of the border) corresponds to an area of 8.08 hectares (113.3% of the intermediate). It is evident how different the extent of a polygon can be when digitized on the internal or external margin of the border.

The final data were checked for topological errors using the topology checker tools available in QGIS.

Map availability

IKFM has been made available, in both raster and vector format, in a WebGIS, as this type of tool allows the efficient distribution of geographic data and even data processing (Federici et al. 2013).

The map is available through the site carta1936.dicam.unitn.it. The data have been released in the Creative Commons licence and will be free to be downloaded and used in the frame of European communication on open data (http://ec.europa.eu/digital-agenda/en/open-data-0), with the obligation of citing the present paper as a reference.

The system was made available on a virtual machine containing both software and data.

The virtual machine was built using VMware as virtualization software and Xubuntu 15.10 64 bit as a guest OS. The WebGIS is based on Geoserver, Tomcat, PHP, OpenLayers, Ext.js, PostgreSQL, PostGIS and GRASS.

The WebGIS shows the IKFM in vector format, using additional maps for the background to help the user identify the current location. The IKFM can be downloaded in the shape file format.

The digital map data set is based on the 276 maps of Carta Forestale del 1936, and additional maps were downloaded from official or free databases that can be found on the web. The Italian state, region and province boundaries were downloaded from the ISTAT site, and maps with the Slovenian and Croatian borders were downloaded from the GADM database of Global Administrative Areas (http://www.gadm.org/). Maps with the main roads, rivers and railways were taken from the www.OpenStreetMap.org site and were processed to reduce the number of geographical entities. Maps with the lakes (category 5.1.2 Inland waters, Water bodies) were derived from Corine Land Cover version 13 (02/2010) using the Italian boundaries.

The data are displayed using the spherical projection EPSG:3857.

The vector IKFM forest data are represented using different colours for each of the 25 subcategories in the map, trying to be coherent with the original category colours (Tab S1).

In the download section of the WebGIS, it is possible to select the maps to download. The system automatically selects the raster maps covering the area in the current visualization window and clips the vector map to the same window.

It is possible to download the whole vector map in a single file for Fuso Ovest (west zone EPSG:3003) and a single file for Fuso Est (east zone EPSG:3004), including Istria.

Tests on the data set

We carried out some basic analysis to verify the data reliability, to check some information about the forest subcategories and to test the possibility of using vector IKFM data.

To emphasize the contribution that this map can offer to ecological studies, it has been used in a study on ecosystems and forest coverage. Therefore we selected a case study in which the modifications to vegetation in the Trentino region, in the North-East of Italy, have been investigated over a period of time spanning almost two centuries (Tattoni et al. 2010, Ciolli et al. 2012). The digital IKFM map was compared to historical cartography, aerial photographs and Natura2000 ecosystem mapping.

Then, area values extracted from the IKFM digital version were compared to original IKFM tables created by hand by Milizia Forestale personnel in 1936, and the results are presented. The IKFM forest cover surfaces were compared to those of IFNI 1985 and INFC 2005 to evaluate whether the results are in line with what is expected and to highlight possible problems and issues.

Finally, some considerations of the forest categories and subcategories are presented, and some IKFM landscape pattern metrics were compared (using Patch Analyst 3.1) among Italian regions in 1936 to evaluate whether the results are reasonable in terms of subcategories quality, quantity and distribution. In our case (vector theme), edge calculations include all the edge on the landscape including boundary edge. In order to set the parameters of the Patch Analyst 3.1 procedure we considered ecologically relevant the differences between different forest subcategories and also relevant the differences between forest/no forest.

Results

The case study we selected (Tattoni et al. 2010, Tattoni et al. 2011, Ciolli et al. 2012) is particularly interesting because reliable historical maps describing the land use in 1800 are available for Trentino (Buffoni et al. 2003). Other maps are available from forest management plans from the 1950s or by photo-interpretation of different series of aerial photographs starting from 1954 and taken in 1973, 1985, 1994, 2000 and 2006.

The information on the total forest cover from the digital IKFM has been successfully compared to the forest situation in 1859 and in 1954, and the results were used to draw conclusions about ecosystem fragmentation, forest expansion and future scenarios (Tattoni et al. 2010, Tattoni et al. 2011, Ciolli et al. 2012, Tattoni et al 2017). To further test whether the IKFM data correctly represent the spatial pattern of the existing forest in 1936, they were also compared to photographic documentation available in touristic images or postcards from the same period (Tattoni et al. 2010).

Forest and land surface

A first apparently obvious use of IKFM data is to collect simple statistics regarding the forest cover total area and macrocategory distribution. The total forest area in the map is approximately 6'000,'000 ha (Tab S2), with 76% broadleaves, 18.4% conifers and 5.6% degraded forest. Overall, mixed forest (broadleaves or conifers) represents 42% of the total forest area. Nevertheless, a comparison of the total values calculated using the IKFM digital data and the summary paper tables that were originally provided by IKFM in 1936, containing the areas of the forest calculated for each administrative region (Province) and the macro regions or zones (north, centre, south and islands), reserves some surprises.

The area values in these tables are not coherent, as the same totals evaluated from different partial subtotals are different, and nor do they agree with the areas computed from the digital map (Tab. 2).

Table 2 - Comparison between areas from IKFM tables measured and calculated by hand in 1936 and areas calculated with GIS from the digital map (Italian boundaries at 1936). The four values should ideally coincide and should be similar to the digitized IKFM, but the sum of the forest calculated for each administrative region (province) and subtotals for macro regions (zones) are not coherent, probably due to the use of manual calculations in 1936. These tables were officially used to produce statistics in the past, and this highlights the importance of the digital version of this document.

Source	Type of sum	Surface ha	difference % compared to shp
IKFM report paper	tables of the provinces	5'519'900	14.0
IKFM report paper	Summary of forest-zones	6'047'460	5.7
IKFM report paper	Summary of forest-province	6'057'462	5.6
Digitized IKFM shp	fuso est with istria + fuso ovest	6'415'473	-

The comparison (Tab S3) of the IKFM - that represents a picture of the Italian forest distribution, cover and surface (Marchetti et al. 2009) in 1936 - with successive situations in 1985 (IFNI 1985) and 2005 (IFNC 2005) shows different situations and trends.

The same regions (Liguria and Trentino-Alto Adige), have the highest values of forest cover percentage in all three surveys. Conversely, the forest cover percentage for Sardinia increases considerably, exceeding the value for Tuscany in the 2005 survey.

Overall, the forest cover increased by 43.9% in 1985 and by 73.6% in 2005 with respect to 1936. The

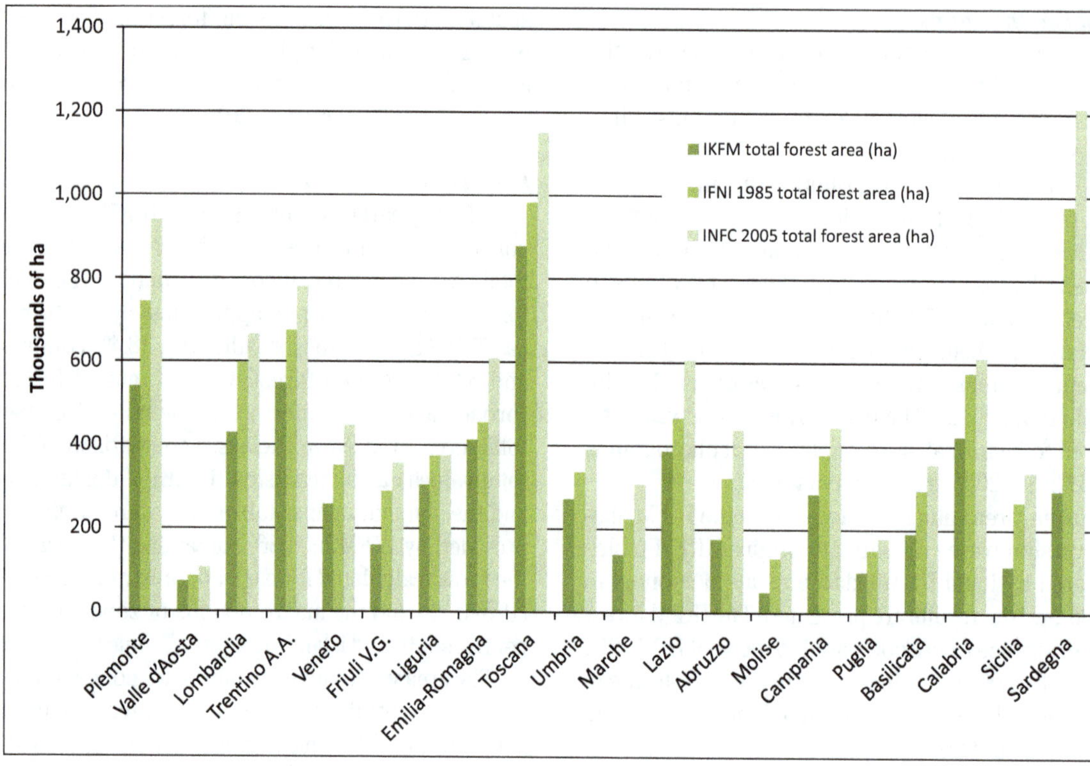

Figure 7 - Forest coverage (hectares) in Italian regions as originally recorded from the analogue IKFM 1936 and two later forest inventories, the IFNI 1985 Inventario Forestale Nazionale [Forest National Inventory] and the INFC 2005 Inventario Nazionale delle Foreste [National Inventory of Forests]. An increasing trend can be observed across regions.

forest cover percentage grew from 20% to 28.8% to 34.7% over the same time span.

While the forest cover increased in every region, the increment ranges from only 22.9% in Liguria to more than 300% in Sardinia (Fig. 7) .

In IKFM, the most common forest species (fig. 8) are 5: Broadleaves: other species or mixed wood, *Quercus robur/Quercus petrea, Castanea sativa, Fagus sylvatica*, and Mixed conifers (*Abies alba, Picea abies, Larix decidua*).

Landscape pattern metrics in 1936

The application of some landscape pattern metrics allows for a comparison of each forest subcategories in 1936 (Tab S4) and the total forest cover in the different regions of Italy in 1936 (Tab S5). For exemplifying, we can consider a landscape analysis of the IKFM by region highlights the results of the forest patch mean size, median size and total number for the Emilia-Romagna and Marche regions. Those two regions showed low values of the forest mean patch size and high total patch numbers; on the opposite the patches in Trentino-Alto Adige were larger and significantly less numerous.

Regarding the patterns of the different forest species, Turkey oak woods usually features large patches with mean patch sizes larger than those of patches of beech forest.

Large patches are also common for Cork oak, chestnut and mixed conifers. Mixed conifers account for a huge surface.

A very evident finding is that stone pine forests feature a high mean shape index.

Discussion

Forest and land surface

As reported before, the area values in Tab. 2 are not coherent.

These differences are possibly due to a series of factors:

a) approximation errors;
b) errors in the evaluation of the surfaces in 1936 (the way they were computed is unknown, but the staff definitely measured the polygons and performed the sum of all the areas by hand);
c) involuntary omissions of some of the polygons;
d) transcription errors on the final document.

It appears that the sum carried out in 1936 measuring by hand the individual polygons and summing them up was not reliable. The different sums (Province, Summary of forests-zones and Summary of forests-province) should reach the same measures of the total surface area, therefore some errors are present in the original paper documents (not in the paper map). It is clear that summing up more than 63'000 polygons by hand could introduce calculation errors. It is therefore highly probable that if some of the forest statistics at the national level calculated in the 1930s and later periods relied on these paper data, these statistics obtained imprecise results. The transformation of the IKFM into vector format makes available a more robust and consistent geo-referenced database that can be analysed using GIS.

The comparison (Tab S3) of the IKFM with successive situations in 1985 (IFNI 1985) and 2005 (IFNC 2005) is obviously affected by approximations and errors, which can be described but not evaluated; the main problems are due to the facts that

a) data have been collected using completely different instruments;
b) instructions given in 1936 to personnel for the survey are unknown;
c) although minimal, transformations from analogue (paper) to digital maps inevitably introduce approximations.

Even with the considerations above, common to other studies that deal with historical forest maps (Kaim et al. 2014), a comparison (Vizzarri et al. 2015) can lead to interesting considerations.

The comparison results (Tab S3) are compatible to the landscape dynamics described in the areas (Falcucci et al. 2006), the general forest cover trend is confirmed in every region (Fig. 7) and the results are in line with what is expected and described and basically confirm the general validity of the IKFM data.

Regarding the reliability of the IKFM forest classification system, some interesting considerations about the forest species (mixing categories and subcategories) in IKFM arise from a simple observation of the distribution of the different forest species (Fig. 8).

The "Broadleaves, other species and mixed wood" category is the most common, but it seems reasonable to suppose that this class includes every forest that the surveyors were not able to classify, so it must be treated with caution. The *Quercus robur/petraea* is also very common but, according to the knowledge of the 1936 forest species distribution, the suspicion is raised that mistakes in the attribution of patches to *Quercus cerris*, *Quercus pubescens* and *Quercus robur/petrea* have occurred. This fact must be taken into account if data are used to compare the present situation, and the different IKFM Quercus species should be grouped into a single one.

Landscape pattern metrics in 1936

The application of some landscape pattern metrics allows for a comparison of each forest subcategories in 1936 (Tab S4) and the total forest cover in the different regions of Italy in 1936 (Tab S5). This comparison may be useful both to understand whether the forest landscape described by the categories and subcategories is plausible based on the peculiarities of the different regions and to determine whether the forest subcategories indicated in the legend can be trusted. In Emilia-Romagna and Marche regions the low values of the forest mean patch size and high total patch numbers can be explained by the fact that these regions are mostly covered by plains and low hills, with widespread agricultural activity and a more fragmented forest environment. In mountainous areas with low population density and agricultural activity, the patches in 1936 were larger and significantly less numerous, describing a forest management approach with large areas (Trentino-Alto Adige).

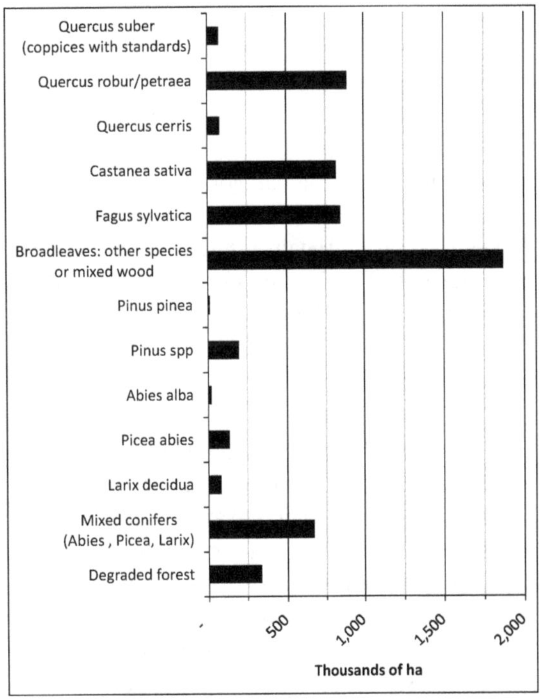

Figure 8 - Forest coverage in hectares according to the original 1936 categories of IKFM.

Regarding the patterns of the different forest species, Turkey oak woods usually features large patches, typical of diffuse forests cultivated wherever possible, with mean patch sizes larger than those of patches of beech forest.

It is hard to tell how much the large surface of Mixed conifers is affected by the very general classification.

The high mean shape index for stone pine forests is mostly due to the fact that their patches have stretched shapes along the coasts.

Considering the above examples, we can say that despite the 1:100'000 scale of the IKFM, the landscape patterns are compatible with the situation in 1936 and contain very descriptive information, at least for some classes, although the generic classification system suggests the need for caution if researchers want to compare these classification system across different periods.

The considered case study clearly demonstrate that with the digital format it is possible to create maps for each forest subcategories that are useful, in assessing forest subcategories distributions or to build custom maps describing the status in 1936 that can be checked against the current situation or historical data. Finally, the use of the digital IKFM in determining the total forest cover is reliable.

Conclusions

The availability of the IKFM makes it possible to fill a gap in the Italian forest mapping timeline, providing accurate, complete and interesting information for researchers who are interested in the study of landscape and ecological changes.

The transformation into vector polygons allows the manipulation of data through the connected database, exploiting its efficiency and flexibility in data querying and processing.

SQL allows a profitable storage of otherwise unavailable information, like the numbers of conifer symbols, although the reliability of these data must be carefully investigated.

The information in the IKFM map has already proven to be useful in reconstructing the total forest cover in past situations at local scale (Geri et al. 2010, Tattoni et al. 2010, Tattoni et al. 2011, Ciolli et al. 2012, Amici et al. 2013, Biasi et al. 2015, Salvati et al. 2015, Tattoni et al 2017) and at country scale (Cammaretta et al. 2017, in the press).

Because the Italian border changed before and after 1936, the data in this map have a super-national value. Some forest areas mapped in the IKFM are presently part of Slovenia, Croatia or France and can now be used to study forest changes in those countries. Moreover, the borders of Italy touch France, Switzerland, Austria, Slovenia and Croatia, and researchers in all these countries may be interested in the historical ecological features on the boundaries, as the borders can have a significant effect on their neighbourhood.

Given the currently available GIS tools and the features of the original material, the manual interpretation and digitization was more productive than a mixed approach based on automatic recognition and a posteriori manual correction. The description of how we dealt with the many technical and practical problems encountered in the creation of this map can be useful for the digitization of similar historical maps in other contexts.

The studies in the historical archive and the tests we carried out highlight validity of the IKFM for GIS processing, although the limits and uncertainties of the original forest classification and thereby of the digitized data must be taken into account by users.

We encourage the research community to perform further studies and comparisons using 1930s historical material at the national but also at the regional and local levels that will help further elucidate the significance of the single forest subcategories adopted for the classification in IKFM.

Acknowledgements

The authors want to thank Michele Bortolotti and the Direzione Sistemi Informativi, Servizi e Tecnologie Informatiche, Università degli studi di Trento for guesting the virtual machine of the WebGIS, Davide Giacomelli, Mario Sulli and Massimo Bianchi for their support. The authors want to thank the Accademia Italiana di Scienze Forestali that guested the Workshop: "La Carta Forestale del Regno d'Italia: 80 anni di evoluzione delle foreste italiane" in Firenze February the 4th 2016 to present the WebGIS see the site: sisef.org/2016/01/18/workshop-la-carta-forestale-del-regno-ditalia-80-anni-di-evoluzione-delle-foreste-italiane-firenze-04022016/ and the SISEF Working Group Gestione Forestale.

References

Aggemyr E., Cousins S.A.O. 2012 - *Landscape structure and land use history influence changes in island plant composition after 100years.* Journal of Biogeography 39 (9): 1645-1656.

Amici V., Rocchini D., Geri F., Bacaro G., Marcantonio M., Chiarucci A. 2012 - *Effects of an afforestation process on plant species richness: A retrogressive analysis.* Ecological Complexity 9: 55-62.

Amici V., Santi E., Filibeck G., Diekmann M., Geri F., Landi S., Scoppola A., Chiarucci A. 2013 - *Influence of secondary forest succession on plant diversity patterns in a Mediterranean landscape,* Journal of Biogeography 40, 2335–2347.

Biasi R., Colantoni A., Ferrara C., Ranalli F., Salvati L. 2015 - *In-between Sprawl and Fires: long-term Forest Expansion and Settlement Dynamics at the Wildland-Urban Interface in Rome, Italy.* International Journal of Sustainable Development and World Ecology 22(6): 467-475.

Bieling C., Plieninger T., Schaich H. 2013 - *Patterns and causes of land change: Empirical results and conceptual considerations derived from a case study in the Swabian Alb, Germany.* Land Use Policy 35: 192-203.

Bracchetti L., Carotenuto L., Catorci A. 2012 - *Land-cover changes in a remote area of central Apennines (Italy) and management directions.* Landscape and Urban Planning 104 (2): 157-170.

Brengola A. 1939 - *La carta Forestale d'Italia. [The Italian Forest Map]* Rivista Forestale Italiana. Milizia Nazionale Forestale. Comando Centrale, Roma, 1-10. ISSN: 1594-686X

Buffoni D., Leoni D., Bortolamedi R. 2003 - *L'eredità cartografica catastale degli Asburgo in forma digitale. [The Habsburgic cadastre cartographic heredity in digital format]* in Atti della 6a Conferenza Nazionale ASITA. Geomatica per l'Ambiente, il Territorio e il Patrimonio Culturale, Perugia, 5-8 November 2002 Vol 1 pp 559-564.

Camarretta N., Chiavetta U., Puletti N., Corona P. 2017 - *Quantitative changes of forest landscapes over the last century across Italy.* Plant Biosystem (in the press) https://doi.org/10.1080/11263504.2017.1407374.

Cimini D., Tomao A., Mattioli W., Barbati A., Corona P. 2013 - *Assessing impact of forest cover change dynamics on high nature value farmland in Mediterranean mountain landscape.* Annals of silvicultural research 37(1):29-37.

Ciolli M., Tattoni C., Ferretti F. 2012 - *Understanding forest changes to support planning: A fine-scale Markov chain approach.* in F Jordán, Se Jørgensen (edited by), Models of the ecological hierarchy from molecules to the ecosphere, Great Britain: Elsevier, 2012:341-359. (Developments in Environmental Modelling; 25)

Corona P., Chirici G., Marchetti M. 2002 - *Forest ecosystem inventory and monitoring as a framework for terrestrial natural renewable resource survey programmes.* Plant Biosystems 136: 69–82

Corona P., Marchetti M., Morgante L., Di Pietro R. 2001 - *Cartografia sperimentale e prodromi di una tipologia dei boschi dell'appennino abruzzese. [Experimental cartography and prelude to a forest type category of Abruzzo Apennine]* Annali Accademia Italiana di Scienze Forestali XLIX-L: 175–241

Falcucci A., Maiorano L., Boitani L. 2006 - *Changes in land-use/land-cover patterns in Italy and their implications for biodiversity conservation.* Landscape Ecology, 22(4): 617–631. doi:10.1007/s10980-006-9056-4

Federici B., Giacomelli D., Sguerso D., Vitti A., Zatelli P. 2013 - *A web processing service for GNSS realistic planning.* Applied Geomatics 5 1: 45-57.

Garbarino M., Lingua E., Weisberg P.J., Bottero A., Meloni F., Motta R. 2013 - *Land-use history and topographic gradients as driving factors of subalpine Larix decidua forests.* Landscape Ecology 28 (5): 805-817.

Geri F., Rocchini D., Chiarucci A. 2010 - *Landscape metrics and topographical determinants of large-scale forest dynamics in a Mediterranean landscape.* Landscape and Urban Planning 95 (1-2): 46-53.

Geri F., La Porta N., Zottele F., Ciolli M. 2016 - *Mapping historical data: Recovering a forgotten floristic and vegetation database for biodiversity monitoring* ISPRS International Journal of Geo-Information, 5 (7) 100. doi: 10.3390/ijgi5070100

Glavan M., Pintar M., Volk M. 2013 - *Land use change in a 200-year period and its effect on blue and green water flow in two slovenian mediterranean catchments-lessons for the future.* Hydrological Processes 27 (26): 3964-3980.

Ianni E., Geneletti D., Ciolli M. 2015 - *Revitalizing Traditional Ecological Knowledge: A Study in an Alpine Rural Community.* Environmental Management, 56 (1): 144-156

IFNI. 1985 - *Inventario forestale nazionale-Sintesi metodologica e risultati. [Forest National Inventory – Methods and results]* MAF – Direzione generale per l'economia montana e per le foreste - Corpo Forestale dello Stato, ISAFA, Trento. pp 462.

INFC. 2007 - *Le stime di superficie 2005 – Prima parte. [Forest cover estimate 2005]* G. Tabacchi, F. De Natale, L. Di Cosmo, A. Floris, C. Gagliano, P. Gasparini, L. Genchi, G. Scrinzi, V. Tosi. Inventario Nazionale delle Foreste e dei Serbatoi Forestali di Carbonio. MiPAF - Corpo Forestale dello Stato - Ispettorato Generale, CRA - ISAFA, Trento. [on line] Available: http://www.infc.it. Accessed May 2017 09

Kaim D., Kozak J., Ostafin K., Dobosz M., Ostapowicz K., Kolecka N., Gimmi U. 2014 - *Uncertainty in historical land-use reconstructions with topographic maps.* Quaestiones Geographicae, 33(3): 55–63 - doi: 10.2478/quageo-2014-0029

Marchetti M., Chiavetta U., Santopuoli G. 2009 - *La cartografia forestale su base tipologica della Regione Abruzzo: dai "prodromi" alla carta forestale dell'Italia centrale. [The forest type cartography of Abruzzo region: from the prelude to Central Italy forest map].* Available: http://www.academia.edu/795865/La_cartografia_forestale_su_base_tipologica_della_Regione_Abruzzo_dai_prodromi_alla_carta_forestale_dellItalia_centrale . Accessed May 2017 09

Marignani M., Rocchini D., Torri D., Chiarucci A., Maccherini S. 2008 - *Planning restoration in a cultural landscape in Italy using an object-based approach and historical analysis.* Landscape and Urban Planning, 84 (1): 28-37.

Martin A., Caro T., Mulder M.B. 2012 - *Bushmeat consumption in western Tanzania: A comparative analysis from the same ecosystem.* Tropical Conservation Science 5 (3): 352-364.

Merler S., Furlanello C., Jurman G. 2005 - *Machine learning on historical air photographs for mapping risk of unexploded bombs.* 13th International Conference on Image Analysis and Processing (ICIAP2005), Lecture Notes in Computer Science, Vol. 3617: 735-742.

Meyer E. 2012 - *Turismo e storia della produzione cartografica: la carta d'Italia del Touring Club italiano al 250.000. [Tourism and history of cartography: the 1:250.000 Italian map of Touring Club Italiano]* Fondazione Luigi Micheletti Centro di ricerca sull'età contemporanea [on line 4 February 2016]. Available: http://www.fondazionemicheletti.eu/contents/documentazione/archivio/Altronovecento/Arc.Altronovecento.06.04.pdf . Accessed May 2017 09

Mikusinska A., Zawadzka B., Samojlik T., Jedrzejewska B., Mikusiński G. 2013 - *Quantifying landscape change during the last two centuries in Bialowieża Primeval Forest.* Applied Vegetation Science, 16 (2): 217-226.

Neteler M., Bowman M.H., Landa M., Metz M. 2012 - *GRASS GIS: A multi-purpose open source GIS.* Environmental Modelling and Software 31: 124-130. doi: 10.1016/j.envsoft.2011. 11.014

Palombo C., Chirici G., Marchetti M., Tognetti R. 2013 - *Is land abandonment affecting forest dynamics at high elevation in Mediterranean mountains more than climate change?* Plant Biosystems, 147 (1): 1-11.

Pezzi G., Maresi G., Conedera M., Ferrari C. 2011 - *Woody species composition of chestnut stands in the Northern Apennines:* The result of 200 years of changes in land use. Landscape Ecology 26 (10): 1463-1476.

Preatoni D.G., Tattoni C., Bisi F., Masseroni E., D'Acunto D., Lunardi S., Grimod I., Martinoli A., Tosi G. 2012 - *Open source evaluation of kilometric indices of abundance.* Ecological Informatics 7(1):35-40. doi: 10.1016/j.ecoinf.2011.07.002

Quantum GIS Development Team 2010, last visited 20/3/2014. Quantum GIS Geographic Information System. Available at http://www. qgis.org

Radicioni F., Stoppini A. 2009 - Datum e coordinate nella geodesia. http://labtopo.ing.unipg.it/files_sito/aurelio/5%20

-%20DATUM%20E%20COORDINATE.pdf, visited 23/10/2017

Renó V.F., Novo E.M.L.M., Suemitsu C., Rennó C.D., Silva T.S.F. 2011 - *Assessment of deforestation in the Lower Amazon floodplain u sing h istorical L andsat M SS/TM i magery.* Remote Sensing of Environment 115 (12): 3446-3456.

Sacchelli S., Zambelli P., Zatelli P., Ciolli M. 2013 - *Biomasfor An open source holistic model for the assessment of sustainable forest bioenergy.* iForest 6:285-293. doi: 10.3832/ifor0897-006.

Salvati L., Biasi R., Carlucci M., Ferrara A. 2015 - *Forest transition and urban growth: exploring latent dynamics (1936-2006) in Rome, Italy, using a geographically weighted regression and implications for coastal forest conservation.* Rendiconti Accademia Nazionale dei Lincei 26(3): 577-585.

Schirpke U., Leitinger G., Tappeiner U., Tasser E. 2012 - SPA-LUCC: *Developing land-use/cover scenarios in mountain landscapes.* Ecological Informatics 12: 68-76.

Sitzia T., Semenzato P., Trentanovi G. 2010 - *Natural reforestation is changing spatial patterns of rural mountain and hill landscapes: A global overview.* Forest Ecology and Management, 259(8): 1354–1362. doi:10.1016/j.foreco.2010.01.048

Tang J., Bu K., Yang J., Zhang S., Chang L. 2012 - *Multitemporal analysis of forest fragmentation in the upstream region of the Nenjiang River Basin, Northeast China.* Ecological Indicators, 23: 597-607.

Tappeiner U., Tasser E., Leitinger G., Tappeiner G. 2007 - *Landnutzung in den Alpen: historische Entwicklung und zukunftige Szenarien,* vol 1. Alpine space man & environment: Die Alpen im Jahr 2020. Innsbruck University Press: 23–39

Tattoni C., Ianni E., Geneletti D., Zatelli P., Ciolli M. 2017 - *Landscape changes, traditional ecological knowledge and future scenarios in the Alps: A holistic ecological approach.* Science of the Total Environment, 579:27-36. DOI: 10.1016/j.scitotenv.2016.11.075

Tattoni C., Ciolli M., Ferretti F. 2011 - *The fate of priority areas for conservation in protected areas: a fine-scale Markov chain approach.* Environmental Management 47(2): 263-278. Springer Verlag. Available: http://www.springerlink.com/content/k532520665515um2/ . Accessed May 2017 09

Tattoni C., Ciolli M., Ferretti F., Cantiani M.G. 2010 - *Monitoring spatial and temporal pattern of Paneveggio forest (northern Italy) from 1859 to 2006.* iForest 3(2): 72-80. - - doi: 10.3832/ifor0530-003

Uuemaa E., Antrop M., Marja R., Roosaare J., Mander Ü. 2009 - *Landscape Metrics and Indices : An Overview of Their Use in Landscape.* Living Reviews in Landscape Research, 3, 1–28. - doi: 10.12942/lrlr-2009-1

Vizzarri M., Chiavetta U., Chirici G., Garfì V., Bastrup- Birk A., Marchetti M. 2015 - *Comparing multisource harmonized forest types mapping: a case study from central Italy.* iForest – Biogeosciences and Forestry, 8(1), 59– 66. doi:10.3832/ifor1133-007

Vota G. 1954 - *I sessant'anni del Touring Club Italiano 1894-1954. [Sixty years of Touring Club Italiano 1894-1954]* Milano, Tci, 1954: 109-117.

Zald H.S.J. 2008 - *Extent and spatial patterns of grass bald land cover change (1948–2000), Oregon Coast Range, USA.* Plant Ecology, 201(2): 517–529. - doi: 10.1007/s11258-008-9511-1

Zambelli P., Gebbert S., Ciolli M. 2013 - *Pygrass: An Object Oriented Python Application Programming Interface (API) for Geographic Resources Analysis Support System (GRASS) Geographic Information System (GIS).* ISPRS International Journal of Geo-Information, (2): 201-219. - doi: 10.3390/ijgi2010201

Zhou W., Huang G., Pickett S.T., Cadenasso M.L. 2011 - *90 Years of Forest Cover Change in an Urbanizing Watershed: Spatial and Temporal Dynamics.* Landscape Ecology 26(5): 645–659. - doi:10.1007/s10980-011-9589-z

Zlinszky A., Timár G. 2013 - *Historic maps as a data source for socio-hydrology: A case study of the Lake Balaton wetland system, Hungary.* Hydrology and Earth System Sciences 17(11): 4589-4606

Tab S 1: VALUES is the ID number of the subcategory, COLOUR is the colour that represents the IKFM category, LABEL is the original IKFM subcategories label, GRASSRGB is the RGB (Red, Green, Blue) value that encodes the colour in GRASS that matches the original colour of the paper map.

VALUE	COLOR	LABEL	GRASSRGB
0	purple – viola	Conifer - RESINOSE	155:80:125
1	purple – viola	Conifer Norway spruce - RESINOSE ABETE ROSSO	155:80:125
2	purple – viola	Conifer Silver fir - RESINOSE ABETE BIANCO	155:80:125
3	purple – viola	Conifer Larch - RESINOSE LARICE	155:80:125
4	purple - viola	Conifer other pines - RESINOSE PINI	155:80:125
5	purple - viola	Conifer Stone pine - RESINOSE PINO DOMESTICO	155:80:125
6	blue - azzurro	Beech high forest - FAGGIO alto fusto	123:169:156
7	blue - azzurro	Beech coppice with standard - FAGGIO ceduo composto	123:169:156
8	blue - azzurro	Beech coppice - FAGGIO ceduo	123:169:156
9	brown - marrone	Sessile oak and English oak high forest - ROVERE E FARNIA alto fusto	150:35:35
10	brown - marrone	Sessile oak and English oak coppice with standard - ROVERE E FARNIA ceduo composto	150:35:35
11	brown - marrone	Sessile oak and English oak coppice - ROVERE E FARNIA ceduo	150:35:35
12	brown - marrone	Turkey oak high forest - CERRO alto fusto	150:35:35
13	brown - marrone	Turkey oak coppice with standard - CERRO ceduo composto	150:35:35
14	brown - marrone	Turkey oak coppice - CERRO ceduo	150:35:35
15	orange - arancione	Cork oak high forest - SUGHERA alto fusto	255:120:15
16	orange - arancione	Cork oak coppice with standard - SUGHERA ceduo composto	255:120:15
17	orange - arancione	Cork oak coppice - SUGHERA ceduo	255:120:15
18	green - verde	Chestnut high forest - CASTAGNO alto fusto	125:160:80
19	green - verde	Chestnut coppice with standard - CASTAGNO ceduo composto	125:160:80
20	green - verde	Chestnut coppice - CASTAGNO ceduo	125:160:80
21	yellow - giallo	Broadleaves: other species or mixed wood high forest - ALTRE SPECIE O MISTI alto fusto	255:190:65
22	yellow - giallo	Broadleaves: other species or mixed wood coppice with standard - ALTRE SPECIE O MISTI ceduo composto	255:190:65
23	yellow - giallo	Broadleaves: other species or mixed wood coppice - ALTRE SPECIE O MISTI ceduo	255:190:65
24	pink - rosa	Degraded forest - BOSCHI DEGRADATI	243:87:125
25	none - nessuno	Not classifiable - NON CLASSIFICABILE	
99	none - nessuno	Island - ISOLA	

Tab S2: Areas for forest subcategories, categories and macrocategories from IKFM vector map (Italian boundaries at 2016), as reported in the original legend.

Forest subcategories	IKFM subcategories		Categories		Broadleaves-Conifers (macrocategosies)	
Name of IKFM subcategories	ha	%	ha	%	ha	%
01 - Quercus suber - high forests	64,080	1.06				
02 - Quercus suber - coppices	10,393	0.17				
03 - Quercus suber - coppices with standards	394	0.01	74,867	1.2		
04 - Quercus robur/quercus petraea - high forests	134,788	2.24				
05 - Quercus robur/quercus petraea - coppices	681,010	11.30				
06 - Quercus robur/quercus petraea - coppices with standards	76,324	1.27	892,122	14.8		
07 - Quercus cerris - high forests	57,990	0.96				
08 - Quercus cerris - coppices with standards	21,667	0.36	79,657	1.3		
09 - Castanea sativa - high forests	481,323	7.98				
10 - Castanea sativa - coppices	323,760	5.37				
11 - Castanea sativa - coppices with standards	13,762	0.23	818,845	13.6		
12 - Fagus sylvatica - high forests	260,912	4.33				
13 - Fagus sylvatica - coppices	550,782	9.14				
14 - Fagus sylvatica - coppices with standards	35,238	0.58	846,932	14.0		
15 - Broadleaves: other species or mixed wood - high forests	146,146	2.42				
16 - Broadleaves: other species or mixed wood - coppices	1,586,647	26.32				
17 - Broadleaves: other species or mixed wood - coppices with standards	137,009	2.27	1,869,802	31.0	4,582,225	76.0
18 - Pinus pinea 10,993	0.18					
19 - Pinus spp 197,222	3.27					
20 - Abies alba 19,345	0.32					
21 - Picea abies 133,820	2.22					
22 - Larix decidua 79,848	1.32					
23 - Mixed conifers (Abies alba, Picea abies, Larix decidua)	669,262	11.10			1,110,490	18.4
24 - Degraded forest	335,588	5.57				
Total	**6,028,303**	**100.00**				

Tab S3: Comparison between areas in IKFM (Italian boundaries at 2016) and in later forest inventories, the IFNI 1985 Inventario Forestale Nazionale [Forest National Inventory] and the INFC 2005 Inventario Nazionale delle Foreste [National Inventory of Forests]. The last three columns report the forest coverage percentage. Calculations were carried out using present Italian boundaries.

	IKFM forest area (ha)	IFNI 1985 forest area (ha)	IFNC 2005 forest area (ha)	Total area from INFC 2005 (ha)	Increment from IKFM to IFNI 1985 %	Increment from IKFM to IFNC 2005 %	IKFM with respect to total area %	IFNI 1985 with respect to total area %	INFC 2005 with respect to total area %
Piemonte	539.759	743.400	940.116	2.539.983	37,7	74,2	21,3	29,3	37,0
Valle d'Aosta	72.111	84.600	105.928	326.322	17,3	46,9	22,1	25,9	32,5
Lombardia	428.426	598.500	665.703	2.386.285	39,7	55,4	18,0	25,1	27,9
Trentino Alto Adige	548.395	675.000	779.705	1.360.687	23,1	42,2	40,3	49,6	57,3
Veneto	258.063	351.000	446.856	1.839.122	36,0	73,2	14,0	19,1	24,3
Friuli V.G.	159.975	289.800	357.224	785.648	81,2	123,3	20,4	36,9	45,5
Liguria	305.291	374.400	375.134	542.024	22,6	22,9	56,3	69,1	69,2
Emilia Romagna	412.981	454.500	608.818	2.212.309	10,1	47,4	18,7	20,5	27,5
Toscana	878.172	982.800	1.151.539	2.299.018	11,9	31,1	38,2	42,7	50,1
Umbria	272.092	336.600	390.255	845.604	23,7	43,4	32,2	39,8	46,2
Marche	137.839	224.100	308.076	969.406	62,6	123,5	14,2	23,1	31,8
Lazio	387.240	466.200	605.859	1.720.768	20,4	56,5	22,5	27,1	35,2
Abruzzo	176.386	322.200	438.590	1.079.512	82,7	148,7	16,3	29,8	40,6
Molise	49.265	129.600	148.641	443.765	163,1	201,7	11,1	29,2	33,5
Campania	285.251	378.900	445.274	1.359.025	32,8	56,1	21,0	27,9	32,8
Puglia	97.335	149.400	179.040	1.936.580	53,5	83,9	5,0	7,7	9,2
Basilicata	190.981	294.300	356.426	999.461	54,1	86,6	19,1	29,4	35,7
Calabria	423.310	576.900	612.931	1.508.055	36,3	44,8	28,1	38,3	40,6
Sicilia	112.223	266.400	338.171	2.570.282	137,4	201,3	4,4	10,4	13,2
Sardegna	293.205	976.500	1.213.250	2.408.989	233,0	313,8	12,2	40,5	50,4
Italy	**6.028.301**	**8.675.100**	**10.467.533**	**30.132.845**	**43,9**	**73,6**	**20,0**	**28,8**	**34,7**

Tab S4: Main landscape pattern metrics by forest subcategories in 1936 according to IKFM (Italian boundaries at 2016).

Subcategory	Class Area (ha)	No. of Patches	Mean Patch Size (ha)	Median Patch Size (ha)	Patch Size Coefficient of Variance	Patch Size Standard Deviation	Total Edge (m)	Edge Density (m ha^{-1})	Mean Patch Edge (m)	Mean Shape Index
01 - Quercus suber - high forests	64,080	237	270.4	63.4	297.8	805.3	1636629	0.271	6906	1.427
02 - Quercus suber - coppices	10,393	95	109.4	27	224.7	245.8	381584	0.063	4017	1.368
03 - Quercus suber - coppices with standards	394	11	35.8	7.5	142.8	51.2	25857	0.004	2351	1.276
04 - Quercus robur/quercus petraea - high forests	134,788	1311	102.8	24.3	484.1	497.8	5548444	0.92	4232	1.449
05 - Quercus robur/quercus petraea - coppices	681,010	14046	48.5	10.9	380.8	184.6	42629141	7.071	3035	1.425
06 - Quercus robur/quercus petraea - coppices with standards	76,324	898	85	24.6	357.3	303.7	4123837	0.684	4592	1.526
07 - Quercus cerris - high forests	57,990	156	371.7	162.5	167.1	621.2	1623030	0.269	10404	1.635
08 - Quercus cerris - coppices with standards	21,667	75	288.9	122.7	156.4	451.7	795405	0.132	10605	1.804
09 - Castanea sativa - high forests	481,323	4310	111.7	21.4	643.8	719	20655668	3.426	4792	1.492
10 - Castanea sativa - coppices	323,760	4824	67.1	10.9	410.9	275.8	16842695	2.794	3491	1.439
11 - Castanea sativa - coppices with standards	13,762	142	96.9	32.9	303.7	294.4	641227	0.106	4516	1.506
12 - Fagus sylvatica - high forests	260,912	740	352.6	47.1	402.8	1420.3	6568624	1.09	8877	1.558
13 - Fagus sylvatica - coppices	550,782	3629	151.8	19.6	394.6	599	21323939	3.537	5876	1.531
14 - Fagus sylvatica - coppices with standards	35,238	162	217.5	109.1	170.7	371.3	1316624	0.218	8127	1.679
15 - Broadleaves: other species or mixed wood - high forests	146,146	1536	95.1	25.7	275.9	262.5	6489060	1.076	4225	1.498
16 - Broadleaves: other species or mixed wood - coppices	1,586,647	18655	85.1	12.7	618.1	525.7	75039181	12.448	4022	1.513
17 - Broadleaves: other species or mixed wood - coppices with standards	137,009	1003	136.6	29.2	320.5	437.8	5359135	0.889	5343	1.562
18 - Pinus pinea	10,993	110	99.9	44.9	155.3	155.2	626028	0.104	5691	1.725
19 - Pinus spps	197,222	1764	111.8	31.5	577.8	646	8267262	1.371	4687	1.485
20 - Abies alba	19,345	201	96.2	31.6	222	213.7	912082	0.151	4538	1.448
21 - Picea abies	133,820	859	155.8	42.8	300.1	467.5	6167029	1.023	7179	1.682
22 - Larix decidua	79,848	933	85.6	30.4	222.3	190.3	4349101	0.721	4661	1.538
23 - Mixed conifers (Abies alba, Picea abies, Larix decidua)	669,262	2632	254.3	7.1	504.4	1282.7	23447560	3.89	8909	1.573
24 – Degraded forest	335,588	4452	75.4	23.5	271.3	204.5	17965281	2.98	4035	1.493
Italy	6,028,304									

Tab S5: Main landscape pattern metrics by region in 1936 according to IKFM (Italian boundaries at 2016).

Regione	Total area (ha)	IKFM with respect to total area %	Total Landscape Area (ha)	No. of Patches	Mean Patch Size (ha)	Median Patch Size (ha)	Patch Size Coefficient of Variance	Patch Size Standard Deviation	Total Edge (m)	Edge Density (m ha^{-1})	Mean Patch Edge (m)	Mean Shape Index
Piemonte	2,539,983	21.3	539,759	7178	75.2	18.5	417	313.6	30,005,754	55.6	4,180	1.56
Valle d'Aosta	326,322	22.1	72,111	382	188.8	38.2	264.8	499.9	3,266,408	45.3	8,551	1.82
Lombardia	2,386,285	18.0	428,426	4827	88.8	20.2	364.6	323.6	22,293,841	52.0	4,619	1.54
Trentino Alto Adige	1,360,687	40.3	548,395	1849	296.6	20.6	467.6	1386.8	20,145,831	36.7	10,896	1.78
Veneto	1,839,122	14.0	258,063	3514	73.4	8.4	705.7	518.3	13,008,625	50.4	3,702	1.46
Friuli V.G.	785,648	20.4	159,975	2953	54.2	12.1	401.8	217.7	9,093,626	56.8	3,079	1.36
Liguria	542,024	56.3	305,291	2806	108.8	24.4	309.6	336.8	13,926,153	45.6	4,963	1.55
Emilia Romagna	2,212,309	18.7	412,981	11991	34.4	6.6	480.1	165.3	30,677,459	74.3	2,558	1.43
Toscana	2,299,018	38.2	878,172	7417	118.4	18.7	630.6	746.7	35,874,774	40.9	4,837	1.50
Umbria	845,604	32.2	272,092	3677	74.0	15.4	498.2	368.7	14,493,127	53.3	3,942	1.50
Marche	969,406	14.2	137,839	4485	30.7	8.5	356.8	109.7	11,493,911	83.4	2,563	1.47
Lazio	1,720,768	22.5	387,240	3737	103.6	20.5	364.7	377.9	18,113,481	46.8	4,847	1.59
Abruzzo	1,079,512	16.3	176,386	1864	94.6	17.1	392.9	371.8	7,520,025	42.6	4,034	1.45
Molise	443,765	11.1	49,265	372	132.4	44	303.8	402.3	1,882,157	38.2	5,060	1.51
Campania	1,359,025	21.0	285,251	1220	233.8	53	386.4	903.4	8,321,351	29.2	6,821	1.52
Puglia	1,936,580	5.0	97,335	1648	59.1	12.6	434.5	256.6	5,018,226	51.6	3,045	1.40
Basilicata	999,461	19.1	190,981	1505	126.9	28.1	324.6	411.9	8,197,905	42.9	5,447	1.58
Calabria	1,508,055	28.1	423,310	1665	254.2	55.4	408.1	1037.5	11,723,711	27.7	7,041	1.52
Sicilia	2,570,282	4.4	112,223	651	172.4	42.8	292.6	504.4	3,300,331	29.4	5,070	1.34
Sardegna	2,408,989	12.2	293,205	906	323.6	112.9	351.2	1136.7	7,115,612	24.3	7,854	1.45
Italy	11,226,157	20.0	6,028,304	62781	96.0	15.7	552.7	530.7	272,734,424	45.2	4,344	1.49

16

Forest tree Marginal Populations in Europe Report on the state of knowledge on forest tree marginal and peripheral populations in Europe

Fulvio Ducci[1*], Kevin Donnelly[2]

Abstract - Marginal/peripheral (MaP) forest populations exist at the edge of species ranges, and may contain unique genetic diversity arising from adaptation to the adverse or unusual conditions within the regions they inhabit. The effects of climatic and other global changes are likely to be increasingly significant, and will affect forest tree species and their marginal populations. Studying adaptive processes in MaP populations is crucial, and of mutual interest for European and non-European countries, to understanding the evolution of species, and for the development of forest genetic resources (FGR) conservation, management strategies, and networks to cope with global change. These populations are threatened not only by global warming, but by various other threats arising from human activities. Because of their millennia-long history of adaptation to environmental changes, FGR growing in Europe may prove invaluable for adapting the forestry sector.

Keywords - Forest genetic resources; forest tree marginal populations; MaPs; marginality; COST Action FP 1202 MaP FGR.

Introduction

Marginal/peripheral (MaP) forest populations exist at the edge of species ranges, and may contain unique genetic diversity arising from adaptation to the adverse conditions in which they live. Because they typically attract less economic interest, peripheral forest tree populations are usually neither recognized nor managed as valuable forest resources (Lindner et al., 2010). Furthermore, conservationists and conservation planners seldon put a high value on peripheral populations, unless they belong to a species that is itself threatened (Leppig and White 2006, Thuiller 2007, Steen and Barrett 2015). Nevertheless, ecologists and geneticists tend to agree with Lesica and Allendorf (1995), that peripheral populations should be considered as valuable for conservation.

Understanding the adaptive processes in MaP populations is crucial and of mutual interest for European and non-European countries in order to safeguard the future of forest ecosystems. Developing conservation and management strategies for the forest genetic resources (FGR) contained within MaP populations is necessary to preserve and adapt European forests to the effects of global change. Because of their millennia-long history of continuous adaptation within rapidly changing environments', FGR extant in MaP populations throughout Europe may prove invaluable for adapting the forestry sector.

However, MaP populations are threatened not only by ongoing climate change but also by disturbances arising from a complex of human activities. Southern Europe represents an ideal region for study, as the effects of climate change on FGR will be greater and occur sooner than in the rest of Europe, and will affect the mesophilic species populations which provided the genetic diversity base of the post-Quaternary recolonisation of the European continent.

It is generally assumed that the present species distribution and genetic variation in Europe derived from Quaternary refuge areas distributed mainly in the southern peninsulas; however, it is likely that other marginal populations survived in central Europe and more northern areas (Tzedakis et al. 2013).

The COST Action FP1202 "*Strengthening conservation: a key issue for adaptation of marginal peripheral populations of forest trees to climate change in Europe (MaP-FGR)*" (http://www.cost.eu/COST_Actions/fps/FP1202) was intended to focus European attention on this particular topic and on

[1] Chair of FPS COST Action FP1202 "Strengthening conservation: a key issue for adaptation of marginal/peripheral populations of forest trees to climate change in Europe (MaP-FGR)" CREA Research centre for Forestry and Wood, Arezzo (Italy)
[2] Vice Chair Centre for Ecology and Hydrology, Edinburgh (United Kingdom)
*fulvio.ducci@crea.gov.it

the adaptive potential of genetic resources growing at the edges of the forest tree species distribution. This potential is today endangered by rapid changes to habitats, caused by factors of both natural and human origin, and needs to be better studied, protected, and valorised in view of the implementation of adaptive strategies.

The idea for the Action was conceived within the framework of the activities of FAO Silva Mediterranea, namely Working Group 4 "*Forest Genetic Resources in the Mediterranean Region*" (http://www.fao.org/forestry/silva-mediterranea/88915/en/), comprised of all Mediterranean countries, where genetic resources must cope with rapid climate change as well as significant changes in land-use patterns.

The Action was then proposed and developed with the contribution of nearly all European countries, and was also extended also to non-COST countries in the Maghreb and the Near East. Additional support was provided by international bodies including FAO, CIHEAM, Bioversity International/Euforgen, EFIMED, and IUFRO. In total, researchers and stockholders from 32 COST, and six COST Near Neighbor Countries attended: Algeria, Austria, Belgium, Bosnia and Herzegovina, Bulgaria, Croatia, Cyprus, Czech Republic, Denmark, Finland, France, FYR Macedonia, Germany, Greece, Hungary, Ireland, Israel, Italy, Latvia, Lebanon, Lithuania, Montenegro, Morocco, Netherlands, Norway, Poland, Portugal, Romania, Serbia, Slovakia, Slovenia, Spain, Switzerland, Tunisia, Turkey, the United Kingdom, and Ukraine.

The main objective of COST Action FP1202 was to generate relevant knowledge on the role and use of MaP populations in the adaptation of forests to global change via a multidisciplinary approach. In particular, the Action addressed the conservation and management of MaP FGR by: (i) compiling information on climate change impacts upon MaP populations, (ii) making information available for the preparation of national and pan-European forest plans and strategies for adaptation and mitigation, (iii) developing criteria for monitoring and conserving FGR, and (iv) sharing results with forest managers. The main results can be found in several documents produced and downloadable at the MaP-FGR web site: http://map-fgr.entecra.it/, or can be supplied by the national experts of the Action Management Committee.

Not all MaP populations are of equal value for adapting forests to climate change (Lesica and Allendorf 1995). Some, as a consequence of maladaptation, may have little value for conservation. Others, particularly those in Southern Europe, may contain unique combinations of genes of high value for mitigation and adaptation, resulting from long term evolutionary and adaptive processes (Hampe and Petit 2005, Fady et al. 2016).

Prioritizing MaP FGR on the basis of their genetic value was one of the objectives of the Action, as identifying high value populations is of direct relevance to conservation and management practice. direct relevance for conservation and management.

Specific objectives in the Action were also:

- to collect, collate, analyze and synthesize information from past and on-going projects related to genetic diversity and impacts of climate change;
- to record and list existing conservation efforts and status, in order to identify gaps and set conservation priorities;
- to analyze and raise awareness on the role of FGR in the adaptation of MaP populations;
- to perform meta-analysis of collected data to identify common trends on the dynamics of genetic diversity in terms of the response to global change;
- to provide recommendations and guidelines for forest managers and national policy makers to conserve and sustainably use MaP FGR for forest adaptation and mitigation of climate change;
- to highlight the potential of MaP populations for the adaptation to climate change in other networks dealing with FGR conservation.

Objectives were targeted by 4 working groups, namely:

WG1: Gathering of already available data and compilation of ecological, genetic, and global change information, and in particular climate effects, relating to FGR and the distribution of MaP populations (Initially led by Julian Gonzalo, and later by Eduardo Notivol and Ricardo Alia).

WG2: Evaluation and analysis of WG1 information, including the development of genetic diversity maps for each species/population, and of methodology for evaluation of FGR diversity; compilation of databases of relevant institutions, genetic material, trials and networks, and identification of gaps in current knowledge. This group also performed meta-analysis of the data to identify common and divergent trends of FGR responses to global change. (Led by Giovanni Giuseppe Vendramin and Paraskevi Alizoti).

WG3: Mainstreaming genetic diversity into sustainable forest management in the context of global change, considering both conservation and use of FGR. (Led by Bruno Fady and Philippos Aravanopoulos).

WG4: Coordination and organization of all networking, database management, training and communication activities: conferences, workshops, training schools, the web toolbox, open access databases, reports, publications, and Short Term Scientific Missions (STSMs). (Initially led by Christophe Besacier, and later by Nicolas Picard and Valentina Garavaglia)

Results of the Action

The main results of the action are reported here, in the papers of this issue, and are also largely available on the Action web site.

When the Action began, there were many ideas, but a consensus had not yet been reached regarding the defining characteristics of MaP populations. Initially, marginality was discussed in terms of geography, then developing to include genetic characteristics, and ecological traits. It has been necessary to define and reconcile the concepts of geographic, ecological, and genetic marginality for both molecular markers and phenotypic traits. Spatialization of data was another important topic addressed in the final stages.

It was also discovered that:

- it is not straightforward to use information available in literature and existing databases, even if already partially consolidated at EU level;
- models of marginality cannot be generalized by species, type of marginality, or by ecological/geographic level;
- when evaluating climatic scenario and other thematic maps, difficulties arise in reproduction of methodology;
- collating geographic, genetic, and phenotypic data is already difficult at the species level, and becomes more difficult when trying to focus attention toward marginal populations;
- it is ultimately easier to use model species to simplify the work of identifying valuable indexes for defining "marginality".

Information on the distribution and ecology of target species was gathered, and meta-databases constructed for both neutral and adaptive genetic variation. Fifteen countries contributed data for 14 species, resulting in a relatively large dataset at the wider European level. These species were: *Abies alba, Fagus sylvatica, Fraxinus excelsior, Fraxinus sp., Picea abies, Pinus halepensis, P. brutia, Pinus nigra, Pinus pinaster, Pinus pinea, Pinus sylvestris, Taxus baccata, Quercus robur* and *Quercus petraea.*

Maps were constructed based upon data collected from the above databases for species with an adequate number of populations across Europe. A niche model approach was used to define the ecological marginality of the populations, and geographical marginality indexes were then established for each of the different species/populations.

Analyses on the patterns of genetic diversity among environmentally or geographically marginal vs core populations were then performed together with the regression of intra-population phenotypic variation and phenotypic population plasticity against five marginality indexes for quantitative/adaptive traits. Based on data collected from the meta-database, several species were identified for use as models based upon to the availability of information for both adaptive genetic variation and molecular markers: *Fagus sylvatica*, *Abies* spp., *Picea abies*, *Pinus halepensis*, and *Pinus brutia.*

Collaborative documents, including an opinion paper (Fady et al. 2016), and a policy brief were produced to reach scientists, policy makers, and other stakeholders. Additional scientific and technical papers were produced by the more than 250 people who had exchanges with the Action initiatives [http://map-fgr.entecra.it/?page_id=74].

Other important outcomes included the compilation of an expert database of marginal and peripheral populations in Europe and in Near Neighbor Countries, along with maps and information describing the state of genetic resources of 24 Mediterranean forest tree species in collaboration with the FAO.

Last but not least, the dissemination of information and training of new scientists has been a concerted effort for all participants: five training schools were provided for PhD students and young researchers in the MaP FGR field, with more than 100 Trainees and 30 Trainers, and more than 60 STSMs funded by the Action.

In terms of gender balance, the majority of participants in the action were female (Figure 1), , from inner to outer ring years considered are 2013, 2015, and 2016.

From the beginning, networking activities have been made available at international level events, including FAO Silva Mediterranea meetings, EFIMED

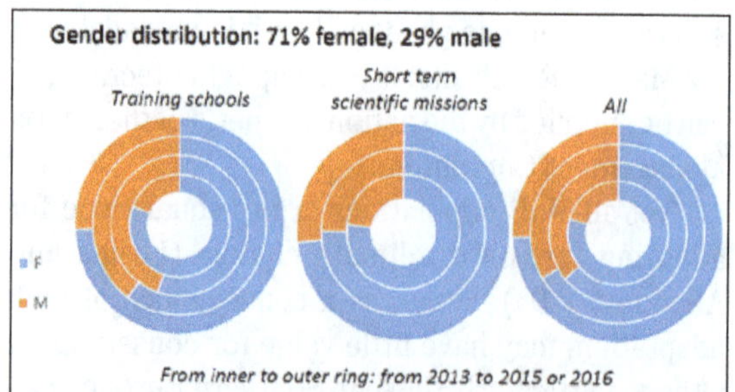

Figure 1 - Gender balance in the Action.

Mediterranean weeks, the Durban World Forestry Congress, FAO COFOs, COP 21, TreeBreedex, Trees-4Future, and IUFRO meetings.

This compilation of a number of national reports represents one of the Action's dissemination routes, and addresses the state of knowledge on conservation and management of MaP FGRs in Europe.

Marginality

Presently, less than 30% of the Earth's land mass is covered by forests. This proportion is steadily decreasing each year, either as a result of active deforestation or due to the effects of global changes.

Table 1 - Forested Areas of the world (https://en.wikipedia.org/wiki/List_of_countries_by_forest_area).

Region	Forested area (km²)	% of land area
World	39,000,000	26.19%
Polar regions (Russia, Scandinavia and North America)	13,800,000	35.38%
Latin America and the Caribbean	9,640,000	45.67%
East Asia and the Pacific	7,332,000	35.18%
Africa	6,500,000	21.80%
Canada and the United States	3,680,000	26.00%
European Union	1,600,000	35.00%
Australia	1,470,832	19.00%
India	778,424	23.68%
West Asia (Middle East and Arabia)	36,600	01.00%

More than 1.6 million km^2 of forested area lies within the European Union, accounting for around 35% of the European continent's total area, and 4% of the global forest area.

In Europe, due to intense and historical anthropogenic pressure, forested areas have been subject to strong fragmentation, and consequently to a progressive and continuous erosion in terms of surface area as well as species composition and genetic diversity. Nowadays, the complex phenomenon of so-called global change tends to exacerbate the effects of this fragmentation, particularly in terms of climatic effects (Brook et al. 2008). Climatic changes are occurring at such a rapid rate that many tree populations are showing evidence of severe stress.

It has been predicted that alleles from populations adapted to warmer climates, in particular those at the leading edge, may flow toward, and promote adaptation in populations growing in cooler conditions, as a consequence of asymmetric gene flow (Aitken et al. 2008, 2016). According to the same authors, populations growing at the edges where factors are changing dramatically will likely face genetic erosion or, under extreme conditions, extirpation. More widespread species, with larger population sizes, are likely to persist and adapt, but are also likely to suffer adaptation lag for several generations. On account of these adaptation lags, interspecific competition may weaken, allowing species to persist under sub-optimal conditions. Species with small populations, fragmented ranges, low fecundity, or suffering decline as a consequence of invasive insects or disease should be candidates for implementing methods of assisted migration.

A consequence of excessive fragmentation is an overall increase in size of forest boundaries which are exposed to limiting factors in the surrounding environment, leading to a progressive reduction in biodiversity, and within-species genetic erosion. However, having undergone relatively long periods of isolation in the past, these fragmented habitats may contain unique combinations of species as well as distinctive genetic and adaptive variability, that may provide the necessary adaptive capacity to safeguard against future uncertainty arising from global change and the ever-increasing anthropogenic pressures on ecosystems (Muillot et al. 2015).

In the Quaternary age, and in the previous eras, climatic changes have occurred at different intensities and over differing time periods (Pinna 1996). In some, climate changes occurred rapidly, over tens of years, and in others slowly, with average variations of only a few degrees over hundreds of years (Pinna 1977). Faster climatic changes have imposed stronger selective pressure on forest populations, most likely driving epigenetic factors and phenotypic plasticity expression (Schlichting 1986, Pigliucci 2001). Today's climate change rates, coupled with anthropogenic modifications, such as changes to land-use, have altered or may prevent gene flow, and can be expected to interfere with adaptation and migration, and likely with the survival of many populations or species (Davis and Shaw 2001).

Indeed, the responses of tree populations to global changes may be too slow to be viable. Moreover, as the population patches resulting from fragmentation are small, they are at an increased risk of a variety of genetic consequences, such as decreased dispersal, genetic erosion, and inbreeding, which are likely to impact upon long-term survival. Remnant populations often contain only a subset of the genetic diversity found in the previously continuous habitat. Adaptive processes that act upon underlying genetic diversity therefore have a smaller pool of fitness-maintaining alleles to act upon in the face of environmental change.

Marginalization *per se* should not be considered as pathological. It should be instead considered simply as a natural aspect of the evolutionary dynamics of a species in relation to the temporal and spatial variation of environmental factors in its natural range. Therefore, a marginal population is part of the species richness. It assumes a pathological meaning when range reduction is excessive, and the speed of change and fragmentation are too high to allow stable adaptive processes.

The COST Action FP1202 effectively summarised some definitions of marginality and marginal tree populations as follows:

- *geographical marginality* is generally used to refer to peripheral populations growing at the edge of their range. Species ranges can shift following environmental, typically climatic changes;
- *ecological marginality* can be observed anywhere within the range of a species, including the core;
- *altitudinally marginal* populations are usually islands of stunted, malformed trees, that with the warming climate could develop into reproductively capable, well-developed individuals, acting as 'nucleus' for further species expansion through seed dispersal;
- geographically marginal populations may themselves be either marginal or disjoint, the former connected to the core population by asymmetric gene flow (Kremer et al. 2012), while the latter are not;
- within the geographically marginal, leading and trailing edge populations can be identified. *Leading edge* populations represent expansion into new territories, their establishment largely determined by long-distance dispersal events. *Trailing edge* populations are those which persist in regions that are becoming unsuitable;
- where unfavourable driving forces occur, the result may be a displacement of the range at the trailing edge; alternatively, a fraction of the populations may persist in locally suitable environments (*stable edges*).

Together with other evolutionary genetic mechanisms, populations can maintain genetic diversity through migration. In continuous habitats, populations encounter few impediments to establishment on suitable sites. In fragmented habitats, suitable sites are separated by barriers and therefore gene flow is inhibited or completely disrupted. Under these circumstances, the capacity of species to integrate the unique genetic information contained within remnant populations is severely limited. Inbreeding increases and the fitness of the population is globally reduced (Hampe and Petit 2005, Fady et al. 2016). However, disrupted habitats and populations can sometimes re-establish connectivity and gene flow, as frequently occurred during the recurrent climate interglacial cycles. These periodically interconnected populations are collectively defined as meta-populations (Hanski 1991).

According to Kremer et al. (2012), theoretical models predict that asymmetric gene flow from larger core populations into those on the periphery may prevent local adaptation in marginal areas. Thus, a large gene flow can homogenize allele frequencies across space, reducing the adaptive divergence of meta-populations.

Following northward postglacial migration of tree species, the remnant forest populations remaining at more southern latitudes have experienced a long period in which to adapt to local conditions, and at the same time have had repeated exchanges with other nearby meta-populations. Because of this, they may have remained sufficiently isolated to fix particular traits while retaining enough variation to be adaptively dynamic. Kremer et al. (2012) provided evidence that, within the life-span of a generation, genes can move over spatial scales larger than the habitat shifts predicted under climate change, and that the northern border of a migrating species, the so-called leading populations, will progressively colonise new available and suitable environments.

Pollen viability and seed dispersal will strongly influence the demographic and genetic structure of leading populations. Austerlitz and Garnier-Géré (2003), modelling migration of outcrossing forest trees, demonstrated the major importance of relatively frequent medium-distance dispersal events over rarer events of much longer distance during the colonization process. In both marginal and leading edges, the survival of species will be dependent on individuals adapted to extreme conditions.

Marginal habitats are also of interest, not only because they may show marked genetic differentiation from populations at the core of a range, but also because of the potential for hybridization among closely related species. Hybridisation can be advantageous for adaptation (Thompson et al. 2009). Differentiated populations of closely related species can produce hybrid individuals of greater fitness than those of the parent species. This has happened in many areas along the Mediterranean peninsulas, which are of high genetic and species diversity. In these areas, the cooler, moister air at high elevation on mountain ranges favoured the survival of mesophilic species throughout the glacial and interglacial periods, and numerous genetic taxa emerged. Peninsulas became established as important refugia for species survival, and engines of speciation and hybridization (Hewitt 2011).

Preserving MaP populations is fundamental in order to protect sources of unique variation, likely to be of great importance for species adaptation to the rapid global changes occurring in the Anthropocenic era (Crutzen 2006, Steffen et al. 2007).

Common characteristics of marginal populations include their spatial isolation from the core (continuous) population and the reduced suitability of the environment as compared to that in the core population. Marginal populations are also usually smaller in size, and may be more prone to extirpation as a result.

According the centre-periphery hypothesis, peripheral populations exhibit lower genetic diversity and greater genetic differentiation as a consequence of small effective population size and greater isolation relative to geographically central populations as confirmed by Eckert et al. (2008), who reviewed this assertion.

On the ground, it can be complex to define marginal and peripheral (MaP) populations in regions where human influence has been significant for centuries or millennia. To do so, we need to understand the geography, ecology, and genetics of populations, as well as the abiotic and biotic drivers responsible for the direction and speed of range shifts.

Because they grow at the edges of distributions, MaP populations may represent 'natural laboratories' for understanding how demography and genetic processes, such as natural selection, shape local adaptation, and if either prevent or facilitate colonization of new habitats (Fady et al 2016).

Approaches to identify existing MaP populations include the use of geographic distances as thresholds when considering geographically marginal populations, while ecological distances (e.g. based on landscape heterogeneity) may highlight ecologically marginal populations.

MaP populations may be distinguished on the basis of genetic characteristics, if genetically distinct from the core population, and the use of genetic tools together with common garden networks is therefore another key approach in their identification. Integration with common garden networks is vital to understanding and improving information on the adaptive value of marginal populations that may be of value for conservation and use in practice (Màtyàs 2007). Moreover, modelling should also be based upon new experiments outside of the current natural ranges. Integration of these trials with genomic information will allow for better prediction of tree responses to climate change (Kremer et al 2012, Alberto et al. 2013).

The key role of marginal populations for maintenance of biodiversity throughout the Quaternary renders them extremely important for the conservation of intra- and interspecific biodiversity in the face of climate change (Hampe and Petit 2005, Fady et al. 2016). Species may also respond relatively rapidly to change via phenotypic plasticity or invocation of an epigenetic response (Bräutigam et al. 2013, Nicotra et al. 2010 (Eriksson, 1996; Eriksson et al., 1993). These mechanisms are particularly important for plants in marginal environmental conditions where phenotypic traits may be canalised to genetic fixation.

Given the potential vulnerability of MaP populations, the degree and extent of their adaptive plasticity the speed with which they may move or become extinct, and their disproportionate contribution to genetic diversity, it is vitally important that we are able to identify them and understand how they are likely to change in response to future climate change (Eckert et al. 2008).

The use of this information in conservation and sustainable management of forest genetic resources will ensure that MaP populations receive the attention they deserve in terms of genetic conservation under climate change.

Main results from the regional compilations

For practical reasons, the area covered by the COST Action has been subdivided into five regions, mainly on the basis of bioclimatic similarity.

There is of course considerable variation within each region, on account of wide latitudinal ranges, as well as differences in climate, soil, and habitat characteristics. This variability is also due to the presence of important mountain ranges which lie in differing orientations, and are responsible for climate variation at a local level. Of these, the Atlas, Cantabrian, Pyrenees, Pontus, Taurus, and Caucasus mountains are oriented in an east-west direction, while the Scandinavian Alps, Alps, Dinaric Alps, Balkans, and Carpathians are mainly oriented in a north-south direction. It is evident that this orographic structure plays an important role in the wind circulation and characteristics of climatic patterns with a strong influence on species distribution.

Altitude is an important factor in gene pool conservation for mesophilic species in southern areas, which host the mountainous habitats suitable for populations belonging to these species. Most of these populations can be classified as *marginal*, and originated from former glacial refugia. Mountains, together with the larger rivers, are in some cases barriers for species migration, but altitude can also compensate for the effect of latitude in habitat distribution, allowing, for example, for the survival of mesophilic species at lower latitudes.

In Europe, the effect of oceanic streams such as the Gulf are important, as they influence the occurrence of habitats in nearly all of western Europe. In the interior of Europe, the oceanic conditions give way to a more continental climate. These continental climates are dominated by northern and eastern winds, while in the south the climatic conditions are strongly influenced by subtropical wind circulation which makes frequent incursions into higher latitudes in the three main European peninsulas (Iberian, Italian, and Balkan).

The changes occurring in recent decades have resulted in a progressive shift to higher mean temperatures at higher latitudes and altitudes, and an increase in the intensity and frequency of extreme events which pose serious problems to species distributions.

Consequently, marginal populations can be seriously endangered by the shift of their ecological envelope, by isolation caused by a reduction in habitability in the surrounding area, as well as by human activities (Zhu et al. 2012).

Human impacts

History and socio-economic evolution have played a significant role in shaping the present-day characteristics of European forests.

According to Eurostat, the population of Europe is around 842 million, living within an area of 26.7 million km^2, in which forests accounted for 1.5 million km^2 in 1990, and 1.6 million km^2 today, with an increase evident over the last twenty-five years (World Bank 2015). Several non-European neighboring countries also participated in the COST Action: Turkey, Israel, Lebanon, Tunisia, Algeria, and Morocco. The total estimated forest area for these countries is around 1.4 million km^2.

According to the World Bank (2006), the rural population in Europe accounts for around 25% of the total, while in the Middle East and North Africa this figure is around 35.5%. Together with the climate, population density is a good indicator of the potential impacts upon habitats and forests. The mean population density in Europe is ~31.6 per km^2, ranging from 16 per km^2 in Norway to 393 per km^2 in the Netherlands. In the Middle East and North Africa, the total population is around 350 million, with an average population density of ~35 per km^2 (World Bank 2006). As countries in these regions typically contain large regions of desert, it should be understood that local population densities will be substantially higher.

The level of economic development should also be considered with regards to the evaluation of potential impacts. People living in temperate areas experience higher living standards and may view forested areas differently than those in the south, where forests are less productive and ecosystems more fragile, but very often overexploited to meet needs of the local economy. In these regions, it is worth remembering the significant impacts to forests resulting from high wood consumption of ancient civilizations. Humans began impacting upon forested regions at southern latitudes at least some 3,000 years prior to northern latitudes. The effects of these impacts are still evident in areas today regarded as biodiversity hot spots. In most countries, the majority of primary forests were cut back over a few centuries to clear land for agriculture, and were strongly modified by plantations and silviculture. In some, intensive forest tree farming accounts for a significant proportion of the forested area and of wood production. It should be acknowledged that protected areas in the EU today cover a relatively large area, almost 21% of the total (EEA 2012), including most of the marginal populations we consider here.

Geographic barriers to gene flow

Five important hotspots of genetic diversity were determined by the geographic structure of the Mediterranean area: Iberia, Italy, the Balkans, the Middle East, and North Africa. These regions served as refugia during glacial periods, providing sources of genetic material for re-colonization during interglacial periods. Orography, climate, and land-use are the main barriers shaping the present genetic structure of biodiversity and species genetics in Europe following the Quaternary period.

Geographic morphology is the main barrier, modifying micro-climatic conditions both at the local level and wider scale, and interacting with another main factor, the climate distribution. As described above, the orientation of mountain ranges has determined the routes of re-colonization available to tree species.

In western Europe, the Atlantic climate represents a barrier due to the effect of the warm Gulf stream. There, the temperate mild Atlantic climate has allowed the post-glacial colonization of mesophilic tree species.

During glaciations, the lower sea level in some areas allowed gene flow to occur between territories which are today disconnected. This is the case for islands such as the Aegean, Great Britain and Ireland, and other territories such as the Balkans and Apennines.

Altitude related barriers occur due to the isotherm distribution, and populations belonging to species requiring colder climates are often isolated on mountain tops. A similar scenario arises from the phenomenon of thermal inversion in some smaller regions, whereby the bottoms of valleys and canyons are unusually cool. In these areas, mesophilic species usually found at higher altitudes may find suitable habitats and remain isolated (Dinuluca et al. 2016, Saracino 2012).

Climatic differences among sites may result in asynchrony in flowering phenology of populations, which may then inhibit gene flow. Indeed, when population ranges are relatively continuous both in altitude and in latitude, flowering phenology is clinal and synchronized. When the range become fragmented a larger asynchrony can occur, and can be a serious barrier to gene flow among the isolated populations (Fuchs et al. 2003).

Together those typically considered, alternative barriers to gene flow should be evaluated. These can arise from silvicultural activities, the introduction of monoculture across large areas, rather than giving preference to a mixed forest system.

Introduced species can be sufficiently invasive as to impede natural regeneration of local autochthonous species. This is the case for Norway spruce in large areas of Europe, or for beech when it stifles other species in southern areas such as silver fir. The same occurs when silviculture is halted, due to an excess of protectionism, in particular, scenarios where clear cutting or coppicing is recurrently needed to clear the ground in order to favour regeneration of some species.

In Mediterranean regions, fire is a basic limiting factor which in some cases has influenced adaptation i.e. for natural regeneration dynamics. The prevention of fires might be a barrier for serotinous species.

In some cases, the forest nursery system itself can became a bottleneck or a barrier when genetic resources are improperly sampled or managed during the nursery breeding processes (Konnert et al. 2015).

Possible future modifications due to climatic change

According to the IPCC (2011), Europe has been warming faster than the global average over the last 30 years, local conditions varying between 2° C and 4° C on average, this variation being more pronounced in winter and at high latitudes (Holtmeier and Broll 2005). Precipitation is forecast to be reduced by 5-15% in northern areas, and by up to 20% in the south. An increase in extreme events, such as storms, long heat waves, and drought is also anticipated. Consequently, more frequent periods of flooding and drought are expected. Drier conditions may cause an increase in the length of the fire season, resulting in an greater risk of forest fires that have increased in frequency and intensity since the 1970s (Schär et al. 2004).

In southern areas of Europe, the African subtropical anticyclone has shifted in its periodical occurrence some several hundred kilometres northward and contributes to an increase in the intensity and frequency of heat waves, the length of dry periods, and other extreme events. As a consequence, climatic instability will increase in southern Europe, and forest fires in these regions may become fiercer.

Desertification risk in southern regions and possible climate tropicalization is predicted in several areas, from Africa to the Middle East, which would result in an acceleration of the water cycle and increased flood risk.

In contrast, in the Atlantic area and at northern latitudes, a longer growing season could have a positive effect. There, phenological records over the last 30 years show earlier springs and later autumns. In general, budburst has occurred consistently earlier for several northern and mesophilic species. This, however, increases the risk of late frost damage that, already an issue at southern latitudes, is expected to become more frequent in the North.

The Mediterranean will probably suffer major changes, affecting the most vulnerable habitats where MaP populations of the major European species grow. Moreover, water scarcity is likely to affect several Mediterranean countries in the coming years with serious social consequences and probable land degradation. Together with forest fires, the harmful effects from natural hazards such as pests and disease may also increase.

Marginal populations in Europe

The species economically most important and widespread in northern Europe, with continuous distributions are *Pinus sylvestris*, *Picea abies*, and *Betula pendula*, and to an extent *Larix decidua*. These are characterized by extensive gene flow and show clinal variation in adaptive traits. Scattered populations, normally associated with a more Atlantic climate, are also present at high latitude, including *Taxus baccata*, *Fagus sylvatica*, *Fraxinus excelsior*, *Tilia cordata*, *Corylus avellana*, *Ulmus glabra*, *Quercus petraea*, *Q. robur*, and *Ilex aquifolium*.

For most of these species, genetic information on marginality exists, showing a general trend of MaP populations to host less variation and exhibit greater differentiation than core populations.

Moving progressively towards the southern latitudes, in the Atlantic region, biodiversity increases dramatically, with a higher number of species for which marginal and peripheral populations have been identified. Among them are *Abies alba*, *Acer monspessulanum*, *Arbutus unedo*, *Castanea sativa*, *Fagus sylvatica*, *Fraxinus excelsior*, *Juniperus turbinate*, *Picea abies*, *Pinus halepensis*, *P. nigra*, *P. pinaster*, *P. pinea*, *P. sylvestris*, *Populus nigra*, *P. tremula*, *Quercus pyrenaica*, *Q. robur*, *Q. petraea*, *Q. pubescens*, *Q. cerris*, and *Q. suber*, *Salix phylicifolia*, *Sorbus torminalis*, and *Ulmus laevis*.

Proceeding towards more continental areas in central Europe, the heterogeneity of the landscape has caused fragmentation of some major species ranges. For example, because of the more xeric conditions, the Pannonian lowland forms a gap in the distribution of many tree species such as the mesophilic scattered hardwoods.

Norway spruce attains its southern limits in the Alps and Carpathians (www.euforgen.org/maps), and the species boundaries for more temperate species such as silver fir occur in northern Bavaria and Thuringia, the Vosges in France, and Carpathians in eastern Europe. This boundary also represents the northern limit for more thermophilic species such as *Sorbus domestica* and *Quercus pubescens*, while *Pinus cembra*, *Betula nana*, and *B. humilis* have marginal edges with respect to their core distribution area in the boreal region. These regions represent the northern range for other hardwoods such as *Castanea sativa*, *Pinus cembra*, *Populus nigra*, *Pyrus pyraster*, *Sorbus torminalis*, *Tilia platyphyllos*, *Ulmus laevis* and more.

The southern edges of *Larix decidua*, *P. abies*, and *Pinus cembra* distributions occur at the Alps, while several species such as *Abies alba*, *Fagus sylvatica* and the most noble hardwoods extend their range towards Apennines, with marginal scattered populations.

Finally, southern areas are rich in biodiversity, especially mountainous regions, due to the high levels of endemism represented by different taxa deriving from remnants of genera nowadays considered typically northern. These may be regarded as remnants of past cooler periods. *Pinus sylvestris* reaches its southern limits in Italy, Spain, and Greece. *Quercus petraea* and *Q. robur* have their fragmented edges in Italy and the Iberian Peninsula.

Many groups of closely related species exist in the southern Europe, which have adaptively diverged i.e. *P. halepensis* and *P. brutia*, several *Acer* and *Fraxinus* species, oaks, Mediterranean firs in northern Africa, junipers, and cedars in Middle East. Moreover, many northern species have trailing edges in the south, and a huge number of marginal habitats occur for more common trees such as *F. sylvatica*, *A. alba*, *Pinus mugo*, *C. sativa*, *P. avium*, *Sorbus* sp., *Populus* sp., *P. abies*, as well as for rarer trees such as *Q. ithaburensis*, *B. pendula*, that exist at their southern limits as marginal populations.

The southernmost area of our network, and the Middle East, represent the northernmost edges of tree species as *Cyperus papyrus* and *Acacia albida*.

P. heldreichii and *Alnus cordata* represent instances of Tertiary species occurring in fragmented populations which are both marginal and isolated at the same time. Unique isolated populations are remnants of ancient ancestral species differentiated, likely as a consequence of isolation, into endemic species such as *A. pinsapo*, *A. numidica*, *A. nebrodensis*, *A. equitrojani*, and *Cupressus dupreziana*.

Current status of protection

In Europe, the majority of the most important legislative initiatives are implemented by individual countries, and a number of protected areas networks already exist that can provide support for identifying, monitoring, and protecting genetic resources growing in MaP populations. In the MCPFE system, as well as according to FOREST EUROPE, UNECE, IUCN, and FAO principia, these protected areas are categorized into "no active intervention", "minimum intervention", and "conservation through active management".

Most European countries have recorded part of their genetic resources into the EUFGIS system database (http://portal.eufgis.org/), and a number of the gene conservation units (GCU) can also be classified as MaP populations. The EUFGIS database contains information on 3,128 units and 102 tree species in 34 countries. These units harbour a total of 3,773 tree populations. EUFGIS can be a useful tool to support countries in the management of MaP FGR as agreed in the context of Forest Europe, the pan-European forest policy process. Similarly, a total of 11,385 sites are recorded into databases of Natura 2000 in the EU territory (https://www.eea.europa.eu/data-and-maps/data/natura-8). However, while Natura 2000 records all habitats, EUFGIS records forest populations.

National Parks and other protected areas (see major information in regional and national reports) encompass the majority of populations defined as marginal. Protection measures and initiatives are typically more consistent among northern European countries than those in the south.

Despite the large human pressure, natural forests have locally been conserved by protected areas networks: these areas vary in size between 0.2% in

Turkey, to 38.2% in Croatia of the total forest area.

In nearly all countries, national registers of forest genetic resources record basic materials including MaP populations. These registers can be a helpful source for the proper use of reproductive materials from MaP populations. Moreover, these registers implement the European rules and the OECD scheme on Forest Reproductive Materials, so that MaP FGR can be of immediate utility when and where needed. Topak (1997) previously prepared a list of basic materials from all Mediterranean basin countries, that can be used as a reference when monitoring the present status of many seed stands, and consequently several MaP populations in the region.

Conclusion

Some authors have shown how fragmentation has been positive for the great endemic diversity that may result, however, fragmentation may also result in isolation, genetic drift, and genetic erosion. Areas of high population density also represent barriers to gene flow, reducing local levels of genetic diversity. Genetic analyses have shown that agricultural land offers ~1,000 times more resistance to gene flow than semi-natural habitats. This indicates that patterns of dispersal are constrained by the spatial configuration of remnant patches of suitable habitat (Ortego et al. 2015). In this context, shifts of isotherms play an important role as they may alter mountain temperate forest ecosystems at too high a rate for adaptation to keep pace, and exacerbate the effects of fragmentation on gene pools.

Forest genetic resources growing in marginal populations can be a valuable source of adapted reproductive materials. Indeed, the use of more adapted genetic materials should be mandatory when selecting species for planting, especially in relation to ecological sustainability and in view of the effects of climate change. In light of this, legislation should be more stringent with regard to certification and traceability, and ensure that the propagation path of material is traceable to the site of plantation.

Providing teaching on MaP forest genetic resources is of primary importance to preserve a genetic heritage which has survived for millennia of natural changes and human pressure. Conserving forest trees genetic resources requires maintaining *in situ* their adaptive potential, their forest community, and their entire forest ecosystem. *Ex situ* methods, such as assisted migration should be used only as the last extreme pragmatic solution.

Scarascia-Mugnozza and Matteucci (2005) listed the following as research priorities for the Mediterranean Forest Research Agenda:

- the impact of climate and land-use change on Mediterranean forest ecosystems' functioning: assessing and monitoring the main physical and biological processes including biodiversity and and the water cycle and the integration of the risk of forest fires in land-use and landscape planning and management;
- policy, economic, and institutional aspects of sustainable provision of forest goods and services;
- forest and woodlands in the context of integrated management of land resources: models and decision systems for optimizing multi-objective and multi-actor management problems;
- monitoring the genetic structure of populations with special attention to the genetic resources of marginal and peripheral populations.

With regards to forestry, the following technical actions would be required, reconciling information on legislation and methods among the Action participants, COST, Neighboring, and Near-Neighboring Countries:

- coordinating actions and strategies among different regions of the pan-European and neighboring regions;
- adaptation measures to be developed combining silviculture with ancillary and more naturalistic measures (e.g. respect for dead biomass and microhabitats in forests);
- pursuing the resilience of the forest ecosystem by strengthening the capacity for natural regeneration;
- implementing forest restoration and restoring the water cycle in degraded habitats and ecosystems;
- maintaining silvipastoral systems where biodiversity could be lost if abandoned;
- extending the due diligence principia and PEFC certification to genetic resources and forest reproductive materials.

These principia are also recognized and adopted as vital by the new ongoing FAO survey of Mediterranean forests.

Genetic variability has to be protected, and in turn, the adaptability of populations to environmental change. Research should be supplemented with data including species inventories, legislation, practical use, and coordination at national and pan-European levels, together with promotion of public awareness on the importance of conservation of the endangered species in forest ecosystems.

Research is called for with an increased emphasis on fine ecophysiology and adaptive traits, genetic variation, and genomics.

Regional reports

The regional reports were assembled as it follows for practical reasons. In some cases the name of a State can be repeated in two different regions. This is due to the occurrence of different climatic features shared by the same country.

Baltic region: Norway, Sweden, Finland, Denmark, Estonia, Latvia, Lithuania, and Poland;

Atlantic region: Belgium, Denmark France, Germany, Ireland, Portugal, Norway, Spain, The Netherlands, and The United Kingdom;

Central Europe region: Austria, Czech republic, Germany, Italy, Poland, Slovakia, Hungary, Ukraine, and Romania;

Southeastern Europe region: Bulgaria, Croatia, Greece; Montenegro, Romania, Serbia, Slovenia, and Ukraine;

Mediterranean region: Croatia, Cyprus, France, Greece, Israel, Italy, Lebanon, Morocco, Portugal, Spain, Tunisia, and Turkey.

References

Aitken SN., Yeaman S., Holliday JA., Wang T., Curtis-McLane S. 2008 - *Adaptation, migration or extirpation: climate change outcomes for tree populations.* Evolutionary Applications 1(1): 95-111. doi:10.1111/j.1752-4571.2007.00013.x.

Alberto FJ., Aitken SN., Alia R., Gonzalez-Martinez SC., Hänninen H., Kremer A., Lefèvre F., Lenormand Th., Yeaman S., Whetten R., Savolainen O. 2013 - *Potential for evolutionary responses to climate change – evidence from tree populations.* Global Change Biology 19: 1645–1661. doi:10.1111/gcb.12181

Austerlitz F., Garnier-Géré PH. 2003 - *Modeling the impact of colonisation on genetic diversity and differentiation of forest trees: interaction of life cycle, pollen flow and seed long-distance dispersal.* Heredity 90: 282–290. doi:10.1038/sj.hdy.6800243

Bräutigam K., Vining KJ., Lafon-Placette C., Fossdal CG., Mirouze M., Marcos JG., Fluch S., Fernández Fraga M., Guevara MA., Abarca D., Johnsen Ø., Maury S., Strauss SH., Campbell MM., Rohde A., Díaz-Sala C., Cervera MT. 2013 - *Epigenetic regulation of adaptive responses of forest tree species to the environment.* Ecology and Evolution 3(2): 399–415. doi:10.1002/ece3.461

Brook BW., Navjot S., Sodhi NS., Corey JA., Bradshaw CJA. 2008 - *Synergies among extinction drivers under global change.* Trends in Ecology and Evolution 23(8): 453–460. doi:10.1016/j.tree.2008.03.011

Crutzen PJ. 2006 - *The Anthropocene: the current human-dominated geological era. Paths of Discovery Pontifical Academy of Sciences,* Acta 18 Vatican City www.pas.va/content/dam/accademia/pdf/acta18/acta18-crutzen.pdf;

Davis MB., Shaw RG. 2001 - *Range Shifts and Adaptive Responses to Quaternary Climate Change.* Science 292 (5517): 673-679. doi:10.1126/science.292.5517.673

Dinulica F., Marcu V., Borz SA., Vasilescu MM., Petritan IC. 2016 - *Wind contribution to yearly silver fir (Abies alba Mill.) compression wood development in the Romanian Carpathians.* iForest 9: 927-936. doi:10.3832/ifor1742-008 [online 2016-10-02]

Eckert CG., Samis KE., Lougheed SC. 2008 - *Genetic variation across species' geographical ranges: the central–marginal hypothesis and beyond.* Molecular Ecology. 17: 1170–1188. doi:10.1111/j.1365-294X.2007.03659.x

EEA 2012 - *Protected areas in Europe - an overview.* EEA Report No 5/2012, Copenhagen. ISBN 978-92-9213-329-0, ISSN 1725-9177 doi:10.2800/55955

Eriksson G. 1996 - *Evolutionary genetics and conservation of forest tree genetic resources.* In: Turok J., Eriksson G., Kleinschmit J., Canger S. (eds) Noble Hardwood Network. Report of the first meeting, 24-21 March 1996, Escherode, Germany. IPGRI, Rome:159-167.

Eriksson G. 1998 - *Sampling for genetic resources populations in the absence of genetic knowledge.* In: Turok J., Collin E., Demesure B., Eriksson G., Kleinschmit J., Rusanen M., Stephan R. (eds). Noble Hardwood Network. Report of the second meeting, 22-25 March 1991, Lourizan, Spain. IPGRI, Rome: 61-75.

Eurostat 2007 - *Total Population as of 01.01.06.* http://ec.europa.eu/eurostat/portal/page?_pageid=1996,39140985&_dad=portal&_schema=PORTAL&screen=detailref&language=en&product=Yearlies_new_population&root=Yearlies_new_population/C/C1/C11/caa10000

Fady B., Aravanopoulos FA., Alizoti P., Mátyás C,von Wühlisch G., Westergren M., Belletti P., Cvjetkovic B., Ducci F., Huber G., Kelleher CT., Khaldi A., Dagher Kharrat M., Kraigher H., Kramer K., Mühlethaler U., Peric S., Perry A., Rousi M., Sbay H., Stojnic S., Tijardovic M., Tsvetkov I., Varela MC., Vendramin GG., Zlatanov T. 2016 - *Evolution-based approach needed for the conservation and silviculture of peripheral forest tree populations.* Forest Ecology and Management 375: 66–75.

Fuchs EJ., Lobo JA., Quesada M. 2003 - *Effects of forest fragmentation and flowering phenology on the reproductive success and mating patterns of the tropical dry forest tree Pachira quinata.* Conservation Biology 17: 149–157. doi:10.1046/j.1523-1739.2003.01140.x

Hampe A., Petit RJ. 2005 - *Conserving biodiversity under climate change: the rear edge matters.* 8 (5): 461-7. doi:10.1111/j.1461-0248.2005.00739.x.

Hanski I.1991 - *Single-species metapopulation dynamics: concepts, models and observations.* Biological Journal of the Linnean Society 42: 17–38. doi:10.1111/j.1095-8312.1991.tb00549.x

Hewitt G M, 2011 - *Mediterranean Peninsulas: The Evolution of Hotspots.* In: Zachos F., Habel J. (eds.). Biodiversity Hotspots. Springer, Berlin, Heidelberg: 123 – 147.

Holtmeier FK., Broll G. 2005 - *Sensitivity and response of northern hemisphere altitudinal and polar treelines to environmental change at landscape and local scales.* Global Ecology and Biogeography, Global Ecol. Biogeogr. 14: 395–41.

Konnert M., Fady B., Gömöry D., A'Hara S., Wolter F., Ducci F., Koskela J., Bozzano M., Maaten T., Kowalczyk J. 2015 - *European Forest Genetic Resources Programme (EUFORGEN). Use and transfer of forest reproductive material in Europe in the context of climate change.* European Forest Genetic Resources Programme (EUFORGEN), Bioversity International, Rome, Italy (E ISBN 978-92-5-107538-8): xvi and 75 p.

Kremer A., Ronce O., Robledo-Arnuncio JJ., Guillaume F., Bohrer G., Nathan R., Bridle JR., Gomulkiewicz R., Klein EK., Ritland K., Kuparinen A., Gerber S., Schueler S. 2012 - *Long-distance gene flow and adaptation of forest trees to rapid climate change.* Ecology Letters 15: 378–392. doi:10.1111/j.1461-0248.2012.01746.x

Leppig G., White JW. 2006 - *Conservation of peripheral plant populations.* California. Madroño 53(3): 264-274.

Lesica P., Allendorf FW. 1995 - *When Are Peripheral Populations Valuable for Conservation?.* Conservation Biology 9: 753–760. doi:10.1046/j.1523-739.1995.09040753.x

Lindner M., Maroschek M., Netherer S., Kremer A., Barbati A., Garcia-Gonzalo J., Seidl R., Delzon S., Corona P., Kolström M., Lexer MJ., Marchetti M. 2010 - *Climate change impacts, adaptive capacity, and vulnerability of European forest ecosystems.* Forest Ecology and Management. 259(4, 5): 698-709. doi:10.1016/j.foreco.2009.09.023

Màtyàs C. 2007 - *What do field trials tell about the future use of forest reproductive material?* In: Koskela J., Buck A., Teissier du Cros E. (eds) "Climate change and forest genetic diversity: implications for sustainable forest management in Europe". Biodiversity International, Rome, Italy: 53-69.

Mouillot D., Bellwood DR., Baraloto C., Chave J., Galzin R. 2013 - *Rare Species Support Vulnerable Functions in High-Diversity Ecosystems.* PLoS Biol 11(5): e1001569. doi:10.1371/journal.pbio.1001569

Myking T., Vakkari P., Skrøppa T. 2009 - *Genetic variation in northern marginal Taxus baccata L. populations. Implications for conservation.* Forestry 82: 5529-539. doi:10.1093/forestry/cpp022

Nicotra AB., Atkin OK., Bonser SP., Davidson AM., Finnegan EJ., Mathesius U., Poot P., Purugganan MD., Richards CL., Valladares F., van Kleunen M. 2010 - *Plant phenotypic plasticity in a changing climate.* Trends in Plant Science, 15, 12: 684-692. doi:10.1016/j.tplants.2010.09.008

Ortego J., Aguirre MP., Noguerales V., Cordero PJ. 2015 - *Consequences of extensive habitat fragmentation in landscape-level patterns of genetic diversity and structure in the Mediterranean esparto grasshopper.* Evol Appl. 8(6): 621–632. doi:10.1111/eva.12273

Pigliucci M. 2001 - *Phenotypic Plasticity: beyond nature and nurture.* The Johns Hopkins Univ. Press, Baltimore and London: 328 pp.

Pinna M. 1977 . *Climatologia. [Climatology].* Ed. UTET, Torino: 442 pp.

Pinna M. 1996 - *Le variazioni del clima : dall'ultima grande glaciazione alle prospettive per il 21° secolo. [The climate changes: since the last glaciation towards persepctives for the 21st century].* Ed. F. Angeli ©, Milano: 214 pp.

Saracino A. 2012 - *Le faggete campane e lucane: gli usi e le tipologie [Beech forests in Campania and Basilicata:uses and types].* In: Sul trattamento delle faggete in Italia: dal metodo scientifico all'empirismo dei nostri giorni *[The silviculture of Beech forests in Italy: from the scientific method to the empiricism of nowadays]* I Georgofili, Quaderni 2012-III, Firenze, Accademia dei Georgofili: 61-72. doi: 10.13140/RG.2.1.2221.7120

Scarascia-Mugnozza G., Matteucci G. 2012 - *Mediterranean forest research: challenges and opportunities in a changing environment.* http://www.enea.it/it/pubblicazioni/pdf-eai/gennaio-febbraio2012/ss-mediterraneanforests.pdf

Schär Ch., Vidale PL., Lüthi D., Frei Ch., Häberli C., Liniger MA., Appenzeller Ch. 2004 -*The role of increasing temperature variability in European summer heatwaves.* Nature 427: 332-336. [www.nature.com/nature]

Schlichting D. 1986 - *The evolution of phenotypic plasticity of Ecology in plants.* Annual Review and Systematic, 17: 667-693.

Steen DA., Barrett K. 2015 - *Should states in the USA value species at the edge of their geographic range?* The Journal of Wildlife Management 79: 872–876. doi:10.1002/jwmg.897

Steffen W., Crutzen PJ., McNeill JR. 2007 - *The Anthropocene: Are Humans Now Overwhelming the Great Forces of Nature.* AMBIO: A Journal of the Human Environment. 36 (8): 614- 621. doi:10.1579/0044-7447(2007)36[614:TAAHNO]2.0.CO;2

Thompson JD., Gaudeul M., Debussche M. 2009 - *Conservation Value of Sites of Hybridization in Peripheral Populations of Rare Plant Species.* Conservation Biology 24 (1): 236-245. doi: 10.1111/j.1523-1739.2009.01304.x

Topak M. 1997 - *Directory of seed sources of the Mediterranean Conifers.* FAO 1997: http://www.fao.org/docrep/006/AD112E/AD112E00.HTM

Tzedakis PC., Emerson BC., Hewitt GM. 2013 - *Cryptic or mystic? Glacial tree refugia in northern Europe.* Trends in Ecology & Evolution 28 (12): 696 -704. http://dx.doi.org/10.1016/j.tree.2013.09.001

World Bank (The), 2006. *Migration and Remittances Factbook-Middle East and North Africa.* http://web.worldbank.org/archive/website01058/WEB/IMAGES/MENA.PDF

World Bank (The), 2015. *The Forest area, European union.* https://data.worldbank.org/indicator/AG.LND.FRST.K2?locations=EU

Zhu K., Woodall CW., James S., Clark JS. 2012 - *Failure to migrate: lack of tree range expansion in response to climate change.* Global Change Biology 18: 1042–1052, doi: 10.1111/j.1365-2486.2011.02571.x

17

Early impact of alternative thinning approaches on structure diversity and complexity at stand level in two beech forests in Italy§

Claudia Becagli[1*], Nicola Puletti[1], Ugo Chiavetta[1], Paolo Cantiani[1], Luca Salvati[2], Gianfranco Fabbio[1]

Abstract - Stand structure, tree density as well as tree spatial pattern define natural dynamics and competition process. They are therefore parameters used to define any silvicultural management type. This work aims to report first data resulting from a silvicultural experiment in beech forests. The objective of the trial is testing the structure manipulation in terms of diversity and the reduction of inter-tree competition of different thinning approaches. Alternative thinning methods have been applied in two independent experimental sites located in the pre-Alps and southern Apennines, in Italy. Specific goals were to: (i) verify the impact early after thinning implementation on forest structure through a set of diversity and competition metrics resulting from a literature review; (ii) the sensitivity of tested indexes to detect effectively thinning manipulation. Main results show the low sensitivity of stand structure indexes and the ability of competition metrics to detect thinning outcome.

Keywords - stand structure, beech forest, thinning, structure diversity, competition, indexes

Introduction

Forest structure plays today a major role in the analysis and management of forest ecosystems (Roberts and Gilliam 1995, Oliver and Larson 1996, Latham 1998, Zenner and Hibbs 2000, Neumann and Starlinger 2001). Both structural diversity and tree competition (Brand and Magnussen 1988) are considered main factors to describe stand dynamics and define ecologically sound and economically sustainable management practices.

Stand structure is defined by tree species composition, tree age, size distribution and spatial patterns, canopy layering, within the forest stand (Husch et al. 2003). Stand structure influences individual tree growth and a number of biotic and abiotic processes (Garcia 2006, Eerikainen et al. 2007, Fox et al. 2007a,b). Silvicultural practices as thinnings and regeneration cuttings, heavily manipulate stand structure (Bailey and Tappeiner 1998) and this impacts wildlife populations (Harrington and Tappeiner 2007, Yamaura et al. 2008), stand dynamics (Saunders and Wagner 2008), tree regeneration dynamics (Getzin et al. 2008) and understory vegetation (Kembel and Dale 2006).

Horizontal/vertical tree density and spatial arrangement, tree species composition and intermingling as well as the range of tree dendrotypes, are usually related to biological diversity, ecological stability, competition processes and forest ecosystem functioning (Sallabanks et al. 2002, Kint et al. 2004).

The analysis of stand structure, tree density and tree species diversity are basic tools to explain both spatial heterogeneity and mechanisms of inter-tree competition (Oliver and Larson 1996, Pommerening 2006). Indeed, structural parameters at the macro, meso, and micro level are reliable indicators of diversity and ability of persistence (Margalef 1982) of forest ecosystems, and thus of the sustainability of the applied management's option (Haber 1982).

Based on these premises, a number of indexes has been developed over the last decades to quantify forest structure (Pommerening 2002). It has also been suggested that different sets of indicators may usefully be used to assess and simulate the effect of silvicultural treatments according to specific research references. Indexes-based literature, however, mainly deals with high forests (Ladermann and Stage 2001, Neumann and Starlinger 2001, Staudhammer and LeMay 2001, Corona et al. 2005, Motz et al. 2010). Few authors analysed the suitability of structural diversity indexes in coppice forests (Fabbio et al. 2006, Becagli et al. 2009). As for studies devoted to competition processes, these were carried out both in naturally originated forests and plantations (Hegyi 1974, Brand and Magnussen 1988, Biging and Dobbertin 1992, Piutti and Cescatti 1997, Cescatti and Piutti 1997, Schroder and von Ga-

§ Funded within the project LIFE+09 ENV/IT000078 ManForCBD

[1] Consiglio per la Ricerca e la sperimentazione in Agricoltura, Forestry Research Centre, Arezzo (CRA-SEL), Italy

[2] Consiglio per la Ricerca e la sperimentazione in Agricoltura, Centro di ricerca per lo studio delle Relazioni Pianta-Suolo, Roma (CRA-RPS), Italy

[*] corresponding author: claudia.becagli@entecra.it

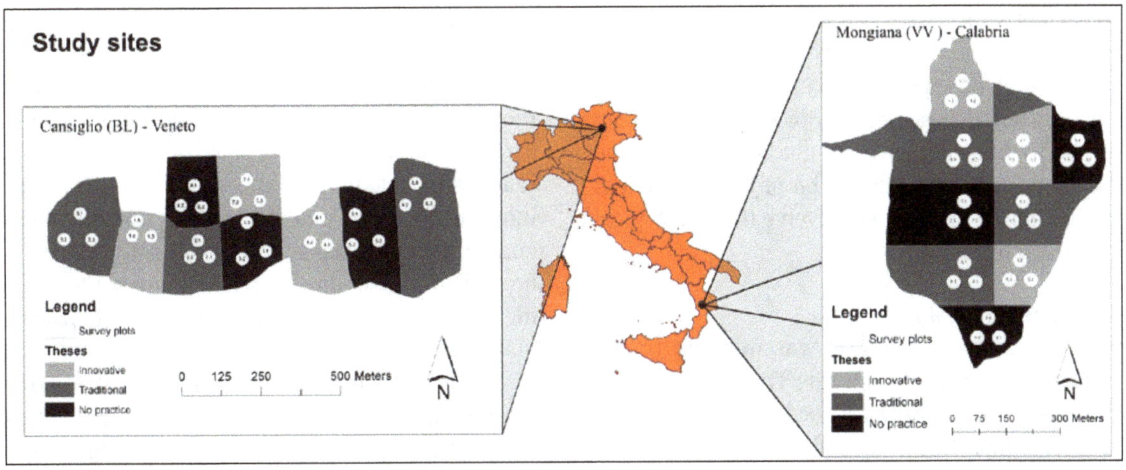

Figure 1- Location of the study areas.

dow 1999, Fichtner et al. 2012, Fichtner et al. 2013).

The knowledge of stand dynamics supports the focus on management goals by (i) predicting the future stand structure and relative pattern; (ii) reducing silvicultural costs, on average and (iii) increasing stand productivity and developing diagnostic criteria for describing, prescribing and marketing decisions for silvicultural operations (Oliver and Larson 1996).

Objective of the analysis is to evaluate the early change in stand structure following thinning practice by a set of indexes and the modification arising from the differently-based silvicultural interventions applied to managed beech forests. A more specific goal is to answer the following question: are all the calculated indexes capable to detect the effect of the intermediate removal and of criteria of intervention on stand structure and tree competition level immediately after their implementation at the tested sites?

Materials and methods

Study sites

The study was carried out in the pre-alpine region (Cansiglio-Veneto) and in southern Apennines (Mongiana-Calabria), two main beech vegetation areas (Fig. 1).

The study area located in the plateau of Cansiglio consists of an even-aged (120 to 145 yrs) compartment. The forest has age-long custom in forest management. Regular applied silviculture along the stand life-cycle consisted of low-mixed thinnings repeated every 20-25 years on the same management unit. Stand regeneration was successfully established following the group shelterwood system. Standing crop is quite uniform resulting from the long-lasting repetition of practices aimed at growing an one-storied structure. Natural variation in site-index makes the only evidence of a different arrangement in the crown layer.

The main management goal in these forests was to get uniform and well-shaped logs for the production of quality timber. Rotation time has been anyway shifting over the last decades and, as in other forest districts, from the customary age of 100-120 years to indefinite ages at now, this better matching the emerging recreational, scenic and mitigation functions besides wood-production goal. Site and climate characteristics are optimal for beech vegetation, these conditions well-supporting the extension of standing crop permanence time (Tab. 1).

The study area established in Mongiana consists of a forest aged about 70 years originated by a set of regeneration cuttings, the group shelterwood system, the clear-cut and clear-cut with reserves. The resulting standing crop is made by a main even-aged stand, where some trees of former cycle are present both grouped and along streams crossing the compartment. The patchy presence of silver fir mother trees and cohorts of regeneration provides spots of compositional and structural diversity. The forest position in the upper part of the mountain system facing the Tyrrhenian sea, the interception of fogs, wet winds and rain, makes physical environment

Table 1 - Site characteristics.

	Cansiglio Forest	Mongiana Forest
Area (ha)	30 - 35	30
Geographical coordinates (UTM-WGS84)	46° 03'N, 12°23' E	38° 30'N, 16°14'E
Altitudinal range (m a.s.l.)	1100 – 1200	1100
Landscape morphology	Gently sloping mountainsides and plains	Patchy succession of small hills (slope 40%), valleys and plateaux
Bedrock	Limestone, Marlstone (Cretaceous Med.- Sup.)	Granite ("Serra and Sila" formation)
Mean Temp °C	5.6	10.1
Max Temp °C (average warmest month)	14.8, August	18.4, July
Min Temp °C (average coldest month)	-4, January	2.2, February
Total Rainfall (mm)	2004	1808

consistent with beech vegetation in this southern Apennines area. Here too, the customary management system based on low to mixed thinnings usually does not open lasting canopy gaps. The random or patchy presence of trees much older than the dominant crop age, makes the physiognomy less regular and to some extent diverse than the typical even-aged beech forest.

Experimental design

An area of 30 hectares was devoted to the management trials at each site. Three theses were identified and randomly assigned in three replicates to each compartment (1 to 9). One thesis is made by the customary thinning practice (T), i.e. the low-mixed thinning as applied by local managers. The second silvicultural option is made by the innovative (for the forests concerned) criterion (I) consisting in a crown thinning at Cansiglio and on the selection of 45-50 trees per ha, with removal of direct competitors at Mongiana. Number of selected trees per unit area was determined on the foreseeable number of mother trees needed to carry out the future regeneration stage. The no-thinning thesis (control C) completes the experimental layout. A cluster of three circular sampling plots (radius 20 m) was established within each compartment according to a systematic design to survey mensurational parameters.

Structural and competition indexes

The impact of silvicultural practice on forest structure following each applied criterion has been assessed by a set of structural diversity and competition metrics resulting from a literature review. Both "spatial" and "non-spatial" indexes were accounted, the term "spatial" being referred to the information needed to get each index. Indexes requiring the calculation of individual tree position were considered as spatial indexes; all the others were considered as non-spatial indexes. To avoid the edge effect, trees at plot border were not considered as references. Selected indexes of stand structure and inter-individual competition are listed in Table 2. The Aggregation index by Clark and Evans (CE) characterizes the horizontal trees' distribution pattern at the stand level by the method of the nearest neighbouring trees. The theoretical range of CE lies between 0 (the maximum aggregation) and 2.1491 indicating a regular, hexagonal distribution pattern. Values about 1 comply with a random distribution. Diameter (TD) and tree height (TH) differentiation indexes provide information on tree size distribution of neighbouring trees. Values range between 0 and 1, with increasing values highlighting higher dbh or tree height variability.

Further non-spatial indexes of tree size diversity are the dbh ($STVI_{dbh}$) and Htot. ($STVIH_{tot}$) variance, They range between 0 and 1. The index is 1 when the parameter (dbh, Htot) distribution is uniform and is 0 when all values are equal or the distribution reaches its maximum bimodality. The advantage inherent to these indexes is their objectiveness compared to class-based indexes. Subjectivity is limited to maximum theoretical values, these being set at 150 cm (dbh) and 40 m (Htot) following Staudhammer and LeMay (2001).

As for tree competition, Hegyi (HG) index, a tree level, spatial indicator where neighbouring trees (competitors) are being selected according to a circle count criterion (Hegyi 1974) using Htot/3 of the reference tree as default distance, was considered. A modified version (HG_{MOD}) was tested too, to take into account larger neighbouring trees lying outside the default distance. Neighbouring trees are being selected here in accordance with an horizontal angle count (Pretzsch 2009).

A second competition index (Schroder and von Gadow 1999) is BAL_{MOD}, a distance-independent competition index derived from the basal area in larger trees (BAL) index combining individual tree's basal area percentile with a relative-spacing stand-density measure (RS_i). BAL_{MOD} assumes value equal to zero for the largest tree in a stand with a basal area percentile equal to 1, assuming that the largest tree provides the maximum growth rate. The minimum growth rate corresponds with the maximum index value. Within this range, the BAL_{MOD} index of trees with the same basal area percentile increases exponentially as RS_i decreases. Assuming constant stand density, the index provides the biologically reasonable effect that a high rank tree has a lower index value than a low-hierarchy tree.

Canopy structure was also assessed, since incoming light is a major driver of inter-tree competition (Kimmins 1987). Two canopy indexes, i.e. crowns coverage (CC) and the sum of crown projections (SCP) were selected to describe canopy structure. CC is a distance-dependent index representing the portion of horizontal space covered by at least one tree crown, theoretically ranging from 0 to 100. SCP, a spatially-independent index, is the sum of all crown areas per ground area unit and is a positive real number ⪆100.

Values before (BT) and after-thinning (AT) were compared also in terms of relative change (RC) of a specific index (i) as follows:

$$RC_i = \frac{(AT - BT)}{BT} \quad [1]$$

BAL_{MOD}, HG and HG_{MOD} indexes are being calculated iteratively for each tree. To assess the ranks of tree population promoted by thinning practice, the first quartile was adopted as the reference value. Hegyi distribution per competition classes was also considered in the analysis.

Non-parametric inference was used to compare the selected indexes among the different silvicultural treatments due to the restricted sample size and the deviation from normality of many variables. Mann-Whitney U statistics was carried out testing for differences between the applied treatments at different probability levels.

Table 2 - List of the stand structure diversity and inter-tree competition indexes.

Index	Equation	Range and reference values
Aggregation index [CE] (Clark and Evans 1954)	$CE = \frac{r_A}{r_E}$ Where: $r_A = \sum_{i=1}^{n} HDist_i$ $r_E = \frac{1}{2}\sqrt{\frac{A}{N}}$ $HDist_i$ = Euclidean distance between *i-th* tree and its nearest neighbour A = plot area N = total number of trees in the plot	Min value = 0 Max value = 2.1491 CE = 1 → completely spatially random distribution CE < 1 → spatially clustered distribution CE > 1 → spatially regular distribution
DBH-Differentiation [TD] (Pommerening 2002)	$TD = \sum_{i=1}^{N} \frac{1}{n} \sum_{j=1}^{n} \left(1 - \frac{\min(DBH_i, DBH_j)}{\max(DBH_i, DBH_j)}\right)$ Where: n = number of the first 4 nearest neighbours DBH = diameter at height = 1.3 m N=total number of trees in the plot	Min value = 0 Max value = 1 TD = 0 → all neighbors have equal DBH TD = 1 → all neighbours have different DBH
Height-Differentiation [TH] (Pommerening 2002)	$TH = \sum_{i=1}^{N} \frac{1}{n} \sum_{j=1}^{n} \left(1 - \frac{\min(H_i, H_j)}{\max(H_i, H_j)}\right)$ Where: n = number of the first 4 nearest neighbours H = total stem height N=total number of trees in the plot	Min value = 0 Max value = 1 TD = 0 → all neighbours have equal height TD = 1 → all neighbours have different height
Diameter diversity based on variance [$STVI_{dbh}$] and Height diversity based on variance [$STVI_{Htot}$] (Staudhammer and LeMay 2001)	$\begin{cases} 1-\left(\frac{S_U^2 - S_k^2}{S_U^2}\right)^{p1} & if\ S_k^2 \le S_U^2 \\ 1-\left(\frac{S_U^2 - S_k^2}{m \cdot S_{max}^2 - S_U^2}\right)^{p2} & if\ S_k^2 \ge S_U^2 \end{cases}$ Where: S_k is the empirical variance (dbh or tree height) S_U is the variance of the univariate uniform distribution (dbh or tree height) S_{max} is the maximum possible variance (when the distribution is maximally bimodal)	Min value = 0 Max value = 1 STVI= 0 → when all values (DBH or Htot) are equal or when DBH (or Htot) distribution is very close to be maximally bimodal STVI = 1 → highest diversity
BAL modified [BAL_{MOD}] (Schroder and von Gadow 1999)	$BALMOD_i = \frac{1 - p_i}{RS}$ Where: p_i is the basal area percentile of the reference tree RS is a relative spacing index (stand level)	Min value = 0 The higher the value the higher the competition the reference tree is subjected to
Hegyi [Hg] (Hegyi 1974)	$Hg_i = \sum_{i=1}^{n} \frac{DBH_j}{DBH_i \cdot HDist_{ij}}$ Where: DBH_i = diameter at breast height of the competitor trees DBH_j = diameter at breast height of the reference tree $HDist_{ij}$ = Euclidean distance between *i-th* tree and its nearest *j-th* neighbor n = number of competitors (all trees that fall within a distance of $1/3 \cdot H_i$)	Min value = 0 The higher the value the higher the competition the reference tree is subjected to
Hegyi modified [Hg mod] (Pretzsch 2009)	$Hg_i = \sum_{i=1}^{n} \frac{DBH_j}{DBH_i \cdot HDist_{ij}}$ Where: DBH_i = diameter at breast height of the competitor trees DBH_j = diameter at breast height of the reference tree $HDist_{ij}$ = Euclidean distance between *i-th* tree and its nearest *j-th* neighbour n = number of competitors (trees identified from an horizontal angle count sample)	Min value = 0 The higher the value the higher the competition the reference tree is subjected to

Results

Main mensurational parameters are reported in Table 3. Values at each site well-depict the differences between, mainly due to stand age and former applied silviculture. Mean values of structural diversity indexes before and after the occurrence of silvicultural practices are reported in Table 4. Thinning operations slightly changed the former values at both sites. CE index moved from random towards regular trees' distribution for both the applied criteria at Cansiglio, whilst it remained quite steady (i.e. random) at Mongiana. As for diameter differentiation (TD), dbh variability was low at Cansiglio and increased slightly at Mongiana (both criteria). As for tree height differentiation (TH), height variability was reduced by customary practice at Cansiglio only. Dbh variance ($STVI_{dbh}$) was kept close to zero by the occurrence of practices at Cansiglio; the same index did not show any significant change following thinnings at Mongiana. Tree height variance ($STVIH_{tot}$) was also very low and kept unchanged (innovative practice) or dropped (customary practice) at Cansiglio, whilst the opposite took place at Mongiana.

Values of relative change (Fig. 2) referred to the set of structural indexes do not exceed 15%, this showing the slight structural differences recorded across the theses.

As for tree competition, the first quartile (Tab. 5) and relative change (Fig.2) were examined. All indexes showed higher values at Mongiana than at Cansiglio before and after the implementation of thinning practices, independently of the applied criterion. All of them proved to be sensitive to the manipulation of stand structure, each one showing a reduction in accordance with the own algorithm. The innovative thesis promoted a higher cut of tree competition than the customary practice at both sites.

The Hegyi distribution pattern per competition classes was also recorded (Fig. 3). The graphical arrangement is similar as for Cansiglio (both theses) and Mongiana (innovative trial), i.e. the increase in low competition classes, the heavy reduction in medium-low competition classes and the zero setting in medium-high competition classes. On the contrary, a much more smoothed pattern across all competition classes was produced by customary practice at Mongiana.

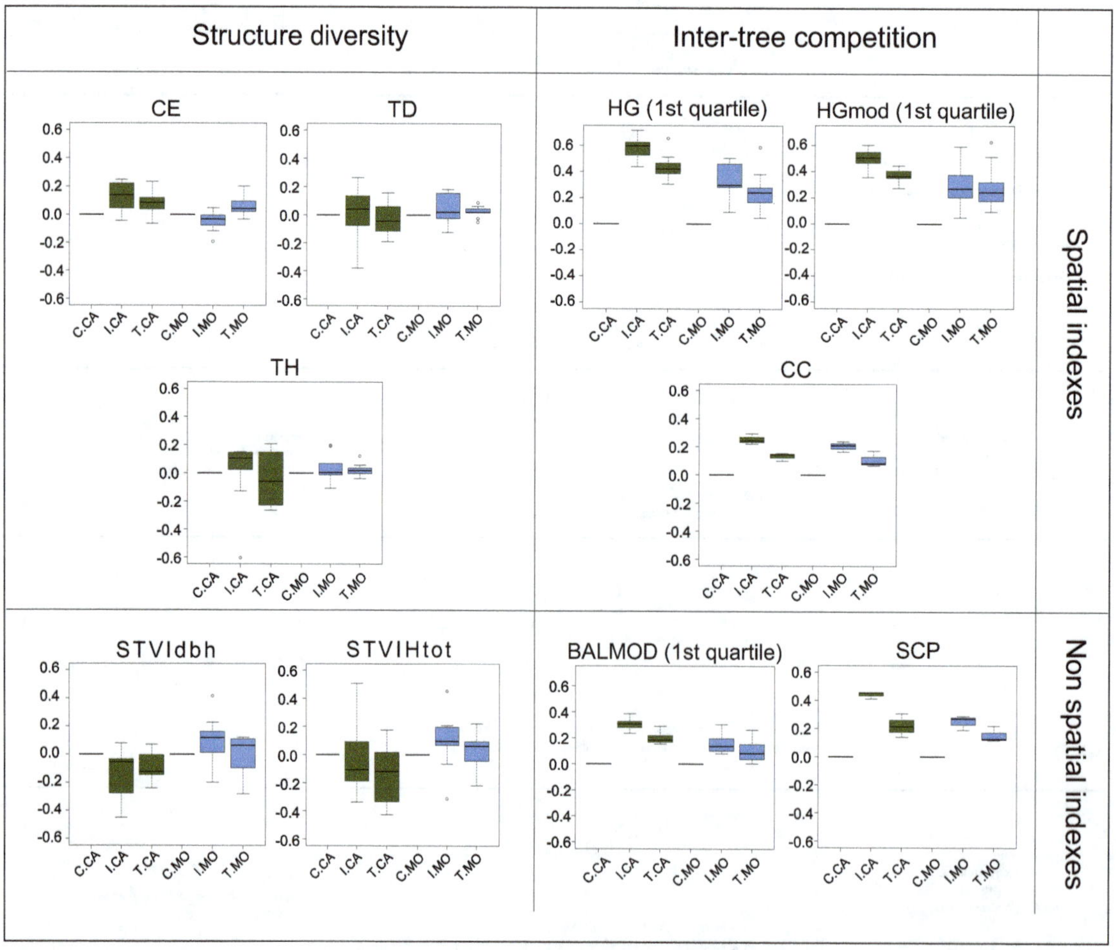

Figure 2 - Relative change of stand structure diversity and inter-tree competition indexes following thinning implementation per forest and thesis.

Table 3 - Main mensurational parameters (±sd) at sites.

Parameters	Before thinning Cansiglio	Before thinning Mongiana	After thinning Cansiglio Innovative	After thinning Cansiglio customary	After thinning Mongiana Innovative	After thinning Mongiana customary
Tree Density	323±65.9	510±130.4	187±9	235±29	409±100	408±168
Basal area m²/ha	40.9±5.2	41.2±7.3	26.3±1.3	30.3±4	33.7±7.7	31.6±3.1
Mean Height (m)	26.6±0.5	23.3±1.5	27±0.7	26.5 ±0.4	23.3±1.5	23.5±2
Mean (DBH) (cm)	40.6±3	32.9±6.1	43.7±2.4	40.5±2.6	33±6.2	33.4±8.2
Dominant DBH (cm)	49±3.6	46.2±6.8	48.1±0.8	46.9±1.2	47±7.9	47.0±7
Dominant Height (m)	27±0.6	26.2±1.2	27±0.1	26.6±0.1	24.2±1.6	26.0±1
Standing Volume (m³)	543±72	497±110.8	358±16	402±60	401±123	376.7±55.3

Table 4 - Mean values of stand structure diversity indexes before and after thinning implementation.

index	Before thinning Cansiglio	Before thinning Mongiana	After thinning Cansiglio Innovative	After thinning Cansiglio customary	After thinning Mongiana Innovative	After thinning Mongiana customary
CE	1.2	1.1	1.4	1.4	1.1	1.2
TD	0.2	0.2	0.2	0.2	0.3	0.3
TH	0.08	0.1	0.09	0.05	0.1	0.1
STVIdbh	0.1	0.2	0.09	0.09	0.3	0.1
STVIHtot	0.1	0.2	0.1	0.03	0.1	0.2

Table 5 - Tree competition indexes: 1stq before and after thinning implementation. Cansiglio = CA; Mongiana = MO.

thesis	BAL_{MOD} before	BAL_{MOD} after	Hegyi before	Hegyi after	$Hegyi_{MOD}$ before	$Hegyi_{MOD}$ after
CA_innovative	0.64	0.45	0.75	0.32	0.97	0.48
CA_customary	0.66	0.53	0.79	0.43	0.91	0.57
MO_innovative	0.96	0.81	1.21	0.79	1.27	0.88
MO_customary	0.88	0.81	1.13	0.80	1.21	0.81

Results of statistical tests are reported in Table 6. A significant difference ($p < 0.01$) is detected as for tree distribution pattern (CE) - innovative thinning thesis - at Cansiglio. A highly significant difference ($p < 0.001$) is present at the same forest for both thinning theses and all tested competition indexes. A significant differences ($p < 0.01$) is also reported by BAL_{MOD} at Mongiana - innovative thesis. Less significant ($p < 0.1$) are the differences recorded by HG and HG_{MOD}. No differences are detected as for the customary thesis at Mongiana.

Finally, mean values of crown-based indexes (Table 7) showed a notable reduction at both sites following thinning practices. Values before, different at each forest, were adjusted unlike. At Cansiglio, SCP dropped drastically with the innovative thesis (-44%) and less with the customary thesis (-19%), whilst, as expected, CC showed a lower reduction (-25% and -13 % respectively). At Mongiana, values before were much lower and, in accordance with the reduced cover before, CC dropped a bit more than SCP in the innovative thesis and followed the same pattern as in Cansiglio as for the customary thesis.

Discussion

The main outcome from the use of stand struc-

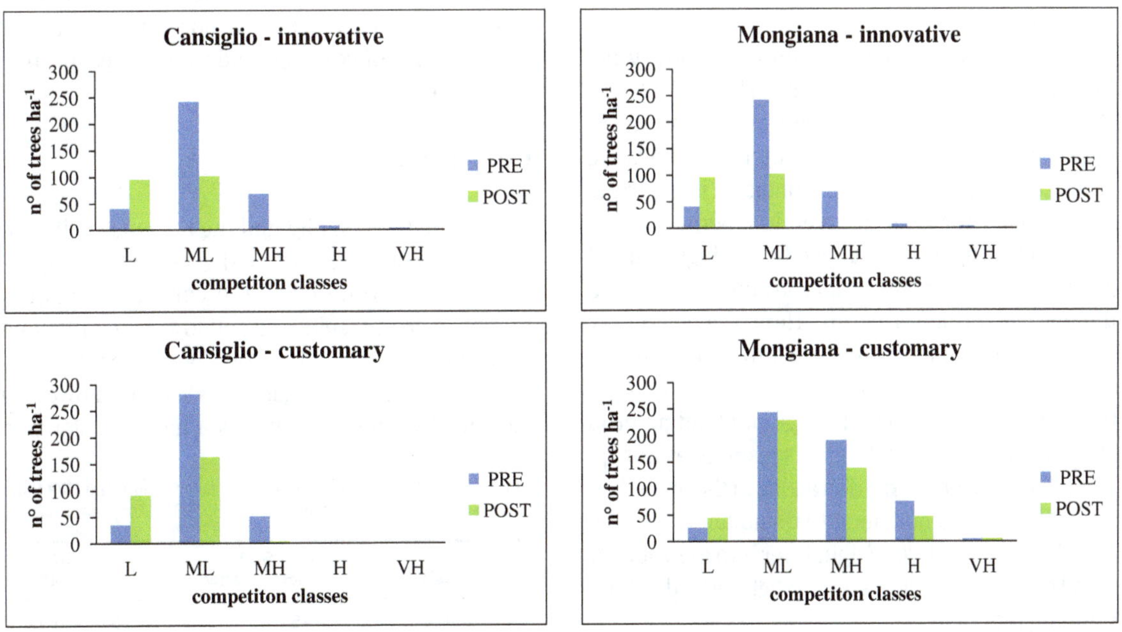

Figure 3 - Tree frequency distribution per competition classes (L: low, ML: medium low, MH: medium high, H: high, VH: very high).

Table 6 - Non-parametric statistics of tested indexes (Cansiglio = CA; Mongiana = MO). Asterisks indicate significant difference at * $p < 0.1$, ** $p < 0.01$, *** $p < 0.001$.

Comparison	Site	test	CE	TD	TH	$STVI_{dbh}$	$STVI_{Htot}$	BAL_{MOD}	Hegyi	$Hegyi_{MOD}$
innovative pre vs innovative post	CA	Mann Whitney U	10**	39	39	23	34	2***	7***	0***
customary pre vs customary post	CA	Mann Whitney U	16	35	39	32	34	1***	1***	0***
innovative pre vs innovative post	MO	Mann Whitney U	28	36	34	36	39	11**	16*	17*
customary pre vs customary post	MO	Mann Whitney U	31	34	39	40	39	25	22.5	19.5

tural indexes to highlight possible, early differences in tree texture and tree sizes was from the Aggregation Pattern index CE. It showed to be sensitive to stand manipulation and able to discriminate the higher heterogeneity of standing crop at the younger site (Mongiana) with respect to the older site (Cansiglio), it being manipulated for a much longer life-span. This sensitivity is well-summarized by the still random trees' horizontal distribution following thinning operations in Mongiana compared with the more regular pattern in Cansiglio. Among the applied theses in Mongiana, customary practice slightly moves the distribution towards regular distribution, given this intervention is performed all over the forest ground.

As for tree diameter differentiation, both spatial (TD) and non-spatial index ($STVI_{dbh}$), highlight similar dbh values in the sample considered. This peculiar feature is linked to the shade-tolerance of beech able to grow quite dense stands made of trees sized likewise. Both indexes do not detect any change as for the condition post-thinning at Cansiglio, whilst an even reduced sensitivity is proved at Mongiana. The typical even-aged structure and the one-storied arrangement of beech forests further promote this attribute.

Important differences are on the contrary detected in tree height variability by TH and $SVTI_{Htot}$ at Cansiglio. Here, values drop following the customary intervention and remain steady after the innovative practice. The steady pre-post condition is recurring at Mongiana for both interventions' criteria. The result in Cansiglio is explained by the higher stand age which ranked tree height arrangement into a thicker one-storied structure. Here, the customary intervention has harvested small to medium-sized stems (i.e. dominated up to a few sub-dominant trees) reducing the height range of standing crop (Roberts and Gilliam 1995). This result does not apply to the innovative criterion acting in the main crop layer made by co-dominant to dominant trees. The less-differentiated tree height layer into the younger stand of Mongiana allows only a slight record of both thinning practices.

Competition indexes (first quartile) are quite responsive - further to the expected reduction - to thinning theses (Brand and Magnussen 1988). They describe properly the targeted criterion of reducing tree competition in the dominant layer (Cansiglio-innovative), and in the neighborhood of a number of selected canopy trees (Mongiana-innovative), as well as the overall competition at stand level (customary thesis) at both sites. Response applies to low up to medium-high competition classes (Piutti and Cescatti 1997, Cescatti and Piutti 1998). Exception is made by customary criterion in Mongiana resulting vice versa into the prevailing reduction of inter-individual competition into medium-high and high classes, i.e. in the most dense patches of the less regular texture characterizing this forest.

Non-parametric statistics confirm the reduced sensitivity of stand structure indexes at this early-post intervention stage. Exception is made for the Aggregation Pattern index CE. The high sensitivity of tree competition indexes is also proved. As for crown-based indexes, the much higher figures of SCP and CC at Cansiglio compared to Mongiana describe a forest stand aged nearly twice Mongiana, much less dense and displaying a quite similar basal area (Fichtner et al. 2013b, Kimmins 1987). That means the establishment of the age-dependent crown interface made up of a more structured layer both in terms of coverage and overlapping, in spite of lower tree density.

Conclusions

The analysis highlighted the response ability of spatial and non-spatial tree diversity and competition indexes to describe the early manipulation following crown, selective and low-mixed thinning practices performed into two beech forests aged differently. The Aggregation index CE showed to be sensitive, moving towards a more regular tree

Table 7 – Crown-based indexes: mean values before and after thinning implementation. Cansiglio = CA; Mongiana = MO.

thesis	SCP before	SCP after	CC before	CC after
CA_innovative	115%	71%	90%	65%
CA_customary	90%	71%	79%	66%
MO_innovative	76%	51%	66%	46%
MO_customary	75%	60%	65%	54%

distribution at the more aged crop, independently of the applied criterion, and maintaining a random pattern at the younger and less homogeneous stand.

A slight or null sensitivity is showed by Diameter and Tree height differentiation, as well as by dbh and tree height variance, i.e. the other spatial and non-spatial tree diversity indexes tested. The condition here analyzed immediately after thinning practice will progressively change over time as a function of tree size differentiation following the applied practice. In fact, thinning makes available higher growing space to released trees, enhances stem radial growth and crown development, reduces symmetrical competition, in the perspective of differentiate further their next growth rate. The use of the above indexes becomes therefore more profitable and their response more consistent at intermediate steps between thinning practice.

All tree competition indexes tested HG, HG_{MOD} and BAL_{MOD}, recorded effectively the change in progress because they do register the different spatial arrangement following the intermediate removal and do not need any growth reaction to be established. As for crown-based indexes, they were able to differentiate thinning criteria much more than what recorded by mensurational parameters e.g. the difference in basal area removed.

Acknowledgements

Study funded within the project LIFE+09 ENV/ IT000078 'Managing Forests for multiple purpose: Carbon, Biodiversity and socio-economic wellbeing' (ManForCBD).

The authors wish to thank the anonymous referees for their helpful comments that contributed to improve the paper.

References

Bailey J.D., Tappeiner J.C. 1998 - *Effects of thinning on structural development in 40- to 100-year-old Douglas-fir stands in western Oregon.* Forest Ecology and Management 108: 99-113.

Becagli C., Amorini E., Fois A., Manetti M.C. 2009 - *Indici strutturali e di competizione per la caratterizzazione di popolamenti cedui di castagno.* Annali CRA-Centro Ricerca per la Selvicoltura 35 (2007-2008): 15-24.

Biging, G. S., Dobbertin M. 1992 - *A comparison of distance-dependent competition measures for height and basal area growth of individual conifer trees.* Forest Science 38 (3): 695-720.

Brand D.G., Magnussen S. 1988 - *Asymmetric, two-sided competition in even-aged monocultures of red pine.* Canadian Journal of Forest Research 18 (7): 901-910.

Cescatti A., Piutti E. 1998 - *Silvicultural alternatives, competition regime and sensitivity to climate in a European beech forest.* Forest Ecology and Management 102: 213-223.

Clark P. J., Evans F.C. 1965 - *Distance to nearest neighbour as a measure of spatial relationships in populations.* Ecology 35: 445-453.

Corona P., D'Orazio P., La Monaca A., Portoghesi L. 2005 - *L'indice di Winkelmass per l'inventariazione a fini assestamentali della diversità strutturale di soprassuoli forestali.* Forest@ (2): 225-232.

Erikäinen K., Miina J., Valkonen S. 2007 - *Models for the regeneration establishment and the development of established seedlings in uneven-aged, Norway spruce dominated forest stands of southern Finland.* Forest Ecology and Management 242: 444-461.

Fabbio G., Manetti M.C., Bertini G. 2006 - *Aspects of biological diversity in the CONECOFOR plots. I. Structural and species diversity of the tree community.* Annali CRA-Centro Ricerca per la Selvicoltura 30, Suppl 2 Special Issue. Aspect of biodiversity in selected forest ecosystems in Italy: 3-16.

Fichtner A., Sturm K., Rickert C., Härdtle W., Schrautzer J. 2013a - *Competition response of European beech Fagus sylvatica L. varies with tree size and abiotic stress: minimizing anthropogenic disturbances in forests.* Journal of Applied Ecology 49: 1306-1315.

Fichtner A., Sturm K., Von Oheimb G., Härdtle W. 2013b - *Crown size-growth relationships of European beech (Fagus sylvatica L.) are driven by the interplay of disturbance intensity and inter-specific competition.* Forest Ecology and Management 302: 178-184.

Fox T.R., Jokela E.J., Allen H.L. 2007a - *The development of pine plantation silviculture in the southern United States.* Journal of Forestry 105: 337-347.

Fox, T.R., Allen H.L., Albaugh T.J., Rubilar R.,. Carlson C.A. 2007b - *Tree nutrition and forest fertilization of pine plantations in the Southern United States.* Southern Journal Applied of Forestry 31 (1): 5-11.

Getzin S.,Wiegand T., Wiegand K., He F. 2008 - *Heterogeneity influences spatial patterns and demographics in forest stands.* Journal of Ecology 96: 807-820.

Harrington T.B., Tappeiner J.C. 2009 - *Long-term effects of tanoak competition on Douglas-fir stand development.* Canadian Journal of Forest Research 39:765-776.

Hegyi F. 1974 - *A simulation model for managing jack pine stands. In: Fries, J., (Ed.). Growth models for tree and stand simulation.* In: Proceedings of IUFRO meeting sr. 01.04. Royal College of Forestry, Stockholm: 74-90.

Husch B., Beers T.W., Kershaw J.A. 2003 - *Forest mensuration* (4th ed.). John Wiley & Sons, New York, 443 p.

Ladermann T.R., Stage A. 2001 - *Effects of competitor spacing in individual - tree indices of competition.* Canadian Journal of Forest Research 31: 2143-2150.

Latham P.A., Zuuring H.R., Coble D.W. 1998 - *A method for quantifying vertical forest structure.* Forest Ecology and Management 104: 157-170.

Kembel W., Dale M.R.T. 2006 - *Within-stand spatial structure and relation of boreal canopy and understory vegetation.* Journal of Vegetation Science 17 (6): 783-790.

Kimmins J.P. 1987 - *Forest Ecology. A foundation for sustainable Forest Management and Environmental Ethics in forestry.* Third edition. The University of British Columbia. Upper Saddley River New Jersey, 531 p.

Kint V., Mohren G.M.J., Geudens G., De Wulf R., Lust N. 2004 - *Pathways of stand development in ageing Scots pine forests.* Journal of Vegetation Science 15: 549-560.

Margalef R. 1982 - *Sono stabili i sistemi naturali?* Economia Montana 3: 17-19.

Neumann M., Starlinger F. 2001 - *The significance of different indices for stand structure and diversity in forests.* Forest Ecology and Management 145 (1-2): 91-106.

Oliver C.D., Larson B.C. 1996 - *Forest Stand Dynamics.* Updated ed., Wiley, New York, 520 p.

Piutti E., Cescatti A. 1997 - *A quantitative analysis of the interactions between climatic response and intraspecific competition in European beech.* Canadian Journal of Forest Research 27 (3): 277-284.

Pommerening A. 2002 - *Approaches to quantifying forest structures.* Forestry 75: 305-324.

Pommerening A. 2006 - *Evaluating structural indices by reversing forest structural analysis.* Forest Ecology and Management 224: 266-277.

Pretzsch H. 2009 - *Forest dynamics growth and yield: From measurement to model.* Berlin: Springer Verlag, 664 p.

Roberts M.R., Gilliam F.S. 1995 - *Patterns and mechanisms of plant diversity in forested ecosystems. implications for forest management.* Ecological Applications 5 (4): 969-977.

Sallabanks R., Riggs R.A., Cobb L.E. 2002 - *Bird use of forest structural classes in grand fir forests of the Blue Mountains, Oregon.* Forest Science 48: 311-321.

Saunders M.R., Wagner R.G. 2008 - *Long-term spatial and structural dynamics in Acadian mixedwood stands managed under various silvicultural systems.* Canadian Journal of Forest Research 38: 498-517.

Schroder J., Von Gadow K. 1999 - *Testing a new competition index for maritime pine in northwestern Spain.* Canadian Journal of Forest Research 29: 280-283.

Staudhammer C.L., LeMay V.M. 2001 - *Introduction and evaluation of possible indices of stand structural diversity.* Canadian Journal of Forest Research 31: 1105-1115.

Yamaura Y., Tojo H., Hirata Y., Ozaki K. 2007 - *Landascape effects in bird assemblages differ between plantations of broadleaved forest in a rural landscape in central Japan.* Journal Forest Research 12: 298-305.

Zenner E.K., Hibbs D.E. 2000 - *A new method for modelling the heterogeneity of forest structure.* Forest Ecology and Management 129: 75-87.

18

Re-use of wastewater for a sustainable forest production and climate change mitigation under arid environments

Maria Cristina Monteverdi[1*], Sara Da Canal[2], Alberto Del Lungo[3], Salvatore Masi[4], Hocine Larbi[5], Paolo De Angelis[2]

Abstract - Over the last decades biotic and abiotic constrains together with human actions are determining a substantial environmental pressure, particularly in dry lands as the south of the Mediterranean region. From very long time, indeed, simultaneous drivers such as demographic growth, climate change and socio-economic factors are weakening the previous homeostasis between human needs and natural resources on the regional scale. Resulting pressures are determining environmental degradation and increase of desertification risk for the arid and semiarid lands. Water quality and availability are both crucial points limiting people well-being and livelihoods in the same context. Scarcity of fresh water and heavy and mismanaged production of wastewater are the main factors affecting water resources. Increasing pollution of soil and ground waters reduces the possibility of sustainable development of local communities with relevant social consequences. The FAO's supporting program in north Africa aims to: a) develop new and cheaper phytotechnologies (e.g. constructed wetland system; innovative treatment system for reuse of waste water for fertigation); b) treat wastewater for water quality protection; c) promote land recovery by means of sustainable multipurpose forestry; d) adopt bioengineering interventions to stop slopes erosion and protect urban, and semi-urban infrastructures; e) create pilot demonstrative areas to test multi-purpose sustainable agroforestry systems. Within this frame, an integrated approach was designed to promote innovative sustainable water management and multipurpose forestry, in order to mitigate the effects of climate change, promote land recovery, and improve the livelihoods of local population. The present paper aims to provide an overview of the FAO project GCP/RAB/013/ITA. Particularly, two pilot studies are shown and discussed.

Keywords - phytotechnologies, arid environment, wastewater, sustainable development, forest and agroforestry systems

Introduction

The Mediterranean Basin with its climatic peculiarities is extended on three continents. It is densely populated (507 million people on 2010) and extremely rich of natural and cultural heritage. It constitutes an “ecoregion”, where both social and economic developments are closely related to somewhat limited natural resources and vulnerable environments (FAO 2013).

After thousands of years of co-evolution between ecosystems and societies, human activities are determining substantial environmental pressure, with a large social-environmental imbalance between the northern and southern regions. Such disparities are exacerbated by climate change effects. These are causing an increase in intensity and frequency of the cyclical period of aridity and of extreme events such as wind duster and water bombs (IPCC 2012). The simultaneous action of biotic, abiotic and socio-economic factors is the main driver of land degradation and of increasing risk of desertification (Bennadji et al. 1998).

In arid and semi arid areas the soil is scarcely productive. In this context, forests can play a crucial role in promoting and restoring the water cycle on the regional scale (David et al. 2011, Matteucci et al. 2011).

Forests in drylands are an invaluable multipurpose resource of fodder for livestock, fuel-wood (e.g. 58% of all the energy consumption in Africa), non-wood and fiber products and biomass (FAO 2013). They also provide a wide range of environmental services (ESs) such as protection from soil erosion, addition of organic matter, increased water retention and biodiversity. ESs are particularly relevant for agricultural activities and desertification control. Afforestation and reforestation activities in arid and semi-arid lands are functional to soil rehabilitation/restoration. Forests increase carbon

[1] Consiglio per la Ricerca e la sperimentazione in Agricoltura, Forestry Research Centre, Arezzo (CRA-SEL), Italy
[2] Department for Innovation in Biological, Agro-food and Forest systems (DIBAF), University of Tuscia, Italy
[3] Forestry Department FAO. Forestry Officer, Project LTO, GCP/RAB/013/ITA
[4] School of Engineering University of Basilicata, Italy
[5] Université de Mascara, Algeria
[*] corresponding author: mcristina.monteverdi@entecra.it

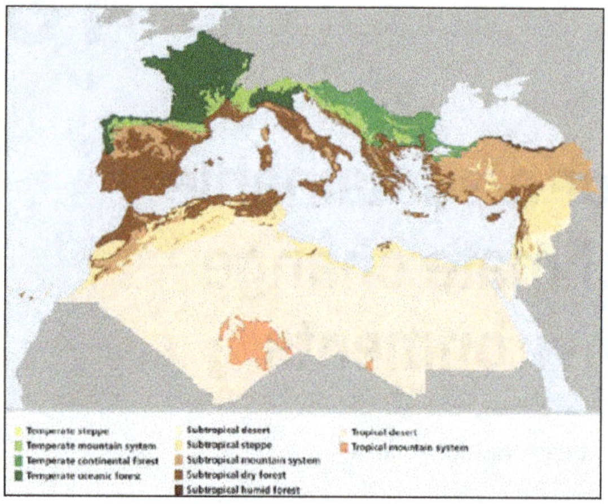

Figure 1 - Steppes and deserts dominates the southern Mediterranean countries. Source: Derived FAO and JRC, 2012 and Iremonger and Gerrand, 2011.

sequestration and storage, generate employment opportunities, provide recreational and landscape value (FAO 2013).

Water quality and availability represent crucial limiting factor for human well-being and livelihoods in arid and semi-arid lands.

The southern Mediterranean countries generally are dominated by steppes and deserts (Fig.1). Such environments are fragile and exposed to degradation and desertification processes.

Northern Africa, is affected by different problems heavily related to the new social and economical development. The population growth of the last decade highlighted the limits of available natural resources. The effects of global change together with new agricultural techniques (e.g. pivot irrigation), unsustainable water and soil management, land use changes led to an overexploitation of natural resources. All this is endangering the fragile balance between human needs and natural resources.

Especially in arid and semi-arid lands, the low availability of fresh water and the mismanaged production of wastewater are main factors affecting use and availability of water resources. Contamination of soil and ground waters biases, finally, any sustainability in the processes of local development.

A contribution to sustainability relies on the use or re-use of non-conventional water resources. Urban waste water can be properly treated and used for forest and agroforestry systems in rural, urban and peri-urban areas. This would increase the availability of wood, firewood, biomass, food and ecosystem services.

Over the last three years FAO Forestry has implemented new activities on the use of non conventional waters in forestry and agroforestry systems of arid zones in some Mediterranean countries. Through a collaboration with the Italian Ministry of Foreign Affairs, FAO launched the GCP/RAB/013/ITA project "*Forest restoration in Algeria, Egypt, Morocco and Tunisia using treated wastewater to sustain smallholders' and farmers' livelihoods*". In this project Scientific Italian Institutions (DIBAF-University of Tuscia, University of Basilicata, National Research Council CNR and Consiglio per la Ricerca e sperimentazione in Agricoltura CRA) in collaboration with the FAO Forestry Department have designed and developed pilot projects in Maghreb countries, innovative methodologies for a sustainable and cheaper use of urban wastewaters.

The project is an important example of participatory approach to the formulation of cooperation projects. Indeed, the project plan directly derives from the needs and requests of the four North African countries involved: Algeria, Egypt, Morocco and Tunisia. These countries specifically requested the Italian expertise to re-use treated wastewater for the restoration of forestry and agro-forestry systems of arid zones.

Project objectives

The main aim was the use of non-conventional waters for irrigation of multipurpose wood plantations for tree crops' production, orchards, fuel-wood and wooden biomass. In particular, the project proposed an integrated approach specifically aimed to: a) reduce pollution of natural water resources (development of new and cheaper phytotechnologies for wastewater treatment); b) mitigate climate change, promote the recovery of degraded areas, reducing contamination risks of ground water and soil; c) improve the living conditions of the local population through forestry and agroforestry sustainable activities, taking into account needs and traditions of local people.

Materials and Methods

The methodologies adopted by the project to treat wastewater are mainly two: constructed wetlands (phytotechnology approach) and a modified version of the standard wastewater treatment system, planned to reduce the energy inputs and preserving the nutrient load useful to the plants (fertigation approach).

Constructed wetlands

Natural wetlands act as bio-filters, removing sediments and pollutants from the water; constructed wetlands are designed to emulate these features. Vegetation in a wetland provides a substratum (roots, stems, and leaves) upon which microorganisms can grow as they break down organic materials. The plants remove about 70% to 90% of pollutants (biotic and abiotic), and provide carbon for microorganism particularly by fine root turnover. Constructed wetlands are largely recognized as effective for primary and secondary urban and agricultural wastewater treatment (Kadlek 1996, US_EPA 1999), and are considered by the World Health Organization (WHO) an effective and cheap way to meet its microbiological guideline for wastewater reuse (WHO 1989). These systems are made up of ponds placed in sequence, filled with specifically selected plants, shrubs and vegetation for their ability to filter impurities from the water. The treated water is then let settle in storage basins. Since the system only uses natural phytotechnology solutions it also appears to be more sustainable and economically-viable for rural communities. The application of this system in Algeria and Tunisia has been designed by researchers of the University of Tuscia, Italy.

Innovative treatment for reuse of waste water for fertigation

Innovative "secondary" treatment for wastewater systems reduces the microbial load to acceptable limits, maximizing the release of useful substances (e.g. Nitrogen and Phosphorus), which can be used in agroforestry systems to provide nutrients and organic matter in the soil. According to this innovative approach a methodology was developed by the University of Basilicata, Italy, after a testing phase on olive trees. From these previous pilot studies, it has emerged that properly controlled modifications of a conventional treatment plant, together with an accurate monitoring scheme which adjusts the provision of organic matter and nutrients to the irrigated soil, is a cost-effective system that maximizes the amount of "good" organic matter available for irrigation, while minimizing health risks.

Main activities supported in the project countries

In response to the specific requests of the participating countries the project is supporting

- In Morocco, the establishment of 10 hectares of palm grove that will be part of the green belt of the town of Marrakech (31°42'3.02"N – 8°3'37.25"W). The green belt will have several functions including agro-silvo-pastoral activities and to act as buffer zone between the main landfill and the urban area. The plantation is irrigated by the innovative treatment for reuse of wastewater for fertigation designed by University of Basilicata, and realized in collaboration with University of Marrakech. Both CNR and CRA have provided further support in the development of the plantation. This plantation will be included into the larger project for the development of "green belt" to protect the city against soil erosion, desertification and to create a functional zone for the development of agroforestry activities. At the same time, the use of fertigation techniques will ensure an increase of soil fertility and carbon storage capacity of the soil.
- In Egypt the project provided the drafting of the first forestry management plan of Serapium forest (30°29'24.41"N -32°13'59.77"E, Ismailia district), which is a 25 years old plantation irrigated with treated wastewater. The Serapium plantation (129 ha) was realized using different species (*Eucalyptus citriodora Hook., E. camaldulensis Dehnh., Casuarina equisetifolia L., Cupressus sempervirens L., Khaya senegalensis* (Desr.) A. Juss, and others) adapted to hot environments and characterised by a high wood productivity. The forest management plan has been drawn taking into account the possibility of making this forest eligible for a possible market for carbon credits. The work is being carried out by the Under-Secretariat for Afforestation and Environment of the Ministry of Agriculture, the Department Forestry University of Alexandria and other relevant Egyptian institutions, under the technical coordination of FAO and with the support of the University of Monaco (Germany) and University of Tuscia (Italy).
- In Tunisia the project activities have been implemented in Ouechtata, a rural village in northern Tunisia, close to the coast (36°57'27.51"N – 9°00'7.29"E). The activities concern the design of a constructed wetland, planned to reduce the pollution of a water reservoir used for agricultural needs, at low cost and in sustainable way. Furthermore, the design of a fertigation system for olive trees and biomass production in Hafouz (inside village of southern Tunisia, 35°38'13.21"N – 9°42'28.44"E) is planned to reduce the cost of management of the already existing wastewater treatment plant. The activities were conducted in close cooperation among the project partners (FAO, University of Tuscia and University of Basilicata) and the national Forestry General Direction.
- In Algeria the FAO project supported the completion of an already existing pilot area in the Oasis of Brézina (33°5'38"N – 1°15'3.40"E). The project was developed in the context of the Italian cooperation (Ministry of Economic Development - MISE, Ministry of Foreign

Affairs - MAE), by the University of Tuscia (IT) and University of Mascara (ALG). The reuse of the wastewater produced by the local community was tested for the irrigation of forestry and agroforestry systems, after treatment by means of the constructed wetland technology. A second activity was the design of constructed wetlands to reduce the pollution of the local river and of the palm plantations, as well to fight the erosion processes in the oasis of Taghit (30°55'3.06"N – 2°2'2.22"W).

Pilot applications of phytotechnologies to realize productive and protective forest systems in southern Mediterranean countries: two case studies.

Phytotechnologies use plants to remediate various media impacted with different types of contaminants (Vangronsveld et al. 2009, Otte and Jacob 2006). In different sites in north Africa, the integrated systems to reduce water and soil pollution have been designed and developed (treating the urban waste water by constructed wetland). At the same time, these systems are planned to promote land rehabilitation (planting trees according to multipurpose scheme and local needs). Below, two pilot applications carried out in Algeria are explained briefly.

First case study

Oases are the main spots of human development in desert areas. Biotic and abiotic constraints together with human actions determine the fragile equilibrium of the ecosystem that is threatened by desertification.

In the oasis of Brézina an integrated, sustainable model for the oasis protection, recovery and development has been recently realized, also considering the advantages related to carbon credits market through the Clean Development Mechanism (CDM) of the Kyoto Protocol. Brézina oasis is located in the pre-Saharan region (800 m a.s.l.), 85 Km south of the county (*wilaya*) town of El Bayadh, Algeria. Brézina is characterized by a pre-Saharan arid climate, with an annual mean temperature of 20 °C; diurnal variation of 11 °C; annual rainfall rarely exceeds 100 mm (NASA 2007); low atmospheric humidity (around 40%); high solar radiation (16.56 MJ $m^{-2}day^{-1}$) and high wind speed. During the April dust storms, the average wind speed grows until 5.9 m s^{-1} (Kasbadji Merzouk 2000). The *Seggueur* ephemeral river, which surrounds the oasis at the southern side (Masini et al. 1988), is the main source of water for the palm plantation and for the traditional agricultural productions, performed under the palm cover of the oasis.

The main problems which afflict Brézina oasis are related to the groundwater depletion resulting from increasing urban population, excessive retention of *Seggueur* river by the dam realized upstream of the oasis, and climate change. All these are causing severe damages of the palm grove and the risk of groundwater contamination due to the outflow of untreated wastewater.

The main goal was to face the challenge from two different perspectives. The first concerns the adoption of a new water management model, inspired to the collection, recycle and reuse of urban wastewater. The second proposes a new agroforestry strategy in order to stimulate both alternative economic chains and environmental restoration. The recent Oasis growth suggested to design a system optimised for the local conditions and easily expandable. For this reason the pilot project was designed as a modular system, able to be replicated in parallel or in series to match increased wastewater flow rate or pollution, respectively. The modular approach implies the definition of a minimum flow rate of wastewater. Such a minimum quantity is collected, treated and, finally, reused through a single process. The wastewater pilot system was designed to fill a good quality of output and to reduce the infiltration and evapotranspiration losses using constructed wetlands. A further target was to build up a simply operated and flexible system, in terms of effluent quality produced (Mara 1999).

Two parallel natural treatment chains compose the system, as illustrated in Figures 2-3-4:

1) CW B is a constructed wetland with horizontal superficial flow, being composed by three cells alternating aerobic (cell 1 and cell 3) to anaerobic process (cell 2) (Fig. 3).
2) CW A is a constructed wetland with horizontal sub superficial flow composed by three separated cells where mainly aerobic processes are triggered (Fig. 4).

Both treatment chains are connected with a storage basin necessary to homogenize the flow rate variation during the day and to provide minimal water storage. For the constructed wetland were chosen two different mixes of plant species (Fig. 5):

- MIX 1: *Phalaris arundinacea* L., *Juncus* spp., *Phragmites* spp., *Typha* spp., optimal for reduced water depth and resistant to short periods of dryness.
- MIX 2: *Myriophyllum* spp., *Thypha* spp., optimal to favor anaerobic processes.

The synergistic effect of hydraulic design, the plant species selection and the related fine root biomass promotion, improve the process of ab-

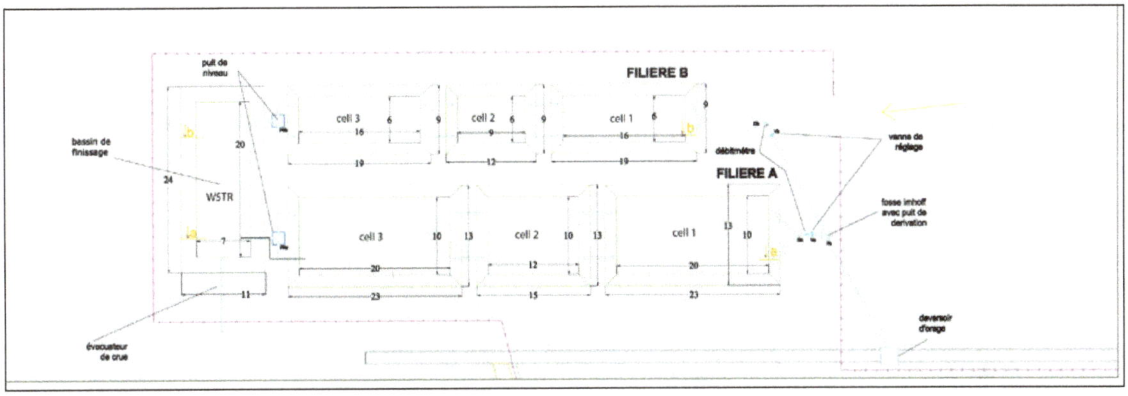

Figure 2 - Oasis of Brézina (Algeria) – waste treatment system: plan.

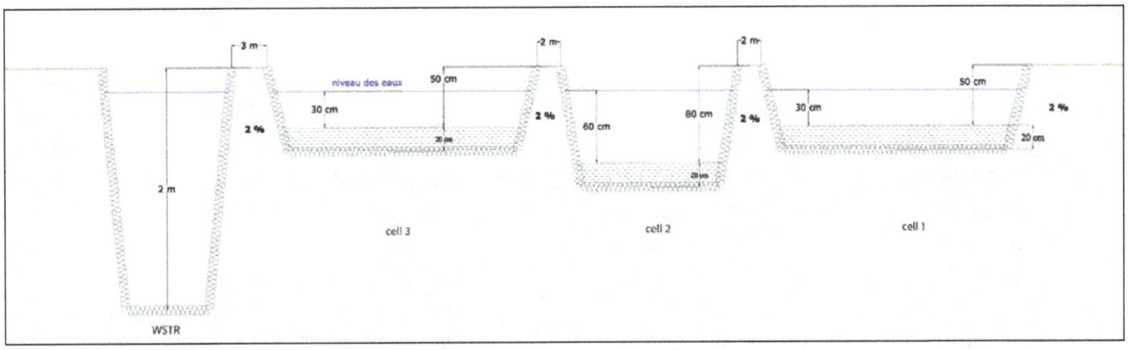

Figure 3 - Oasis of Brézina (Algeria) – CW B section: a constructed wetland with horizontal superficial flow. Cell 1 is designed to provide preliminary sedimentation with a water level up to 30 cm, the vegetation is emergent with a density of 90%. Cell 2 is designed to provide anaerobic digestion of Cell 1 effluent, water level never exceed 60 cm and the vegetation is submerged with a density of 20%. Cell 3 is designed to complete the water treatment of Cell 2 effluent trough aerobic process. Water level is up to 30 cm and vegetation is emergent with a density of 90%. WSTR (Water Storage Treatment Reservoir).

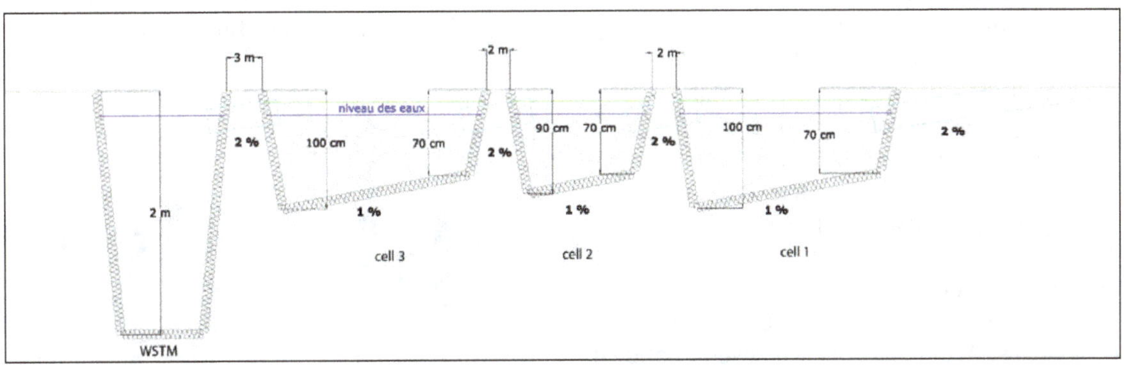

Figure 4 - Oasis of Brézina (Algeria) – CW A section: a constructed wetland with horizontal sub superficial flow composed by three separated cells where mainly aerobic processes are triggered. WSTR (Water Storage Treatment Reservoir).

sorption, sedimentation and inactivation of major pollutant as BOD5 (biochemical oxygen demand of wastewater during decomposition occurring over a 5-day period), TSS (total suspended solids), nitrogen, phosphorus and *Escherichia coli*.

The treated wastewater is efficiently reused into the forest agroforestry plots, using drip irrigation. The forest specie selection maximizes both environmental and economical improvement through marginal land restoration, no-food products, and potential revenue from carbon sequestration (Fig. 6).

In the forest plantation four experimental systems were tested.

Figure 5 - Oasis of Brézina (Algeria) – view of the constructed wetlands in October (2013).

Figure 6 - Oasis of Brézina (Algeria) –agroforestry testing plots: species (wood and forage), densities (biomass) genotypes evaluation.

1. *Tamarix* spp. collection to compare different species and populations for plant breeding.
2. *Tamarix* spp. plantation for biomass production under short rotation regime to evaluate productivity, inter-annual growth rates, wood quality also for bioethanol production, environmental benefits.
3. Plantation of different species with a relevant ecological and economical value to produce forage, wood and non-wood products (e.g., *Argania spinosa* L., *Pistacia atlantica* Desf., *Olea europaea* L.). The trees were planted at a distance of 6 m, on 7 m spaced rows; the inter row will be planted with forage intercrop species: *Atriplex* spp., *Medicago arborea* L., and *Opuntia ficus-indica* (L.) Mill.
4. Experimental comparison on field for genotype evaluations of *Populus* spp. in order to select suitable plant material for forestry plantations in arid and semi-arid lands.

Second case study

Along with the objective of preserving oasis ecosystems, Algeria has opened several national programs for environmental protection and local development, as the one for Taghit Oasis. This program is focused on the identification of measures able to protect the environmental resources and cultural heritage, promoting a sustainable development of a vulnerable area, transformed by human activities for centuries, at present under the pressure of new societal drivers. From this perspective, the oasis of Taghit represents a small-scale experiment to test new approaches for environmental protection in the frame of the sustainable development of the Oasis.

In Taghit the project aims to reduce the impact of wastewater on the principal river, where the palm cultivation is managed in combination with agro-pastoral activities. In fact, as consequence of an increasing wastewater discharge and of increasingly frequent events of flooding, the fertile riparian area results polluted and the palm trees damaged. Taghit is located in sub-Saharan area, near the Moroccan border, the western edge of the Grand Erg Occidental (GEO) (623 m a.s.l.). In the Figures 7 and 8, the main climate characteristics of the oasis are represented.

Although almost 98% of urban wastewater is collected regularly, the main source of pollution is represented by a poor sustainable wastewater management. Currently, wastewater generated from 5 villages located around the oasis Taghit (Zaouia Fougania, Thaghit, Berrabi, Bhakti, Zaoui Tehtania /

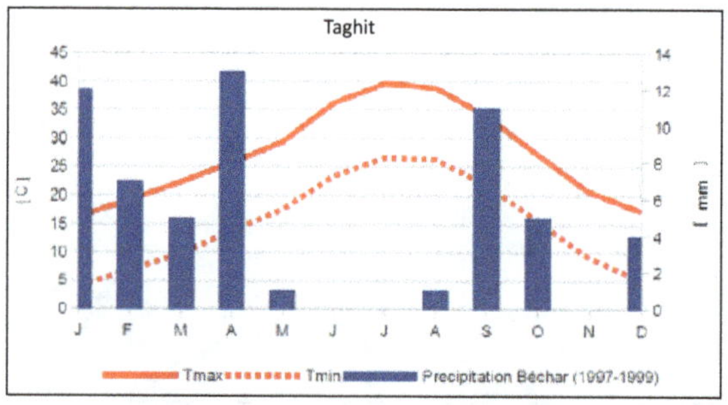

Figure 7 - Taghit (Algeria) – annual average, maximum and minimum temperatures, and annual rainfall distribution.

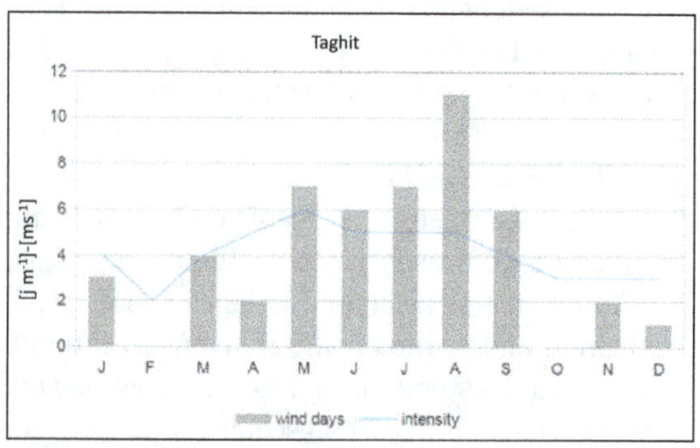

Figure 8 - Taghit (Algeria) – wind regime during the year.

Brika) are released downstream of urban areas and upstream of the palm stands. Final destination of the wastewater stream is the bed of the river Zousfana. Based on qualitative and quantitative characteristics of wastewater produced by each village, constructed wetlands were specifically designed (Fig. 9). After a careful analysis of the characteristics and problems of Taghit Oasis, different possible reuse of the treated wastewater for reforestation and revegetation were also designed, and summarized as follows.

1. Realization of multipurpose irrigated forestry and agroforestry systems

Given the fragility of the oasis ecosystems and climate extremes that characterize them, the fundamental priority is to promote the enhancement of local biodiversity and forest genetic resources (FGR). Indeed, the selection of suitable species for the realization of multifunctional production systems is strongly required. Collection and comparative test of species and provenances is an essential tool for the selection and improvement of forest trees and for starting a process of conservation and management of Forest Genetic Resources. In the specific case of the oasis of Taghit, the implementation of a pilot comparative plantation (species/provenances) was proposed, aiming to assess the variability of the main adaptive traits responsible of the individual adaptability, to sustain the growth and carbon sequestration and to reverse the ecological degradation of fragile ecosystems. Species of high ecological and economic interest (e.g., *Populus alba* L., *P. euphratica* Oliv., etc), will be tested in the experimental plantation (Fig.10).

2. Consolidation of embankment by woody species in order to stop the erosion and to protect the urban infrastructures

Part of the treated wastewater can be used for irrigation of naturalistic engineering works such as: consolidation of embankment, roadsides, riverbanks and landslide slopes. The use of woody species (e.g. *Nerium oleander* L., *Olea europaea var sylvestris* L., *Capparis spinosa* L., *Tamarix* spp.) in such applications allows to increase the effectiveness of anti-erosion and consolidation, reducing water and wind erosion as well as landslides occurrence. Thus, such applications increase the safety of peri-urban and urban infrastructures. In addition to the technical and functional effects mentioned above, works of this kind are also important for other purposes. These bioengineering works, by using native species of the different stages of the dynamic series of potential vegetation, can favour the reconstruction or trigger of the natural potential at the landscape scale. Bioengineering provides cost-effective structures as alternatives (e.g. slope fascines) in respect to traditional structures (e.g. retaining concrete head wall). These are usually more expensive and characterised by higher envi-

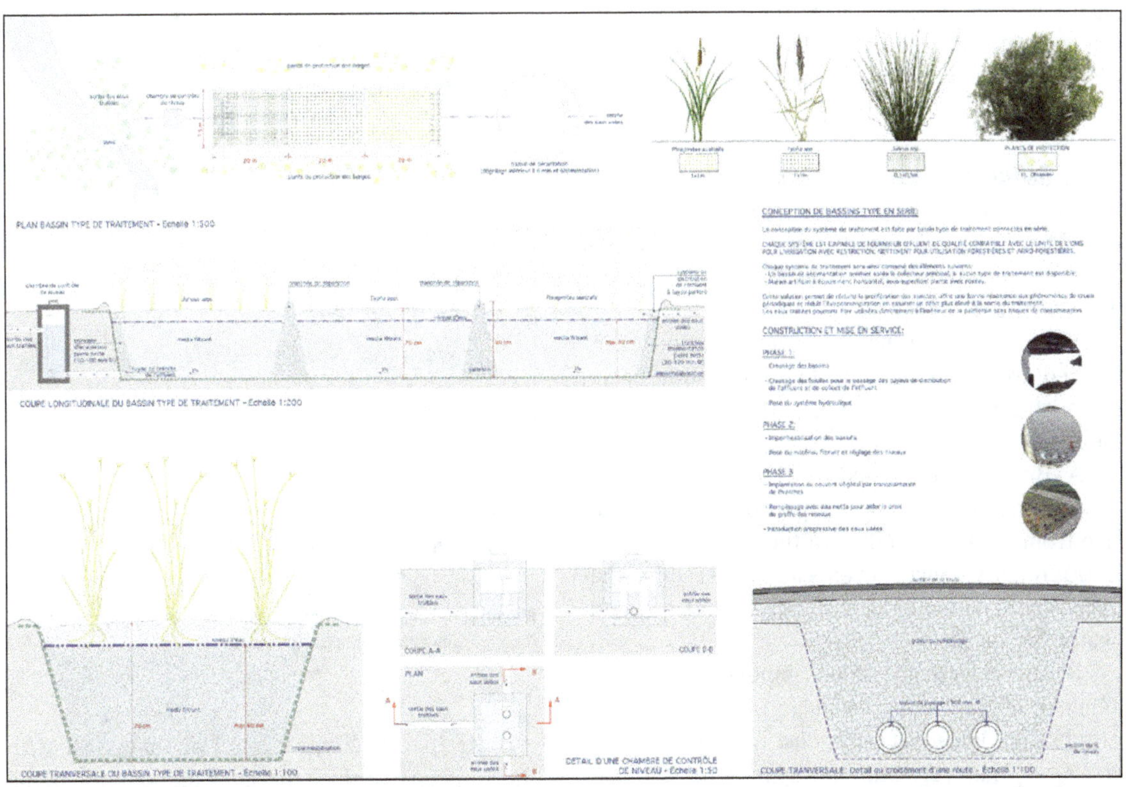

Figure 9 - Taghit (Algeria) – scheme of treated wastewater: plan and sections.

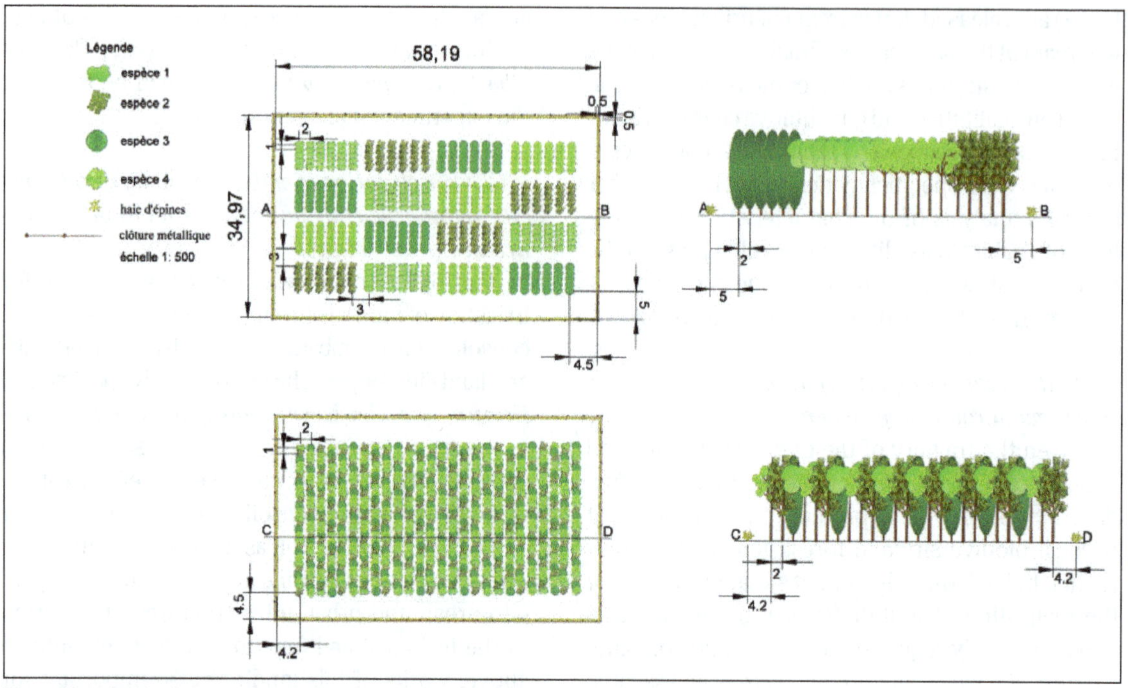

Figure 10 - Taghit (Algeria) – test plots for field trial of trees species irrigated with treated wastewater.

ronmental impacts. Moreover, the green structures provide social and economic incomes and benefits, as their construction and maintenance determine employment, improved environmental quality and environmentally sound management of natural and non-conventional resources (e.g. treated wastewater). Given the degradation and instability of most of road slopes in Taghit, stabilizing interventions were planned, based on the use of local herbaceous plants and shrubs and on rescue irrigations with treated water. In case of superficial instability of the slopes, realization of small steps with wattle fences and fascines was considered.

3. Reducing the environmental impact of the basins of primary treatment (lagoons, settling basins), through the use of vegetated strips of herbaceous species, aromatic shrubs and trees with use of treated waste water for irrigation

The project goal has been the creation of irrigated green belts surrounding the lagoon basin in construction at Taghit (Fig. 11), which are structured as:

A. Buffer zone, formed by herbaceous species (e.g. *Mesembryanthemum* spp.) to limit the damage from possible flooding of the pond caused by extraordinary climatic events and thus reduce the risk of contamination of surface and underground water resources.

Figure 11 - Plantation of aromatic plants and shrubs to reduce the impacts of the treatment basin.

B. After a service corridor for the ordinary maintenance of the lagoon basine, a belt of aromatic shrubs (e.g. *Artemisia herba-alba* Turra, *Lavandula antineae* Maire, *Rosmarinus officinalis* L., *Rosa* spp., *Astragalus armatus* Willd., *Helichrysum* spp., *Atriplex halimus* L., etc.) was envisaged with the purpose to reduce the smell air pollution of the primary treatment system.

C. A trees or shrubs windbreak is finally expected (e.g. *Nerium oleander* L., *Tamarix africana* Poir., *Popolus* spp.) that, in addition to the protective function of the lagoon basin to the possible silting effects, will preclude the view from road. The use of species such as N. *oleander* characterized by intense colored blooms, also determines a pleasant visual effect contributing to the aesthetic landscaping, a basic element given the huge tourist interest of Taghit Oasis.

Conclusions

Wastewater harvesting and treatment are required to guarantee a sustainable development of dry lands. Treated urban wastewater is an unconventional but relevant resource, allowing to reserve fresh water uses for more essential needs.

The actual human and economic development produces widespread availability of wastewaters even in areas with scarce surface and ground water. Recycling wastewaters reduces the risk of pollutants seeping into natural habitats and of contaminations of drinking water wells. Such recycle has beneficial environmental effects on soils and water bodies. It can improve carbon storage and water retention in dry soil.

Treated wastewater can be safely used for irrigation of wood plantations, non-food and energy crops. Recycling makes available fresh water resources for the essential needs of people living in dry areas. Moreover, fertigation methodologies, such as those selected for the project, reduce the energy costs of up to 50% compared with traditional systems, allowing the recovery of nutrients to enrich soil fertility.

The use of non-conventional water resources promote reclamation of marginal land, reduction of wind erosion, amelioration of soil conditions, economic stimulation for timber and non-timber products as well as local manpower requirements for new management systems. A new and sustainable agro-forestry management of arid and degraded land is required to guarantee new economic benefits to the local people and to conserve and recover the arid and semiarid lands. The integrated approach proposed in the FAO project can be considered a "*win win*" solution for restoration and conservation of dry lands, mitigation of global climate change and for a sustainable development on the local scale.

In addition, the use of partially treated wastewaters, in forestry and agroforestry activities can provide valid water supply to support the production of timber and firewood, the creation of windbreaks and thus preserve and protect the productivity of agricultural lands. This unconventional resource can contribute to ensure food for local communities, and helps to combat desertification. All this concurs to the achievements of some of "The Millennium Development Goals" established by the United Nations, as the objective 1, 7 and 8 "eradicate poverty and hunger, ensure environmental sustainability, and develop a global partnership for development".

Acknowledgements

The activities reported in the present work were conducted with the support of:

Italian Ministry of Foreign Affair, Italian Ministry for Education, Research and University, Bilateral agreement Italy-Algeria - PRGR 2009;

Italian Ministry for Economic Development, Italian Office for Foreign Trade, Italian Conference of University Rector project DESERTFOR;

FAO project "Forest Restoration in Algeria, Morocco and Tunisia Using Treated Waste Water to Sustain Smallholders' and Farmers' Livelihoods" - GCB/RAB/013/ITA.

The authors also thanks for their support:

- the team of the University of Mascara (Algeria): A. Miloudi, M. Benslimane, B. Boumediene, Z. Souidi;
- the team of the University of Tuscia (Italy): Perugini L. Chiani, R. Valentini
- the Directors and the teams of the Forest General Directions of Algeria and Tunis;
- the local authorities of Brézina and Taghit.

A special thank to Mr. Abdelkader Benkeira and Ms. Sabrina Rachedi (DGF Algeria), for their collaboration for the development of the project in Taghit, and to Dr. Marco Lauteri for useful discussions.

The authors wish to thank the anonymous referees for their helpful comments that contributed to improve the paper.

References

Bennadji A., Bennadji H., Bounaga N., Cheverry C. 1998 - *Béni-Abbés ou le dépérissement d'une palmeraie.* Sécheresse 9 (2): 131-137.

David J.S., Bellot J., Birot Y., David T.S. 2011 - *Water Fluxes in Forests.* In: Birot Y., Gracia C., Palahi M. (eds). "Water for forests and people in the Mediterranean: a challenging balance." What Science Can Tell Us 1: 32-36. Helsinki, European Forest Institute.

FAO & JRC 2012 - *Global forest land-use change 1990–2005*, by E.J. Lindquist, R. D'Annunzio, A. Gerrand, K. MacDicken, F. Achard, R. Beuchle, A. Brink, H.D. Eva, P. Mayaux, J. San-Miguel-Ayanz, H-J. Stibig. FAO Forestry Paper N. 169. FAO and European Commission Join Research Centre. Rome, FAO.

FAO 2013 - By various authors. *State of Mediterranean Forests 2013.* [Online]. Available: http://www.fao.org/docrep/017/i3226e/i3226e.pdf [2014]

FAO 2013 - By various authors . *Planted forests are a vital resource for future green economies.* Summary Report of the 3rd International Congress on Planted Forests based on three scientific workshops held in Bordeaux (France), Dublin (Ireland) and Porto (Portugal), and one plenary meeting held in Estoril (Portugal) from May 15th to 21st, 2013. [Online]. Available: http://www.fao.org/forestry/37902-083cc16479b4b28d8d4873338b79bef41.pdf [2014]

IPCC 2008 - By various authors. B.C. Bates, Z.W. Kundzewicz, S. Wu, J.P. Palutikof, Eds., 2008: *Climate Change and Water.* Technical Paper of the Intergovernmental Panel on Climate Change, IPCC Secretariat, Geneva, 210 p.

IPCC 2012 - By various authors. *Managing the Risks of Extreme Events and Disasters to Advance Climate Change Adaptation.* A Special Report of Working Groups I and II of the Intergovernmental Panel on Climate Change. C.B. Field, V. Barros, T.F. Stocker, D. Qin, D.J. Dokken, K.L. Ebi, M.D. Mastrandrea, K.J. Mach, G.-K. Plattner, S.K. Allen, M. Tignor, P.M. Midgley (eds.). Cambridge University Press, Cambridge, UK, and New York, NY, USA, 582 p.

Iremonger S., Gerrand A.M. 2011 - *Global ecological zones for FAO forest reporting, 2010.* Unpublished report. Rome, FAO.

Kadlec R.H, Knight R.L. 1996 - *Treatment Wetlands.* CRC Press, New York.

Mara D.D, Paerson H. W. 1999 - *A Hybrid waste stabilization pond and wastewater storage and treatment reservoir system for wastewater reuse for both restricted and unrestricted crop irrigation.* Water Resources 33 (2) : 591-594.

Masini L., Dalia S., Ori G.G., Bianconi P., Visentini C. 1988 - *Oued Saura: un fiume effimero nel Sahara nord-occidentale.* Giornale di Geologia 501 (2): 177-184.

Matteucci G., Vanclay J., Martin Vide J. 2011 - *Do Forest areas influence Rainfall Regime?* In: Birot Y., Gracia C., Palahi M. (eds). "Water for forests and people in the Mediterranean: a challenging balance." What Science Can Tell Us 1: 32-36. Helsinki, European Forest Institute.

Moreau S. 2005 - *Conservation de la biodiversité et gestion durable des ressources des sites de Mergues (M'sila), Oglet Ed Daira (Naama) et Taghit (Bechar) : Plan de gestion du site de Taghit-Gui.*

NASA 2007- Atmospheric Science data Center.

Otte Marinus L., Jacob Donna L. 2006 - *Constructed wetlands for phytorimediation, rhizofiltration, phytostabilisation and phytoextraction.* In : Phytoremediation Rhizoremediation, M. Mackova et al. (eds), Springer : 57-67. DOI: 10.1007/978-1-4020-4999-4_5

Salt David E., Blaylock M., Kumar Nanda P.B.A., Dushenkov V., Ensley Burt D., Chet I., Raskin I. 1995 - *Phytoremediation: A Novel Strategy for the Removal of Toxic Metals from the Environment Using Plants Nature Biotechnology* 13: 468 – 474.

US-EPA 1999 - Manual: *Constructed wetlands treatment of municipal wastewater.* EPA/625/R-99/010: Cincinnati, Ohio.

Vangronsveld J., Herzig R., Weyens N., Boulet J., Adriaensen K., Ruttens A., Thewys T., Vassilev A., Meers E., Nehnevajova E., van der Lelie D., Mench M. 2009 - *Phytoremediation of contaminated soils and groundwater: lessons from the field.* Environmental Science and Pollution Research 16: 765–794. DOI: 10.1007/s11356-009-0213-6

WHO 1989 - *Health Guidelines for Use of Wastewater in Agriculture and Aquaculture.* Technical Report Series, 778 p.

Kasbadji Merzouk N. 2000 - *Wind energy potential of Algeria, Renewable Energy* 21: 553-562.

19

Exploring forest infrastructures equipment through multivariate analysis: complementarities, gaps and overlaps in the Mediterranean basin

Sofia Bajocco[1*], Elisabetta Raparelli[1], Francesco Patriarca[1], Giovanni Di Matteo[1], Pierfrancesco Nardi[1], Luigi Perini[1], Luca Salvati[2], Giuseppe Scarascia Mugnozza[3]

Abstract - The countries of the Mediterranean basin face several challenges regarding the sustainability of forest ecosystems and the delivery of crucial goods and services that they provide in a context of rapid global changes. Advancing scientific knowledge and fostering innovation is essential to ensure the sustainable management of Mediterranean forests and maximize the potential role of their unique goods and services in building a knowledge-based bio-economy in the region. In this context, the European project FORESTERRA ("Enhancing FOrest RESearch in the MediTERRAnean through improved coordination and integration") aims at reinforcing the scientific cooperation on Mediterranean forests through an ambitious transnational framework in order to reduce the existing research fragmentation and maximize the effectiveness of forest research activities. Within the FORESTERRA project framework, this work analyzed the infrastructures equipment of the Mediterranean countries belonging to the project Consortium. According to the European Commission, research infrastructures are facilities, resources and services that are used by the scientific communities to conduct research and foster innovation. To the best of our knowledge, the equipment and availability of infrastructures, in terms of experimental sites, research facilities and databases, have only rarely been explored. The aim of this paper was hence to identify complementarities, gaps and overlaps among the different forest research institutions in order to create a scientific network, optimize the resources and trigger collaborations.

Keywords - FORESTERRA, infrastructures, multidimensional scaling, Mediterranean countries

Introduction

The countries of the Mediterranean basin, as well as those of other geographical areas with a Mediterranean climate, are facing problems regarding the sustainability of forest ecosystems and the provision of essential goods and services, in a context of rapid global change (Lindner 2010, Lafortezza et al. 2013).

The Mediterranean forest regions cover areas typically characterized by high intrinsic diversity in several inter-related aspects: ecological, environmental, cultural, economic, social and historical (Scarascia Mugnozza et al. 2000, Fabbio et al. 2003). For this reason it is important to advance scientific knowledge and promote innovation, which is essential to ensure the sustainable management of Mediterranean forests, maximizing the potential role of their goods and services in building a knowledge-based bio-economy of the region (Croitoru 2007, Merlo and Rojas 2000). The FP7 European Project FORESTERRA ("Enhancing FOrest RESearch in the MediTERRAnean through improved coordination and integration") has the objective to strengthen scientific cooperation on Mediterranean forest research through the integration and harmonization of research, infrastructure and existing databases through an ambitious transnational framework to reduce the fragmentation of existing forest research and maximize its impact (http://www.foresterra.eu/).

There are a number of regional organizations and research projects that collect data and compile information relevant to Mediterranean forests and related issues, such as FAO Silva Mediterranea Committee, AIMF, Biodiversity International, Blue Plan, CIHEAM, CIRCE of INGV, CMA and SEL of CRA, EFIMED, FOREST EUROPE, JRC (in particular EFFIS for forest fires), MMFN, OFME, UNECE, URFM of INRA, WWF, LFCC process, etc. (FAO 2011). Yet, each of these entities looks at this issue from different perspectives by collecting data in response to specific questions and needs or with a geographic focus which only embeds a part of the

[1] Consiglio per la Ricerca e la sperimentazione in Agricoltura, Unità di ricerca per la climatologia e la metereologia applicate all'agricoltura, Roma (CRA-CMA), Italy
[2] Consiglio per la Ricerca e la sperimentazione in Agricoltura, Centro di ricerca per lo studio delle Relazioni Pianta-Suolo, Roma (CRA-RPS), Italy
[3] University of Tuscia, Dipartimento per la Innovazione nei sistemi Biologici, Agroalimentari e Forestali, Viterbo (UNITUS -DIBAF), Italy
[*] corresponding author: sofia.bajocco@entecra.it

Mediterranean rim. The resulting evidence is that forest research in the Mediterranean is poorly exchanged and unequally developed (Saket 2002). In fact, while forest research is well developed in some European countries like France, Spain and Italy, its development is significantly lower in other countries where few resources are allocated to forest research (Palahi et al. 2008, Scarascia Mugnozza et al. 2000). Such gaps are particularly evident when dealing with infrastructures. In this perspective, the identification of Mediterranean countries whose level of forest research development and infrastructure equipment is different provides an opportunity to enhance the general level and to improve the quality of forest research in the area as a whole (Turner et al. 2003).

To the best of our knowledge, the equipment and availability of infrastructures, in terms of experimental sites, research facilities and databases, have only rarely been explored. Other scientific disciplines tried to analyze such issue (e.g. Kennedy et al. 2008), but no attention has been paid to forest research infrastructures so far.

According to the European Commission, research infrastructures are facilities, resources and services that are used by the scientific communities to conduct research and foster innovation (http://eurofed.stis.belspo.be/eurofed.asp?id=300;100;400&lang=en). They include: major scientific equipment (or sets of instruments); knowledge-based resources such as collections, archives or scientific data; e-infrastructures, such as data and computing systems and communication networks; and any other infrastructure of a unique nature essential to achieve excellence in research and innovation.

Research infrastructures play an increasing role in the advancement of knowledge and technology and of their exploitation (Archibugi and Pietrobelli 2003). Research infrastructures help structuring the scientific community and play a key role in the construction of an efficient research and innovation environment by offering high quality research services to users from different countries, by attracting young people to science and by networking facilities towards open, interconnected, data-driven and computer-intensive science and engineering.

In this perspective, within the framework of FORESTERRA project, the aim of this paper is twofold: (i) making a survey of the types of forest infrastructures available across the Mediterranean basin and (ii) identifying complementarities, overlaps and gaps among the different Mediterranean countries in terms of forest infrastructures, in order to create a scientific network, optimize the resources and trigger future collaborations.

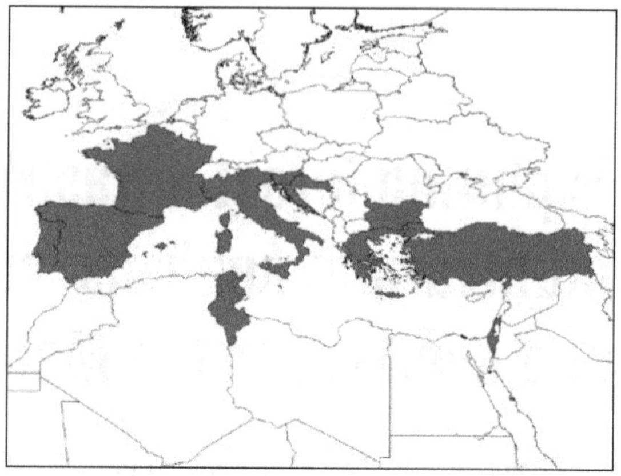

Figure 1 - Location of the FORESTERRA countries analyzed.

Study area

The Mediterranean basin extends for 3,800 km from east to west, starting from the head end of Portugal to the shores of Lebanon, and about 1,000 km from north to south, from Italy to Morocco and Libya. In the European Union, seven Member States are included in the Mediterranean region, some only partially (France, Portugal, Italy, Spain), wholly others (Greece, Malta, Cyprus).

All areas of the Mediterranean region are home to site-specific wild animals and plants, with a large number of species not detectable anywhere else in the world. The rate of endemism is exceptionally high: over half of the 25,000 flowering plants identified to date in the region, equivalent to about 10% of all known plants on Earth, is made up of endemic species. Mediterranean is thus among the areas of the world with the highest biodiversity (Mittermeier et al. 2004, Underwood et al. 2009).

Data

The FORESTERRA questionnaires on infrastructures were generated by reviewing and amending similar initiatives carried out previously (e.g. ERANET-ARIMNET and JPI-FACCE). A Mediterranean forestry research framework database was hence generated from the information gathered from the questionnaires received from eleven Mediterranean countries of the FORESTERRA Consortium: Spain, France, Italy, Turkey, Portugal, Tunisia, Bulgaria, Slovenia, Croatia and Greece (Fig. 1).

We detected nine-teen types of infrastructures (Tab. 1) covering the forest research activities of the FORESTERRA countries analyzed. We excluded from further analysis the infrastructures common to all countries: Experimental plantation & Forest research site for long-term monitoring; Forest Genetics and Biotechnology Lab; Forest Nursery, Rhizo/

Table 1 - List of forest infrastructures recorded through the questionnaires.

INFRASTRUCTURES
1. Arboretum, Forest genetic trial, Seed bank & Seed lab
2. Biomass, Economics & Wood Technology labs
3. Botanical garden & Botany lab
4. Carbon flux towers
5. Eco-physiology lab
6. Environmental, GIS and Forest Geomatics labs
7. Experimental plantation & Forest research site for long-term monitoring
8. Fire ecology labs
9. Forest Biochemistry lab
10. Forest Genetics and Biotechnology lab
11. Forest Hydrology lab
12. Forest Nursery, Rhizo/Phytotrons & Micropropagation lab
13. Forest Protection lab (entomology and pathology)
14. Meteorological station, lab & field equipments for climate change studies
15. Mobile Lab Unit
16. Silviculture & Forest biometry lab
17. Soil lab
18. Stable isotopes & Chemistry labs
19. Wood harvesting, Mechanization, Trasportation & Ergonomics lab

Phytotrons & Micropropagation lab; Meteorological station, lab & field equipments for climate change studies; Stable isotopes & Chemistry labs.

Methodology

A data matrix X was created having dimension i, j where i is the number of different types of research infrastructure while j is the number of considered countries. The matrix element x_{ij} was set up to one if the j-th country owns the i-th infrastructure, or to zero when such infrastructure is not present in the country.

The row vector x_i represents the distribution of the i-th infrastructure i within countries. The column vector x_j represents the infrastructural endowment of the country. A country owning all (none) infrastructure would have a vector x_j having all "1" (or "0").

The analysis was carried out in three steps. In the first step we analyzed the simple ordering of countries by number of infrastructures owned and of infrastructures by number of countries in which it is distributed. Then, we analyzed the relationships between pairs of countries in terms of mutual complementarities of infrastructures. In the last step we made use of multidimensional scaling technique to determines groups of countries similar by infrastructural equipment (Salvati and Zitti 2009). This allows to highlight common patterns of specialization.

The simple ordering of countries and infrastructures (step 1) was obtained by respectively summing the elements of the correspondent column or row vectors:

$$C_j = {}_i\Sum x_{ij} \qquad R_i = {}_j\Sum x_{ij}$$

In the step 2, we created the square matrixes I having dimension jxj in which each element represents a relationship between a pair of countries. Such relationship is expressed by the mean of a complementarity index between the two countries $j1$ e $j2$. We created three matrixes that correspond to three different complementarity indices: a general complementarity index I^1, a specific complementarity index I^2 and a relative specific complementarity index I^3. The general complementarity index I^1 represents the number of different infrastructures that are owned by one and only one of the two countries divided by the total number of different infrastructures I:

$$I^1_{j1,j2} = [{}_i\Sum |x_{ij1} - x_{ij2}|] / I$$

It takes maximum value 1 when the two countries are perfectly complementary without overlapping; it takes the minimum value of 0 when the countries own exactly the same kind of infrastructures. This index is symmetric (i.e. the value does not change inverting the two countries $j1$ and $j2$) and thus the matrix I^1 is symmetric. We can interpret this index as the overall result of a partnership between the two countries.

The specific complementarity index I^2 represents the number of infrastructures missing in the country J^1 and owned by the country J^2, divided by the total number of different infrastructures J:

$$I^2_{j1\,j2} = [{}_i\Sum (1-x_{ij1})\, x_{ij2}] / I$$

It takes the maximum value 1 when the country j^1 has no infrastructure while country J^2 has all the kind of infrastructures considered; it assumes the minimum value 0 when the countries have the same kind of infrastructures. It can be considered as a proxy for the effect of the partnership between the two countries from the point of view of country $j1$.

The relative specific complementarity index I^3 represents the number of infrastructure missing in the country J^1 and owned by the country J^2, divided by the number of infrastructures missing to the country J^1:

$$I^3_{j1,j2} = [{}_i\Sum (1-x_{ij1})\, x_{ij2}] / [{}_i\Sum (1-x_{ij1})]$$

It takes the maximum value 1 when the country J^2 has all the infrastructures missing to the country J^1. It takes the minimum value 0 when the country J^2 has none of the infrastructures missing in country J^1. It can be considered as the relative effect of the partnership between the two countries from the point of view of country J^1.

Finally, to determine a two-dimensional clas-

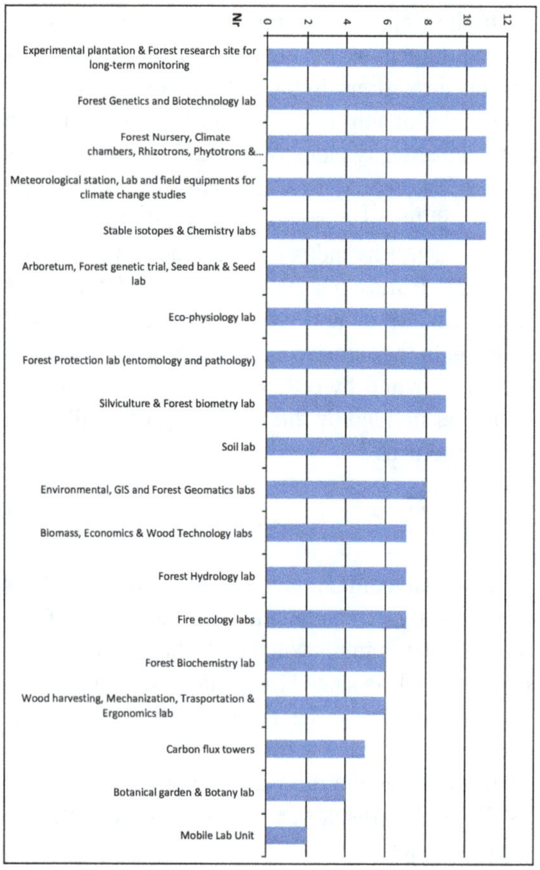

Figure 2 - Frequency graph of the infrastructures belonging to the forest research institutes of the FORESTERRA countries.

Table 2 - Ranking of countries by endowment of infrastructures.

FORESTERRA countries	Nr. of infrastructures
Italy	18
Turkey	17
Spain	17
France	17
Greece	16
Slovenia	15
Israel	14
Portugal	13
Croatia	11
Tunisia	9
Bulgaria	6

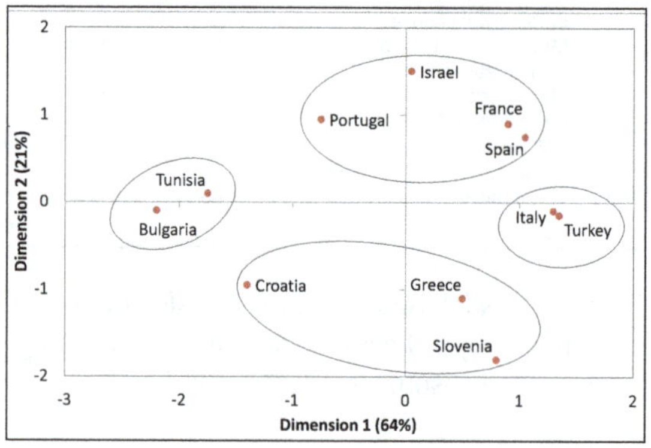

Figure 3 - Scatter plot of the MDS analysis.

sification of the infrastructural systems among the different countries, we performed a Multi-Dimensional Scaling (MDS) analysis. MDS is a means of visualizing the level of similarity of individual cases of a dataset. It refers to a set of related ordination techniques used in information visualization, in particular to display the information contained in a distance matrix. An MDS algorithm aims to place each object in N-dimensional space such that the between-object distances are preserved as well as possible. Each object is then assigned coordinates in each of the N dimensions. The number of dimensions of an MDS plot N can exceed 2 and is specified a priori. Choosing N=2 optimizes the object locations for a two-dimensional scatterplot (Borg and Groenen 2005, Honarkhah and Caers 2010).

Starting from the matrix X, we set up the square matrix of scalar products $P=X^TX$ having dimension JxJ (which corresponds to the matrix of the indices I^1 multiplied by J). We decided to apply an Euclidean metric to the matrix of distances between countries.

We applied to this matrix the dimensional reduction by finding sequentially the vectors that minimize a loss function of the distances from the points d.

Results and Discussion

The analysis allowed to identify the geographical distribution of the forest infrastructures across Mediterranean countries, identifying the most and the less common infrastructure (Fig. 2) and the most and less endowed country (Tab. 2).

The MDS analysis in two dimensions had a data fitting of about 85% (depending on Mardia's Fit Measure) with 64% in the first dimension and 21% of the second, respectively (Fig. 3). The first dimension summarizes the infrastructure that are substantially more common (greater weight to more variables present). Essentially, it repeats the ranking by number of infrastructure. While Turkey and Italy are almost overlapping and are well represented by the first dimension, the second dimension shows two models of specification in the countries immediately following in the list: on the one hand, Slovenia, Greece and Croatia, on the other Israel, Portugal , France and Spain. Bulgaria and Tunisia are confirmed to be poorly endowed by general infrastructure, and not equipped of specific infrastructure. The intermediate position on this dimension of the countries most endowed is compatible with the presence in these countries of both types of specializations.

Looking at the original data (not shown here), it is possible to identify the specific infrastructures that characterize the groups of countries classified by the second dimension: Carbon Flux Towers; Forest Biochemistry lab; Forest Nursery, Rhizo/Phytotrons & Micropropagation lab; Stable isotopes & Chemistry labs, for the countries at the top of the figure; Botanical garden & Botany lab; Eco-physiology lab; Arboretum, Forest genetic trial, Seed

Table 3 - Matrix of indices of general complementarity (number of complementary infrastructures on total infrastructure).

	Italy	Turkey	Spain	France	Greece	Slovenia	Israel	Portugal	Croatia	Tunisia
Turkey	15%	-								
Spain	20%	15%	-							
France	30%	25%	10%	-						
Greece	40%	25%	40%	40%	-					
Slovenia	35%	30%	45%	55%	45%	-				
Israel	40%	35%	20%	20%	60%	65%	-			
Portugal	45%	40%	35%	35%	55%	60%	25%	-		
Croatia	65%	70%	55%	55%	55%	50%	55%	50%	-	
Tunisia	65%	60%	55%	55%	55%	60%	45%	20%	40%	-
Bulgaria	80%	75%	70%	70%	60%	75%	50%	35%	35%	25%

Table 4 - Matrix of indices of specific complementarity (number of infrastructures obtained in the partnership by the row country on total infrastructure).

	Italy	Turkey	Spain	France	Greece	Slovenia	Israel	Portugal	Croatia	Tunisia	Bulgaria
Italy	-	5%	10%	5%	10%	5%	5%	0%	5%	0%	0%
Turkey	10%	-	10%	5%	5%	5%	5%	0%	10%	0%	0%
Spain	20%	15%	-	5%	15%	20%	0%	0%	5%	0%	0%
France	15%	10%	5%	-	15%	15%	0%	0%	5%	0%	0%
Greece	30%	20%	25%	25%	-	20%	25%	15%	10%	5%	0%
Slovenia	30%	25%	35%	30%	25%	-	30%	20%	10%	10%	10%
Israel	35%	30%	20%	20%	35%	35%	-	5%	15%	5%	0%
Portugal	45%	40%	35%	35%	40%	40%	20%	-	20%	0%	0%
Croatia	60%	60%	50%	50%	45%	40%	40%	30%	-	15%	5%
Tunisia	65%	60%	55%	55%	50%	50%	40%	20%	25%	-	5%
Bulgaria	80%	75%	70%	70%	60%	65%	50%	35%	30%	20%	-

Table 5 - Matrix of indices of relative specific complementarity (number of infrastructures obtained in the partnership by the row country on total infrastructure missing to the same country).

	Italy	Turkey	Spain	France	Greece	Slovenia	Israel	Portugal	Croatia	Tunisia	Bulgaria
Italy	-	50%	100%	50%	100%	50%	50%	0%	50%	0%	0%
Turkey	67%	-	67%	33%	33%	33%	33%	0%	67%	0%	0%
Spain	100%	75%	-	25%	75%	100%	0%	0%	25%	0%	0%
France	75%	50%	25%	-	75%	75%	0%	0%	25%	0%	0%
Greece	100%	67%	83%	83%	-	67%	83%	50%	33%	17%	0%
Slovenia	86%	71%	100%	86%	71%	-	86%	57%	29%	29%	29%
Israel	88%	75%	50%	50%	88%	88%	-	13%	38%	13%	0%
Portugal	82%	73%	64%	64%	73%	73%	36%	-	36%	0%	0%
Croatia	92%	92%	77%	77%	69%	62%	62%	46%	-	23%	8%
Tunisia	87%	80%	73%	73%	67%	67%	53%	27%	33%	-	7%
Bulgaria	89%	83%	78%	78%	67%	72%	56%	39%	33%	22%	-

bank & Seed lab for the countries in the lower part.

In conclusion, we can classify the forest research infrastructures equipment of the analyzed countries in four groups:

1) countries characterized by the presence of nearly all the infrastructure, consisting on Italy and Turkey;
2) countries with only a few general infrastructure (Tunisia and Bulgaria);
3) a "West Europe and Israel" model with large occurrence of Carbon Flux Towers, Forest Biochemistry lab, Stable isotopes and Microscopy Lab & Chemistry labs, composed of Israel, which is the most characteristic country of the group, and three Western Mediterranean countries (Portugal, France and Spain);
4) a "Balkan" model with large occurrence of Botanical garden & Botany lab, Eco-physiology lab and Arboretum, Forest genetic trial, Seed bank & Seed lab, composed of Slovenia, Greece and Croatia (which has less infrastructure but more specific than group 2).

Consequently, as also suggested by the complementarity indices analysis (Tab. 3, Tab. 4 and Tab. 5), while for the group 2 is necessary to develop agreements with most endowed countries (group 1 but also some in groups 3 and 4), bilateral partnerships could be fruitful in particular for groups 3 and 4 considering the different pattern of specialization. The unique bilateral partnership that allows full coverage are those integrating Italy (which also has the specific infrastructure of the two specialized models) with France or Greece. Agreements among more than two countries should include members of all the four groups.

Conclusions

In a framework of scientific and technical forest research, this study may help to: (i) foster coordination and integration of forest research activities at European level; (ii) endorse the multidisciplinary approach to address new research challenges through closer cooperation between different research fields (silviculture, ecology, physiology, climatology, molecular biology, pathology, wood technology, etc.); (iii) increase scientific and technical excellence; (iv) make easier the access to complementary research facilities and expertise between national and international institutions, optimizing the use of technical resources; (v) create support studies on the multifunctionality of the European forests.

Acknowledgements

The research leading to these results has received funding from the European Community's Seventh Framework Programme [FP7/2007-2013] under grant agreement n° [291832] - FORESTERRA. This Publication reflects only the authors views and the European Community is not liable for any use that may be made of the information contained herein.

The authors wish to thank the anonymous referees for the careful review.

References

Archibugi D., Pietrobelli C. 2003 - *The globalisation of technology and its implications for developing countries. Windows of opportunity or further burden?* Technological Forecasting & Social Change 70: 861–883.

Borg I., Groenen P. 2005 - *Modern Multidimensional Scaling: theory and applications* (2nd ed.). New York: Springer-Verlag, 207–212. ISBN 0-387-94845-7

Croitoru L. 2007 - *How much are Mediterranean forest worth?* Forest Policy and Economics 9: 536-545.

Fabbio G., Merlo M., Tosi V. 2003 - *Silvicultural management in maintaining biodiversity and resistance of forest in Europe – the Mediterranean region.* Journal of Environmental Management 67: 67-76.

FAO 2011 - *State of Mediterranean Forest (SoMF).* FAO Rome, Italy Press: 9-11

Honarkhah M., Caers J. 2010 - *Stochastic Simulation of Patterns Using Distance-Based Pattern Modeling.* Mathematical Geosciences 42: 487–517.

Kennedy A., Khoja T.A.M., Abou-Zeid A.H., Ghannem H., IJsselmuiden C. 2008 - *National health research system mapping in 10 Eastern Mediterranean countries.* La Revue de Santé de la Méditerranée orientale 14: 502-517.

Lafortezza R., Sanesi G., Chen J. 2013 - *Large-scale effects of forest management in Mediterranean landscapes of Europe.* iForest 6: 331-335.

Lindner M., Maroschek M., Netherer S., Kremer A., Barbati A., Garcia-Gonzalo J., Seid R., Delzon S., Corona P., Kolstroma M., Lexer M.J., Marchetti M. 2010 - *Climate change impacts, adaptive capacity, and vulnerability of European forest ecosystem.* Forest Ecology and Management 259: 698-709

Merlo M., Rojas Briales E. 2000 - *Public goods and externalities linked to Mediterranean forests: economic nature and policy.* Land Use Policy 17: 197-208

Mittermeier R.A., Gil P.R., Hoffmann M., Pilgrim J., Brooks T., Mittermeier C.G., Lamoreux J., Da Fonseca G.A.B. 2004 - *Hotspots Revisited: Earth's Biologically Richest and Most Endangered Terrestrial Ecoregions.* University of Chicago Press for Conservation International.

Palahi M., Mavsar R., Gracia C., Birot Y. 2008 - *Mediterranean forests under focus.* International Forestry Review 10: 676-688.

Saket M. 2002 - *Gaps in national-level information on forests and trees in developing countries.* Unasylva 53: 24-27.

Salvati L., Zitti M. 2009 - *The environmental 'risky' region: identifying land degradation processes through integration of socio-economic and ecological indicators in a multivariate regionalization model.* Environmental Management 44: 888-899.

Scarascia-Mugnozza G., Oswald H., Piussi P., Radoglou K. 2000 - *Forests of the Mediterranean region: gaps in knowledge and research needs.* Forest Ecology and Management 132: 97–109.

Turner K., Paavola J., Cooper P., Farber S., Jessamy V., Georgiou S. 2003- *Valuing nature: lessons learned and future research directions.* Ecological Economics 46: 493– 510.

Underwood E. C., Viers J. H., Klausmeyer K. R., Cox R. L., Shaw M. R. 2009 - *Threats and biodiversity in the Mediterranean biome.* Diversity and Distributions 15: 188–197.

20

Assessing impact of forest cover change dynamics on high nature value farmland in Mediterranean mountain landscape

Dora Cimini[1], Antonio Tomao[1*], Walter Mattioli[1], Anna Barbati[1], Piermaria Corona[2]

Abstract - A general trend of scrub encroachment and natural forest expansion (old-field succession) has been occurring in the past fifty years in Mediterranean mountain areas. While this phenomenon enhances a series of ecological processes and environmental services, it may represent a threat for biodiversity conservation when occurring on High Nature Value (HNV) extensive farmland. These open areas, created by traditional farming systems, have been recognized as a critically important European biodiversity resource. Understanding the drivers of Land Use and Land Cover Change (LUCC) and their implications on the dynamics of forest and HNV farmland habitats plays a crucial role in biodiversity conservation in protected areas. In this perspective, the case study here presented aims to model recent (1989-2008) LUCC in a National Park of the Central Apennines (Italy) and to produce a mid-term forecast (2008-2020). In the past twenty years, 57% of the land uptake by shrubland and 46% by forest has been occurring on former HNV farmland. Mid-term projections (2008-2020) confirm further HNV farmland decline. Localization of HNV farmland habitats vulnerable to change is a valuable decision-making tool to tradeoff in protected areas the conservation of traditional landscapes with the increase of forest and shrubland areas. In this perspective, concrete socio-economical and silvicultural measures are outlined to preserve these socially and economically-fragile open habitats.

Keywords - forest dynamics, Land Use and Land Cover Changes, spatial drivers, High Nature Value farmland

Introduction

Mountain environments of Southern Europe host significant biodiversity associated to a landscape mosaic, characterized by a high proportion of forest and low-input farmland habitats, like extensively grazed uplands, alpine meadows and pastures. While the key role of forests in biodiversity conservation is widely acknowledged, it is only recently that open habitats created by traditional farming systems have been recognized as a critically important European biodiversity resource (Baldock et al. 1993, Beaufoy et al. 1994), gaining the title of high nature value (HNV) farmland (EEA 2006, Paracchini et al. 2008). HNV farmland designates areas where agriculture supports, or is associated with, either a high species and habitat diversity or the presence of species of European conservation concern, or both. In fact, many grassland and heaths listed in the Annex I of the Habitats Directive depend on, or are associated with, extensive agricultural practices, like priority habitats 6210* (Semi-natural dry grassland and scrubland *facies* on calcareous substrates) and 6230* (Species-rich *Nardus* grassland on siliceous substrates in mountain areas). About half of Europe's endemic species depends on grassland and declining trends in the population of farmland birds and butterflies depend very much on the loss of HNV farmland (EEA 2010).

The landscape mosaic of forest and open habitats associated to extensive agriculture was relatively stable in Southern Europe until the 1950s. Since then, rapid changes in land use (intensification of agriculture, urbanisation, land abandonment and movement towards urban areas) has led to the widespread collapse of the socio-economic systems that supported these economically fragile farming systems.

The decrease of agricultural and pasture activities has determined the progressive colonization of grassland and agricultural lands by shrubland and forest. Moreover, the abandonment of management of trees outside forest in agricultural areas resulted in their spatial aggregation and the creation of new forest patches. In Mediterranean areas this general trend of scrub encroachment and natural forest expansion has been clearly documented in the past fifty years (Tatoni et al. 1994, Debussche et al. 1996, Bonet 2004, Pugnaire et al. 2006, Corona et al. 2008, Barbati et al. 2013). This phenomenon, also known as old-field succession, on one hand enhances a series of ecological processes and environmental

[1] University of Tuscia, Department for Innovation in Biological, Agro-food and Forest systems (DIBAF), Viterbo, Italy
[2] Consiglio per la Ricerca e la sperimentazione in Agricoltura, Forestry Research Centre (CRA-SEL), Arezzo, Italy
[*] corresponding author: antonio.tomao@unitus.it

services, but on the other hand may represent a threat for biodiversity conservation, when occurring on HNV farmland. The disappearance of cultural landscapes with a consequent increase of woodland cover is partially causing the decline of open habitats and associated species in the Mediterranean Basin (Finck et al. 2002, Mazzoleni et al. 2004, Bracchetti et al. 2012).

Understanding the drivers of Land Use and Land Cover Change (LUCC) and their implications on the dynamics of forest and HNV farmland habitats plays a crucial role in biodiversity conservation in protected areas. In this context, the development of robust models to project future LUCC might represent a valuable decision-making supporting tool to outline natural habitats vulnerable to change and to target land planning measures. According to this perspective, in this paper we present a case study in the Gran Sasso and Monti della Laga National Park (PNGSML), a high conservation value area, containing 32 Sites of Community Importance (SCI; Habitats Directive 92/43/EEC). The whole area is also recognized as Special Protection Area (SPA; Birds Directives 79/409/EEC).

The objectives of this study are: (i) to quantify the LUCC in the PNGSML focusing on recent dynamics of forest and shrubland expansion (period 1989-2008) and the related impacts on HNV farmland; (ii) to understand driving forces of the LUCC in the investigated area; (iii) to apply a LUCC model to predict future land cover scenario in order to identify areas and habitats most prone to LUCC in the upcoming decades.

Materials and methods

Study area and land cover history

PNGSML was established in 1995 and has become one of the largest protected areas in Italy. It spans over 143,300 ha in the Apennines mountains of central Italy (42° 10' N - 42° 50'N, 13° 10' E - 13° 16' E). Its elevation ranges from 225 to 2,908 m above sea level with an average elevation of 1,300 m above sea level (Fig. 1).

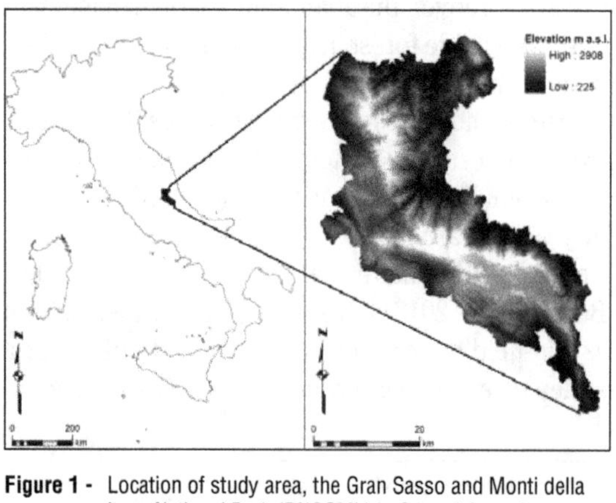

Figure 1 - Location of study area, the Gran Sasso and Monti della Laga National Park (PNGSML) in Central Apennines.

Table 1 - Park zoning of PNGSML

	Park zoning	Proportion of total PNGSML area (%)
A	strict nature reserve	11
B	general reserve with traditional sustainable use of natural resources by residents	41
C	protected landscape where new tourism, farming, pastoral and forestry activities may be permitted	39
D	economic and social promotion areas	9

Woodlands can be classified into six forest types: (i) beech forest (the dominant tree vegetation in the area), (ii) oak forest, (iii) hornbeams forest, (iv) chestnut forest, (v) mixed forest (which includes *Abies alba* remnants), (vi) riparian forest. HNV habitats are mainly represented by grassland notably priority habitats (Dir. 92/43/CEE) 6210* (Semi-natural dry grassland and shrubland *facies* on calcareous substrates) and 6230* (Species-rich *Nardus* grassland). According to the National Law on protected areas in Italy, Park zoning represents the main planning tool to pursue management objectives and to reconcile potential conflicts through a regulation of land use activities: zones vary from the strict level of protection (A Zone) to areas supporting sustainable land use activities (D Zone) (Tab. 1).

In the case of PNGSML, promoting local sustainable development becomes a key management issue also for the conservation of grassland, maintained by traditional pastoral activity. Like in other areas of the Apennines, the socioeconomic structure has been considerably altered after World War II. The population of mountain villages dropped greatly up to the first half of 1960s (Catorci 2007), while the inhabitants in the town of L'Aquila increased until 1970 and then stabilized. Today, the number of inhabitants living inside the PNGSML boundaries is 12% of the total population (137,635) of the 44 municipalities of the Park (ISTAT 2011).

Multi-temporal database and ancillary data

A summary of datasets processed for the study area is reported in Tab. 2. LUCC in the period 1989-2008 has been analyzed on the basis of high resolution land cover maps (minimum mapping unit: 0.5 ha) derived by visual interpretation of orthophotos images. Three broad land cover classes were adopted for mapping land cover in the considered period (1989-2008): (i) forest, (ii) shrubland, (iii) non-forest area. Forest and shrubland were classified according

Table 2 - Statistics of land use and land cover datasets used in the case study.

Year	Land cover datasets	Minimum mapping unit (ha)
1989	Land cover map from digital B/W orthophoto (TERRAITALY 1988/89)	0.5
1990	CLC	25
2006	CLC	25
2008	Land cover map from digital color) orthophoto (TERRAITALY 2008	0.5
2010	Land cover map from digital color orthophoto (Abruzzo Region)	0.5

to FAO/FRA2000 definition (UN/FAO 2001); non-forest class includes agricultural land, fallow land, grassland, bare ground and rocks, and settlements.

Besides the two maps at the year 1989 and at the year 2008 over the whole PNGSML territory, a third reference map, used as independent validation set, was also derived by the same method, to classify the land cover at 2010 for L'Aquila province territory included in PNGSML. To detect links between LUCC processes and the loss of HNV farmland, the Corine Land Cover (CLC) datasets (years 1990 and 2006) have been used to assess, on a coarser spatial scale, the distribution of HNV farmland in the PNGSML at the beginning and at the end of the reference period adopted for analyzing LUCC (1989-2008), following the methodology proposed by Paracchini et al. (2008).

The multi-temporal land cover databases were integrated with georeferenced ancillary information (Tab. 3). Spatial information on physiographic attributes (elevation, slope, aspect) was derived from the Digital Elevation Model (DEM, 20 m geometrical resolution). Road network was drawn from official cartographic sources provided by the PNGSML, urban areas were extracted from 2006 CLC map. As census observations indicate that population density in the study area was nearly stable during the investigated period (ISTAT 2011), inhabitants density at Municipality level for 1989 was used as reference data in this study. All spatial data were transformed to raster format with 20 m spatial resolution to match the DEM raster grid.

Forest cover change and future scenario

The integrated module of IDRISI Taiga Land Change Modeler (LCM) (Clark Labs, 2009) was used to model LUCC in the investigated period (1989-2008) and to predict a future scenario according to the steps showed in Fig. 2.

The LCM aggregates a Multi-layer Perceptron (MLP) and Markov Chains Analysis (MCA) for time prediction. The MLP neural network algorithm returns potential transition maps of each land cover class (Eastman 2009) modelling all transitions at once. The potential transition model serves as the basis for the future scenario of forest cover using a MCA. This process calculates the conditional probability of the state of a system at a given time (time t), based on the state of the system at the previous time (time t_0, with $t_0 < t$) (Reddy et al. 2009).

We assumed that (i) future LUCC dynamics (2008-2020) follows the same trend of the past twenty years (1989-2008), (ii) the changes in forest/shrubland cover most likely occur in areas close and/or adjacent to areas affected by recent changes. LCM has been run twice in this study: first to produce a short-term LUCC forecast (2010) and second to produce a mid-term forecast (2020).

The accuracy of the land cover forecast map at year 2010 was assessed by comparison with the reference map at the year 2010 for the L'Aquila province territory. Accordingly, a matrix of classification errors, Overall Accuracy (OA) index and Kappa Index of Agreement (KIA) were calculated; the Accuracy of User (AU) and the Producer (AP) were also estimated for each class (Congalton 1991).

Predictions of LUCC scenario for the year 2020 were processed in a hard prediction map to show the estimated spatial distribution of land cover classes according to results of the MCA.

Drivers of LUCC

The spatial pattern of areas affected by forest and shrubland expansion was analysed according to the set of independent spatial drivers (Tab. 1 and Tab. 3): (i) physiographic variables (aspect, elevation, slope); (ii) demographic variable (population density); (iii) distance variables, i.e. Euclidean distances between areas affected by LUCC and elements such as roadway infrastructure, town/village centres and forest edges; (iv) PNGSML zones.

A change/persistence density analysis was carried out to identify the spatial drivers most related to the spatial pattern of forest and shrubland expansion. The method is based on spatial overlay of change/persistence maps with the set of spatial drivers. Change/persistence maps were derived as

Table 3 - Statistics of spatial drivers datasets used in the case study.

Spatial drivers	Source	Unit	Mean	Coefficient of variation
slope	DEM	%	42.1	0.66
elevation	DEM	m	1,281	0.31
aspect	DEM	cosine	0.03	23.6
		sine	-0.004	177
population density	ISTAT	inhab. ha^{-1}	0.32	1.5
distance from town/village center	CLC	m	4,759	0.64
distance from road	PNGML	m	777	0.81
distance from forest edge	2008 land cover map	m	149	2.23

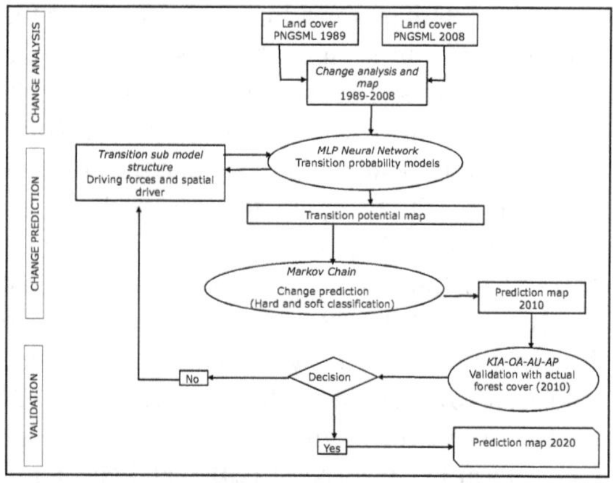

Figure 2 - Flow chart of the LUCC analysis.

Boolean raster pictures plotting the non-forest areas at the year 1989 converted to forest or shrubland at the year 2008 (change) and the remaining stable ones (persistence). The spatial overlay of the change/persistence maps with the geospatial ancillary information allowed to calculate the density of change or persistence by single spatial drivers, as follows:

$$d = \frac{r_i}{\sum_{i=1}^{n} r_i} \quad [1]$$

where d is density of change, r_i is the ratio between the area occupied by change and the whole surface in the i-th class of the considered independent variable and n is the total number of classes. Such an analysis allows comparing, by independent spatial drivers, the relative proportion of areas affected by changes and the remaining stable ones.

Results

LUCC and loss of HNV farmland

Land cover change in the past twenty years occurred in 4,871 ha of the PNGSML territory, corresponding to 3.4% of total area (Tab. 4). Forest spreading is the most relevant LUCC process, occurring on the 2.1% of the area; shrubland encroachment occurred on 1.2%. Forest spreading is mainly due to the evolution of areas covered by shrubland in 1989 into forest habitats (1.5% of the territory), thus causing a slight decline of shrubland habitats (-0.34%). Non-forest land cover classes affected by forest and shrubland expansion in the period 1989-2008 are identified by spatial overlay of the high resolution 2008 land cover map with 1990 CLC map (Tab. 5). A significant proportion of non/forest areas colonized by shrubland and forest (57% and 46%, respectively) were located in HNV farmland in 1990, to a large extent represented by grassland.

Table 4 - Transition matrix of LUCC from the year 1989 to the year 2008 (values expressed as percentage of the PNGSLM territory).

Land cover class		2008 shrubland	forest	non-forest
1989	*shrubland*	7.30	1.44	0.08
	forest	0.01	50.83	0.01
	non-forest	1.17	0.69	38.47
	total change	-0.34	2.11	-1.77

Table 5 - Land uptake by shrubland and forest in the period 1989-2008 by HNV and other land use classes.

	Land cover class	Shrubland ha	Shrubland % of total land uptake	Forest ha	Forest % of total land uptake
	olive groves	1.1	0.07	2.8	0.29
	pastures	17.2	1.03	23.8	2.39
	complex cultivation patterns	12.2	0.73	10.6	1.06
HNV farmland	lands principally occupied by agriculture	165.7	9.90	122.6	12.30
	natural grassland	749.4	44.80	290.4	29.14
	moors and heatland	2.3	0.14	3.5	0.35
	total of HNV land use classes	*947.9*	*56.66*	*453.7*	*45.54*
	discontinous urban fabrics	0.3	0.02	-	-
	arable lands	67.6	4.04	44.2	4.44
	broad-leaved forests	371.6	22.21	370.6	37.20
	coniferous forests	15.8	0.95	13.4	1.34
Other land use classes	mixed forests	7.6	0.46	30	3.01
	transitional woodland-shrubs	234.4	14.01	79.6	7.99
	bare rocks	6.2	0.37	0.04	0.00
	sparsely vegetated areas	21.5	1.29	4.7	0.48
	water bodies	0.05	0.00	-	-
	total of other land use classes	*725.1*	*43.34*	*542.5*	*54.46*

Characterization of forest cover change according to spatial drivers

Change/persistence density statistics by environmental spatial drivers are shown by Fig. 3 and Fig. 4. Forest and shrubland expansion occurred mainly on elevations below 1,000 m, with a low slope gradient (< 40%). Distance-related variables show that change was higher near the edge of forest (≤ 500 m) and near the roads (≤ 500 m). Aspect and proximity to urban settlements have a less clear effect. However, much change (40%) occurs in the neighbourhood of villages (distinctively, within the first 3 km). Concerning functional zoning of PNGSML, the areas corresponding to strict and integral protection (zone A and zone B) are more stable compared to the C and D zones.

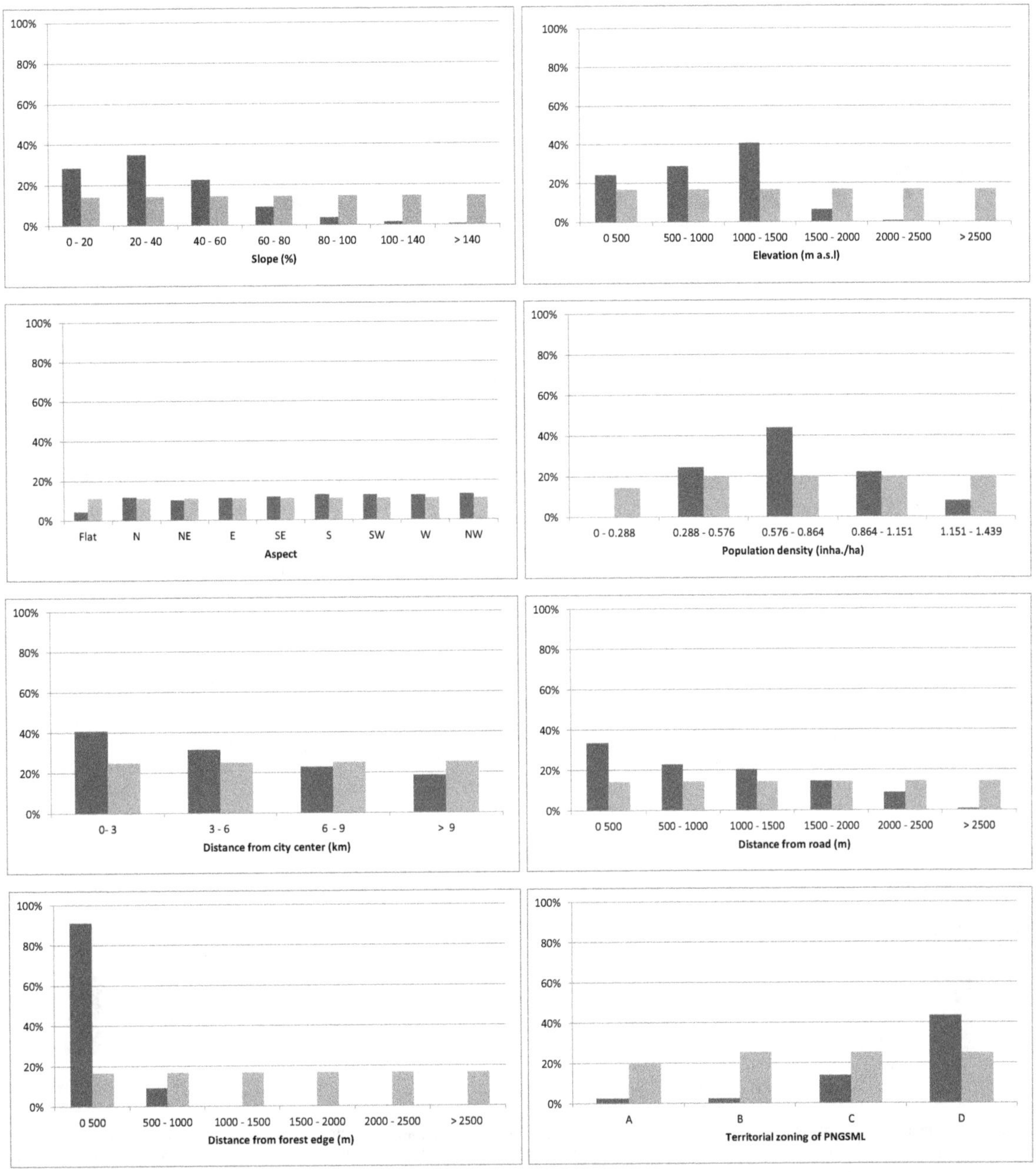

Figure 3 - Density (y-axis, values expressed as percentage) of change from non-forest areas to shrubland (dark grey) and no-change areas and other transitions (light grey) with respect to the considered spatial drivers. Data for the continuous variables (elevation, slope, aspect, density, distance from town/village center, distance from road and from forest edge) are shown by class intervals (x-axis).

Location and forest cover scenario

The transition probability matrix of LUCC in the period 2008-2010 is shown in Tab. 6. The diagonal cells represent the probability for a given land cover class to remain stable, whereas the off-diagonal values indicate the probability of a change from one land cover class to another. Results of the accuracy assessment of the model prediction at year 2010 are highly satisfactory (KIA and OA index > 0.99), as reported in Tab. 7.

The PNGSML areas prone to change cover class have been identified (Fig. 5) by projecting the transition potential observed in the period 1989-2008 into the future (year 2020), based on the transition probability matrix shown in Tab. 8. The projected landscape is characterized by an increase in forest areas and a persistence of shrubland (Tab. 9). Most non-forest land projected to be colonized by shrubland and forest in the period 2008-2020 is localized on areas currently covered by HNV farmland, 42% and 36% respectively (Tab. 10).

Discussion

LUCC analysis in the PNGSML territory reveals a general trend from 1989 to 2008 characterized by a continuous increase in forest cover, confirming the trend documented over Mediterranean Europe

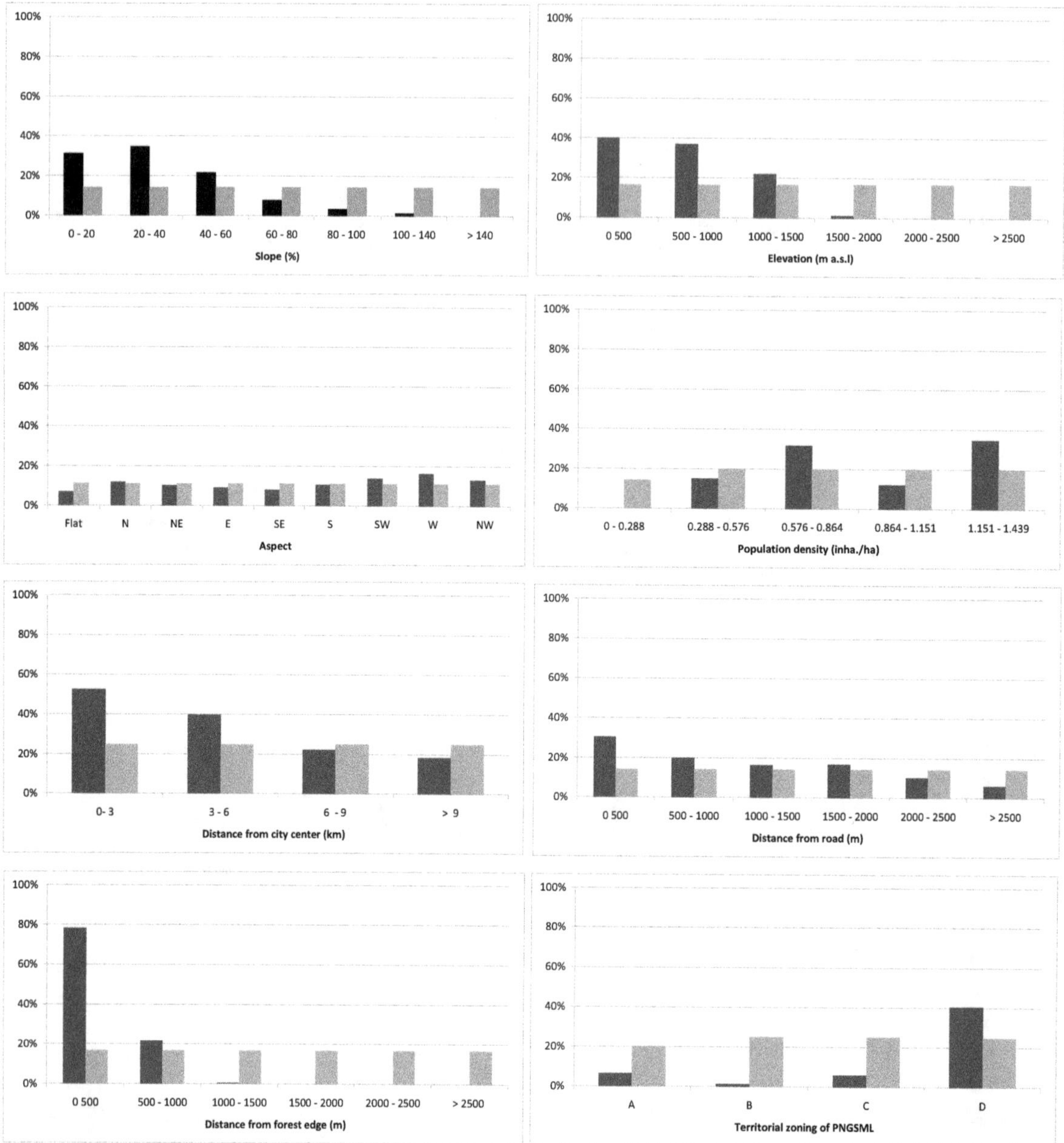

Figure 4 - Density (y-axis, values expressed as percentage) of change from non-forest areas to forest (dark grey) and no-change areas and other transitions (light grey) with respect to the considered spatial drivers. Data for the continuous variables (elevation, slope, aspect, density, distance from town/village center, distance from road and from forest edge) are shown by class intervals (x-axis).

since the early 1950s (Moreira et al. 2011). These observations indicate that a period of 19 years is sufficient for tree canopy closure from scattered trees in farmland and for the evolution of shrubland into forest. On one hand, this value is compatible with those reported by Bracchetti et al. (2012); on the other hand, our results also suggest that the considered period is sufficient for forest vegetation to colonize the most suitable sites.

Natural forest expansion produces complex and far reaching effects on rural environment: it brings production of direct and indirect benefits, mainly related to the reduction of water runoff and soil erosion (Garcia Ruiz et al. 1996, Tasser et al. 2003), the control of sediment yield (Boix-Fayos et al. 2008), the improvement of soil properties (Seeber and Seeber 2005), and the increase of presence of large vertebrates (Falcucci et al. 2007). Therefore, an enlargement of forest land is usually expected to provide the increase of ecosystem services. Conversely, this phenomenon can cause a sensible decrease of HNV farmland, including high value grassland as in this case study, which are extremely important to preserve rare plants and feeding habitats of many birds, and to maintain landscape

Table 6 - Markov transition probabilities for the period 2008-2010 derived by the 1989-2008 change analysis.

Land cover class	shrubland	forest	non-forest
shrubland	0.976	0.023	0.001
forest	0.000	0.999	0.001
non-forest	0.004	0.001	0.995

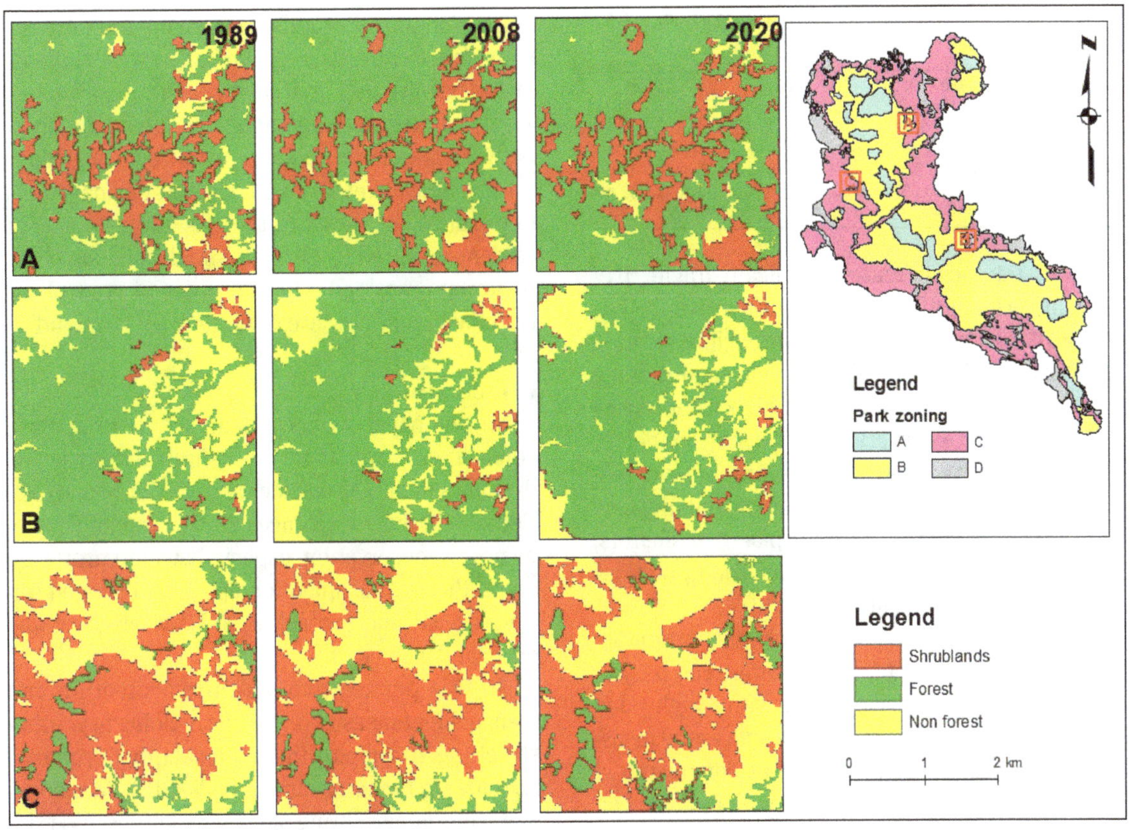

Figure 5 - Land cover maps at the year 1989 (observed), 2008 (observed) and 2020 (predicted) in three sample areas of the PNGSML.

diversity too (Bracchetti et al. 2012). In Mediterranean mountain landscapes, especially within areas designated for nature conservation, a tradeoff between the increase of forest and shrubland and the conservation of traditional landscapes is needed. In this perspective, the present study provides a methodological framework to spatially assess the localization of open natural habitats vulnerable to change. It represents a valuable decision-making support tool to better establish tradeoffs between forest expansion and HNV farmland protection.

Analyzing drivers of change and land cover general trend, we found that the increase in wooded areas is located mainly in mountain areas where a greater decrease in farming and pastoral activities has taken place, producing the same effect highlighted in other studies (Moreira et al. 2011). This phenomenon significantly decreases over 2,000 m a.s.l., confirming that in Central Apennines this altitude can be considered as a threshold for forest cover evolutionary dynamics. Distance from the mature woods is highly related to cover change. These edges are, in fact, the most susceptible to change (Matlack 1994, Oikonomakis and Ganatsas 2012). Also areas near roads are significantly affected by re-naturalization phenomena linked to the increase of the woods and scrubs: this may be explained, at least partially, by the fact that roads and other infrastructures (power lines, canals, railways) can

Table 7 - Summary of Kappa Index of Agreement (KIA) and User's (AU), Producer's (AP), and Overall (OA) accuracies for the map predictions at the year 2010.

Land cover class	AU	AP	KIA	OA
shrubland	0.977	0.999		
forest	0.998	0.999	0.999	0.995
non-forest	1.000	0.995		

Table 8 - Markov transition probabilities for the period 2008-2020 derived by 1989-2008 change analysis.

Land cover class	shrubland	forest	non-forest
shrubland	0.89	0.109	0.001
forest	0.00	1.00	0.00
non-forest	0.02	0.01	0.97

be considered as potential corridors for some generalist species (Spellerberg 1998), above all when their maintenance works are reduced or suspended.

The observed spatial drivers are not expected to change, so that the trend of HNV farmland loss, due to forest and shrubland expansion, might not be reversed, as confirmed by the mid-term scenario for 2020. Park zoning, that is the most important management and planning tool for habitat conservation, cannot provide protection against the loss of HNV farmland, as documented by the high level of HNV farmland land uptake within those zones with a low degree of protection (C and D zones). Concrete efforts are therefore required to counteract this phenomenon: specific approaches must be adopted to

Table 9 - Forest cover values at the years 1989 (observed), 2008 (observed) and 2020 (predicted).

Land cover class	1989		2008		2020	
	ha	%	ha	%	ha	%
shrubland	12,643	8.82	12,157	8.48	13,067	9.12
forest	72,867	50.85	75,897	52.96	76,579	53.44
non-forest	57,794	40.33	55,250	38.56	53,659	37.44

maintain the socially and economically-fragile HNV farming systems (EEA 2010). In this perspective, PNGSML has supported this approach, activating an European LIFE+ project to help maintaining and enhancing the presence of high value grassland by concrete conservation measures aimed at the sustainability of pastoral activities in mountain areas. This goal could be achieved by building new infrastructures such as watering places or shelters, but also by training farmers about good practices for pasture management (Felli 2013).

Moreover, silvicultural practices able to contain the expansion of shrubland and forest can be proposed. Cutting of invasive shrubs within high value farmland and grassland could be a successful measure to contrast grassland decline, as demonstrated in several LIFE+ projects in Europe (Toland et al. 2008). Thinning of woodlands near open areas within the forest can also create link-areas where spreading of pasture and grassland species is promoted. This practice can help enhancing the connectivity between grassland by linking many fragmented open areas (Seravelli et al. 2005). A careful management of forests near grassland, the most prone to change as demonstrated by this case study, can also help restoring grassland edges in order to control the colonization of new areas within HNV farmland.

Table 10 - Land uptake by shrubland and forest in the period 2008-2020, by HNV and other land use classes.

		Shrubland		Forest	
	Land cover class	ha	% of total land uptake	ha	% of total land uptake
HNV farmland	olive groves	0.3	0.03	5.5	0.37
	pastures	3.1	0.27	26.7	1.79
	complex cultivation patterns	1.7	0.15	63.4	4.24
	lands principally occupied by agriculture	16.3	1.43	185.2	12.38
	natural grassland	448.9	39.33	263.4	17.61
	moors and heatland	8.3	0.72	-	-
	total of HNV land use classes	*478.6*	*41.93*	*544.2*	*36,39*
Other land use classes	discontinous urban fabrics	0.05	0.00	0.7	0.05
	industrial areas	-	-	0.1	0.01
	arable lands	9.7	0.85	65.5	4.38
	broad-leaved forests	202.5	17.75	628.8	42.03
	coniferous forests	5.7	0.50	20.7	1.38
	mixed forests	3.4	0.30	9.6	0.64
	transitional woodland-shrubs	195.9	17.16	197.6	13.21
	bare rocks	61.1	5.36	4.7	0.31
	sparsely vegetated areas	178.8	15.67	22.8	1.52
	water bodies	5.5	0.48	1.1	0.07
	total of othe land use classes	*662.7*	*58.07*	*951.6*	*63.61*

Conclusions

In this paper we have demonstrated the close connection between forest and shrubland expansion dynamics in the last decades and HNV farmland loss. Because of the importance of HNV farmland in biodiversity conservation, especially within protected areas, the establishment of a tradeoff between forest increase and protection of HNV farmland and grassland is needed.

In this perspective, specific protection and management measures, such as building new infrastructures to support the sustainability of pastoral activities in mountain areas (Felli 2013), occasional and continuous cutting of invasive shrubs and thinning of forests near HNV grassland (Seravelli et al. 2005, Toland et al. 2008) could represent effective tools for the protection of these landscape mosaic elements strongly influenced by human traditional activities.

Under a more general framework, distinctive concern should be paid at territorial planning level to reduce the loss of biodiversity and habitats of traditional landscapes, that in Italy and in the Mediterranean mountains result from traditional agricultural and silvicultural activities of thousands of years of human presence.

Aknowledgements

The authors want to thank the anonymous reviewers for the helpful revision of the paper.

References

Baldock D., Beaufoy G., Bennett G., Clark J. 1993 - *Nature Conservation and New Directions in the Common Agricultural Policy.* Report for the Ministry of Agriculture, Nature Management and Fisheries, the Netherlands. Institute for European Environmental Policy, HPC, Arnhem, 224 p.

Barbati A., Corona P., Salvati L., Gasparella L. 2013 - *Natural forest expansion into suburban countryside: Gained ground for a green infrastructure?.* Urban Forestry & Urban Greening 12: 36 - 43. http://dx.doi.org/10.1016/j.ufug.2012.11.002

Beaufoy G., Baldock D., Clark J. 1994 - *The nature of farming. Low intensity farming systems in nine European countries.* Report IEEP/WWF/JNCC, London, Gland, Peterborough, 66 p.

Boix-Fayos C., de Vente J., Martinez-Mena M., Barbera G.G., Castillo V. 2008 - *The impact of land use change and check-dams on catchment sediment yield.* Hydrological Processes 22 (25): 4922–4935. http://dx.doi.org/10.1002/hyp.7115

Bonet A. 2004 - *Secondary succession of semi-arid Mediterranean old-fields in south-eastern Spain: insights for conservation and restoration of degraded lands.* Journal of Arid Environments 56: 213–233. http://dx.doi.org/10.1016/S0140-1963(03)00048-X

Bracchetti L., Carotenuto L., Catorci A. 2012 - *Land-cover changes in a remote area of central Apennines (Italy) and management directions.* Landscape Urban Planning 104: 157–170. http://dx.doi.org/10.1016/j.landurbplan.2011.09.005

Catorci A. 2007 - *Lineamenti storico-economici del paesaggio pastorale maceratese [Historical and economic features of the pastoral landscape in the Macerata area].* Braun-Blanquetia 42: 39–45.

Clark Labs 2009 - *IDRISI Taiga* 16.02. Clark Labs, Worcester, MA, 342 p.

Congalton R. 1991 - *A review of assessing the accuracy of classifications of remotely sensed data.* Remote Sensing of Environment 37: 35–46

Corona P., Calvani P., Mugnozza G.S., Pompei E. 2008 - *Modelling natural forest expansion on a landscape level by multinomial logistic regression.* Plant Biosystems 142 (3): 509–517. http:// dx.doi.org/10.1080/11263500802410850

Debussche M., Escarré J., Lepart J., Houssard C., Lavorel S. 1996 - *Changes in Mediterranean plant succession: old-fields revisited.* Journal of Vegetation Science 7: 519–526. http://dx.doi.org/10.2307/3236300

Eastman R. 2009 - *IDRISI Taiga: Guide to GIS and Image Processing.* Worcester, MA, 342 p.

EEA 2006 - *Progress towards halting the loss of biodiversity by 2010.* EEA Report No 5/2006. Luxembourg: Office for Official Publications of the European Communities, 100 p.

EEA 2010 - *Assessing biodiversity in Europe - the 2010 report.* EEA Report No 5/2010. Luxembourg: Office for Official Publications of the European Communities, 58 p. http://dx.doi.org/10.2800/42824

Falcucci A., Maiorano L., Boitani L. 2007 - *Changes in land-use/land-cover patterns in Italy and their implications for biodiversity conservation.* Landscape Ecology 22: 617–631. http://dx.doi.org/10.1007/s10980-006-9056-4

Felli G. 2013 - *Notiziario LIFE PRATERIE.* Ente Parco Nazionale del Gran Sasso e Monti della Laga. [Online] Available: http://www.lifepraterie.it/pdf/1-NL-Praterie.pdf [2013, December 31]

Finck P., Riecken U., Schröder E. 2002 - *Pasture landscapes and nature conservation. New strategies for preservation of open landscapes in Europe.* In: "Pasture landscapes and nature conservation", Redecker B., Finck P., Härdtle W., Riecken U., Schröder E. Eds., Springer, Berlin, 1–14.

Garcia Ruiz J.M., Lasanta T., Ruiz Flano P., Ortigosa L., White S., Gonzalez C., Martì C. 1996 - *Land-use changes and sustainable development in mountain areas: A case study in the Spanish Pyrenees.* Landscape Ecology 11 (5): 267–277.

ISTAT 2011 - *Bilancio demografico anno 2011 e popolazione residente dal 9 ottobre al 31 dicembre.* National Institute for Statistics. [Online] Available: http://demo.istat.it/bil20111009/index02.html [2013]

Matlack G.R. 1994 - *Vegetation dynamics of the forest edge–trends in space and succesional time.* Journal of Ecology 82: 113–123.

Mazzoleni S., di Pasquale G., Mulligan M., di Martino P., Rego F. 2004 - *Recent dynamics of the Mediterranean vegetation and landscape.* Chichester, UK: Wiley & Sons Ltd, 320 p. http://dx.doi.org/10.1002/0470093714.fmatter

Moreira F., Viedma O., Arianoutsou M., Curt T., Koutsias N., Rigolot E., Barbati A., Corona P., Vaz P., Xanthopoulos G., Mouillot F., Bilgili E. 2011 - *Landscape-wildfire interactions in southern Europe: implications for landscape management.* Journal of Environmental Management 92: 2389-2402. http://dx.doi.org/10.1016/j.jenvman.2011.06.028

Oikonomakis N., Ganatsas P. 2012 - *Land cover changes and forest succession trends in a site of Natura 2000 network (Elatia forest), in northern Greece.* Forest Ecology and Management 285: 153-163. http://dx.doi.org/10.1016/j.foreco.2012.08.013

Paracchini M.L., Petersen J.E., Hoogeveen Y., Bamps C., Burfield I., van Swaay C. 2008 - *High Nature Value Farmland in Europe. An estimate of the distribution patterns on the basis of land cover and biodiversity data.* European Commission, Joint Research Centre, Institute for Environment and Sustainability, 87 p.

Pugnaire F.I., Luque M.T., Armas C., Gutierrez L. 2006 - *Colonization processes in semi arid Mediterranean old-fields.* Journal of Arid Environments 65: 591–603. http://dx.doi.org/10.1016/j.jaridenv.2005.10.002

Reddy C.S., Rao L.R.M., Pattanaik C., Joshi P.K. 2009 - *Assessment of large scale deforestation of Nawarangpur district, Orissa, India: a remote sensing based study.* Environmental Monitoring and Assessment 154: 325–335. http://dx.doi.org/10.1007/s10661-008-0400-9

Seeber J., Seeber G.U.H. 2005 - *Effects of land-use changes on humus forms on alpine pastureland (Central Alps Tyrol).* Geoderma 124 (3–4): 215–222. http://dx.doi.org/10.1016/j.geoderma.2004.05.002

Seravelli M., Papini F., Samaden S., Venturi V. 2005 - *Gli interventi realizzati dalle comunità montane del Casentino e del Pratomagno.* In: "Conservazione delle praterie montane dell'Appennino toscano". Atti del Convegno finale del progetto Life Natura NAT/IT/7239, Borchi S. Ed., Comunità Montana del Casentino, Arti Grafiche Cianferoni, Stia (AR): 36-61.

Spellerberg I.F. 1998 - *Ecological effects of roads and traffic: a literature review.* Global Ecology and Biogeography 7: 317-333. http://dx.doi.org/10.1046/j.1466-822x.1998.00308.x

Tasser E., Mader M., Tappeiner U. 2003 - *Effects of land use in alpine grasslands on the probability of landslides.* Basic and Applied Ecology 4 (3): 271–280. http://dx.doi.org/10.1078/1439-1791-00153

Tatoni T., Magnin F., Bonin G., Vaudour J. 1994 - *Secondary succession on abandoned cultivation terraces in calcareous Provence. Vegetation and soil.* Acta Oecologica-International Journal of Ecology 15: 431–447.

Toland J., Jones W., Eldridge J., Thorpe E., O'hara E. 2008 - *LIFE and Europe's grasslands: restoring a forgotten habitat.* Office for Official Publications of the European Communities, 56 p.

UN/FAO 2001 - *Global Forest Resouces Assessment - Main Report.* [Online]. Available: http://www.fao.org/forestry/site/fra2000report/en. [2013]

21

Modelling Crown-Stem Diameters Relationship for the Management of *Tectona grandis* Linn f. Plantation in Omo Forest Reserve, Western Nigeria

Friday Nwabueze Ogana[1*]

Abstract - Crown-stem diameter relationship is important for sustainable forest management. Effective forest management requires routine inventory data of tree crops for decision making. Therefore, in this study, crown-stem diameter relationship was modelled for the management of *Tectonia grandis* L. f. plantation in Nigeria. The dataset consists of 1,919 trees measured from 35 sample plots. Three methods including ordinary least square (OLS), quantile regression (QR) and linear mixed model (LMM) were used to model the relationship between crown diameter (CD) and diameter at breast height (DBH). The model was used to estimate the limiting density, growing space and projection area. The result shows that the model from LMM was more suitable than OLS and QR. It explained 0.622 of the variation in CD with root mean squared error of 0.736. The study also shows e.g., that *T. grandis* trees of 25 cm DBH would occupy an equivalent area 36.32 m^2 and to avoid inter-tree competition the stand density should not exceed 275 Nha^{-1}. And if the plantation is to be thinned and clear-felled when mean DBH is 50 cm, density remaining after thinning would be 115 Nha^{-1} with average growing space of 9.3 m. This information would help in the management of the plantation.

Keywords - crown diameter; linear mixed model; ordinary least square; quantile regression; teak.

Introduction

Crown is an important part of trees which is required for photosynthesis and reproduction. The leaves on the crown absorb sunlight in the form of radiant energy for photosynthesis (Sharma et al. 2016). Photosynthesis is essential for the survival of tree crops. The larger the tree crown, the more foliage for photosynthesis; in consequence, more potential for carbon fixation (Zarnoch et al. 2004, Yang and Huang 2017). The condition of the crown gives information on the health status of the tree, because the crown often shows the first sign of deterioration (Zarnoch et al. 2004). Tree crowns are also required for bird nest and perching (Deng et al. 2003). Crown characteristics such as crown length and crown width/diameter, are routinely measured for quantification and better understanding of tree production efficiency, stability and vigor (Fu et al. 2017). They are also important for the prediction of "growth responses in spacing, thinning and fertilizer trials" (van Laar and Akca 2007). Crown characteristics with other growth variables have been used for forest growth and yield prediction (Gonzalez-Benecke et al. 2014, Fu et al. 2016).

Crown diameter (CD) has been defined as the "average of the widest axis of the crown and its perpendicular axis" (Zarnoch et al. 2004). It is a determinant of crown surface area, volume, crown form index, crown thickness index, linear crown index and crown spread ratio (van Laar and Akca 2007). CD has also been used to estimate canopy cover (Gill et al. 2000) and for crown profile analysis (Marshall et al. 2003, Raptis et al. 2018). Crown diameters are difficult to measure on the field, especially in forest stand with irregular edges (Kershaw et al. 2017, Fu et al. 2017). Thus, CD is often obtained as photo-derived variable from remote sensing (Gering 1995, Kershaw et al. 2017). Also, forest managers have depended on the ability to predict crown diameter from diameter at breast height/stump diameter based on a simplified relationship (e.g., Bechtold 2003, Foli et al. 2003, Hemery et al. 2005, Sönmez 2009, Pretzsch et al. 2015). This relationship can be established either through the method of ordinary least squared (OLS), linear mixed model (LMM) or quantile regression (QR)

Classic OLS has been recognized as one of the best methods of estimating conditional mean of response variables for a given value of predictor vari-

1 University of Ibadan, Ibadan, Nigeria - Department of Social and Environmental Forestry
Opposite Agbowo Shopping Complex, Off Sango-Ojoo Expressway Ibadan 200284
Nigeria

*Corresponding author: ogana_fry@yahoo.com

Table 1 - Descriptive statistics of the fitting (75%) and validation data sets (25%).

Variables	Fitting data (n = 1439)				Validation data (n = 480)			
	Mean	Max	Min	SD	Mean	Max	Min	SD
DBH (cm)	17.9	38.8	6.0	5.47	17.8	37.8	7.9	5.10
Height (m)	16.7	25.6	7.1	3.49	16.7	26.5	6.6	3.55
CD (m)	5.7	10.9	2.9	1.19	5.7	10.5	2.2	1.19
CPA (m^2)	27.4	94.1	6.6	11.45	27.3	87.4	3.8	11.59
BA (m^2)	0.027	0.118	0.002	0.01	0.027	0.112	0.004	0.01

N = 1919 trees; CD = crown diameter; CPA = crown projection area; BA = basal area; SD = standard deviation

able (Zhang et al. 2005). In OLS, the estimates of the parameters of CD-DBH relationship are obtained by minimizing the sum of squared residuals (Hao and Naiman 2007). It is however, most reliable only if the assumption of normality, homoscedasticity and independence are met (Sun et al. 2017). The problem of independence is overcome in linear mixed model. Most inventory data used for modelling are hierarchical or nested in nature i.e., tree in plot; such data structure is effectively modelled using mixed-effect approach. This approach considered both within and between-plots variability. Thus, because of the fixed and random parameters in linear mixed model, the covariate effect and heterogeneity between plots are better explained (De-Miguel et al. 2012, Sun et al. 2017). The quantile regression (QR) is another approach that is gaining wide application in forestry. In QR, the problem of optimization is solved by minimizing the sum of absolute residuals (Zhang et al. 2005). It was introduced by Koenker and Bassett (1978). And since then, several researchers have applied this method to forest resources management because of its relative flexibility and robustness (e.g., Zhang et al. 2005, Gao et al. 2007, Mehtatalo et al. 2008, Ducey and Knapp 2010, Bohora and Cao 2014, Gao et al. 2016, Sun et al. 2017, Raptist et al. 2018, Ozcelik et al. 2018). QR provides detailed estimation to different parts of the dependent/response distribution without restriction on the assumption of the error term (Zhang et al. 2005, Sun et al. 2017).

Tectona grandis L. f. (teak) is a tropical hardwood species belonging to the family of Verbenaceae. It originated from the Asia-Pacific and can be found in some tropical countries including Brazil, Costa Rica, Kenya, Panama, Togo and Nigeria (Sousa et al. 2011). It was introduced in Nigeria to augment other slow-growing indigenous tree species and has been widely cultivated across length and breadth of the country. *T. grandis* is an important timber species that is fast-growing, it is of excellent anatomical and physical properties (e.g., it is durable, resistant to termite, easy of working, etc.) (Chukwu and Osho 2018). *T. grandis* can be used for different purpose because of its excellent properties and as such, there is high market demand for its products (Miranda et al. 2011). Thus, the need for its management. Effective forest management requires routine inventory data of the tree crops for decision making. The knowledge of crown diameter-DBH relationship can be used to determine tree spacing, stand density, basal area per ha, thinning regimes etc. (Hemery et al. 2005, Pretzsch et al. 2015). This information is relevant for sustainable forest management. Therefore, the main purpose of this study was to develop crown diameter-DBH relationship for the management of *T. grandis* plantation in Nigeria using different modelling methods.

Methodology

Data

The *Tectona grandis* L. f. data set from Omo Forest Reserve was used in this study. The plantation occupies an area of 139,100 ha and its lies between Latitudes 6°35′ to 7°03′N and Longitudes 4°09′ to 4°40′E of Ogun State, Nigeria (Chukwu and Osho 2018). The data set consists of 1919 trees measured from 35 sample plot of 25 x 25 m size. Diameter at breast height (DBH), total height and crown diameter (CD) were measured to the nearest 0.1 cm, 0.1 m and 0.3 m, respectively. CD was measured as the linear distance between edges of the tree crown in a north-south and east-west directions. The mean value of north-south and east-west measures was recorded as the crown diameter. The data set was randomly split into two groups – 75% (1.439) and 25% (480). The 75% was used for model fitting while the 25% was used for model validation. The descriptive statistics of the fitting and validation data sets are presented in Table 1.

Crown diameter (CD)-DBH relationship

Several studies have established linear relationship between tree crown diameter (CD) and diameter at breast height (DBH) for different species (Gering et al. 1995, Foli et al. 2003, Hemery et al. 2005, Adesoye and Ezenwenyi 2014, Yang and Huang 2017, Raptis et al. 2018). Furthermore, Dawkins

(1963) showed that the relationship between crown diameter and DBH can be effectively represented by straight lines. In this study, a scatterplot of crown diameter and DBH also showed a linear relationship with positive intercept (Fig. 1).

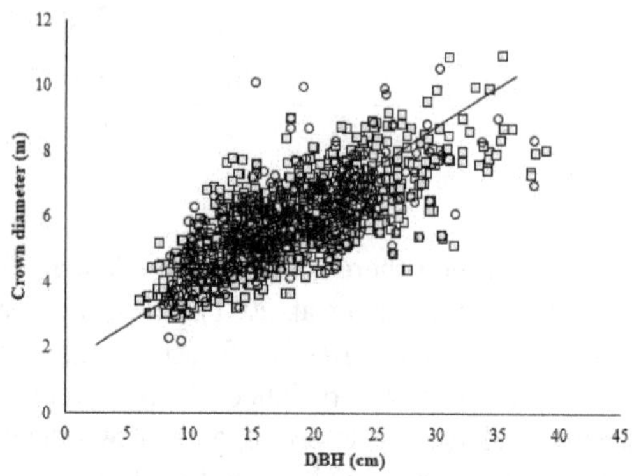

Figure 1 - Scatterplots of the relationship between crown diameter and DBH for the fitting data (squares) and validation data (circles).

Thus, a linear relationship was considered in this study. The model expression is given by:

$$CD_{\kappa i} = b_0 + b_1 DBH_{\kappa i} + \epsilon_{\kappa i} \quad (1)$$

Where CD = crown diameter (m), DBH = diameter at breast height (cm), and = intercept and slope, respectively, = error term which is assumed to be normal and independent with a zero mean and constant variance i.e., . The subscript i and k = individual tree i of plot k.

Fitting methods

Three fitting methods were considered in this study including ordinary least square (OLS), quantile regression (QR) and linear mixed model (LMM).

Ordinary Least Square (OLS)

This method estimates the conditional mean of the response variable i.e., crown diameter given a fixed value of another variable. The OLS estimated the parameters (and) in Eq.1 by minimizing the sum of squared residuals of the function (Hao and Naiman 2007). It is however, most reliable only if the assumptions of normality, homoscedasticity and independence are met (Raptist et al. 2018).

Quantile Regression (QR)

Similar equation form (Eq. 1) was applied to estimate the τth crown diameter quantile:

$$\hat{y}_\tau(CD) = b_0 + b_1\, DBH \quad (2)$$

Where is the estimated value of the τth quantile of crown diameter at diameter of breast height (DBH), the parameters and from the quantile regression were obtained by minimizing the sum of absolute residual expressed as:

$$\hat{\beta}_\tau = argmin_{\beta \epsilon R^2} \sum_{i=1}^{n} \rho_\tau (y_i - x'_i \beta) \quad (3)$$

Where = parameter and ; τ = quantiles. Different quantiles were used including 0.3, 0.45, 0.5, 0.55 and 0.60 were considered.

Linear Mixed Model (LMM)

In contrast to OLS and QR, the linear mixed model accounts for both within and between plot variations. LMM has both fixed and random parameters. Two model forms were considered in this study. In the first LMM form (hereafter called LMM1), the intercept (a) was assumed to vary between plots while the slope (b) coefficient was fixed, expressed as:

$$CD_{\kappa i} = (b_0 + a_\kappa) + b_1 DBH_{\kappa i} + \epsilon_{\kappa i} \quad (4)$$

Where is the random component. The covariance between trees i and i' of plot k is given by:

$$cov(a_\kappa + \epsilon_{\kappa i}, a_\kappa + \epsilon_{\kappa i'}) = cov(a_\kappa, a_\kappa) + cov(a_\kappa, \epsilon_{\kappa i'}) + cov(a_\kappa, \epsilon_{\kappa i}) + cov(\epsilon_{\kappa i}, \epsilon_{\kappa i'}) = var(a_\kappa) = \sigma_a^2 \quad (5)$$

In the second LMM form (hereafter called LMM2), both the intercept and slope were assumed to vary between plots (Eq. 6).

$$CD_{\kappa i} = (b_0 + a_\kappa) + (b_1 + c_\kappa) DBH_{\kappa i} + \epsilon_{\kappa i} \quad (6)$$

Where and are the random component, i.e., plot effects which have multivariate normal distributions with zero mean and variance-covariance matrix (D) defined as:

$$D = var\begin{pmatrix} a_k \\ c_k \end{pmatrix} = \begin{pmatrix} var(a_K) & cov(a_K, c_K) \\ cov(a_k, c_k) & var(c_K) \end{pmatrix} \quad (7)$$

Other parameters and variables are previously defined. Analysis was carried out in R (R Core Team 2017).

Model assessment and validation

In this study, both quantitative and qualitative analyses were used to assess the different modelling methods. Quantitative assessment includes the use of adjusted coefficient of determination (R^2_{adj}), efficiency index (EI), root mean square error (RMSE), mean absolute bias (MAB), relative mean absolute error (RMA) and Akaike Information Criterion (AIC). Methods with high R^2_{adj} and low EI, RMSE, MAB, RMA and AIC were regarded as good. Quali-

Table 2 - Estimated parameters from ordinary least square (OLS), quantile regressions (QR) and linear mixed models (LMM) and their fit indices.

Methods						†R2adj	‡EI	‡RMSE	‡MAB	‡RMA (%)	‡AIC
OLS	2.929*	0.159*				0.526	0.472	0.824	0.642	11.21	3533
QR											
30%	2.529*	0.156*				0.390	0.608	0.935	0.724	13.69	3644
45%	2.822*	0.157*				0.512	0.486	0.836	0.647	11.57	3589
50%	2.889*	0.159*				0.525	0.473	0.825	0.641	11.27	3600
55%	2.991*	0.158*				0.524	0.474	0.826	0.646	11.17	3627
60%	3.012*	0.163*				0.510	0.489	0.838	0.661	11.23	3669
Mixed											
LMM1	2.881*	0.161*	0.768	0.302		0.597	0.401	0.760	0.586	10.21	3414
LMM2	2.939*	0.159*	0.750	0.578	0.033	0.622	0.377	0.736	0.571	10.01	3385

* means significant at 5%; † means the larger the value the better the method; ‡ means the smaller the value the better the method; Joint test of equality of slopes for the quantiles = 0.342

tative assessment was done by visual observation of the residual graphs of the different methods for the validation data set.

$$R^2_{adj} = 1 - \frac{(n-1)\sum_{i=1}^{n}(y_i-\hat{y}_i)^2}{(n-p)\sum_{i=1}^{n}(y_i-\bar{y}_i)^2} \quad (8)$$

$$EF = \frac{\sum_{i=1}^{n}(yi-\hat{y}_i)^2}{\sum_{i=1}^{n}(y_i-\bar{y}_i)^2} \quad (9)$$

$$RMSE = \sqrt{\frac{\sum_{i=1}^{n}(y_i-\hat{y}_i)^2}{n}} \quad (10)$$

$$MAB = \frac{\sum_{i=1}^{n}|y_i-\hat{y}_i|}{n} \quad (11)$$

$$RMA(\%) = \frac{\sum_{i=1}^{n}\left|\frac{y_i-\hat{y}_i}{\hat{y}_i}\right|}{n} x100 \quad (12)$$

$$AIC = -2\log lik + 2p \quad (13)$$

Where: n = sample size, p = number of parameters; = average tree height; Y_i is the observed value and is the theoretical value predicted by the model. The models were also validated using independent data set (i.e., test data).

Results And Discussion

Crown diameter-stem diameter model

The estimated parameters and fit indices from ordinary least squared (OLS), quantile regression (QR) and linear mixed model (LMM) for modelling the relationship between crown diameter (CD) and DBH of *T. grandis* are presented in Table 2. The results show that all methods had positive intercept which lied between the range of 2.5 and 3.5. This is expected for CD-DBH relationship in most tree species. Hemery et al. (2005) reported positive intercepts (0.6 – 2.8) for CD-DBH relationship in 11 temperate species. Although in some exceptional cases the intercept could be negative which is akin to the Dawkins's "type 3 behaviour". For example, Foli et al. (2003) reported a nonsignificant negative intercept for *Entandrophragma angolense* C. DC. among other tropical tree species investigated. The positive intercept has management implication as it suggests that "the stand basal area could be allowed to increase towards maturity because CD-DBH ratio decreases with stem size" (Foli et al. 2003). The slope values for OLS, median regression (QR at 50%) and linear mixed model with random intercept and slope (LMM2) were approximately the same (0.159 up to 3 decimal places). In addition, QR provides detailed information on the prediction of the response variable compared to OLS. For example, if multiple quantiles were to be considered, 0.05 to 0.95 say, the different/similar slope values from these quantiles will give an indication of heteroscedasticity/homoscedasticity (Koenker and Hallock 2001). In this study, the slope values of the five quantiles indicated homoscedasticity. Also, the joint test of equality of slopes for the five quantiles was not significant at 5% level (p = 0.432)

The quantitative assessment of the methods showed that LMM2 had the highest R^2_{adj}, and least EI, RMSE, MAB, RMA and AIC values. Furthermore, the loglikelihood ratio tests of LMM2, LMM1 and the base model (LMM2 vs LMM1, LMM2 vs OLS and LMM1 vs OLS) were significant at 5% level (not documented). OLS and median regression were relatively the same. The performance of linear mixed model is not unexpected. This method has both fixed and random effects parameters which account for covariate effects (DBH) and heterogeneity be-

Table 3 - Evaluation of the methods using validation data set.

Methods	EI	RMSE	MAB	RMA (%)
OLS	0.579	0.9078	0.6927	12.10
Quantiles				
30%	0.716	1.0091	0.7480	14.19
45%	0.593	0.9186	0.6934	12.42
50%	0.580	0.9087	0.6910	12.16
55%	0.581	0.9095	0.6973	12.05
60%	0.597	0.9216	0.7127	12.11
Mixed Model				
LMM1	0.507	0.8493	0.6350	11.07
LMM2	0.482	0.8279	0.6239	10.89

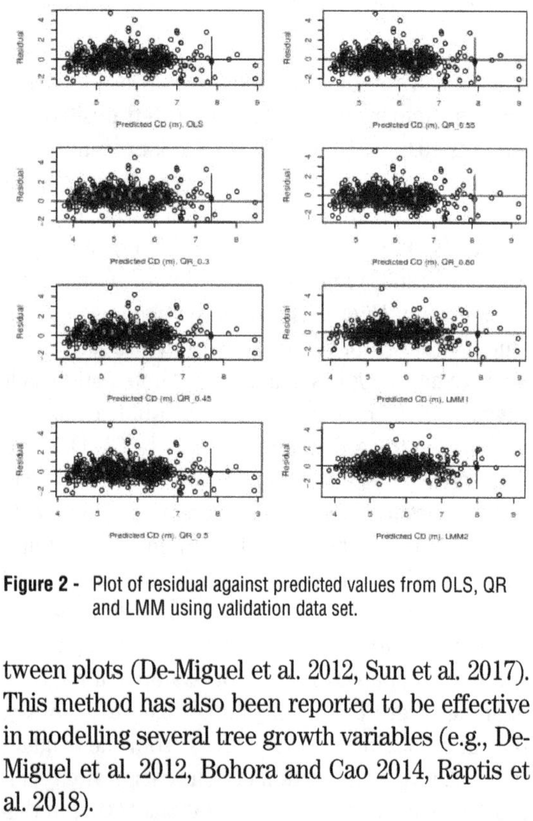

Figure 2 - Plot of residual against predicted values from OLS, QR and LMM using validation data set.

tween plots (De-Miguel et al. 2012, Sun et al. 2017). This method has also been reported to be effective in modelling several tree growth variables (e.g., De-Miguel et al. 2012, Bohora and Cao 2014, Raptis et al. 2018).

The performance of LMM2 was still better than other methods with respect to the validation data set (Table 3). Its efficiency index (EI), root mean square error (RMSE), mean absolute bias and relative mean absolute error (RMA) values were the least. Furthermore, residual trend was assessed by plotting the residuals and predicted CD in 10 classes using validation data set (Fig. 2). Thin and thick lines corresponding to the class-specific standard deviation and 95% confidence interval, respectively were added to the class means. The red colours are thick lines that do not intersect the horizontal line and as such, indicate poor prediction. Only in the first class did LMM performed poorly while OLS and QR (50) predicted poorly in the 4 and 5th classes. In addition, the thin lines did not show heteroscedasticity as function of CD, as such the constant variance assumption was not violated.

Application of CD-DBH relationship

One important application of CD-DBH relationship to forest management is the determination of sustainable thinning regimes/schedules. Hemery (2005) asserted that such thinning schedules are usually built on average diameter (DBH) of the stand to be realized. The estimated parameters from ordinary least square (OLS), quantile regression (0.5) (QR) and linear mixed models (LMM) with random intercept and slope were further applied to prescribe thinning regime for the *T. grandis* plantation in Omo Forest Reserve. Table 4 shows the estimated crown diameter (CD), CD-DBH ratio (R), crown area (CPA), number of trees per ha (10000/CPA), stand basal area (10000 x $(\pi/4)/R^2$) and the average spacing required (Sp). From the table, different thinning regimes can be considered. For example, if an average DBH of 25 cm is considered for the *T. grandis*, the expected crown diameter would be 6.9 m, 6.9 m and 6.8 m, predicted from OLS, QR and LMM methods, respectively. Estimations from LMM would be addressed here since it performed relatively better than OLS and QR methods. Thus, given an average crown diameter of 6.8 m, it would occupy an equivalent area 36.32 m^2. And to avoid inter-tree-crown competition (for space, light, etc.) in the stand when DBH is 25 cm, the stand density (N tree per ha) should not exceed 275 Nha^{-1}. This is the limiting density needed for complete canopy (Foli et al. 2003).

Furthermore, supposing the trees are thinned and clear-felled when mean DBH is 50 cm, the expected crown diameter would be 10.5 m and the area occupied would be 86.59 m^2. The stand density remaining after thinning would be 115 Nha^{-1} with an average growing space of 9.3 m. The growing space was obtained from the square root of the ratio of 1

Table 4 - Stand characteristics of the mahoe plantation at LCS in each inventory, showing mean values with standard errors when applicable.

	Ordinary Least Square (OLS)						Quantile (0.5) regression (QR)						Linear Mixed Model (LMM)					
DBH cm	CD m	R	CPA m2	N ha-1	G ha-1	Sp m	CD m	R	CPA m2	N ha-1	G ha-1	Sp m	CD m	R	CPA m2	N ha-1	G ha-1	Sp m
10.0	4.5	45.00	15.90	628	3.88	4.0	4.5	45.00	15.90	628	3.88	4.0	4.6	46.00	16.62	601	3.71	4.1
15.0	5.3	35.33	22.06	453	6.29	4.7	5.3	35.33	22.06	453	6.29	4.7	5.3	35.33	22.06	453	6.29	4.7
20.0	6.1	30.50	29.22	342	8.44	5.4	6.1	30.50	29.22	342	8.44	5.4	6.1	30.50	29.22	342	8.44	5.4
25.0	6.9	27.60	37.39	267	10.31	6.1	6.9	27.60	37.39	267	10.31	6.1	6.8	27.20	36.32	275	10.62	6.0
30.0	7.7	25.67	46.57	214	11.92	6.8	7.7	25.67	46.57	214	11.92	6.8	7.5	25.00	44.18	226	12.57	6.7
35.0	8.5	24.29	56.75	176	13.31	7.5	8.5	24.29	56.75	176	13.31	7.5	8.3	23.71	54.11	184	13.97	7.4
40.0	9.3	23.25	67.93	147	14.53	8.2	9.3	23.25	67.93	147	14.53	8.2	9.0	22.50	63.62	157	15.51	8.0
45.0	10.1	22.44	80.12	124	15.6	9.0	10.0	22.22	78.54	127	15.91	8.9	9.8	21.78	75.43	132	16.56	8.7
50.0	10.9	21.80	93.31	107	16.53	9.7	10.8	21.60	91.61	109	16.83	9.6	10.5	21.00	86.59	115	17.81	9.3
55.0	11.7	21.27	107.51	93	17.36	10.4	11.6	21.09	105.68	94	17.66	10.3	11.3	20.55	100.29	99	18.60	10.1
60.0	12.5	20.83	122.72	81	18.1	11.1	12.4	20.67	120.76	82	18.38	11.0	12.0	20.00	113.10	88	19.63	10.7
65.0	13.3	20.46	138.93	71	18.76	11.9	13.2	20.31	136.85	73	19.04	11.7	12.8	19.69	128.68	77	20.26	11.4
70.0	14.1	20.14	156.15	64	19.36	12.5	14.0	20.00	153.94	64	19.63	12.5	13.5	19.29	143.14	69	21.11	12.0
75.0	14.9	19.87	174.37	57	19.89	13.2	14.8	19.73	172.03	58	20.18	13.1	14.3	19.07	160.61	62	21.60	12.7
80.0	15.7	19.62	193.59	51	20.4	14.0	15.6	19.50	191.13	52	20.65	13.9	15.0	18.75	176.71	56	22.34	13.4
85.0	16.5	19.41	213.82	46	20.85	14.7	16.4	19.29	211.24	47	21.11	14.6	15.7	18.47	193.59	51	23.02	14.0
90.0	17.3	19.22	235.06	42	21.26	15.4	17.2	19.11	232.35	43	21.51	15.2	16.5	18.33	213.82	46	23.38	14.7
95.0	18.1	19.05	257.30	38	21.64	16.2	18.0	18.95	254.47	39	21.87	16.0	17.2	18.11	232.35	43	23.95	15.2

CD = crown diameter, CPA = crown projection area, N= number of tree per ha, G = basal area per ha; R = CD/DBH(m); Sp = Spacing

hectare of an area to the survival number of trees per ha (Hafley and Buford 1985). However, if the stand will be thinned to a final crop and clear-fell when the mean DBH of the forest stand is 75 cm (this is possible if the objective of management is for timber production), it would occupy an area of 160.61 m^2. The number of stems per ha remaining would be 62 Nha^{-1}. The estimated growing space in Table 4 was based on two assumptions – circular tree crown and non-overlapping crown. Foli et al. (2003) stated that such assumptions are possible for well-managed monoculture stand. The *T. grandis* plantation used in this study is a monoculture stand that has been managed for more than a decade specifically for timber production.

T. grandis tree is a fast-growing tropical hardwood species that grows to a height of 39.6 m and DBH of more than 50 cm in well-managed undisturbed stand (Robertson 2002). The average DBH of *T. grandis* data used in this study was 17.9 cm and the predicted CD was 5.7 m (estimated from LMM2). The estimated stocking density (N), basal (G) and growing space (Sp) for the stand would be 307 Nha^{-1}, 7.75 m^2ha^{-1} and 5.7 m, respectively. Other applications of crown diameter-DBH relationship include determination of tree spacing/planting distance for agroforestry project, free growth, managing shelterwood systems, etc. In agroforestry for example, planting distance is predetermined from the commencement of the project. This distance can be determined from the CD-DBH model as the average growing space. The growing space requirement for different sizes were estimated (see Table 4).

Hemery et al. (2005) applied the OLS approach to prescribe thinning decision for *Fraxinus excelsior* L. They estimated 65 Nha^{-1} for the stand after the trees have been thinned to final crop and clear-felled when the mean DBH was 65 cm. The OLS performed relatively well in this study. However, this approach tends to violate the assumption of independence and ignores the plot-crown diameter variation. Thus, predictions from such approach may need to be applied with caution. Also, Zhang et al. (2005) reported poor self-thinning line with OLS method. In the case of linear mixed model (LMM2), the within plot variation is adequately accounted for and as such, provides reliable prediction. The quantile regression was equally good as its predictions were comparable to OLS and LMM2. One advantage of quantile regression is that it provides estimation to any part of a response distribution without restriction on the assumption of the error term. It also provides robustness and is insensitive to outliers (Hao and Naiman 2007, Sun et al. 2017). The quantile regression has been found to be effective in modelling stem diameter percentiles and stand density index (Mehtätalo et al. 2008, Ducey and Knapp 2010).

Conclusion

This study has assessed the relationship between tree crown diameter and diameter at breast height (CD-DBH) of *T. grandis* stand in Omo Forest Reserve, Nigeria using different modelling methods. Of the methods considered, linear mixed model had the best fit. However, when prediction at different percentiles of the crown diameter is of interest, then quantile regression is recommended. The CD-DBH relationship was used to derive different thinning schedules based on average diameter. The limiting stocking per ha, basal area per ha and growing space were defined. This information would help in the sustainable management of the *T. grandis* plantation in the Forest Reserve.

References

Adesoye P.O., Ezenwenyi J.U. 2014 - *Crown diameter prediction models for* Tectona grandis *Linn f. Omo forest reserve.* Journal of Forestry Research and Management 11: 72-87.

Bechtold W. A. 2003 - *Crown-diameter prediction models for 87 species of stand-grown trees in the eastern United States.* Southern Journal of Applied Forestry 27(4): 269-278.

Bohora S.B., Cao Q.V. 2014 - *Prediction of tree diameter growth using quantile regression and mixed-effects models.* Forest Ecology and Management 319: 62-66.

Chukwu O., Osho J.S.A. 2018 - *Basal area-stump diameter models for* Tectona grandis *Linn.f. stands in Omo Forest Reserve, Nigeria.* Journal of Forest and Environmental Science 34(2): 119-125.

Dawkins H.C. 1963 - *Crown diameters: their relation to bole diameter in tropical forest trees.* Commonwealth Forestry Review 42: 318-333.

De-Miguel S., Mehtatalo L., Shater Z., Kraid B., Pukkala T. 2012 - *Evaluating marginal and conditional predictions of taper models in the absence of calibration data.* Canadian Journal of Forestry Research 42: 1383-1394.

Deng W.H., Wei G., Mei Z.G. 2003 - *Nest and roost habitat characteristics of the grey-faced buzzard in northeastern China.* Journal of Raptor Research 37(3): 228-235.

Ducey M.J., Knapp R.A. 2010 - *A stand density index for mixed species forest in the northeastern United States.* Forest Ecology and Management 260: 1613-1622.

Foli E.G., Alder D., Miller H.G. Swaine M.D. 2003 - *Modelling growing space requirements form some tropical forest tree species.* Forest Ecology and Management 173:79-88.

Fu L., Sharma R.P., Hao K., Tang S. 2017 - *A generalized interregional nonlinear mixed-effects crown width model for Prince Rupprecht larch in northern China.* Forest Ecology and Management 389: 364-373.

Fu Y., Lv F., Zhu G., Lv Y. 2016 - *Construction of crown width model for single tree of natural* Larix principis-rupprechtii *Forest.* Forest Resource Management 5: 65-70 (in Chinese with English abstract).

Gao H.L., Bi H.Q., Li F.R. 2007 - *Modelling conifer crown profiles as nonlinear conditional quantiles: An example with planted Korean pine in northeast China.* Forest Ecology Management 398: 101-115.

Gao H.L., Dong L.H., Li F.R. 2016 - *Maximum density-size line for Larix olgensis plantations based on quantile regression.* Chinese Journal of Applied Ecology 27: 3420-3426.

Gering L.R. 1995 - *The relationship of diameter at breast height and crown diameter for species groups in Hardin County, Tennessee.* Southern Journal of Applied Forestry 19 (4): 177-184.

Gill S.J., Biging G.S., Murphy E.C. 2000 - *Modelling conifer tree crown radius and estimating canopy cover.* Forest Ecology and Management 126: 405-416.

Gonzalez-Benecke C.A., Gezan S.A., Samuelson L.J., Cropper W.P., Ludec D.J., Martin T.A. 2014 - *Estimating* Pinus palustris *Mill. tree diameter and stem volume from tree height, crown area and stand-level parameters.* Journal of Forestry Research 25(1): 43-52.

Hafley W.L., Buford M.A. 1985 - *A bivariate model for growth and yield prediction.* Forest Science 31: 327-247.

Hao L., Naiman D. 2007 - *Quantile regression. In Statistics for Social Science and Behavourial Science.* Sage Publications, Inc: London, UK.

Hemery G.E., Savill P.S., Pryor S.N. 2005 - *Applications of the crown-stem diameter relationship for different species of broadleaved trees.* Forest Ecology and Management 215: 285-294.

Kershaw J.A., Ducey M.J., Beers T.W., Hush B.M. 2017 - *Forest mensuration.* 5th Ed. Chichester, UK; Hoboken, NJ; John Wiley & Sons. 632 p.

Koenker R., Bassett G. 1978 - *Regression quantiles.* Econometrica 46: 33-50.

Koenker R., Hallock K. 2001 - *Quantile regression.* Journal of Economic Perspective 15: 143-156.

Marshall D.D., Johnson G.P., Hann D.W. 2003 - *Crown profile equations for stand-grown western hemlock trees in northeastern Oregon.* Canadian Journal of Forestry Research 33: 2059-2066.

Mehtätalo L., Gregoire T.G., Burkhart H.E. 2008 - *Comparing strategies for modelling tree diameter percentiles from remeasured plots.* Environmetrics 19: 529-548.

Miranda I., Sousa V., Pereira H. 2011 - *Wood properties of teak (*Tectona grandis *L.f.) from a mature unmanaged stand in East Timor.* Journal Wood Science 57: 171-178.

Ozcelik R., Cao Q.V., Trincado G., Nilsum G. 2018 - *Predicting tree height from tree diameter and dominant height using mixed-effect and quantile regression models for two species in Turkey.* Forestry Ecology and Management 419 (420): 240-248.

Pretzsch H., Biber P., Uhl E., Dahlhausen J., Rotzer T., Caldentey J., Koike T., Van Con T., Chavanne A., Seifert T., Et A.L. 2015 - *Crown size and growing space requirement of common tree species in urban centres, parks, and forests.* Urban Forestry and Urban Gardening 14: 466-479.

R Core Team 2017 - *R: A language and environment for statistical computing.* R Foundation for Statistical Computing, Vienna, Austria. [Online] Available: http://www.R-project.org/ [2007, June 30]

Raptis D., Kazana V., Kazaklis A., Stamatious C. 2018 - *A crown width-diameter model for natural even-aged black pine forest management.* Forests 9: 610-629.

Robertson B. 2002 - *Growing Teak in the Top End of the NT.* Ag note 812 No G26.

Sharma R.P., Vacek Z., Vacek S. 2016 - Individual tree crown width models for Norway spruce and European beech in Czech Republic. *Forest Ecology and Management* 366: 208-220.

Sonmez T. 2009 - *Diameter at breast height-crown diameter prediction models for* Picea orientalis *(L.) Peterm.* African Journal of Agricultural Research 4(3): 215-219.

Sousa V.B., Cardoso S., Quilho T., Pereira H. 2011 - *Growth rate and ring width variability of teak,* Tectona grandis

(Verbenanceae) in an unmanaged forest in East Timor. Revista Biologia Tropical 60 (1): 483-494.

Sun Y., Gao H., Li F. 2017 - *Using linear mixed-effects models with quantile regression to simulate the crown profile of planted* Pinus sylvestris *var.* Mongolica *trees*. Forests 8: 446 – 463.

Van Laar A., Akça A. 2007 - *Forest Mensuration*. Dordrecht, Netherlands, Springer. 389 p.

Yang Y., Huang S. 2017 - *Allometric modelling of crown width for white spruce by fixed- and mixed-effects models*. The Forest Chronicle 93: 138-147.

Zarnoch S.J., Bechtold W.A., Stoke K.W. 2004 - *Using crown condition variables as indicators of forest health*. Canadian Journal of Forestry Research 34: 1057-1070.

Zhang L., Bi H., Gove J.H., Heath L.S. 2005 - *A comparison of alternative method of estimating the self-thinning bound ary line*. Canadian Journal of Forestry Research 35: 1507-1514.

22

High-frequency seasonal variation of leaf fluorescence and reflectance in Mediterranean forest species under natural environmental conditions

Sabrina Raddi[1*], Federico Magnani[2]

Abstract - The strategies of two evergreen Mediterranean tree species (*Arbutus unedo* L. and *Quercus ilex* L.) to face annual temperature variability and extreme event responses has been monitored in continuum by leaf fluorescence and reflectance. The effect of chilling spells and the accumulation of cold days with chilling and freezing temperatures on winter photo-inhibition was discussed in terms of modulation capacity for fluorescence parameters to daily temperature and irradiance conditions. Throughout the winter *Q. ilex* resulted less photoinhibited than *A. unedo*, showing a more dynamic response in all fluorescence parameters and a higher non-photochemical quenching capacity. These characteristics effectively act to maintain a higher electron transport capacity in *Q. ilex* than in *A. unedo*.

Keywords - *Quercus ilex* L.; *Arbutus unedo* L.; photosynthesis; photochemistry; temperature stress.

Introduction

Gross primary production (GPP), the amount of atmospheric carbon fixed through the process of photosynthesis by living biomass, is the largest flux of CO_2 between the atmosphere and the Earth surface (Ciais et al. 2013), and has a key role as a primary driver of terrestrial productivity. A clear understanding of the adjustment of photosynthetic processes to surrounding conditions is therefore crucial in order to understand C sequestration and forest growth, and its response to the environment.

For plants, the adjustment of photosystem II (PSII) to environmental conditions is of vital importance in order to maintain an optimal photosynthetic electron flow to the Calvin cycle. This may be achieved through the dissipation of energy excess as heat, so minimizing the reactive oxygen species generated by surplus light (Horton et al. 1996, Logan et al. 2014).

Some of these adjustments occur at a rapid-time scale and are reversible in the range from minutes to hours, tracking the changes of light conditions (following for example sunflecks or temporary shading by clouds or overlapping branches). Other modifications occur at longer time-scale, involving also changes in foliar biochemical contents within days, weeks or months. In response to severe meteorological events or biotic stresses, or more commonly to seasonal changes in light intensity, temperature and water or nutrient availability, slowly reversible changes occur in order to acclimate photosynthesis to on-going environmental conditions (Demmig-Adams et al. 2014). This double (short- and long-term) response should be accounted for in order to infer carbon assimilation from fluorescence parameters and environmental variables.

Chlorophyll fluorescence was discovered at the end of the XVII century (Müller 1874), as the capacity to re-emit at longer wavelength part the energy absorbed. It has since been investigated and quantified in order to understand the functioning of the photosystems and their regulatory mechanisms (Kautsky 1931). Chlorophyll molecules in plant leaves are organized into macromolecular complexes (called photosynthetic units), containing many chlorophyll molecules and reaction centers (RC). After a photon is captured by pigment molecules in RC, that act as antenna, the absorbed photon ('exciton') can move among the chlorophyll molecules of the photosynthetic unit (Engel et al. 2007). The exciton can have one of following fates occurring at different rates: (a) it can be used by photosystem II (PSII) to take an electron from water, producing

1 DAGRI, Università degli Studi di Firenze, Florence (Italy)
2 DISTAL, Alma Mater - Università di Bologna, Bologna (Italy)

*Corresponding author: * sabrina.raddi@unifi.it

oxygen; (b) it can undergo radiationless decay; (c) it can be re-emitted as a fluorescent photon, or (d) it can be quenched by regulated non-photochemical trapping centers (NPQ). When the potential flux of electrons exceeds the capacity to use them in photochemistry, NPQ activity increases, balancing supply with demand and resulting in diagnostic changes in fluorescence yield. In order to preserve the photosynthetic energy balance, partitioning of excitation energy between photochemical quenching (PQ), fluorescence and non-photochemical quenching (NPQ) processes is constantly adjusted in the leaf (Papageorgiou and Govindjee 2004).

Even if fluorescence is only a small part of the total changes in yield (1% - 2%), it is therefore useful to investigate the photosynthetic process and the fate of absorbed energy (for review of method and parameters see i.a. Lichtenthaler et al. 1986, Maxwell and Johnson 2000, Baker 2008, Pessarakli 2016).

Following several decades of laboratory and field studies, today passive measurements of solar induced chlorophyll fluorescence (SIF) can be used to estimate photosynthesis and vegetation GPP from space (Frankenberg and Berry 2018). SIF can be assessed from the filling-in of the spectrally wide atmospheric oxygen absorption bands at wavelengths around 687 nm and 760 nm (Davidson et al. 2003, Moya et al. 2004). Several sensors already have the capability to estimate SIF from space by means of high-spectral-resolution interferometers. Estimates at large spatial scale were first achieved by SCIAMACHY on Japanese GOSAT launched in 2009 (30×60 km^2, Frankenberg et al. 2011, Joiner et al. 2011, 2012) and EUMETSAT MetOp satellites launched in 2006 and 2012 with 0.5° resolution, later refined to 0.05° by Duveiller and Cescatti (2016) model. More recently, information has been obtained at a more detailed scale both by NASA OCO-2 (nadir 1.3 × 2.25 km^2, Frankenberg et al. 2014) and European Space Agency (ESA) TROPOMI sensor on the Sentinel 5 Precursor satellite (7 × 3.5 km^2, Köhler et al. 2018), with the aim of quantifying photosynthetic activity and GPP globally (Sun et al. 2018, Zhang et al. 2018). Following the results of SIFLEX (*Solar Induced Fluorescence EXperiment*) campaigns, ESA will launch in 2022 the *Fluorescence Imaging Spectrometer* (FLORIS) sensor onboard the FLEX satellite, with global coverage and a spatial resolution of just 300 m. Such efforts are highly relevant to get a better understanding of carbon and water cycles, in particular to reduce uncertainties in the estimate of GPP, to determine the cause-effect relationships between environmental factors and vegetation carbon sequestration, and to assess the anthropogenic influences on vegetation functioning (Magnani et al. 2009, Mohammed et al. 2014, ESA 2015).

As part of a research program started in 2013 on the diurnal and seasonal acclimation of photosystem II (PRIN project «*Effects of Global Change on Productivity and Radiative Forcing of Italian Forests*»), the chlorophyll fluorescence of two evergreen Mediterranean forest species (*Arbutus unedo* L. and *Quercus ilex* L.) has been monitored throughout the year. Two short-term (one month) trials were also carried out to evaluate the effects on photosynthetic parameters and spectral reflectance of water stress and nutrient supply. The evaluation of the response of coexisting Mediterranean species, such as *Q. ilex* and *A. unedo* – two species evolved under the Mediterranean climate (Quézel 1985, Blondel and Aronson 1999) - to the variability in meteorological conditions has important implications for climate-driven range dynamics and vegetation shifts (Liu et al. 2018).

The main aim of this study has been to describe the different strategies of evergreen Mediterranean tree species to face temperature variability during the year and their response to extreme events. This is particularly relevant in order to assess the acclimation and resilience capacity of these species in a climate change scenario, with a predicted increase in frequency and intensity of extreme climatic conditions (IPCC 2014). The Mediterranean area is actually considered a primary climate change hotspot (Giorgi 2006), because of the above average warming (Guiot and Cramer 2016). This could drive other ecologically important variables in complex or even counterintuitive ways with episodic or nonlinear responses (Jackson et al. 2009, Connor et al. 2018). Intra-annual chlorophyll fluorescence variability has been investigated together with some key aspects related to the response to severe winter conditions (low winter temperatures and frost days).

Materials and methods

Plant material and instrumentation

Well-irrigated five-year-old plants of *Arbutus unedo* L. (strawberry tree) and *Quercus ilex* L. (holm oak) about 1-2 m in height were grown outdoor in full sun in large (40 L) pots (70% peat, 30% pumice and fertilized by 120 g pot^{-1} slow release fertilizer Osmocote Exact Standard 12-14 months) by a commercial nursery of the Pistoia nursery district. Plants were then acclimated for two months to the experimental conditions in the Florence University nursery (Florence, 43.79 N, 11.17 E; 60 m a.s.l.). Chlorophyll *a* fluorescence and reflectance were measured in each species for more than a year, starting in September 2014 for strawberry trees (measurement period: 365

days) and in September 2015 for holm oak (measurement period: 631 days). Measurements were carried out on healthy, fully expanded leaves from the last growth flush in the upper third of the crown. In order to minimize shadows, leaves were randomly chosen among those of the horizontal, South-facing and in full-light subset. Pulse-amplitude modulated fluorescence was monitored every 10 minutes by a multi-channel MONITORING-PAM fluorometer system (Heinz Walz GmbH, Effeltrich, Germany) with six MONI-head/485 fluorometers connected to a single data acquisition system (MONI-DA). The MONI-head delivers measuring light and saturating pulses to the sample through a window that transmits radiation in the range of 400–750 nm, situated at one end of the probe cylinder. Both the dim measuring light and the saturating light pulse (lasting 0.8 s with an intensity of more than 4000 μmol m^{-2} s^{-1}) were emitted by a blue LED source. The MONI-head employs PIN photodiodes to measure pulse-amplitude modulated (PAM) chlorophyll fluorescence at wavelengths longer than 625 nm. Using a photodiode protected by near-infrared filters, the MONI-head simultaneously measures the radiation (400-700 nm range) reflected by a 1.3 x 0.7 cm area of an optically diffuse Teflon sheet, 1 mm thick, mounted at the edge of the leaf clip, providing an estimate of incoming PAR (Photosynthetically Active Radiation). Air temperature is also measured inside the instrument head.

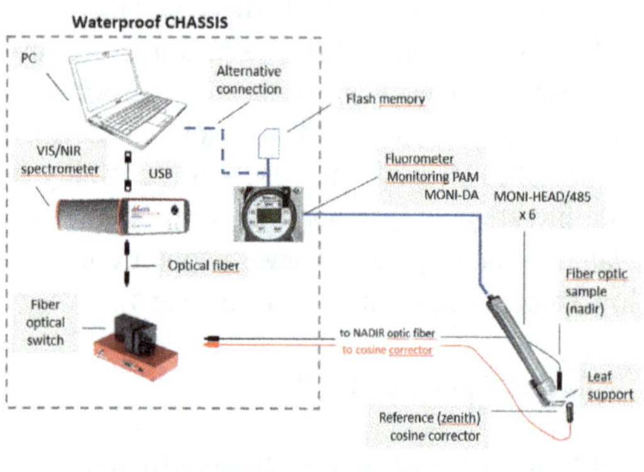

Figure 1 - Schematic representation for foliar fluorescence and reflectance measurement system. Fluorescence by Monitoring-PAM (Walz, Effeltrich, Germany) equipped with a MONI-DA acquisition system and MONI-HEAD/485 fluorometers. Reflectance VIS/NIR specrometers (Avan-Spec-2048-2) with optical fibers connected to (FOS-2-Inline, Avantes Eerbeck, The Netherlands). Further details in text.

Leaf reflectance was also measured on the same leaves through a couple of optical fibers (P200-UV-VIS, Ocean Optics Inc. FL, USA) per sample, with one fiber pointing nadir to the same leaf area sampled by the fluorescence probe with a 28° field-of-view, while the other was equipped with a cosine corrector (CC3, Ocean Optics Inc. , FL, USA) so as to acquire a hemispherical reference signal. The latter was placed vertically on the same plane of the MONI-head leaf holder (as shown in Figure 1), in order to sense a comparable light environment to the leaf sample. Both fibers were connected through a fiber optical switch (FOS-2-Inline, Avantes, Eerbeck, The Netherlands) to a VIS/NIR spectrometer (AvanSpec-2048-2, Avantes), operated through a dedicated software for the acquisition of reflectance spectra (both leaf and reference) every 2 minutes.

Computation of environmental and fluorescence parameters

Among the environmental variables, daily mean temperature (T_{day} in °C), daily minimum and maximum temperature (T_{min} and T_{max}, resp., in°C), and daily mean and maximum value for photosynthetically active radiation (PAR_{day} and PAR_{max}, μmol m^{-2} s^{-1}, resp.) were computed from long-term measurements. The number of frost days (T_{min} < 0°C) in each month and in the entire year (from 1 September to 31 August) was also calculated. Chilling and freezing periods were defined as days with T_{min} ≤ 15°C and ≤ 0°C, respectively (Yadav 2010).

The temporal changes in photosynthetic processes were assessed from the following chlorophyll *a* fluorescence parameters computed after Porcar-Castell (2011).

- The operating quantum yield of photochemistry in photosystem II, φ_P, was estimated as: $(F_m' - F_t)/F_m'$ (Genty et al. 1989), where F_m' and F_t are the maximum and instantaneous fluorescence in light condition, respectively.

- The daily maximum quantum yield of PSII, F_v/F_m, defined as $(F_m - F_o)/F_m$ (Kitajima and Butler 1975), where F_m and F_o are the maximum and minimum fluorescence of a dark-adapted leaf, respectively. A value of F_v/F_m ≥ 0.75 was deemed optimal and indicative of no sustained photoprotection/photoinhibition; for each month, the fraction of days above this threshold was also computed (ND_{opt}).

- Heat dissipation by PSII was estimated by the Stern-Volmer non-photochemical quenching (NPQ) parameter, computed as NPQ = $(F_m/F_m') - 1$ (Bilger and Björkman 1990). Sustained NPQ (NPQ_S) resulting from the overnight maintenance of photoprotection under stress conditions, was estimated as $(F_{mR}/F_m') - 1$ (Porcar-Castell 2011), where F_{mR} is the reference maximum value of F_m measured in the year during the night. In this long-term approach, the requirement of maintaining the same leaf area portion under observation during the season is essential for NPQ_S determination (Logan et al. 2007).

- The sustained photochemical quenching, qL_S, is the fraction of functional reaction centers (Porcar-Castell 2011) and it was computed as:

$$qL_S = (1/F_o' - 1/F_m')/(1/F_{oR} - 1/F_{mR})$$

- The electron transport rate ETR (µmol m^{-2} s^{-1}) at the leaf level is estimated as:

$$ETR = \alpha \cdot \beta \cdot PAR \cdot \varphi_P$$

where α, β, and PAR are leaf absorbance (set to $\alpha = 0.83$), energy partitioning to PSII (set to $\beta = 0.5$), and photosynthetic photon flux density (in µmol photon m^2 s^{-1}) and φ_P is the quantum yield of photochemistry in PSII (electron/photon; Baker 2008). ETR_{max} is the maximum electron transport rate under saturating light and was estimated by fitting a non-quadratic hyperbolic model to the diurnal response of ETR to PAR.

Foliar reflectance

Leaf reflectance at each wavelength (R_{wl}) was calculated as the ratio between spectral radiance reflected by the leaf surface and the radiance received by the surface itself (R_{wl} = Sample$_{wl}$ / Reference$_{wl}$).

The photochemical reflectance index (PRI) was originally developed by Gamon et al. (1992) to estimate the de-epoxidation of violaxanthin to zeaxanthin in the xanthophyll cycle, and as such was proposed to estimate photosynthetic light use efficiency at different scales, from leaf to canopy (for review see Garbulsky et al. 2011). The xanthophyll (VAZ) cycle is responsible for the development of NPQ, and acts as a photoprotective mechanism, preventing the oxidative damage of the photosynthetic apparatus when exposed to high light. Violaxanthin (V), prevailing in the dark, is converted upon illumination to antheraxanthin (A) and then to zeaxanthin (Z). The reaction begins when an excess of light induces a lowering of chloroplast lumen pH; Z binds to PSII proteins causing a conformational change that quenches the excess of energy as heat (Havaux and Niyogi 1999, Krause and Jahns 2004, Dall'Osto et al. 2010). In the dark, Z reverts back to V. Since A and Z show a higher absorption at 531 nm than V, reflectance decreases and a lower PRI is observed in high light (relative to low light) conditions (Middleton et al. 2012). Following Gamon et al. (1997), PRI was estimated as $(R_{531} - R_{570})/(R_{531} + R_{570})$, where R_{531} indicates reflectance at 531 nm (the waveband of the xanthophyll signal), and R_{570} indicates reflectance at 570 nm (the reference waveband).

Results

Environmental conditions

Environmental conditions were typical for the site, with an average annual temperature of 16.5°C and more than 50% of the days with high irradiance (PAR_{max} > 1000 µmol m^{-2} s^{-1}) distributed from March to October (Figure 2). Over the measurement period, absolute minimum temperatures were -4°C, -6°C and -7°C for the three winter seasons, respectively (Figure 3); colder minimum temperatures were therefore experienced by *Q. ilex* than *A. unedo* over the monitoring period, which could partly explain observed differences among species in fluorescence parameters. Frost days occurred over a period of four-five months from November to February with 33, 16, 39 frost days in the three years. January was the coldest month, accounting for 36%, 63% and 46% of frost days in each year. Several consecutive frost days were observed on *A. unedo* in 2014/15 in two periods (at the end of December and of January), and on *Q. ilex* in 2015/16 at mid-January and 2016/17 during a prolonged intermittent period from early December to the end of January, with freezing temperature in about 50% of the days. Leaves emitted on lateral shoots at mid-November in *Q. ilex* (in both years) were successively killed by these frost events. Low chilling temperatures (0°C < T_{min} ≤ 8°C) were also observed from September to May, well after leaf unfolding in early April for *Q. ilex* and end of January for *A. unedo*. The highest annual temperatures occurred in July, with the hottest two weeks in the second half of the month, when high T_{max} and low air relative humidity (RH < 25%) induced strong vapor pressure deficits (vpd > 4 kPa).

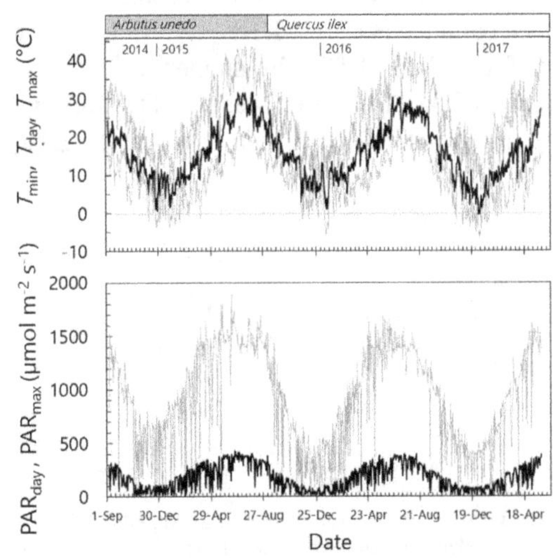

Figure 2 - Above. Daily temperature (*T*, °C): ***Tmin*** and ***Tmax*** (gray), ***Tday*** (black). Horizontal line, ***Tday*** = 0 °C. Below. Photosynthetic active radiation (PAR, µmol m^{-2} s^{-1}): PAR-***max*** (gray), PAR***day*** (black). Measurements on ***Arbutus unedo*** L. until 6-Sep 2015, then on ***Quercus ilex*** L.

Table 1 - Average ± standard deviation of fluorescence parameters over the course of the study. ***Fv/Fm*** = maximum fluorescence yield. ETR***max*** = maximum electron transport rate (μmol $m^{-2} s^{-1}$), NPQ_S = sustained non photochemical quenching, qL_S = fraction of functional reaction centers. Different letters indicate significant differences within species (***p*** < 0.05). Italic bold, seasons without significant differences (***p*** ≥ 0.05) between species (Wilcoxon (1945) rank sum test with continuity correction).

Season	*Fv/Fm*	ETR*max*	NPQ_S	qL_S
	Arbutus unedo **L.**			
spring (AMJ)	0.728 ± 0.038 b	47 ± 10 b	0.87 ± 0.56 b	0.58 ± 0.12 b
summer (JAS)	***0.744 ± 0.029 a***	***57 ± 8 a***	0.46 ± 0.25 c	0.51 ± 0.08 c
autumn (OND)	0.742 ± 0.048 a	36 ± 14 c	0.78 ± 0.61 b	0.57 ± 0.07 b
winter (JFM)	0.650 ± 0.049 c	31 ± 9 d	2.64 ± 0.89 a	0.76 ± 0.09 a
	Quercus ilex **L.**			
spring (AMJ)	0.742 ± 0.041 a	63 ± 18 a	1.24 ± 0.45 c	0.43 ± 0.05 b
summer (JAS)	***0.743 ± 0.032 a***	***59 ± 13 a***	0.51 ± 0.21 d	0.33 ± 0.04 c
autumn (OND)	0.690 ± 0.064 b	50 ± 16 b	1.84 ± 1.33 b	0.39 ± 0.05 d
winter (JFM)	0.636 ± 0.120 c	39 ± 16 c	4.56 ± 4.28 a	0.49 ± 0.11 a

Relationship between environmental and physiological parameters

Pooling the whole experimental period for the two species, significant relationships were detected between mean daily temperature (T_{day}) and fluorescence parameters (F_v/F_m: R = 0.63, R^2_{adj} = 0.40; NPQ_S: R = -0.56, R^2_{adj} = 0.31; ETR_{max}: R = -0.58, R^2 = 0.34; qL_S: R = -0.30, R^2_{adj} = 0.10; all P < 0.001; n_{obs} = 996). A quadratic function increased model prediction for F_v/F_m, NPQ_S and ETR_{max} (but not qL_S) to R^2_{adj} values of 0.61, 0.42 and 0.40, respectively (P < 0.001, n_{obs} = 996). The relationship between fluorescence parameters and radiation (PAR_{max}) showed significant (P < 0.001, n_{obs} = 996) but lower R values than reported for temperature, with the best linear prediction found with ETR_{max} (R = 0.53, R^2_{adj} = 0.28), followed by F_v/F_m (R = 0.40, R^2_{adj} = 0.17), NPQ_S (R = -0.40, R^2_{adj} = 0.16) and qL_S (not significant).

Species differed within each season for fluorescence parameters (F_v/F_m, ETR_{max}, NPQ_S and qL_S) except for F_v/F_m and ETR_{max} in summer (Wilcoxon sum rank test P > 0.05). In each season, NPQ_S was higher and qL_S lower in *Q. ilex* than in *A. unedo*. ETR_{max} was also higher in *Q. ilex* in all of the seasons except summer, while F_v/F_m was higher in *A. unedo* in autumn and winter, but lower in spring (Table 1).

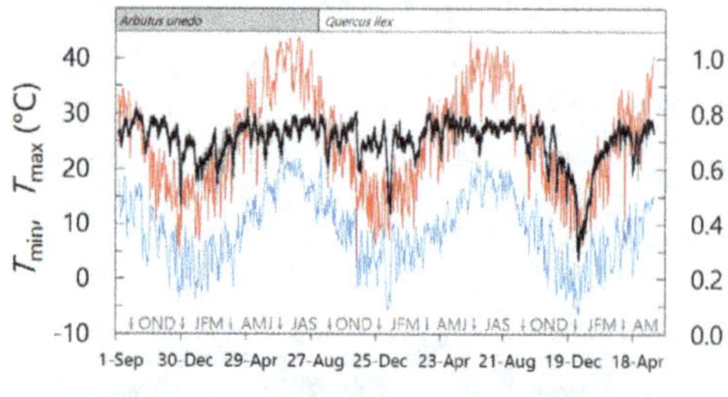

Figure 3 - Daily measurements of ***Fv/Fm*** (black) from Sep 2014 to May 2017. Seasons are identified by the capital letter of the months (OND, October November December; JFM, January February March; AMJ, April May June; JAS, July August September). Concurrent changes in air temperature (***T***, °C; daily maximum, in red, and minimum, in blue) are presented as a reference. In the upper part of the graph the measurement period for ***A. unedo*** and ***Q. ilex*** is indicated.

Temperature response dynamics

The time-course of temperature and F_v/F_m for both species is presented in Figure 3. In each season, F_v/F_m appears to rapidly acclimate to temperature changes, mainly decreasing in response to chilling spells and recovering when temperatures rose again. F_v/F_m acclimated to temperature more rapidly (in just a few days) in *A. unedo* than *Q. ilex* (often more than a week).

In mild winters, such as 2015 and 2016, the two species showed similar F_v/F_m minimum values (*A. unedo*, 0.48; *Q. ilex*, 0.45). In *A. unedo* minimum values occurred both after 7 consecutive frost days at the end of December 2014 and 2 days after the year absolute T_{min} = -4°C in January. *Q. ilex* showed the minimum value in F_v/F_m after 10 consecutive frost days at the end of January 2016 (F_v/F_m = 0.48) and again at the beginning of January 2017 (F_v/F_m = 0.28). The severe frosts of winter 2017 had a particularly severe effect and were followed by a long period of photo-inhibition lasting more than two months until spring (Figure 3). This could be the result of the accumulation of cold days with chilling and freezing temperatures: even if T_{min} values of the 10 days before the F_v/F_m minimum were similar in 2016 and 2017 (P > 0.8), temperatures averaged over a longer period (20-40 days) showed colder (P < 0.05) conditions in 2017.

The dynamics of the recovery to optimal F_v/F_m values (F_v/F_m ≥ 0.75) are also worth considering. In autumn (both species) or in winter (only *Q. ilex*), F_v/F_m recovered to its optimal value following warm days (T_{day} > 8°C), while later in the season (April-May) a higher temperature was required (T_{day} of about 14°C). In late spring and early summer (June until mid-July) both species became sensitive to high temperatures, which induced a drop of F_v/F_m below 0.7 in association with the first T_{max} peaks (and drops of RH below 30%). From the end of the summer (mid-September) both species responded again to drops in temperature, resulting in a reduction in F_v/F_m (Figure 3). Overall, in both species F_v/F_m followed T_{day} changes according to a quadratic function (*Q. ilex*, $R^2 = 0.65$; *A. unedo*, $R^2 = 0.53$, $P < 0.001$). While similar values of F_v/F_m were observed above $T_{day} = 10$°C, chilling (or freezing) temperatures ($T_{day} < 10$°C) induced a higher depression in F_v/F_m in *Q. ilex* than in *A. unedo*, with F_v/F_m values below 0.5 and 0.6 at $T_{day} = 0$°C, respectively (Figure 4).

The fraction of days with sub-optimal F_v/F_m values (< 0.75) within each month and season is shown in Figure 5. The frequency of days with sub-optimal F_v/F_m progressively increased in *Q. ilex* from spring (48%) to winter (91%), while *A. unedo* progressively decreased this frequency from winter (100%) to autumn (48%). *Q. ilex* (but not *A. unedo*) still showed optimal F_v/F_m values in favorable winter days (JFM), but never in December. In early summer (July) both species showed an increase in the frequency of days with sub-optimal F_v/F_m values, extended also to June in *A. unedo* (Figure 5).

Considering other fluorescence parameters, low temperatures resulted in an increase in sustained non-photochemical quenching (NPQ_S), which reached the maximum values in winter ($NPQ_s > 4$), while in summer never exceeded 1.2. PRI peaked at 0.03 at temperatures between 14 and 21°C, decreasing at both higher and lower temperatures to values of -0.10 and -0.15, respectively. Only a weak quadratic relationship with T_{day} was observed, however ($R^2_{adj} = 0.28$). As expected (Porcar-Castell et al. 2012), a strong negative linear relationship was observed at seasonal scale between NPQ_S and PRI (R = -0.78, $R^2_{adj} = 0.621$, $P < 0.001$). An overall linear and positive relationship between T_{day} and ETR_{max} was also observed in both species, which was stronger in *A. unedo* than in *Q. ilex* (*A. unedo*: R = 0.800, R^2 = 0. 632; *Q. ilex*: R = 0.571, $R^2_{adj} = 0.325$; $P < 0.001$, $n_{obs.}$ = 365; 631, resp.), with no difference in slope (P > 0.05), but a 70% higher intercept in *Q. ilex*.

Response to solar radiation

Both fluorescence parameters and PRI showed a lower correlation with radiation than with temperature, except for ETR_{max} (overall, $R^2_{adj} = 0.57$; *A. unedo*: $R^2_{adj} = 0.59$; *Q. ilex*: $R^2_{adj} = 0.29$, $P < 0.001$; Figure 4). It should be noted, however, that PAR_{max} and T_{day} were linearly correlated (R = 0.755, $R^2_{adj} = 0.569$, P < 0.000), making the interpretation of results more dif-

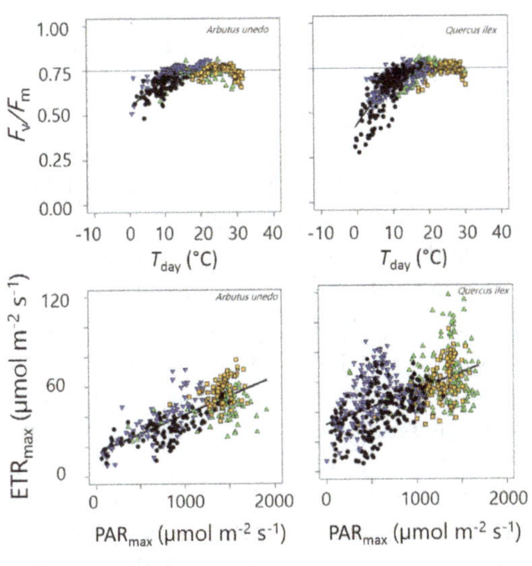

Figure 4 - Above. Relationship between ***Tday*** (°C) and ***Fv/Fm: Arbutus unedo***, y = -0.00045·x² + 0.0198x + 0.538 (R² = 0.53, p < 0.000, nobs. = 365); ***Quercus ilex*** L., y = -0.00071x² + 0.0304ˣ + 0.4348 (R² = 0.65, P < 0.000, ***nobs.*** = 630). Below. Relationship between PARmax (μmol m⁻² ˢ⁻¹) and ***ETRmax*** (μmol m-2 s-1): ***Arbutus unedo*** L., y = 0.025 x + 15.96 (R² = 0.59, P < 0.000, nobs. = 365); ***Quercus ilex*** L. , y = 0.023 x + 31.77 (R² = 0.29, p < 0.000, ***nobs.*** = 630). Black circles (winter), green triangles (spring), orange squares (summer), blue down-pointing triangles (autumn).

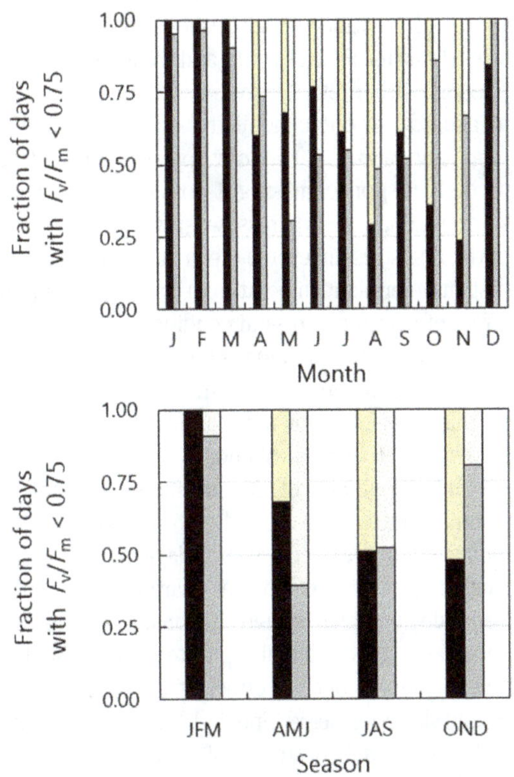

Figure 5 - Fraction of days in each month (above) or season (below) showing photoprotection (*Fv/Fm* < 0.75) (***A. unedo*** in black; ***Q. ilex*** in gray). Complement to 1 (colored bars) is ***NDopt***, the fraction of days with optimal ***Fv/Fm*** (≥ 0.75).

ficult. As already reported for temperature, regression lines of ETR_{max} as a function of PAR_{max} for the two species were parallel (no significant slope difference) and had a higher intercept (almost double) in oak. High irradiance did not appear to have a lasting photo-inhibitory effect. In *A. unedo*, measurements of F_v/F_m, NPQ_S and qL_S taken after days with saturating light conditions (PAR > 1000 μmol m^{-2} s^{-1} in spring and summer; PAR > 500 μmol m^{-2} s^{-1} in autumn and winter) did not differ from the overall seasonal average. *Q. ilex* showed a higher sensitivity to high solar radiation, as saturating light conditions resulted in higher F_v/F_m (and lower NPQ_S) values in autumn and winter but no significant deviation from the overall average was observed in spring and summer.

Discussion

Long-term observations of chlorophyll fluorescence and reflectance has allowed the investigation at daily scale of the annual dynamics of leaf photochemistry several fluorescence parameters and reflectance indices in response to temperature and light environment in two co-existing Mediterranean evergreen species (Gasparini and Tabacchi 2011). Although our experimental setup did not allow a truly paired comparison between species, it provided useful information on their response to temperature stresses differing in severity and frequency. In comparison with previous studies which periodically measured leaf photochemistry on a monthly basis (e.g. Aranda et al. 2005), our approach could track its daily response, making it possible to determine seasonal changes in stress sensitivity, the outcome of the first stressful events (i.e. first autumn frosts or first summer high temperatures), and the cumulative effect of cycles of stressful conditions. On the other hand, had the research focus been on the evaluation of variability of physiological traits at population level, this time-intensive monitoring approach would not have been be the best choice.

Under cool and cold Mediterranean climate, hot summer periods (associated with drought and high radiation), but also cold winter days, might be challenging for evergreen species (Larcher 2000, 2003). Generally, summer drought is considered the most important growth-limitating factor under Mediterranean-climate, but the effect of low temperatures should also considered (Specht 1981, Öquist and Huner 2003). A variety of structural, anatomical and functional traits help plants to face stressful conditions (Cherubini et al. 2003, Bussotti et al. 2015, Niinemets and Keenan 2014, Gratani et al. 2018), and the role of photoprotection should not be overlooked.

Photosynthesis requires fine regulation mechanisms to effectively cope with environmental variability in radiation, temperature, water and nutrient availability. The balance between growth and protection is essential for assuring survival in perennial life-form. Thus, a competitive strategy should assure a high rate of electron transport to the Calvin cycle under favorable conditions and avoid damages by hazardous reactive oxygen species formation by dissipating any excess energy as heat through NPQ. The non-photochemical quenching of excitation energy relies on physiological processes regulated at different timescales from pico-seconds to weeks (Garcia-Plazaola et al. 2017, Porcar-Castell et al. 2014). The decrease (or down-regulation) in F_v/F_m, sometimes referred to as 'photoinhibition', is often the result of the maintenance overnight of a sustained photo-protective state, in response to light levels in excess of the Calvin cycle capacity. In reality, down-regulation of F_v/F_m may be caused both by an increase in sustained NPQ and by damaged (photo-inactivated) reaction centers (Porcar-Castell et al. 2014).

The F_v/F_m reduction associated with a temperature decrease and the ensuing recovery indicate the high modulation capacity of this parameter in order to ensure an adequate photoprotection and flexible acclimation of photosynthesis processes throughout the year. Only under the first hot days in late spring/early summer did F_v/F_m reverse its response to temperature, showing an enhanced photo-protection/photo-inhibition state in response to a temperature (and vpd) increase (Lange et al. 1985, Ripullone et al. 2009, Martínez-Vilalta et al. 2003). Under such conditions, high non-stomatal limitations to photosynthesis (Grassi et al. 2009) and growth inhibition (Campelo et al. 2018) are not always linked to a reduction in carbon assimilation (Crescente et al. 2002). In the present study, under conditions of full water availability both species had rather constant values of F_v/F_m and ETR_{max} over the summer. This is consistent with previous studies, which for many Mediterranean evergreens under field conditions reported a maximum carbon assimilation over an air temperature range of 10-20°C, with the higher limit determined by the onset of summer drought (Flexas et al. 2014, Gratani et al. 2018). In both species examined in the present study, a high thermostability at short-term time scale has also been observed for F_v/F_m over a wide range of leaf temperatures, from 5°C to 48°C (Méthy et al. 1997, Bilger et al. 1987). In response to stress evergreen species change both their foliar biochemical contents (increasing antioxidants and xanthophyll pigments; Verhoeven 2014) and the composition of PS II antenna but also increase energy dissipation through cyclic electron transport (Öquist and Huner 2003), in particular during the chilly/frosty winter period. In our study, the strong reduction of F_v/F_m was mainly related to the increase in sustained non-photochemical quenching during the winter (NPQ_S, R = -0.85). This latter was strongly associated to PRI, suggesting that also seasonal changes in xanthophyll (VAZ) pools or xanthophyll de-epoxidation state occurred in both species (Baraldi et al. 2008, Ripullone et al. 2011, Corcuera et al. 2005; Camarero et al. 2012). In their de-epoxidated state (Z+A) xanthophylls prevent damages to the thylakoid membranes, quenching excess energy through heat (NPQ_s increase) and thus reducing electron flow in parallel with Calvin cycle limitations (ETR_{max} reduction). García-Plazaola and colleagues (2017) observed in *Q. ilex* an emergency mechanism associated with the lutein cycle (LxL), which supplements the VAZ cycle in response to prolonged environmental stresses, both in winter and in summer. Although *A. unedo* had a lower content of lutein than holm oak (García-Plazaola et al. 2004), in the former species the LxL cycle was involved in photo-protection from salinity-induced water stress over the summer (Fusaro et al. 2014). Regulatory mechanisms via non-photochemical quenching could allow photosynthesis (and eventually growth) in seasons when water is abundant but temperatures are sub-optimal (Miller 1982), and thus reward the construction and maintenance costs. In evergreens, greater resistance to frost and freezing-induced cavitation may be obtained by thicker leaves (Lo Gullo & Salleo 1993), but higher allocation of structural carbohydrates and nitrogen to cell wall than to photosynthetic apparatus (González-Zurdo et al. 2016) adds construction costs at the expenses of carbon gain (van Ommen Kloeke et al. 2012). In both species, ETR_{max} and F_v/F_m values were still high during the autumn (even if lower than in spring and summer), consistent with the high carbon assimilation capacity found in this season (Gratani and Varone 2004, Beyschlag et al. 1987, Ogaya and Peñuelas 2003). In autumn (and winter, if temperatures are not too low) a dynamic response to temperature and light allows evergreen species to effectively exploit these seasons for carbon assimilation (Oliveira and Peñuelas 2004, Ripullone et al. 2009). This is particularly relevant, since in autumn under natural conditions photosynthesis could still drive growth, in terms of new shoots flushing and elongation, stem enlargement (Susmel et al. 1976, Gratani et al. 1992, Martin 2012), or mobilizing processes to storage reserves (Canadell et al. 1999, Zavala et al. 2011). In both species, a minor annual peak in stem growth is sometimes observed during the autumn (Martin 2012, Campelo et al. 2018, Castell et al. 1994), and brachyblasts can be emitted (Orsham 1989, Pereira et al. 1987); although a strong variability among years and sites exists, related to environmental conditions (Milla et al. 2010). Winter ETR_{max}, even if reduced in comparison with other seasons, still responded dynamically to radiation and temperature, taking advantage of warm sunny days, adjusting dynamically together with the other physiological parameters (F_v/F_m, NPQ_S and PRI), similarly to what observed by Zhang (2017). Oliveira and Peñuelas (2004) reported a positive relationship between F_v/F_m and light-saturated carbon assimilation capacity (A_{sat}) in winter, maintained as long as leaves are not subjected again to favourable conditions that enhanced F_v/F_m more rapidly than photosynthetic capacity. In the present study, the positive correlation between F_v/F_m and ETR_{max} was maintained also in the response to the lowest T_{min} (-7°C), but after this frost event F_v/F_m recovered more rapidly in response to the temperature increase than ETR_{max}, which showed a lag of 10 days until also daily maximum temperature increased. The positive electron transport capacity (ETR_{max}) observed at sub-freezing temperatures in both species, if not quenched through photochemistry or cyclic electron transport, might induce damaging side reactions, exposing leaves to the risk of oxidation by reactive oxygen species (Murchie and Harbinson 2014, Verhoeven et al. 2018). Low winter temperatures and frost events in particular (Aranda et al 2005, Sperlich et al. 2014) can thus limit the survival of evergreen species not only at high latitudes or altitudes (Nichol et al. 2019, Williams et al. 2003), but also under Mediterranean climate (Alessi et al. 2018). For surviving and at the same time maintaining growth (and thus a high competitive capacity) in harsh environments, evergreen species have developed acclimation mechanisms to cope with cold spells, so limiting the damages induced by low temperatures in combination with high irradiance (and eventually low water availability by frozen soil, Sutinen et al. 2001). This allows the plant to maintain a positive carbon assimilation both in the summer, also in presence of high leaf temperatures (Gratani et al. 2000, Grassi et al. 2009), and in warm autumn days, so prolonging the period of active photosynthesis. In association with other traits (e.g. capacity to recover from drought- or frost-induced xylem cavitation; Salleo et al. 1997, Cherubini et al. 2003), this could result in a selective advantage over deciduous species (Blondel et al. 2010, Barbeta et al. 2019), if temperatures are not too extreme (Gentilesca et al. 2017, Pollastrini et al. 2018).

Conclusions

From an ecological point of view, the two evergreen species displayed slightly different strategies; under the sub-humid Mediterranean climate at the study site, both species showed the highest photoinhibition in winter, with a high temperature dependence (Fernández et al. 2008) but with different strategies and acclimation capacity. On average, *A. unedo* was more photoinhibited than *Q. ilex* throughout the winter, despite the less negative temperatures experienced, and to a lesser extent also in spring and autumn, maintaining a chronic photoinhibition for at least 3 months. *Q. ilex* has a more dynamic response, with a high plasticity (range/max) in all fluorescence parameters, and appears to have a higher non-photochemical quenching capacity in all seasons, complemented by a lower fraction of open reaction centers. These characteristics effectively maintain a high electron transport capacity in *Q. ilex*, and protect the photosystems by dissipating the excess energy as heat and rapidly reducing the fraction of functional reaction centers. In the summer, under optimal water conditions, the differences between *Q. ilex* and the more thermophilous *A. unedo* become not significant.

Acknowledgments

This work was funded by MIUR-PRIN grant No. 2012E3F3LK "Global change effects on the productivity and radiative forcing of Italian forests: a novel retrospective, experimental, and prognostic analysis". We are indebted to Alberto Pierguidi (DAGRI, University of Florence) for technical assistance during the experiment.

References

Alessi N., Wellstein C., Spada F., Zerbe S. 2018 - *Phytocoenological approach to the ecology of* Laurus nobilis *L. in Italy.* Rendiconti Lincei. Scienze Fisiche e Naturali 29: 343–354. doi: 10.1007/s12210-018-0677-8

Aranda I., Castro L., Alía R., Pardos J.A., Gil L. 2005 - *Low temperature during winter elicits differential responses among populations of the Mediterranean evergreen cork oak* (Quercus suber). Tree Physiology 25: 1085–1090. doi: 10.1093/treephys/25.8.1085

Baker N.R. 2008 - *Chlorophyll fluorescence: A probe of photosynthesis in vivo.* Annual Review of Plant Biology 59: 89–113. doi: 10.1146/annurev.arplant.59.032607.092759

Baraldi R., Canaccini F., Cortes S., Magnani F., Rapparini F., Zamboni A., Raddi S. 2008 - *Role of xanthophyll cycle-mediated photoprotection in* Arbutus unedo *plants exposed to water stress during the Mediterranean summer.* Photosynthetica 46: 378-386. doi: 10.1007/s11099-008-0069-x

Barbeta A., Camarero J.J., Sangüesa-Barreda G., Muffler L., Peñuelas J. 2019 - *Contrasting effects of fog frequency on the radial growth of two tree species in a Mediterranean-temperate ecotone.* Agriculture and Forest Meteorology 264: 297-308. doi: 10.1016/j.agrformet.2018.10.020

Beyschlag W., Lange O.L., Tenhunen J.D. 1987 – *Photosynthese und Wasserhaushalt der immergrünen mediterranean Hartlaubpflanze* Arbutus unedo *L. im Jahreslauf am Freilandstandort in Portugal. II. Charakteristische Gaswechselparameter von CO_2-Aufnahme und Transpiration.* Flora 179: 399-420. doi: 10.1016/S0367-2530(17)30275-X

Bilger W., Björkman O. 1990 - *Role of the xanthophyll cycle in photoprotection elucidated by measurements of light-induced absorbance changes, fluorescence and photosynthesis in leaves of* Hedera canariensis. Photosynthesis Research 25: 173-185. doi: 10.1007/BF00033159

Bilger W., Schreiber U., Lange O.L. 1987 - *Chlorophyll fluorescence as an indicator of heat induced limitation of photosynthesis in* Arbutus unedo *L.* In: "Plant Response to Stress". J.D. Tenhunen, F.M. Catarino (Eds.) Springer Berlin Heidelberg: 391–399 doi: 10.1007/978-3-642-70868-8_24

Blondel J., Aronson J. 1999 - *Biology and Wildlife of the Mediterranean Region.* Oxford University Press, New York. 328 p. ISBN 0198500351

Blondel J., Aronson J., Bodiou J.-Y., Boeuf G. 2010 - *The Mediterranean Region: Biological Diversity in Space and Time.* 2nd Edition, Oxford University Press, New York. 376 p. ISBN 978-0-19-955799-8

Bussotti F., Pollastrini M., Holland V., Brüggemann W. 2015 - *Functional traits and adaptive capacity of European forests to climate change. Review.* Environmental and Experimental Botany 111: 91–113. doi: 10.1016/j.envexpbot.2014.11.006

Camarero J.J., Olano J.M., Arroyo Alfaro S.J., Fernández-Marín B., Becerril J.M., García-Plazaola J.I. 2012 - *Photoprotection mechanisms in* Quercus ilex *under contrasting climatic conditions.* Flora 207: 557-564. doi: 10.1016/j.flora.2012.06.003

Campelo F., Gutiérrez E., Ribas M., Sánchez-Salguero R., Nabais C., Camarero J.J. 2018 - *The facultative bimodal growth pattern in* Quercus ilex – *A simple model to predict sub-seasonal and inter-annual growth.* Dendrochronologia 49: 77-88. doi: 10.1016/j.dendro.2018.03.001

Canadell J., Djema A., López B., Lloret F., Sabaté S., Siscart D., Gracia C.A. 1999 - *Structure and dynamics of the root system.* In: "Ecology of Mediterranean Evergreen Oak Forests". F. Rodà, J. Retana, C.A. Gracia, J. Bellot (Eds.) Springer-Verlag, Berlin and Heidelberg: 47–59. doi: 10.1007/978-3-642-58618-7_4

Castell C., Terradas J., Tenhunen J.D. 1994 - *Water relations, gas exchange, and growth of resprouts and mature plant shoots of* Arbutus unedo *L. and* Quercus ilex *L.* Oecologia 98: 201-211. doi: 10.1007/BF00341473

Cherubini P., Gartner B.L., Tognetti R., Bräker O.U., Schoch W., Innes J.L. 2003 - *Identification, measurement and interpretation of tree rings in woody species from mediterranean climates.* Biological Review 78: 119–148. doi: 10.1017/S1464793102006000

Ciais P., Sabine C., Bala G., Bopp L., Brovkin V., Canadell J., Chhabra A., DeFries R., Galloway J., Heimann M., Jones C., Le Quéré C., Myneni R.B., Piao S., Thornton P. 2013 - *Carbon and Other Biogeochemical Cycles.* In: "Climate Change 2013: The Physical Science Basis. Contribution of Working Group I to the Fifth Assessment Report of the Intergovernmental Panel on Climate Change". T.F. Stocker, D. Qin, G.-K. Plattner, M. Tignor, S.K. Allen, J. Boschung, A. Nauels, Y. Xia, V. Bex, P.M. Midgley (Eds.) Cambridge University Press, Cambridge, United Kingdom and New York, NY, USA: 465–570. doi:10.1017/CBO9781107415324.015

Connor N., Overpeck J.T., Allen J.R.M., Anderson J.M., Betancourt J.L., Binney H.A., Brewer S., Bush M.B., Chase B.M., Cheddadi R., Djamali M., Dodson J., Edwards M.E., Gosling W.D., Haberle S., Hotchkiss S.C., Huntley B., Ivory S.J., Kershaw A.P., Kim S.-H., Latorre C., Leydet M., Lézine A.-M., Liu K.-B., Liu Y., Lozhkin A.V., Mcglone M.S., Marchant R.A., Momohara A., Moreno P.I., Müller S., Otto-Bliesner B.L., Shen C., Stevenson J., Takahara H., Tarasov P.E., Tipton J., Vincens A., Weng C., Xu Q., Zheng Z., Jackson S.T. 2018 - *Past and future global transformation of terrestrial ecosystems under climate change.* Science 361: 6405: 920-923. doi: 10.1126/science.aan5360

Corcuera L., Morales F., Abadía A., Gil-Pelegrín E. 2005 - *Seasonal changes in photosynthesis and photoprotection in a* Quercus ilex *subsp. ballota woodland located in its upper altitudinal extreme in the Iberian Peninsula.* Tree Physiology 25: 599–608. doi: 10.1093/treephys/25.5.599

Crescente M.F., Gratani L., Larcher W. 2002 - *Shoot growth efficiency and production of* Quercus ilex *L. in different climates.* Flora 197: 2-9. doi: 10.1078/0367-2530-00007

Dall'Osto L., Cazzaniga S., Havaux M., Bassi R. 2010 – *Enhanced photoprotection by protein-bound vs free xanthophyll pools: a comparative analysis of chlorophyll b and xanthophyll biosynthesis mutants.* Molecular Plant 3: 576-593. doi: 10.1093/mp/ssp117

Davidson M., Berger M., Moya I., Moreno J., Laurila T., Stoll M.-P., Miller J. 2003 - *Mapping photosynthesis from space - a new vegetation-fluorescence technique.* ESA Bulletin 116: 34-37. ISSN 0376-4265

Demmig-Adams B., Koh S.-C., Cohu C. M., Muller O., Stewart J.J., Adams III W. W. 2014 - *Non-photochemical fluorescence quenching in contrasting plant species and environments. Chapter 24.* In: B. Demmig-Adams, G. Garab, W. W. Adams III, Govindjee (Eds.) *Non-Photochemical Quenching and Energy Dissipation in Plants, Algae and Cyanobacteria.* Advances in Photosynthesis and Respiration 40. Springer Dordrecht, Heidelberg, New York and London: 531-554. doi: 10.1007/978-94-017-9032-1_24

Duveiller G., Cescatti A. 2016 - *Spatially downscaling sun-induced chlorophyll fluorescence leads to an improved temporal correlation with gross primary productivity.* Remote Sensing of Environment 182: 72–89. doi: 10.1016/j.rse.2016.04.027

Engel G.S., Calhoun T.R., Read E.L., Ahn T.-K., Mančal T., Cheng Y.-C., Blankenship R.E., Fleming G.R. 2007 - *Evidence for wavelike energy transfer through quantum coherence in photosynthetic systems.* Nature 446: 782–786. doi: 10.1038/nature05678

ESA (European Space Agency) 2015 - *Report for Mission Selection: FLEX.* ESA SP-1330/2 (2 volume series), European Space Agency, Noordwijk, The Netherlands. 197 p. ISBN 978-92-9221-428-9

Fernández M., Alejano R., Domínguez L., Tapias R. 2008 - *Temperature controls cold hardening more effectively than photoperiod in four Mediterranean broadleaf evergreen species.* Tree and Forestry Science and Biotechnology 2: 43-49.

Flexas J., Diaz-Espejo A., Gago J., Gallé A., Galmés J., Gulías J., Medrano H. 2014 - *Photosynthetic limitations in Mediterranean plants: A review.* Environmental and Experimental Botany 103: 12-23. doi: 10.1016/j.envexpbot.2013.09.002

Frankenberg C., Berry J. 2018 - *Solar Induced Chlorophyll Fluorescence: Origins, Relation to Photosynthesis and Retrieval.* In: "Reference Module in Earth Systems and Environmental Sciences: Comprehensive Remote Sensing". Elsevier, Oxford: 143-162. ISBN: 978-0-12-803220-6

Frankenberg C., Butz A., Tooc G.C. 2011 - *Disentangling chlorophyll fluorescence from atmospheric scattering effects in O_2 A-band spectra of reflected sun light.* Geophysical Research Letters 38: L03801. doi: 10.1029/2010GL045896

Frankenberg C., O'Dell C., Berry J., Guanter L., Joiner J., Köhler P., Pollock R., Taylor T.E. 2014 - *Prospects for chlorophyll fluorescence remote sensing from the Orbiting Carbon Observatory-2.* Remote Sensing of Environment 147: 1–12. doi: 10.1016/j. rse.2014.02.007

Fusaro L., Mereu S., BrunettI C., Di Ferdinando M., Ferrini F., Manes F., Salvatori E., Marzuoli R., Gerosa G., Tattini M. 2014 - *Photosynthetic performance and biochemical adjustments in two co-occurring Mediterranean evergreens,* Quercus ilex *and* Arbutus unedo, *differing in salt-exclusion ability.* Functional Plant Biology 41: 391–400. doi: 10.1071/FP13241

Gamon, J.A., Peñuelas J., Field C.B. 1992 - *A narrow-waveband spectral index that tracks diurnal changes in photosynthetic efficiency.* Remote Sensing of Environment 41: 35–44. doi: 10.1016/0034-4257(92)90059-S

Gamon J.A., Serrano L., Surfus J.S. 1997 - *The photochemical reflectance index: an optical indicator of photosynthetic radiation use efficiency across species, functional types, and nutrient levels.* Oecologia 112: 492–501. doi: 10.1007/s004420050

Garbulsky M.F., Peñuelas J., Gamon J., Inoue Y., Filella I. 2011 - *The photochemical reflectance index (PRI) and the remote sensing of leaf, canopy and ecosystem radiation use efficiencies: A review and meta-analysis.* Remote Sensing of Environment 115: 281–297. doi: 10.1016/j.rse.2010.08.023

García-Plazaola J.I., Hernández A., Fernández-Marín B., Esteban R., Peguero-Pina J.J., Verhoeven A., Cavender-Bares J. 2017 - *Photoprotective Mechanisms in the Genus* Quercus *in Response to Winter Cold and Summer Drought.* In: "Oaks Physiological Ecology. Exploring the Functional Diversity of Genus Quercus L." E. Gil-Pelegrín, J.J. Peguero-Pina, D. Sancho-Knapik (Eds.). Tree Physiology 7. Springer International Publishing AG 2017: 361-391. doi: 10.1007/978-3-319-69099-5_11

García-Plazaola J.I., Hormaetxe K., Hernández A., Olano J.M., Becerril J.M. 2004 - *The lutein epoxide cycle in vegetative buds of woody plants.* Functional Plant Biology 31: 815–823. doi: 10.1071/FP04054

Gasparini P., Tabacchi G. (Eds.) 2011 - *L'Inventario Nazionale delle Foreste e dei serbatoi di Carbonio. INFC 2005. Secondo inventario forestale nazionale italiano. Metodi e risultati.* Ministero delle Politiche Agricole, Alimentari e Forestali, Corpo Forestale dello Stato. Consiglio per la Ricerca e la Sperimentazione in Agricoltura, Unità di ricerca per il Monitoraggio e la Pianificazione Forestale, Edagricole, Milano. 653 p. ISBN 978-88-506-5394-2

Gentilesca T., Camarero J.J., Colangelo M., Nolè A., Ripullone F. 2017 - *Drought-induced oak decline in the western Mediterranean region: an overview on current evidences, mechanisms and management options to improve forest resilience.* iForest - Biogeosciences and Forestry 10: 796-806. doi: 10.3832/ifor2317-010

Genty B., Briantais J.-M., Baker N.R. 1989 - *The relationship between the quantum yield of photosynthetic electron transport and quenching of chlorophyll fluorescence.* Biochimica et Biophysica Acta 990, 87–92. doi: 10.1016/S0304-4165(89)80016-9

Giorgi F. 2006 - *Climate change hot-spots.* Geophysical Research Letters 33: L08707. doi:10.1029/2006GL025734

González-Zurdo P., Escudero A., Babiano J., García-Ciudad A., Mediavilla S. 2016 - *Costs of leaf reinforcement in response to winter cold in evergreen species.* Tree Physiology 36: 273–286. doi: 10.1093/treephys/tpv134

Grassi G., Ripullone F., Borghetti M., Raddi S., Magnani F. 2009 - *Contribution of diffusional and non-diffusional lim-*

itations to midday depression of photosynthesis in Arbutus unedo *L.* Trees 23: 1149–1161. doi: 10.1007/s00468-009-0355-7

Gratani L., Marzi P., Crescente M.F. 1992 - *Morphological adaptions of* Quercus ilex *leaves in the Castelporziano forest.* Vegetatio 99-100: 155-161. doi: 10.1007/BF00118221

Gratani L., Pesoli P., Crescente M.F., Aichner K., Larcher W. 2000 - *Photosynthesis as a temperature indicator in* Quercus ilex *L.* Global and Planetary Change 24: 153–163. doi: 10.1016/S0921-8181(99)00061-2

Gratani L., Varone L. 2004 - *Adaptive photosynthetic strategies of the Mediterranean maquis species according to their origin.* Photosynthetica 42: 551-558. doi: 10.1007/S11099-005-0012-3

Gratani L., Varone L., Crescente M.F., Catoni R., Ricotta C., Puglielli G. 2018 - *Leaf thickness and density drive the responsiveness of photosynthesis to air temperature in Mediterranean species according to their leaf habitus.* Journal of Arid Environments 150: 9–14. doi: 10.1016/j.jaridenv.2017.12.007

Guiot J., Cramer W. 2016 - *Climate change: The 2015 Paris Agreement thresholds and Mediterranean basin ecosystems.* Science 354: 465–468. doi: 10.1126/science.aah5015

Havaux M., Niyogi K.K.1999 - *The violaxanthin cycle protects plants from photooxidative damage by more than one mechanism.* Proceeding of the National Academy of Science USA 96: 8762-8767. doi: 10.1073/pnas.96.15.8762

Horton P., Ruban A.V., Walters R.G. 1996 - *Regulation of light harvesting in green plants.* Annual Review of Plant Physiology and Plant Molecular Biology 47: 655–684. doi: 10.1146/annurev.arplant.47.1.655

IPCC (Intergovernmental Panel on Climate Change) 2014. *Climate Change 2014: Synthesis Report. Contribution of Working Groups I, II and III to the Fifth Assessment Report of the Intergovernmental Panel on Climate Change.* R. Pachauri and L. Meyer (Eds.). Geneva: IPCC. ISBN 978-92-9169-143-2

Jackson S.T., Betancourt J.L., Booth R.K., Gray S.T. 2009 - *Ecology and the ratchet of events: Climate variability, niche dimensions, and species distributions.* doi: PNAS 106: 19685–19692 doi: 10.1073pnas.0901644106

Joiner J., Yoshida Y., Vasilkov A.P., Yoshida Y., Corp L.A., Middleton E.M. 2011 - *First observations of global and seasonal terrestrial chlorophyll fluorescence from space.* Biogeosciences 8: 637–651. doi: 10. 5194/bg-8-637-2011

Joiner J., Yoshida Y., Vasilkov A.P., Middleton E.M., Campbell P.K.E., Yoshida Y., Kuze A., Corp L.A. 2012 - *Filling-in of near-infrared solar lines by terrestrial fluorescence and other geophysical effects: simulations and space-based observations from SCIAMACHY and GOSAT.* Atmospheric Measurements Techniques 5: 809–829. doi: 10.5194/amt-5-809-2012

Kautsky H. 1931 - *Energie-Umwandlungen an Grenzflächen,* IV. Mitteil. : H. Kautsky and A. Hirsch: *Wechselwirkung zwischen angeregten Farbstoff‐Molekülen und Sauerstoff.* European Journal of Inorganic Chemistry 64: 2677–2683. doi: 10.1002/cber.19310641017

Krause G.H., Jahns P. 2004 - *Non-photochemical energy dissipation determined by chlorophyll fluorescence quenching: characterization and function.* In: "Chlorophyll a Fluorescence: A Signature of Photosynthesis. Advances in Photosynthesis and Respiration, Volume 19". G.C. Papageorgiou and Govindjee (Eds.). Springer, The Netherlands: 463–495. doi: 10.1007/978-1-4020-3218-9_18

Kitajima M., Butler W.L. 1975 - *Quenching of chlorophyll fluorescence and primary photochemistry in chloroplasts by dibromothymoquinone.* Biochimica et Biophysica Acta 376: 105–115. doi: 10.1016/0005-2728(75)90209-1

Köhler P., Frankenberg C., Magney T.S., Guanter L., Joiner J., Landgraf J. 2018 - *Global retrievals of solar-induced chlorophyll fluorescence with TROPOMI: First results and intersensor comparison to OCO-2.* Geophysical Research Letters 45: 10456–10463. doi: 10.1029/2018GL079031

Lange O.L., Tenhunen J.D., Beyschlag W. 1985 - *Effects of humidity during diurnal courses on the CO2- and light-saturated rate of net* CO_2 *uptake in the sclerophyllous leaves of Arbutus unedo.* Oecologia 67: 301–304. doi: 10.1007/BF00384305

Larcher W. 2000 - *Temperature stress and survival ability of Mediterranean sclerophyllous plants.* Plant Biosystems 134: 279–295. doi: 10.1080/11263500012331350455

Larcher W. 2003 - *Physiological Plant Ecology* (4th Ed.), Springer-Verlag Berlin, Germany. 514 p. ISBN 978-3-540-43516-7

Lichtenthaler H.K., Buschmann C., Rinderle U., Schmuck, G. 1986 - *Application of chlorophyll fluorescence in ecophysiology.* Radiation and Environmental Biophysics 25: 297–308. doi: 10. 1007/BF01214643

Liu D., Ogaya R., Barbeta A., Yang X., Peñuelas J. 2018 - *Long-term experimental drought combined with natural extremes accelerate vegetation shift in a Mediterranean* Holm oak *forest.* Environmental and Experimental Botany 151: 1-11. doi: 10.1016/j.envexpbot.2018.02.008

Lo Gullo M.A., Salleo S. 1993 - *Different vulnerabilities of* Quercus ilex *L. to freeze - and summer drought-induced xylem embolism: an ecological interpretation.* Plant, Cell & Environment 16: 511-519. doi: 10.1111/j.1365-3040.1993.tb00898.x

Logan B.A., Adams III W.W., Demmig-Adams B. 2007 - *Avoiding common pitfalls of chlorophyll fluorescence analysis under field conditions.* Functional Plant Biology 34: 853-859. doi: 10.1071/FP07113

Logan B.A., Demmig-Adams B., Adams W.W., Bilger W. 2014 - *Context, quantification, and measurement guide for non-photochemical quenching of chlorophyll fluorescence. Chapter 7.* In: "Non-Photochemical Quenching and Energy Dissipation in Plants, Algae and Cyanobacteria". Advances in Photosynthesis and Respiration 40. B. Demmig-Adams, G. Garab, W.W. Adams III, Govindjee (Eds.) Springer Dordrecht, Heidelberg, New York and London: 187–201. doi: 10.1007/978-94-017-9032-1_7

Magnani F., Olioso A., Demarty J., Germain V., Verhoef W., Moya I., Van der Tol C. 2009 - *Assessment of Vegetation Photosynthesis through Observation of Solar Induced Fluorescence from Space.* Final Report for the European Space Agency under ESTEC Contract No. 20678/07/NL/HE; ESA, Paris, France. 256 p.

Martin N. 2012 -*Ajustements fonctionnels du chêne vert* (Quercus ilex *L.) à la sécheresse à différentes échelles temporelles: Incidences sur la modélisation des processus.* PhD Thesis Université de Montpellier II, 364 p.

Martínez-Vilalta J., Mangirón M., Ogaya R., Sauret M., Serrano L., Peñuelas J., Piñol J. 2003 - *Sap flow of three co-occurring Mediterranean trees under varying atmospheric and soil water conditions.* Tree Physiology 23: 747–758. doi: 10.1093/treephys/23.11.747

Maxwell K., Johnson G.N. 2000 - *Chlorophyll fluorescence: A practical guide.* Journal of Experimental Botany 51: 659–668. doi: 10.1093/jexbot/51. 345. 659

Mèthy M., Gillon D., Houssard C. 1997 - *Temperature-induced changes of photosystem II activity in* Quercus ilex *and* Pinus halepensis. Canadian Journal of Forest Research 27: 31–38. doi: 10.1139/x96-127

Middleton E.M., Huemmrich K.F., Cheng Y.-B., Margolis H.A. 2012 - *Spectral bioindicators of photosynthetic efficiency and vegetation stress. Chapter 12.* In: "Hyperspectral Remote Sensing of Vegetation". P.S. Thenkabail, J.G. Lyon, A.

Huete. CRC Press, Taylor & Francis Group, Boca Raton: 265-288. ISBN: 9780429192180

Milla R., Castro-Díez P., Montserrat-Martí G. 2010 - *Phenology of Mediterranean woody plants from NE Spain: Synchrony, seasonality, and relationships among phenophases.* Flora 205: 190–199. doi: 10.1016/j.flora.2009.01.006

Miller P.C. 1982 - *Environmental constraints to vegetation form in Mediterranean type ecosystems.* Ecologia Mediterranea 8: 411–416. ISSN : 0153-8756

Mohammed G.H. et al. 2014 - *2012 FLEX/Sentinel-3 Tandem Mission Photosynthesis Study,* ESA Contract No 4000106396/12/NL/AF, Study Final Report. www.flex-photosyn.ca/Reports/PS-Study_FINAL_REPORT_Full_Report_(Public).pdf

Moya I., Camenen L., Evain S., Goulas Y., Cerovic Z.G., Latouche G., Flexas J., Ounis A. 2004 - *A new instrument for passive remote sensing. 1. Measurements of sunlight-induced chlorophyll fluorescence.* Remote Sensing of Environment 91: 186–197. doi: 10.1016/j.rse.2004.02.012

Müller N. J. C. 1874 - *Untersuchungen über die diffusion der atmosphärischen Gase und die Gasausscheidung unter verschiedenen Beleuchtungsbedingungen.* Jahrbucher für Wissenschaftliche Botanik 9: 36–49.

Murchie E.H., Harbinson J. 2014 - *Non-Photochemical Fluorescence Quenching Across Scales: From Chloroplasts to Plants to Communities. Chapter 25.* In: "Non-Photochemical Quenching and Energy Dissipation in Plants, Algae and Cyanobacteria". Advances in Photosynthesis and Respiration 40. B. Demmig-Adams, G. Garab, W. W. Adams III, Govindjee (Eds.). Springer Dordrecht, Heidelberg, New York and London: 553-582. doi: 978-94-017-9032-1_25

Nichol C.J., Drolet G., Porcar-Castell A., Wade T., Sabater N., Middleton E.M., MacLellan C., Levula J., Mammarella I., Vesala T., Atherton J. 2019 - *Diurnal and seasonal solar induced chlorophyll fluorescence and photosynthesis in a boreal scots pine canopy.* Remote Sensing 11: 273. doi: 10.3390/rs11030273

Niinemets Ü., Keenan T. 2014 - *Photosynthetic responses to stress in Mediterranean evergreens: Mechanisms and models.* Environmental and Experimental Botany 103: 24–41. doi: 10.1016/j.envexpbot.2013.11.008

Ogaya R., Peñuelas J. 2003 - *Comparative seasonal gas exchange and chlorophyll fluorescence of two dominant woody species in a* Holm Oak *Forest.* Flora 198: 132–141. doi: 10.1078/0367-2530-00085

Oliveira G., Peñuelas J. 2004 - *Effect of winter cold stress on photosynthesis and photochemical efficiency of PSII of the Mediterranean* Cistus albidus *L. and* Quercus ilex *L.* Plant Ecology 175: 179–191. doi: 10.1007/s11258-005-4876-x

Öquist G., Huner, N.P.A. 2003 - *Photosynthesis of overwintering evergreen plants.* Annual Review of Plant Biology 54: 329–355. doi: 10.1146/annurev.arplant.54.072402.115741

Orsham G. 1989 - *Plant Pheno-morphological Studies in Mediterranean Type Ecosystems.* Kluwer Academic Publishers Group, Dordrecht, The Netherlands. 404 p. doi: 10.1007/978-94-010-7897-9

Papageorgiou G. C., Govindjee 2004 - *Chlorophyll a Fluorescence: A Signature of Photosynthesis.* In: "Advances in Photosynthesis and Respiration". Volume 19. Springer, Dordrecht, The Netherlands. 818 p. ISBN 978-1-4020-3218-9

Pereira J.S, Beyschlag G., Lange O.L., Beyschlag W., Tenhunen J.D. 1987 - *Comparative phenology of four Mediterranean shrub species growing in Portugal.* In: "Plant Response to Stress: Functional Analysis in Mediterranean Ecosystems". J.D. Tenhunen, F.M. Catarino, O.L. Lange, W.C. Oechel (Eds.). Springer Verlag, Berlin Heidelberg, New York, London, Paris, Tokyo: 503-514. doi: 10.1007/978-3-642-70868-8_33

Pessarakli M. 2016 - *Handbook of Photosynthesis.* CRC Press, Taylor and Francis Group. 846 p. ISBN 9781482230734

Pollastrini M., Bussotti F., Iacopetti G., Puletti N., Mattioli W., Selvi F. 2018 - *Forest tree defoliation and mortality in Tuscany (central Italy) connected to extreme drought and heat wave in the 2017 summer: a preliminary report.* Geophysical Research Abstracts 20, EGU 2018-9958. doi: 10.13140/RG.2.2.32745.39526

Porcar-Castell A. 2011 - *A high-resolution portrait of the annual dynamics of photochemical and non-photochemical quenching in needles of* Pinus sylvestris. Physiologia Plantarum 143: 139–153. doi: 10.1111/j.1399-3054.2011.01488.x

Porcar-Castell A., Garcia-Plazaola J.I., Nichol C.J., Kolari P., Olascoaga B., Kuusinen N., Fernández-Marín B., Pulkkinen M., Juurola E., Nikinmaa E. 2012 - *Physiology of the seasonal relationship between the photochemical reflectance index and photosynthetic light use efficiency.* Oecologia 170: 313–323. doi: 10.1007/s00442-012-2317-9

Porcar-Castell A., Tyystjärvi E., Atherton J., van der Tol C., Flexas J., Pfündel E.E., Moreno J., Frankenberg C., Berry J.A. 2014 - *Linking chlorophyll a fluorescence to photosynthesis for remote sensing applications: mechanisms and challenges.* Journal of Experimental Botany 65: 4065–4095. doi: 10.1093/jxb/eru191

Quézel P. 1985 - *Definition of the Mediterranean region and the origin of its flora.* In: "Plant Conservation in the Mediterranean Area". Gómez-Campo C. Eds. Springer, Dordrecht, The Netherlands: 9-24. ISBN 978-90-6193-523-0

Ripullone F., Borghetti M., Raddi S., Vicinelli E., Baraldi R., Guerrieri M.R., Nolè A., Magnani F. 2009 – *Physiological and structural changes in response to altered precipitation regimes in a Mediterranean macchia ecosystem.* Trees 23: 823–834. doi: 10.1007/s00468-009-0323-2

Ripullone F., Rivelli A.R., Baraldi R., Guarini R., Guerrieri R., Magnani F., Peñuelas J., Raddi S., Borghetti M. 2011 - *Effectiveness of the photochemical reflectance index to track photosynthetic activity over a range of forest tree species and plant water statuses.* Functional Plant Biology 38: 177–186. doi: 10.1071/FP10078

Salleo S, Nardini A., Lo Gullo M.A. 1997 – *Is sclerophylly of Mediterranean evergreens an adaptation to drought?* New Phytologist 135: 603–612. doi: 10.1046/j.1469-8137.1997.00696.x

Specht R.L. 1981 - *Primary production in Mediterranean climate ecosystems regenerating after fire.* In: "Ecosystems of the World II: Mediterranean-type Shrublands". F. di Castri, D.W. Goodall and R.L. Specht (Eds.). Elsevier Scientific Publishing Company, Amsterdam, The Netherlands: 257–267. doi: 978-0444418586

Sperlich D., Chang C.T, Peñuelas J., Gracia C., Sabaté S. 2014 - *Foliar photochemical processes and carbon metabolism under favourable and adverse winter conditions in a Mediterranean mixed forest, Catalonia (Spain).* Biogeosciences 11: 5657–5674. doi:10.5194/bg-11-5657-2014

Sun Y., Frankenberg C., Jung M., Joiner J., Guanter L., Köhler P., Magney T. 2018 - *Overview of Solar-Induced chlorophyll Fluorescence (SIF) from the Orbiting Carbon Observatory-2: Retrieval, cross-mission comparison, and global monitoring for GPP.* Remote Sensing of Environment 209: 808–823. doi: 10.1016/j.rse.2018.02.016

Susmel L., Viola F., Bassato G. 1976 - *Ecologia della lecceta del Supramonte di Orgosolo (Sardegna Centro-orientale).* Cedam, Padova, Italy. 261 p.

Sutinen M.-L., Arora R., Wisniewski M., Ashworth E., Strimbeck R., Palta J. 2001. *Mechanisms of frost survival and freeze-damage in nature.* In: "Conifer Cold Hardiness".

F.J. Bigras, S.J. Colombo (Eds.). Springer, Dordrecht, The Netherlands: 89-120. doi: 10.1007/978-94-015-9650-3_4

van Ommen Kloeke A.E.E., Douma J.C., Ordoñez J.C., Reich P.B., van Bodegom P.M. 2012 - *Global quantification of contrasting leaf life span strategies for deciduous and evergreen species in response to environmental conditions*. Global Ecology and Biogeography 21: 224–235. doi: 10.1111/j.1466-8238.2011.00667.x

Verhoeven A. 2014 - *Sustained energy dissipation in winter evergreens*. New Phytologist 201: 57–65. doi: 10.1111/nph.12466

Verhoeven A., García-Plazaola J.I. , Fernández-Marín B. 2018 - *Shared mechanisms of photoprotection in photosynthetic organisms tolerant to desiccation or to low temperature*. Environmental and Experimental Botany 154: 66-79. doi: 10.1016/j.envexpbot.2017.09.012

Wilcoxon F. 1945. *Individual comparisons by ranking methods*. Biometrics Bullettin 1: 80-83. https://www.jstor.org/stable/3001968

Williams E.L., Hovenden M.J., Close D.C. 2003 - *Strategies of light energy utilization, dissipation and attenuation in six co-occurring alpine heath species in Tasmania*. Functional Plant Biology 30: 1205 – 1218. doi: 10.1071/FP03145

Yadav S.K. 2010 - *Cold stress tolerance mechanisms in plants. A review*. Agronomy for Sustainable Development 30: 515–527. doi: 10.1051/agro/2009050

Zavala M.A., Espelta J.M., Caspersen J., Retana J. 2011 - *Interspecific differences in sapling performance with respect to light and aridity gradients in Mediterranean pine-oak forests: implications for species coexistence*. Canadian Journal of Forest Research 41: 1432–1444. doi: 10.1139/X11-050

Zhang C. 2017 – *The photochemical reflectance index (PRI) as an indicator of changes in photosynthetic dynamics and gross primary productivity in response to climate change*. PhD Thesis, Universitat Autònoma de Barcelona, 231 p. ISBN: 9788449075704

Zhang Y.-J., Xiao X., Zhang Y., Wolf S., Zhou S., Joiner J., Guanter L., Verma M., Sun Y., Yang X., Paul-Limoges E., Gough C.M., Wohlfahrt G., Gioli B., van der Tol C., Yann N., Lund M., de Grandcourt A. 2018 - *On the relationship between sub-daily instantaneous and daily total gross primary production: Implications for interpreting satellite-based SIF retrievals*. Remote Sensing of Environment 205: 276–289. doi: 10.1016/j.rse.2017.12.009

A narrow size diameter class model for tree growth and yield simulation in a mahoe (*Talipariti elatum* (SW.) Fryxell, *Malvaceae*) plantation in Puerto Rico

Jimena Forero Montaña[1*], Jess K. Zimmerman[2], Thrity Vakil[3], Mark Nelson[3,4], Marìa Eglee Perez[5], Joaquin Medin[6]

Abstract - Forest management planning requires forest growth models that provide a reliable way to forecast growth, calculate yield, and examine the effects of silviculture. Here we used a size stem frequency approach based on differential equations to developed a growth simulator system to guide the management of mahoe (*Talipariti elatum* (SW.) *Fryxell*) in a small plantation in Puerto Rico. We estimated the optimal harvest age using the Faustmann model, and evaluated alternative harvesting schemes using sensitivity analyses. The growth simulator system predicted that tree volume at first rises quickly, reaches a maximum value, and then decreases because of a lack of natural regeneration in the plantation. Thus, harvesting must be followed by replanting and large and healthy trees must be maintained as seed sources. The optimal harvest age was 43 years for a discount rate of 2.5% and 19 years for 5% interest. Analysis for alternative management schemes based on selective harvesting of the largest tress showed that 5%, 15%, or 30% of the trees ≥54.5 cm dbh could be harvested in cycles of 1, 5 or 10 years respectively, without drastically decreasing the basal area. The sustainability of these management schemes will depend on the costs of management, as well as the responses of mahoe to selective harvesting.

Keywords - forest management modelling; optimal harvest age; selective harvesting; system dynamics.

Introduction

Forest plantations help to meet the global demand for wood and have been increasingly established in the tropics to cope with the process of land degradation (Parrotta el al. 1997, Lamb et al. 2005, ITTO 2009,) and advance rural development (ITTO 2009). In the Caribbean island of Puerto Rico, for example, more than 38,200 hectares of degraded lands were reforested during the twentieth century prior to 1980 (Birdsey and Weaver 1982) and reforestation efforts have continued over the last four decades. Systematic trials included more than 100 exotic timber tree species, but only 19 produced satisfactory growth, and merely nine were later extensively planted (Francis 1995). Records of early growth and survival provided good information of the range of sites where these species can succeed (Francis 1995). However, the overall successes and the current area of forest plantations are not known. Moreover, few plantations are currently harvested or managed, and to our knowledge tree growth and yield models for planning management have been not developed for these timber species. Developing methods to more thoroughly assess the sustainability of these forest plantations is therefore clearly needed.

Forest management planning requires growth models to forecast forest growth, examine the effects of management, and calculate yield of different management strategies (Vanclay 1994, Pretzsch 2009). Whole stand models have been traditionally used to model growth and yield in pure even-aged forest stands such as plantations when the main focus is the production of a single product in one rotation (Vanclay 1994, Pretzsch 2009). The stand level approach uses stand parameters such as tree density (No trees/ha), basal area, and volume to predict growth. Whole stand models require the lowest resolution and level of abstraction of forest models, but consequently provide rather general information of the future stand structure (Vanclay 1994, Pretzsch 2009). On the other hand, single tree models which use individual trees as the basic unit of modelling provide detailed information of stand structure (Vanclay 1994, Palahí et al. 2003, Pretzch 2009). Tree

1 University of Puerto Rico Rio Piedras, Faculty of Natural Sciences, Department of Biology , San Juan Puerto Rico
2 University of Puerto Rico Rio Piedras, Environmental Sciences, San Juan Puerto Rico
3 Tropic Ventures Research and Education Foundation, Patillas, Puerto Rico
4 Institute of Ecotechnics (US/UK), Santa Fe, New Mexico, United States
5 University of Puerto Rico Rio Piedras, Mathematics, San Juan, Puerto Rico
6 University of Puerto Rico in Bayamon, Physics, Bayamon, Puerto Rico

*Corresponding author: forero.jimena@gmail.com

growth models with finer resolution, however, are susceptible to higher estimation errors, which may compromise the estimation accuracy desired for forest management decision making (Ferraco Scolforo et al. 2019). Stand oriented growth models can reveal some aspects of stand structure by considering steam number frequencies in diameter classes (Vanclay 1994, Pretzch 2009). These models predict not only stand mean values, but also changes in the diameter frequency distribution as result of growth, removal, and mortality (Pretzch 2009). The forest growth models described above represent different approaches in describing stand development. None of these models can be replaced entirely by one another, each modeling approach provides specific knowledge of forest growth, yield, and stand structure (Vanclay 1994, Pretzch 2009). Therefore, the aim and purpose of a model, as well as the stage of knowledge about the system considered determine the degree of complexity necessary or possible and its temporal and spatial resolution (Pretzsch 2009). In this study we used a size frequency approach based on differential equations to model tree growth and yield in mahoe (*Talipariti elatum* (SW.) Fryxell, Malvaceae), a valuable hardwood tree native to Cuba (Adams 1971) and Jamaica (Little et al. 1977), in a 25 year old enrichment plantation to guide its management in a subtropical wet secondary forest in Puerto Rico. The modeling approach was a compromise resolution between whole stand models and individual tree models. The system breaks down the stand into many narrow size classes of 1 cm and therefore permits simulations of growth with detailed information of stand structure (Muller-Landau et al. 2006), but keeping the model conceptually and operationally simple. Tree growth was modeled over a nine-year period using nonlinear equations and the optimal time for harvesting was calculated using the Faustmann (1849) formula to combine biological and economic risks (Chang 1998, Buongiomo 2001, Buongimo and Gilles 2003). Sensitivity of the optimal management schedule to changes in the discount rate, timber prices, and harvesting costs was analyzed. Finally, alternative management schemes based on selective harvesting of the largest trees were examined using sensitivity analysis.

Methods

Study area description

The study was conducted at Las Casas de la Selva (here after LCS), 376 ha of private land located on the south flank of the Cordillera de Cayey in the municipality of Patillas, Puerto Rico. The topography is mountainous and characterized by steep slopes that vary from 300 to 600 m in elevation. The forest is classified as subtropical wet forest in the Holdridge life zones system and is locally known as "tabonuco forest", with *Dacryodes excelsa* Vahl. being a dominant species. Annual rainfall averages 3,000 mm and annual temperature 22 °C (Nelson et al. 2010).

Las Casas de la Selva is owned and operated by Tropic Ventures, a joint venture between Global Ecotechnics Corporation and Decisions Team Inc. Before 1983, when Tropic Ventures acquired the land, it was mainly used for traditional cultivation of coffee, and a smaller part was cleared for grazing cattle. Between 1984 and 1989, valuable introduced hardwood species such as mahoe (*T. elatum*), and mahogany (*Swietenia macrophylla* x *S. mahagoni*), some native hardwood species, and Caribbean pine (*Pinus caribaea* Morelet) were planted on approximately 78 ha of the property (Nelson et al. 2011). The plantation of mahoe was mainly established on nine hectares of abandoned pastures. All mahoe trees were planted in a few months but the specific date of planting of each tree is not known. The initial stand density was 330 saplings ha^{-1} with 10 m spacing from line to line and 3 m spacing within the lines. During planting, larger native hardwoods were left untouched when they were present in the tree lines (Nelson et al. 2011). The plantation currently has reached full canopy closure with 38 associated species and a total density of 571 trees ha^{-1} with a basal area of 37.43 m^2 ha^{-1} (Forero-Montaña 2015). Most of the associated trees are native species (95%) and comprise 61% of the stems > 10 cm dbh, but in general these trees are small and do not compete directly for light with mahoe. Indeed, mahoes had outgrown the other species and represent 49% of the trees and 72% of the total basal area (Forero-Montaña 2015).

Table 1 - Stand characteristics of the mahoe plantation at LCS in each inventory, showing mean values with standard errors when applicable.

Year	Number of stems ha^{-1}	Mean dbh (cm)	Mean total height (m)	Basal area (m^2 ha^{-1})	Volume (m^3 ha^{-1})
2003	285	22.09 (0.48)	16.04 (0.24)	11.82	112
2011	278	27.5 (0.62)	19.35 (0.29)	19.27	207

Forest inventories

In 2003, a 1.2 ha permanent plot was established to monitor tree growth and survival in 342 mahoes ≥ 3.5 cm dbh. Diameter at breast height (1.3 m) (dbh), total height, and commercial height were measured on each mahoe using standardized methodology (Nelson et al. 2011). A second inventory was carried out between 2010 and 2011. Tab. 1 summarizes the characteristics of the stand in each inventory.

Data analyses

Tree growth simulator system

The growth simulator system was implemented in STELLA 9.0.9 (Richmond 2001). The basic structure of the model consisted of 66 one-cm diameter classes ranging from 9.5 to 54.5 cm that correspond to the stocks, which are connected by flows or processes that change the number of trees in each diameter class during a time step. The main processes of yield include ingrowth, outgrowth, mortality, and harvesting. The model was formulated as a set of differential equations that altogether change the number of trees in Diameter $Class_i$ with the form:

$$\frac{\Delta(DiameterClass_i)}{\Delta t} = \frac{\Delta(Ingrowth)}{\Delta t} - \frac{\Delta(Outgrowth)}{\Delta t} - \frac{\Delta(Mortality)}{\Delta t} - \frac{\Delta(Harvesting)}{\Delta t} \quad (1)$$

The time step Δt was set as a discrete interval of one year. The differential equations of each process were mathematically disaggregated into sub-functions (control functions) to facilitate the model formulation. The distribution of tree numbers into the diameter classes at t_0 was taken from the first inventory. The tree numbers from t_0 to time t_n were obtained by stepwise integration of the differential equations in each time step. The new tree numbers in each size class were used as input variables for the control functions in the subsequent step. The major components of the model were implemented in STELLA using a design were ingrowth, outgrowth, mortality, and harvesting are the parameters to be set:

Ingrowth

Ingrowth or recruitment is the number of trees entering into the smallest stem diameter class in each year. Natural regeneration of mahoe has not occurred in the LCS plantation. Seedlings emerge in the forest but die apparently because they are not shade tolerant. Thus, recruitment into the first class was simulated assuming that harvesting must be followed by replanting. Stand density was reduced from 330 trees ha^{-1} to 285 trees ha^{-1} during the ~25 years since the plantation was established. Therefore, we considered it conservative to assume that 80% of the planted trees survive and enter into the smallest size class. In addition, because there is no information on growth of mahoe saplings, it was assumed that the planted trees that survive enter into the smallest diameter size class (3.5 cm) in a period of five years. For the other diameter classes outgrowth from the smaller size class represented ingrowth into the next larger diameter class.

Outgrowth

Outgrowth is the number of trees moving from each diameter class to the next higher diameter class in each year. The largest class serves as a container for very large trees and therefore its outgrowth is null. Outgrowth was defined as:

$$Outgrowth = Diameter\ Class_i \times Transition\ rate_i \quad (2)$$

where the transition rate is the probability that a surviving tree moves from its Size $class_i$ to the next larger Diameter $class_{i+1}$ in one time step. The transition rate was calculated as:

$$Transition\ rate_i = \frac{1}{\alpha} \times Id_i \quad (3)$$

where a is the width of Diameter $class_i$, defined as 1 cm in this study, and Id_i is the estimated diameter increment for an average tree in the *Size* $class_i$. Diameter increment is a function dependent on tree size and stand density, and it was modeled using nonlinear equations by means of the methods explained below in the diameter increment model section.

Mortality

The number of dead trees in Diameter $class_i$ in each year was defined as:

$$Mortality = Diameter\ Class_i \times Mortality\ rate_i \quad (4)$$

where mortality rate is the probability of a tree in Diameter $class_i$ dying in a year. It is usually a function of tree size and total basal area because higher probabilities of mortality typically occur at smaller sizes and in crowded stands. Mortality rate, however, was very low during the study period, only nine trees died, and mortality was independent of tree size. Therefore, annual probability of mortality was introduced in the growth simulator system as a constant value to randomly kill 0.003% (9/342 over nine years) of the trees in each time step.

Harvesting

Harvesting is defined as the number of live trees

in Diameter class$_i$ harvested per area in each cutting cycle. The system was first formulated to simulate growth in absence of harvesting with the aim of estimating the optimal harvest age or time for harvesting all mahoes followed by replanting applying the principles of the Faustmann formula for even-aged stands. Then, alternative management schemes were simulated to explore other possible scenarios for the management of the plantation that allow for periodic revenue without having to wait a long period to harvest all the trees (i.e., clear cutting). With this aim the model was constructed to simulate selective harvesting of the largest diameter size (i.e., dbh ≥54.5 cm) trees to provide revenues, which also enables release of small trees and faster growth of a now less crowded stand. We explored numerous scenarios with different cycles and levels of harvesting to find a management scheme that produces the highest sustained yield without reducing the residual basal area to lower levels than the initial value in 2003 (i.e., 12 m^2 ha^{-1}) with the aim of maintaining a reserve of trees that allow adequate growth. For all the scenarios we assumed that harvesting is followed by replanting to replace both the trees that are harvested and the trees that die by natural causes, thus maintaining the number of trees in equilibrium. To be as realistic as possible, harvesting started at the 16th year of the simulations corresponding to the state of the stand at the first inventory.

Fig. 1 shows the complete system dynamics diagram of the tree growth simulation system as it was implemented in STELLA. The diagram consists of an array that includes all diameter classes with the connected flows in one, the relationships shown in Fig. 1 are in effect for each diameter class. The complete set of model equations and variables as shown in Fig. 1 are listed in Anx. 1.

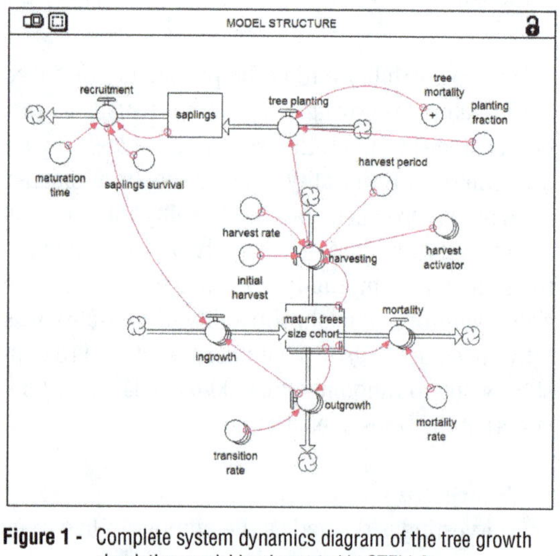

Figure 1 - Complete system dynamics diagram of the tree growth simulation model implemented in STELLA.

Diameter increment model

Growth calculations were performed for trees that were measured in both inventories without large or evident mistakes. Trees with negative dbh increment and extreme outliers were excluded from the analyses. Annual diameter increment was calculated as:

$$\frac{(dbht1-dbht2)}{(date\ 1-date2)} \times 365 \quad (5)$$

where subscripts refer to successive forest inventories and time is measured in days. Traditional analyses of growth rates are based on linear models that assume constant absolute growth rate. However, the rate of biomass accumulation, almost universally slows as plants grow (Paine et al 2012). This decrease in growth results from a combination of factors that affect plants ontogeny such as accumulation of non-photosynthetic biomass in the form of stems and roots, self-shading of leaves, and decline in local concentrations of soil nutrients (South 1995). Nonlinear models provide the best method to account for slowing growth during plant ontogeny (Paine et al 2012). In this study, we used the nlsList function of the nlme package in the R statistical language to fit independent nonlinear functions to model tree growth of mahoe using the Gauss–Newton algorithm (Pinheiro and Bates 2000). Mean annual diameter growth rates within 1 cm dbh classes were calculated to facilitate the calibration of the growth simulator system. A power law function was used to model the relationship between diameter increment and initial size to allow for the assumption of slowing growth during tree ontogeny (Paine et al. 2012).

Height static equation for volume estimation

Analysis of the height data revealed obvious and large errors in the measurements of the first inventory that did not allow for the estimation of a height or volume increment model. Therefore, a static height model was constructed using the measurements of the second inventory. The relationship between total height and dbh was also modeled using a power law function to allow for the assumption of slowing height growth during tree ontogeny. This allometric equation served to calculate total height from dbh and was incorporated into the growth simulator system at the end of each time step to estimate mean volume of each diameter class using the relationship:

$$Tree\ volume = 0.5 \times basal\ area \times total\ heigh \quad (6)$$

Model validation and harvesting optimization

Once the growth simulator system was calibrated, its forecasting accuracy was assessed by comparing the observed size distribution in the second inventory with the distribution predicted by the model in the year ninth of the simulations using a Two-sampled Kolmogorov-Smirnov test (Crawley 2013). In addition, long-term steady states predicted by the model were also checked for quantitative plausibility of the model.

The specific age of harvesting that generates the maximum sustained yield depends on annual volume increment and socio-economic factors such as wood prices, costs of management, and discount rate, which account for the costs of waiting to realize profits. The problem of optimal harvest age for the management of even-aged stands such as plantations involves choosing the optimal time for clear-cutting followed by replanting and subsequent development of a new even-aged stand (Nautiyal 1988, Buongiomo 2001).

Using discounted cash flow techniques Faustmann (1849) provided a method to integrate biological and economic risks to the problem of land valuation to decide when to cut the forest (Chang 1998, Buongiomo 2001, Bungimo and Gilles 2003). The harvest age that maximizes the land value can be found using the following mathematical expression:

$$FEV = \frac{pV_t - C}{1 + r^t - 1} - C \quad (7)$$

where FEV is the forest expectation value that includes the value of the land plus trees, p is the net price of wood per unit of volume at age t, C is costs of replanting, and r is discount rate. For even-aged stands the net present value is basically formed by a continuous periodic series of harvesting revenues minus costs of harvesting at the end of every rotation of t years.

Given that the aim of this study was to use discounted cash flow techniques and sensitivity analysis to explore possible scenarios for management, rather than assess the absolute economic value of the LCS mahoe plantation, the capital value of the land was neglected and for simplicity it was assumed that future wood prices and costs remain constant or increase regularly in the future. The forest expectation value of different harvesting cycles was calculated for various combinations of costs, revenues and discount rates. In 2003, Tropic Ventures started selling mahoe in niche markets for expensive tropical hardwoods in the United States at a relatively stable price of US \$10 per board foot (0.002 m^3). A previous evaluation at LCS determined a cost of production of US\$ 3 dollars per board foot (Vakil 2005 unpublished data). So, we included a net wood price of US \$7 per board foot and a base cost of US\$ 25.000 ha^{-1} for replanting after clear cutting the whole stand. This corresponds to a full-time salary for a worker with minimum federal wage (US\$ 7.25) for less than two years in Puerto Rico. We explored the sensitivity of the optimal harvest age to a 1% annual increment in wood prices and a 20% increment in the costs of reforestation on the basis of the Puerto Rican Senate project Law 1195 of September 24 2014, which seeks to gradually increase the minimum wage from US\$ 7.25 to \$ 15 hr^{-1} in 2025.

Results

Annual stem diameter increment model

Annual stem diameter increment and size (dbh) exhibited a positive relationship with clear deviations from linearity and high individual variability. The best model had the form:

$$F(x) = \alpha \times dbh^{\beta} \quad (8)$$

with $\alpha = 0{,}043 \pm 0{,}009$ (1 standard error) and $\beta = 0{,}77 \pm 0{,}06$ (1 standard error, n= 38) and explained 86% of the variation (residual standard error 0.082, df= 36). The exponent value indicated that diameter increment progressively decreases as stem diameter increases (Fig. 2).

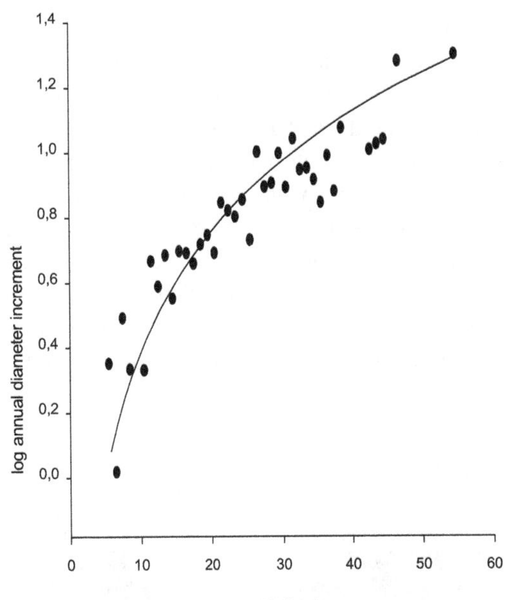

Figure 2 - Data and fitted power law function for the relationship between mean annual dbh increment and 1-cm diameter size classes for mahoe.

Height static equation for volume estimation

Total height and dbh exhibited a positive relationship (Fig. 3). The power law function that best fitted the data had the form:

$$F(x) = \alpha \times dbh^{\beta} \quad (9)$$

with $\alpha = 5{,}61 \pm 0{,}69$ and $\beta = 0{,}37 \pm 0{,}03$ (n=41) (residual standard error 2.389, df=43) and explained 79% of the variation. The value of the exponent indicates that tree height slowly decreased as dbh increases (Fig. 3).

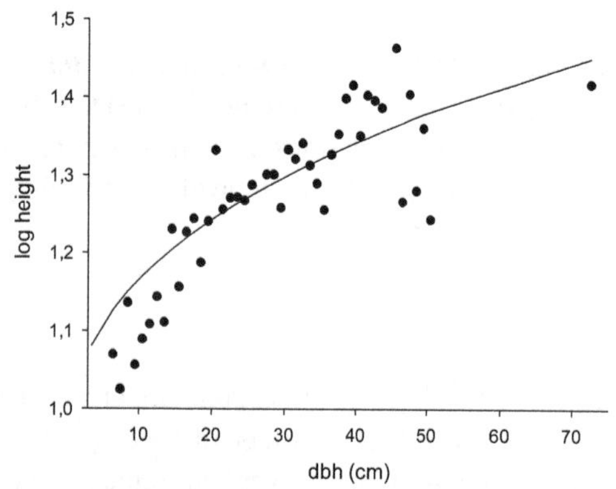

Figure 3 - Data and fitted power law function for the relationship between total height and stem diameter at breast height (dbh) for mahoe.

Model validation

The growth simulator system predicted a total volume of 173 m^3 ha^{-1} in the ninth year of the simulations, while the observed mahoe volume in 2011 was 207 m^3 ha^{-1}. The growth simulator system likely underestimated tree volume increment because the simulations accumulated the largest trees in the 54.5 cm diameter class, while mahoe can grow to more than 70 cm. However, the predicted size and tree volume distributions in the ninth year of the simulations were not statistically different to the observed number of trees (D= 0.09, P> 0.05) and observed volume (D=0.15, P>0.05) in each size category in 2011 (Fig. 4). These results indicate that the growth simulator system provides accurate predictions for mahoe growth and yield at LCS.

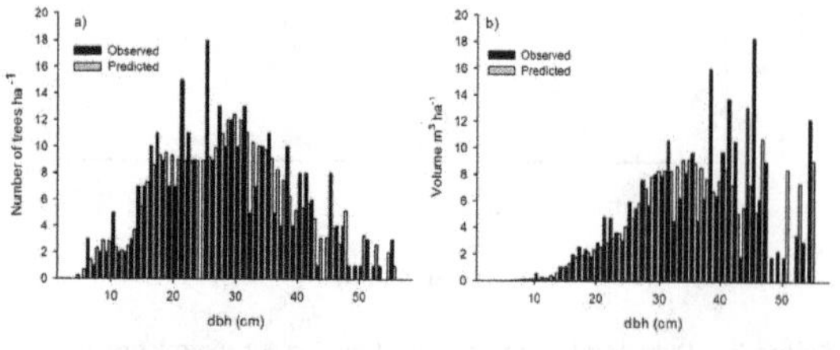

Figure 4 - Comparison between observed and predicted distributions by the tree growth simulator system in the year ninth of the simulations. a) Tree size distribution, b) Tree volume distribution.

Optimal harvest age

The model predicts that in absence of management tree volume will initially increase quickly, reach a maximum value of 714 m^3 ha^{-1} at the age of ~80 years, stay relatively constant for few years, and then progressively decline because natural regeneration is absent in the plantation (Fig. 5).

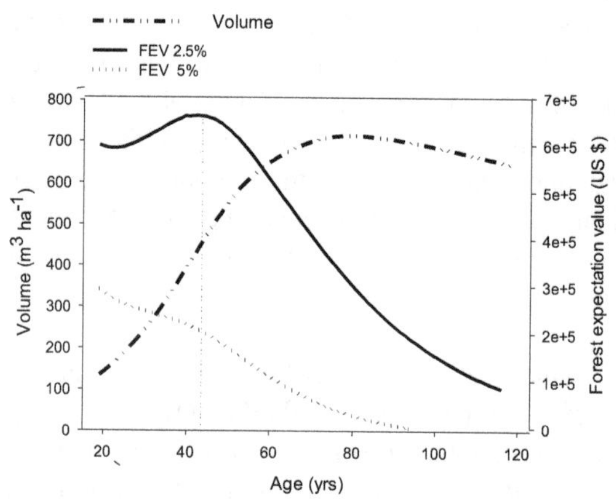

Figure 5 - Predicted annual volume increment for 100 years (right) and forest expectation value (FEV) (left) showing optimal harvest age for 2.5% and 5% discount rates. Notice different scales in the y axis.

The harvest age that maximizes the land expectation value, when the net price of wood was \$7 per board foot, was 43 years with a discount rate of 2.5% and 19 years with a discount rate of 5%, and yielded a volume of 445 m^3 ha^{-1} and 134 m^3 ha^{-1}, respectively (Fig. 5). In contrast, one-time changes in both wood prices and costs of reforestation modified the FEV, but did not change the optimal harvest age. A 1% annual increment in wood prices prolonged the optimal harvest age to 47 years with a 2.5% discount rate, but did not change the harvest age with a 5% discount rate. A 20% annual increment in the costs of reforestation decreased the optimal harvest to 19 years with a 2.5% discount rate, but it did not alter the optimal harvest age with a 5% discount rate (Tab. 2).

Selective harvesting schemes

The growth simulator system showed that 5%, 15%, or 30% of the trees ≥54.5 cm dbh could be harvested in cycles of 1, 5 or 10 years, respectively, without severely decreasing the basal area (Fig. 6). The

Table 2 - Optimal harvest age and forest expectation value for different wood prices, costs of reforestation, and discount rates.

r	p (US$ BF)	C (US$ ha^{-1})	FEV (US $ ha^{-1})	Optimal harvest age (yr)	Yield (m^3 ha^{-1})
	7	25'000	557'633	43	445
	10.5	25'000	817'926	43	445
2.5	7	50'000	494'679	43	445
	8.96	25'000	684'917	*47	509
	7	25'000	478'857	*19	134
	7	25'000	168'196	19	134
	10.5	25'000	751'345	19	134
5	7	50'000	127'430	19	134
	7	25'000	168'196	*19	134
	7	25'000	168'196	*19	134

Net wood price (p), costs of reforestation (C), forest expectation value (FEV), board foot (BF). Asterisk correspond to the optimal harvest age for 1% annual increment in wood prices and 20% increment in costs of replanting starting with a US $7 BF and a US $ 25,000 ha^{-1} respectively at 19 years of age.

dynamics of the residual basal area follows a similar trend to volume increment without harvesting. As result the harvested volume initially increases, reaches a maximum value, then progressively declines, but it recovers after some years because the model assumes that planting takes place regularly to replace the trees that die by both harvesting and natural causes (Fig. 6). The annual harvesting scheme involves harvesting between one to a maximum of four trees per year, the five year cycle scheme starts harvesting one tree and reaches a maximum of 15 trees at the age of 65 years, finally the ten year cycle scheme starts harvesting 2 trees and reaches a maximum of 31 trees at the age of 65 years. The sustainability of these management schemes will depend on mahoe responses to partial canopy opening, as well as on the costs of management, as we discuss below.

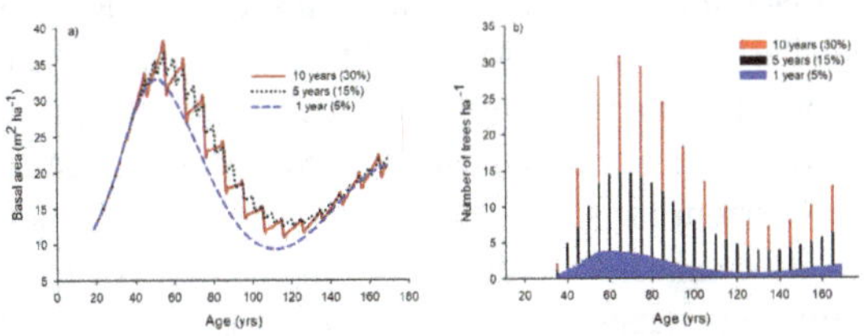

Figure 6 - a) Residual basal area and b) Number of trees harvested in each cycle for the alternative management schemes based on selective harvesting of trees ≥ 54.5 cm dbh.

Discussion

In this study we developed a narrow size diameter class model based on differential equations for tree growth and yield simulation in mahoe . Given the initial condition of the stand, the model predicted the state of the stand after given years with the alternative assumed prescriptions. Our approach has low complexity and is useful for users that have a small data set, thus it is easy to use for forest management planning. We examined two management possibilities: first, we allow the trees to grow and clear cut the whole stand when it reaches the optimal harvest age using the Faustmann formula. And second, we explored alternative harvesting schemes using sensitivity analysis based on selective harvesting of the largest mahoes, which would allow a steadier revenue stream, while small promised trees are released.

The optimal harvest age was 43 years for a discount rate of 2.5% and 19 years for a discount rate of 5%. Thus, given that the plantation is currently 25 years old under a high interest scenario the plantation must be harvested as soon as possible. The analysis of alternative scenarios for selective harvesting suggests that 5%, 15%, or 30% of the mahoes ≥54.5 cm dbh could be harvested in cycles of 1, 5 or 10 years respectively, without severely decreasing the basal area. The sustainability and economic viability of these management schemes will depend on the costs of both harvesting and replanting, and ultimately on the responses of mahoe to selective harvesting.

Our data indicate that mahoe is shade intolerant and we do not know to what extent it could regenerate under partial canopy opening. Moreover, due to its rapid growth rate, mahoe is a high nutrient demanding tree (Kihlgren-Smith 2000). Therefore, it is vital to maintain soil productivity after harvesting. Cleaning, weeding, and soil fertilization are labor intensive operations that can be especially expensive in the tropics where soils are susceptible to erosion and nutrient loss, and where large canopy openings can trigger the growth of vines and grasses that obstruct regeneration (Wadsworth 1997). Thus, a management scheme should be selected to maintain the stand basal area low enough to maximize growth, but at the same time avoid severe canopy and soil disturbance to ensure adequate regeneration (Hutchinson 1993, Wadsworth 1997). It is important to avoid removing only the most valuable trees of the stand leaving behind mostly damaged diseased or genetically inferior trees with little potential for future growth. On the other hand, sick trees or tress with defects such as poor form, low forks, or too many large branches, which will never produce a high-quality saw log, should be eliminated (Hutchinson 1993). Trees of undesired species that are inhibiting growth of mahoes with high potential should be eliminated too. If two healthy vigorous mahoes are competing with one another, one should be harvested (Hutchinson and Wadsworth 2006, Wadsworth and Zweede 2006).Moreover, given that natural regeneration is currently absent at LCS any management scheme that is adopted must leave large and healthy trees as seed sources.

The modeling approach of this study can help to answer key questions of forest management planning. However, the deterministic nature of both the forest growth model and the classic Faustmann analysis , limits its accuracy to predict tree growth, estimate the optimal harvest age, and examine different silvicultural options. The classic Faustmann analysis considers that tree growth, wood prices, and costs remain constant over time or can be predicted with certainty. These assumptions, however, are not generally true. Variation in tree growth, wood prices, costs of reforestation, and discount rates, are the rule rather than the exception (Chang 1998, Buongiomo 2001).

Storms and hurricanes are important stochastic processes that shape tree growth in the secondary forests of Puerto Rico (Uriarte et al. 2004, 2009). Thus, consideration of natural risks is essential to improve our ability to model tree growth and guide management of mahoe in Puerto Rico. Montecarlo modelling techniques have been used to include natural risks such as storms and insect attacks in forest optimization models (Dieter 2001). This technique, however, requires the incorporation of empirical survival probabilities and specific information on the volume that can be harvested after a perturbation. This is information that we don't have for mahoe yet. Nonetheless, when risk is considered, the forest expectation value tends to decrease and the optimal harvest age becomes shorter (Dieter 2001). So, risks of storms would shorten the optimal harvest age of mahoe at LCS.

Our analysis showed that annual increments in wood prices tend to delay the optimal harvest age, while annual increments in the costs of reforestation and harvesting tend to shorten it. Prices respond to changing supply-demand balances that are not always predictable. Indeed, wood markets are continually undergoing dynamic changes (Bumgardner et al. 2014). Currently, there is a trend towards a reduction on the demand for tropical hardwoods in the international market because some species have been substituted by other woods or materials. Moreover, environmental concerns about tropical wood production represent a major challenge for exporters of tropical wood products. There is the expectation, however, that the implementation of the European Union Timber Regulation in 2013 would increase the demand for certified tropical woods (Bumgardner et al. 2014). The emergence of niche markets for fine tropical hardwoods in the USA represents a good commercial opportunity for mahoe, which is perhaps the only commercial hardwood in the world that has a blue heartwood color. Mahoe is now in low supply and it is only seen in small sizes and quantities in the market. Thus, prices of mahoe might likely increase in the near future.

Costs of management and production would likely change mainly in response to changes in labor expenses. For example, if the costs of replanting increase by 20% every year, in line with the Project Law 1195 of the Senate of Puerto Rico, the optimal harvest age of mahoe will decrease to 19 years with a discount rate of 2.5%, indicating that under this scenario the plantation would be not economically viable unless the prices of wood rise enough to cover the costs of management. Nevertheless, given the current fiscal deficit of the government of Puerto Rico salaries are likely to remain stable or even could decrease, which will prolong or not change the optimal harvest age of mahoe in the short term.

Conclusions

The model approach of this study made the best possible use of the existing data. The model is conceptually and operationally simple and can address strategic questions for the sustainable management of mahoe at our study area. The model approach was chosen on base of the existing data and the component functions were grounded on existing theories of forest growth. However, direct experimentation on selective harvesting is required to evaluate mahoe responses to partial canopy opening. Moreover, in order to reliably predict outcomes for different harvesting strategies for sustainable management, it is necessary to add to our database information on growth and survival of mahoe saplings, as well as estimate the real costs of selective harvesting and reforestation at LCS. Finally, risks of tropical storms and hurricanes is another aspect that must be considered to improve the prediction accuracy of the model. This will require evaluation of hurricane effects on vegetation structure and growth dynamics to determine survival probabilities and gather specific information on the volume that could be harvested after a perturbation.

Acknowledgements

This work was supported by the Center for Applied Tropical Ecology and Conservation (CREST-CATEC) under the Grant HRD 0734826, and the Luquillo LTER Program under the Grant DEB-1546686 to University of Puerto Rico. We thank support by The Earthwatch Institute and the work of Earthwatch volunteer groups from 2001-2012 in the collection of the data on tree growth at LCS. We also thank Dr. Fumero-Cabán and two anonymous reviewers for their valuable comments and criticism, which improved the manuscript.

References

Adams C.D. 1971 - The *blue Mahoe and other bush: an introduction to the plant life in Jamaica*. McGraw-Hill. Singapore. 159 p.

Birdsey R.A, Weaver P.L. 1987 - *Forest area trends in Puerto Rico*. New Orleans (LA): United States Department of Agriculture Forest Service Southern Forest Experiment Station. Research Note SO 331.

Buongiomo J. 2001 - *Generalization of Faustmann's formula for stochastic forest growth and prices with markov decision process models*. Forest Science 47(4): 466-471.

Buongiomo J., Gilless J.K. 2003 - *Decision methods for forest resource management. Academic Press San Diego*.439 p.

Bumgardner M., Johnson S., Luppold W., Maplesden F., Pepke E. 2014 - *Markets and market forces for lumber*. In: "The global forest sector: changes, practices and prospects" Hansen E., Panwar R., Vlosky R., Ed. CRC Press: 41-76.

Crawley M.J. 2013 - *The R Book*. Second Edition. John Wiley and Sons, Ltd London. 975 p.

Chang S.J. 1998 - *A generalized Faustmann model for the determination of optimal harvest age*. Canadian Journal of Forest Research 28: 652-659.

Dieter M. 2001 - *Land expectation values for spruce and beech calculated with Monte Carlo modelling techniques*. Forest Policy and Economics 2: 157-166.

Faustmann M. 1849 - *On the determination of the value which forest land and immature stands pose for forestry*. In: "Martin Faustmann and the evolution of discount cash flow" Gane M. Ed., Commonwealth Forestry Institute, Oxford Institute Paper 42, 1968.

Ferraco Scolforo H., McTague J.P., Burkhart H. Campoe O., Stape J.L., Roise J. 2019 - *Eucalyptus growth and yield system: Linking individual-tree and stand-level growth models in clonal Eucalypt plantations in Brazil*. Forest Ecology and Management 432:1-16.

Forero-Montaña J. 2015 - *Potential of subtropical secondary forests for sustainable forestry in Puerto Rico*. Doctoral dissertation, University of Puerto Rico, Río Piedras, San Juan, PR.

Francis J.K. 1995 - *Forest plantations in Puerto Rico*. In: "Tropical Forest: Management and Ecology" Lugo A. E. and Lowe C. Ed., Springer-Verlag New York: 210-223.

Hutchinson I.D. 1993 - *Puntos de partida y muestreo diagnóstico para la silvicultura de bosques naturales del trópico húmedo*. Colección Silvicultura de Bosques Naturales 7. CATIE Turrialba. 32 p.

Hutchinson I.D., Wadsworth F.H. 2006 - *Efectos de la liberación en un bosque secundario de Costa Rica*. Recursos Naturales y Ambiente 46: 155-160.

International Timber Organization (ITTO) 2009 - *Encouraging industrial forest plantations in the tropics: Report of a global study*. Technical Series 33. 141 p.

Kihlgren-Smith J. 2000 - *An Assessment of blue mahoe (Hibiscus elatus) growth using topographic, soil, and litter properties in St. Vincent, West Indies*. Master Thesis, the University of Calgary. Calgary Alberta.

Lamb D., Erskine P.D., Parrota J.A. 2005 - *Restoration of degraded tropical forests lands*. Science 316: 1628-1632.

Little E.L. Jr., Woodbury R.O., Wadsworth F.H. 1977 - *Trees of Puerto Rico and the Virgin Islands, second volume*. Agricultural Handbook 449. Washington, DC; U.S. Department of Agriculture, Forest Service. 596 p.

Muller-Landau H.C., Condit R., Chave J., Thomas S.C., Bohlman S.A., et al. 2006 - *Testing metabolic ecology theory for allometric scaling of tree size, growth and mortality in tropical forests*. Ecology Letters 9: 575-588.

Nautiyal J.C. 1988 - *Forest economics: Principles and applications*. Canadian Scholar's Press Inc. Toronto. 569 p.

Nelson M., Silverstone S., Reiss K.C., Burrows P., Joglar R., Robertson M., Vakil T. 2010 - *The impact of hardwood line planting on tree and amphibian diversity in a secondary subtropical wet forest of southeast Puerto Rico*. Journal of Sustainable Forestry 29: 503-516.

Nelson M., Silverstone S., Reiss K.C., Vakil T., Robertson M. 2011 - *Enriched secondary subtropical forest trough line-planting for sustainable timber production in Puerto Rico*. Bois et forests des Tropiques 309 (3): 51-61.

Palahí M., Pukkala T., Miina J., Montero G. 2003 - Individual-tree growth and mortality models for Scots pine (*Pinus sylvestris* L.) in north-east Spain. Annals of Forest Science: 60: 1-10.

Paine C.E.T., Marthews T.R, Vogt D.R., Purves D., Rees M., Hector A., Turnbull L.A. 2012 - *How to fit nonlinear plant growth models and calculate growth rates: an update for ecologists*. Methods in Ecology and Evolution 3(2): 245-256.

Parrotta J.A., Turnbull J.W., Jones N.1997 - Catalyzing *native forest regeneration on degraded tropical lands*. Forest Ecology and Management 99: 17.

Pinheiro J., Bates D. 2000 - *Mixed-effects models in S and S-PLUS*. Springer Verlag New York. 528 p.

Pretzsch H. 2009 - *Forest dynamics, growth and yield*. Heidelberg Springer, Berlin, Germany. 664 p.

Richmond B. 2001 - *An introduction to systems thinking*. Stella Software. High Performance Systems Inc. Hannover. 165 p.

South D.B. 1995 - *Relative growth rates: a critique.* South African Forestry Journal 173: 43-48.

Uriarte M., Canham C.D., Thompson J., Zimmerman J.K. 2004 - *A maximum-likelihood, spatially-explicit analysis of tree growth and survival in a hurricane-driven tropical forest.* Ecological Monographs 71: 591-614.

Uriarte M., Canham C.D, Thompson J., Zimmerman J. K. 2009 - *Understanding natural disturbance and human land use as determinants of tree community dynamics in a subtropical wet forest: Results from a forest simulator.* Ecological Monographs 79: 423-443.

Vanclay J.K. 1994 - *Modelling forest growth and yield: applications to mixed tropical forests.* CAB International, Wallingford, UK. 312 p.

Wadsworth F.H 1997 - *Forest production for tropical America.* U.S. Department of Agriculture, Forest Service Washington DC. 563 p.

Wadsworth F.H., Zweede J.C. 2006 - *Liberation: Acceptable production of tropical forest timber.* Forest Ecology and Management 233: 45–51.

24

Biomass Expansion Factor (BEF) and Carbon Stock for Brant's Oak (*Quercus brantii* Lindl.) Forests of West-Iran

Yaghoub Iranmanesh[1*], Hormoz Sohrabi[2], Khosro Sagheb Talebi[3], Sayed Mohsen Hosseini[2], Heidari Safari Kouchi Abouzar[4]

Abstract - Investigating a tree's biomass can provide basic information about forest carbon stock. The Biomass Expansion Factor (BEF) is a variable for estimating carbon stock of forests. The aim of this study was to analyse the Above Ground Biomass (AGB) allocation, developing the BEF and carbon stock for two vegetation forms of Brant's Oak (*Quercus brantii* Lindl.) based on forest inventory data. BEF is defined as the ratio of AGB to crown volume variables. The study data were taken from 30 trees that include 16 individual trees with single stem and 14 coppice shoots located in West-Iran. The trees selected were felled and separated into different components including: bole, main branches, lateral branches, twigs and leaves. The fresh weight of the trees was obtained with a portable hanging scale and several samples were taken from each tree component. The results of this study showed significant differences between component biomass proportions of the two vegetation forms of Brant's Oak trees and determined average biomass and carbon content of the forms studied. We also conclude that BEF of Brant's Oak could be improved by applying crown variables. According to the results, BEFs are tree-size dependent variables. Finally, this study indicates that age-dependent BEFs cannot be applied to conditions where stand development deviates from the conditions that in which the BEFs were developed.

Keywords - canopy volume; carbon; coppice; open forest; single stem; Zagros.

Introduction

CO_2 is one of the main greenhouse gases and the main reason for the global warming issue. Widespread concern about global warming and climate change has brought about an international agreement to reduce the amount of emission of this gas into the atmosphere. Under the United Nations Framework Convention on Climate Change (UNFCCC), most of world's countries are required to prepare a national inventory of greenhouse gas emissions and sequestration. Forest trees are the major terrestrial carbon pool and their ability to sequester carbon in their tissues has captured the interest of the world's governments. Estimating the biomass of trees is useful to assess forest structure and its condition (Chavé et al. 2003), carbon stocks and carbon sequestration in their biomass components and it is an indicator of whole site productivity (Návar 2009).

Currently, the methods used to calculate the biomass and carbon stock of trees are different (IPCC 2003). There are two approaches for estimating the above-ground biomass (AGB) of trees: a direct approach using allometric equations, and an indirect approach using biomass expansion factors (BEF). The indirect method is based on factors developed at stand level, and cannot be used to estimate biomass of individual trees (IPCC 2003). On the other hand, one of the methods used to convert field measurements of trees (forest inventory data) to stand biomass values is based on BEF (Soares and Tome 2004). These are mostly based on forest inventory information, by transforming the diameter, height or volume data into biomass estimates (Somogyi et al. 2006).

Sharp et al. (1975) were probably the first persons to use a constant BEF to estimate forest biomass. In their study a BEF of 2.0mg/m^3 was used to calculate the forest biomass in North Carolina, USA, based on forest inventory data. However, other studies indicated that the BEF is not constant (Guo et al. 2009). Since BEF is easier to use than biomass equations, the former is preferred (Johnson and Sharpe 1983). The aim of using such a factor is to take advantage of many tree-volume functions that are already available (West 2009). BEF application may vary in different projects. In some studies, single default values are often used, such as Kauppi et al. 1995. However, these factors may depend on the species, growth phase and site conditions (Satoo and Madgwick 1982). Therefore, calculations of BEF under specific conditions are to be preferred

1 Research Division of Natural Resources, Chaharmalhal and Bakhtiari Agricultural and Natural Resources Research and Education Center, AREEO, Shahrekord, I.R. Iran
2 Faculty of Natural Resources and Marine Science, University of Tarbiat Modares, Noor, I.R. Iran
3 Research Institutes of Forest and Rangelands, AREEO, Tehran, I.R. Iran
4 Faculty of Natural Resources, University of Guilan, Sowme'eh sara, I.R. Iran

(Lehtonen 2004). Muukkonen (2007) also made a compilation from several published studies and introduced some generic equations for volume and biomass in Europe.

In most European countries, greenhouse gas inventories are largely based on converting tree volume data from National Forest Inventories (NFI) to biomass using BEFs (Teobaldelli et al. 2009). However, BEF should be developed locally; otherwise, biased biomass estimation will be made (Lethonen 2005). NFI data can be processed in the tree level in order to provide accurate estimates of stand biomass. However, BEFs are still needed in published NFI results in which biomass is not estimated and as a complement of growth models that do not include biomass predictions (Soares and Tome 2012). In some research, BEF is defined as the ratio of the total AGB of trees to the biomass of the commercial timber. One example of this type of approach is the study by Levy et al. (2004) about BEF for coniferous tree species in Britain. In their study, BEF varied between 1.04 and 2.32, with a mean of 1.43 and tree height accounted for 45% of the variance in BEF in a logarithmic regression. The BEF was defined as, BEF=W/V, where W(mg) is AGB (including leaves, twigs, branches and stem) of the trees and V(m^3) contains the volume of sellable woody parts of the trees.

Chhabra et al. (2002) used BEF as a function of stock volume growth. Lehtonen et al. (2004) described stand-level BEFs, by converting stem volume to tree-component biomass (foliage, branches, stem wood, bark, stump, coarse roots and small roots). Lehtonen et al (2007) also suggested BEF estimation is an uncertain approach and found it an age-dependent approach. It is also sensitive to the dependencies among errors. Most of the uncertainty in estimating BEFs is related to uncertainty in applied biomass and volume models. Peichl and Arain (2007) measured AGB of forest trees in different ages and found that individual trees' BEFs for leaves, branches and roots change with the stand age. So, they developed experimental functions to relate those factors to the tree's age. Pajtik et al. (2008) presented the allometric equations and BEFs for young Norway (*Pice abies* L.) spruce trees (less than 10 years old) from natural regenerations in Slovakia. In other studies, Soares and Tome (2012) provided BEF for *Eucalyptus globules* Labill. stands in Portugal. They analysed changes in BEF by stand variables. Strong relationships were observed between BEF and stand age, basal area, volume and total height.

Brant's Oak (*Quercus brantii* Lindl.) is the main tree species in West-Iran. It is the dominant forest species in all of the southern zone of the Zagros area in the Irano-Turanian phytogeographical region. Although Brant's oak is important in ecological terms and carbon sequestration discussions, there are no available studies about biomass or carbon allocation or other similar studies related to this species. Because of the lack of general biomass functions for Brant's Oak in West-Iran, the aim of this study is to develop BEF for above-ground components in two common vegetation forms of Brant's Oak in the Zagros region of Iran. These equations could be applied to estimate the amount of carbon stored in this type of forest. These kinds of studies are able to help forest managers to estimate the stored carbon in Brant's oak stands with different components using forest inventory data.

Materials and Methods

Site description

The study was conducted in the Zagros region in West-Iran, which covers a vast area of the Zagros Mountains that is classified as semi-arid and open forests. The area selected for this study with a surface of 90ha is located between 50° 59′ 00″-50° 59′ 54″ E and 31° 14′ 20″-31° 15′ 24″ N in Chaharmahal and Bakhtiari province (Fig. 1). The mean annual rainfall of the area is 567mm and the mean annual temperature is 15.5°C. The main soil types of the study area are clay and clay-loam.

Brant's Oak is one of the most important tree species of Iran's western forests with an area of 3,500,000ha that covers the Zagros Mountains and makes a vast distributed pure and mixed oak community from 1,000 to 2,000m above sea level and extended from the North-West to the South-West of the country. These oak stands mixed with *Juniperus excelsa* M. Bieb at higher altitudes and with *Amygdalus scoparia* Spach., *Pistacia Atlantica* Desf. and *Acer monspessulanum* L. at lower altitudes. Due to human impact, only 7% of the oak forests is considered as high forests (trees with single stem), while the remaining 93% is coppice stands.

Tree selection

Thirty, one-hectare sample plots were established in the study area (Fig. 1). In each plot diameter at breast height (DBH), total height and crown diameter of all trees were measured. Trees were selected based on random sampling. Therefore 11 classes for individual trees with single stem based on DBH and 7 classes for coppice shoots based on crown diameter were identified with an equal proportion of trees in each class. Then, one or two of the trees were selected in each class. Thus 30 trees including 16 single-stem trees and 14 coppice

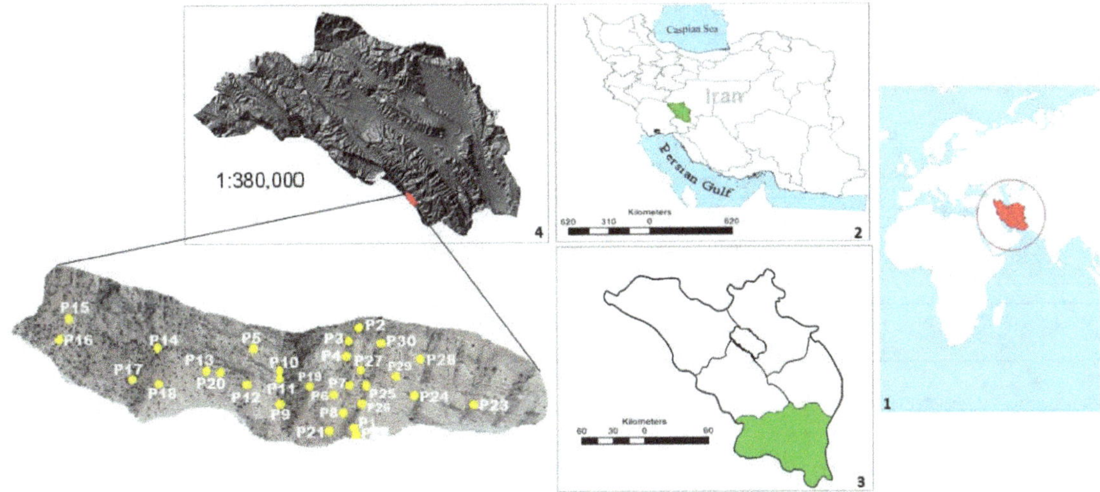

Figure 1 - Study area and location of sample plots.

shoots were selected to fell. Before felling, DBH, total height, crown diameter, crown height and shoot density of the trees selected were measured immediately.

Felling the trees and sampling procedures

Field work took place over 4 weeks in July and August, before leaves fall in autumn. For this purpose, sampled trees were felled and separated into different components including: trunk, main branches (diameter > 5cm), lateral branches (diameter between 1 and 5cm), twigs (diameter < 1cm) and leaves. All diameters were measured over the bark. Then fresh weights of all compartments were measured directly in the field using a portable spring scale (with 0.5kg accuracy).

Sample processing

The fresh weights of all felled tree components were measured. Thereafter, we randomly selected 30 leaves, 10 twigs (with 20cm length), 5 sample discs of branches with different sizes and 1 disk (with 5cm thickness) from the trunk (Losi et al. 2003).

The fresh weight of samples was measured immediately in the field. Afterward, to prevent any change in the quality of the samples, all labelled samples were transported to the nearest lab inside special bags. Then, all samples were dried at 80°C for 24 hours until the weight of samples became stable.

Equation (1) was used to determine dry weight of the components (Heidari Safari Kouchi et al. 2017).

$$WDc = \frac{WFc * WDs}{WFs} \qquad (1)$$

Where: WDc is the dry weight of each component of the tree, WFc is the fresh weight of each tree, WDs is the dry weight of each sample and WFs is the fresh weight of each sample.

Biomass expansion factor calculation

Paying attention to the growth form of Persian oak species in the Zagros habitat, trees trunks are not high and cylindrical. Actually, the tree trunk does not have a recognisable shape. Also, the volume of the trunk is not noticeable against the biomass of the crown (main branches, branches, twigs and leaves). In other words, the tree crown is important in biomass calculations. This is visible in Figure 2. So, we used crown volume instead of the oak trunk in the calculations.

Canopy volume was calculated by Crown form factor (C_f) Value, using equation (2) (Forrest Frank 2010):

$$C_v = (C_f) \times (C_{th}) \times (A_{amcs})^2 \qquad (2)$$

Where: C_v is the crown volume, C_f is the crown form factor and A_{amcs} is the average maximum crown spread.

Finally, a BEF for each form of the trees was calculated by using equation (3).

$$BEF = \frac{(\Sigma(AGB)_i / X_i)}{n} \qquad (3)$$

Where: AGB is above-ground biomass of each tree. X_i in this study is crown volume (Cv, which could be different in other studies depending on the study species characteristics) and n is the number of trees used to calculate the BEF (Lehtonen et al. 2004).

Table 1 - Biomass, carbon fraction and carbon content of two forms of Brant's Oak in different components (mean ± S.E.).

	Component	Biomass (kg)	Carbon Fraction (%)	Carbon (kg)
Individuals with single stem	Leaf	14.7 ± 2.8	47.1	6.9 ± 1.3
	Twig	15.6 ± 2.6	48.1	7.5 ± 1.2
	Lateral Branches	89.4 ± 25.5	48.6	43.4 ± 12.4
	Main Branches	115.6 ± 24.8	48.4	55.9 ± 19.6
	Bole	96.4 ± 23.0	48.6	46.8 ± 11.2
	Stump	28.2 ± 8.1	48.6	13.7 ± 3.9
	Total	359.9 ± 82.9		174.2 ± 46.3
Coppice Shoots	Leaf	9.8 ± 2.2	47.1	4.6 ± 1.0
	Twig	7.8 ± 1.6	48.1	3.7 ± 0.8
	Lateral Branches	19.9 ± 3.7	48.6	9.7 ± 1.8
	Main Branches	38.4 ± 9.1	48.4	18.5 ± 4.4
	Bole	54.6 ± 14.7	48.6	26.5 ± 7.1
	Stump	13.7 ± 3.9	48.6	6.7 ± 1.9
	Total	144.2 ± 34.0		69.8 ± 16.5

Calculation of the carbon stock

Carbon percent (C_c %) of the samples was obtained by burning the samples in an electric kiln and the carbon stock of the tree components investigated by using equation (2) (Heidari Safari Kouchi et al. 2017).

$$Wc = \frac{w_{dc} * C_c\%}{100} \quad (4)$$

Where: Wc is the weight of carbon for each component and W_{dc} is the dry weight of each component of the trees.

Data analysis

Normality of data was tested with the Kolmogorov-Smirnov test. Root mean square error (RMSE) and bias were also used to evaluate the goodness of model fit. To compare the observed biomass with predicted biomass, paired sample t-tests were used. Significance between the means was evaluated at the ϱ= 0.05 and 0.01 probability levels. All computations were performed using the statistical software SPSS-22.

Results

Results are presented separately for individual trees with single stem and coppice shoots. Table 1 shows a summary of the descriptive statistics of the variables analysed. The mean of trees biomass expressed as dry mass in individual trees with single stem is calculated as about 359.9kg in comparison to the coppice shoots with 144.2kg. Measured carbon fraction was between 47.1% and 48.6% for the whole dry-component samples. Carbon content of individual trees with single stem and coppice shoots were 174.2kg and 69.8kg respectively.

The tree biomass is normally divided into the above-ground components including stump, bole,

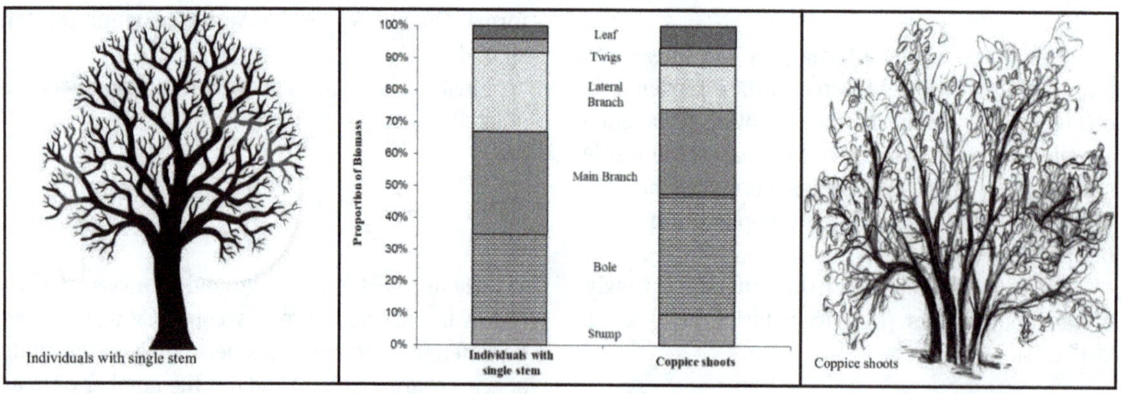

Figure 2 - Proportion of different components biomass in two vegetation forms of Brant's oak.

Table 2 - Test statistics of the BEFs based on Crown Volume from individual trees with single stem. Bias refers to the mean of differences between observed and predicted diameters in percentage (%) from mean biomass. RMSE refers to the root mean squared error. * refers to < 0.05 and ns to non-significant.

Component	Actual Biomass (kg)	Predicted Biomass (kg)	Sig.	BEF	%Bias	%RMSE
Leaf	14.7 ±2.8	20.8 ±5.1	*	0.15	41.56	110.51
Twig	15.6 ±2.6	23.2 ±5.7	*	0.17	49.11	113.23
Lateral Branches	89.4 ±25.5	72.9 ±18.1	ns	0.53	-18.34	52.38
Main Branches	115.6 ±24.8	83.2 ±20.7	ns	0.60	-28.06	80.12
Bole	96.4 ±23	100.8 ±25	ns	0.73	4.45	29.34
Total AGB	359.9 ±82.9	325.7 ±80.9	ns	2.37	-9.51	22.81

branches and foliage. The maximum biomass in individual trees with single-stem components was calculated as 115.6kg for main branches, while in coppice-shoot components tree bole showed the maximum biomass with 54.6kg (Table 1). According to the results, the ratio of branch biomass (main and lateral) in individual trees with single stem was more than that of in coppice shoots, whereas the proportion of stump and trunk biomass in coppice shoots was more than those in individual trees with single stem (Fig. 2). The canopy biomass including foliage and branches was heavier than the trunk and stump biomass for both growth forms of oak trees. The amount of foliage biomass, including twigs and leaves, is less than other components (Fig. 1).

According to Fig. 3 the foliage biomass in large trees is less than those in small and medium size trees. On the other hand, the trees' AGB was increased by tree-size increasing but trees' non-woody part biomass (foliage) to the total AGB ratio decreased with tree-size increase for both tree forms of Brant's Oak.

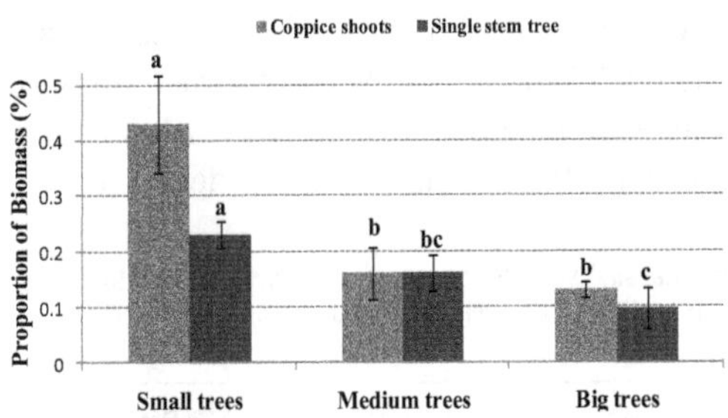

Figure 3 - Proportion of trees non-woody (foliage) parts biomass to the total above-ground biomass. The bars are standard errors and the different letters over the columns indicate significant differences ($\varrho < 0.05$).

Also, calculating the AGB for two vegetation forms of oak trees were surveyed separately. The Kolmogorov-Smirnov normality tests indicated that the BEF for all components was distributed normally.

Individual trees with single stem

For all components of the trees, BEFs were calculated based on crown volume. The BEFs, bias and RMSE of the biomass estimated for all components of Brant's Oak trees are summarised in table 2. The BEF for the total AGB of individual trees with single stem was 2.37kg/m^{3}. According to BEFs, the correlations of tree-component biomass with the different quantitative variables were weaker in comparison to the crown volume. On the other hand, the correlations between the component biomass and crown volume were higher than other variables. Also, the most observed difference between the results was related to the foliage (leaves and twigs). The BEFs describing AGB of components, especially foliage showed higher RMSEs and Bias than those for other tree components. On the other hand, for the non-woody components, the BEFs did not show enough accuracy (Table 2). The paired sample t-test exam results, of BEFs ratio based on crown volume, showed no significant difference between the estimating method and actual biomass measurement in individual trees with single stem, except the foliage.

Coppice Forests

The calculation of BEFs for Coppice shoots of Brant's Oak trees showed that composite variables of "Crown diameter x crown height x shoot density" provided the most acceptable models to estimate the trees biomass. In BEF calculations based on this variable, the observed biomass of all components of oak trees did not have significant difference with its estimation by predicted biomass (Table 3). The BEFs for the total biomass of coppice shoots was 7.2kg/m^{3}. The RMSE and bias of BEFs developed for AGB of coppice shoots were less than those for trees with single stem.

The relationship between BEFs and diameter at breast height (DBH) of individual trees with single stem is shown in Figure 4 (part a), while Figure 4 (part b) shows the relationship between BEFs and crown diameter for coppice shoots. The values of BEFs increase with increasing the size of AGB. These relations are consistent for tree boles with single stem and twig and leaf in two vegetation forms of Brant's Oak, but not consistent for branches. The highest values of the BEFs were found for AGB depending on the size of the tree, while the lowest values were found for foliage. Similarly, there is a relationship between the variability of BEFs between trees and tree size. Generally, the accuracy of estimated BEFs for all component biomass increased with increase of tree size. The results of relationship of BEFs with other tree variables like total height and crown height is weaker in comparison with DBH and crown-diameter variables.

Discussion

Biomass partitioning

Related theories to carbohydrate allocation rules can be used to quantify biomass of trees in various scales, e.g. tree, stand and region (Lethonen 2005). However, the partitioning of carbohydrates for tree growth of different biomass pools and their respiration is not adequately known (Lacointe 2000). The optimality concept (Hari et al. 1990) states that trees allocate carbohydrates in such way that they maximise their annual photosynthesis with restriction of carbohydrate amount. According to the theory of allocation priority, trees also perform carbohydrate allocation prioritisation. Oliver and Larson (1996) affirmed that priority of carbohydrate allocation is given to maintenance respiration, which is followed by fine root and foliage production, flowering, height growth and diameter growth, although recognising that the order of the priority may change temporarily.

These concepts lead to the assumption that the biomass proportions of trees are dependent on various environmental conditions, genetics and also depended on tree age (Lethonen 2005). Forest trees usually compete with their neighbours for essential resources, e.g. sunlight, water, and nutrients (Simon and Edmund 2000). Plants always adjust their above- and below-ground structure and biomass to environmental changes. The results of this study showed that the relative proportion of the bole biomass in the coppice shoots is more than other components and also more than those in individual trees with a single stem, whereas the canopy biomass in coppice shoots is smaller than those in trees with a single stem. Nevertheless, it would be expected that the bole biomass would be different in oaks which are coppiced from those in single stem oaks, due to differences in architecture. Coppice shoots have different stems because of their sprouting type.

Another important factor for biomass partitioning in tree species is tree age. With the increase in age, the proportion of stem wood in the biomass becomes more obvious (Peichl and Arain 2007, Nogueira et al. 2008, Sanquetta and Silva 2011). Wang et al. (2011) explained that the relative contribution of canopy-part (living branches and foliage) biomass decreased with increasing tree size. In our study branch biomass (main and lateral branches) had a significant role in AGB of oak trees especially in high (single stem) forests. The proportion of canopy biomass increased with increasing tree size except the foliage biomass. These findings are in accordance with some research that explained how in mature trees the rate of stem growth decreases relative to that of foliage (Waring and Schlesinger 1985), while biomass accumulation in stem increases (Wieser 2007). This is also consistent with the results of Kantola and Makela (2006), which showed that in young Norway spruce the proportion of branch biomass increases and needles decrease with increasing tree height inside the spread crowns. On the other hand, trees growing in open spaces tended to have widespread crowns and large biomass in branches and leaves in contrast to the tree's bole. Kantola and Makela (2006) found that Norway spruce trees initially allocate most percentage of their biomass to branches, while allocation to foliage decreases.

Table 3 -Test statistics of the BEFs based on "Crown diameter × crown height × shoot density" variables of coppice shoots. Bias refers to the mean of differences between observed and predicted diameters in proportional terms (%) percent from the mean biomass. RMSE refers to the Root mean squared error.

Component	Real Biomass (kg)	Predicted Biomass (kg)	Sig.	BEF	%Bias	%RMSE
Leaf	9.8 ± 2.1	10.1 ± 2.1	ns	0.6	2.86	31.92
Twig	7.8 ± 1.6	8.7 ± 1.8	ns	0.5	12.46	31.08
Lateral Branches	18.6 ± 3.7	18.6 ± 3.9	ns	1.1	0.06	31.82
Main Branches	38.1 ± 9.1	29.8 ± 6.3	ns	1.8	-21.92	42.87
Bole	56.2 ±14.6	44.6 ± 9.4	ns	2.6	-20.57	49.66
Total AGB	144.1 ± 34.1	122 ± 25.7	ns	7.2	-15.34	34.80

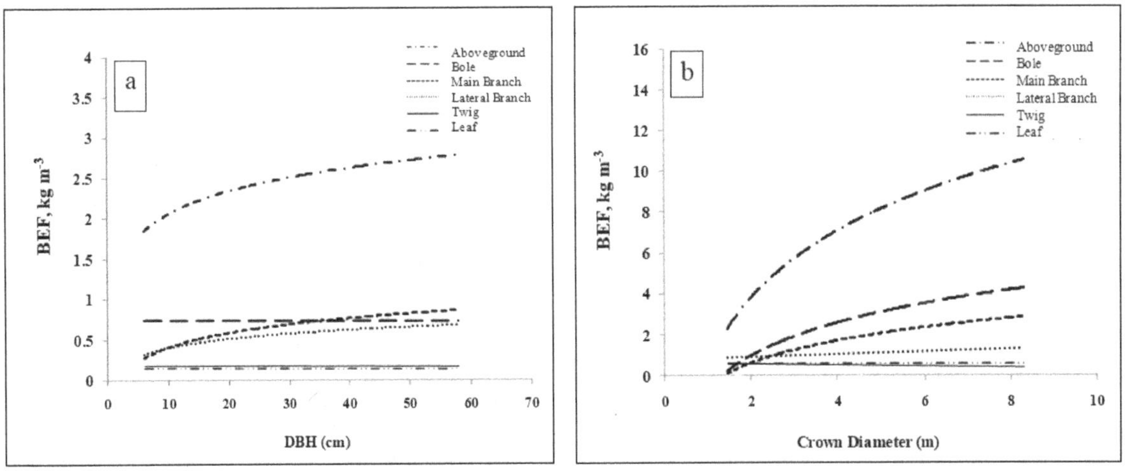

Figure 4 - Relationship between BEFs and DBH for individual with single stem (a) and crown diameter for coppice shoots (b) of Brant's Oak as a function of tree size.

Biomass expansion factor

In this study we used crown variable in BEF ratio, because of the tree-crown importance in semi-arid Mediterranean Brant's Oak measurements. The crown of trees encompasses the main proportion of AGB in two vegetation forms of Brant's Oak.

The results indicated that BEFs are tree-size dependent (Fig. 3). The main and lateral branch BEFs increase with tree size. On the other hand, for the BEFs, the trend is less steep although a constant increase is also recorded. Lehtonen et al. (2004) observed that BEF for total AGB increased in beech (*Fagus sylvatica* L.) and pine trees (*Pinus* spp.) but diminished in oak and spruce trees with incremental DBHs. These relationships are consistent for bole, twig and leaf biomass.

The increasing trend for BEFs leads to the conclusion that with increasing tree size the proportion of crown volume decreases compared to the proportion of the AGB. The increasing trend of BEFs is a result of changes in crown density with changes in tree size. These findings are in accordance with Návar (2009) who described BEFs for tree components on different sizes. The results obtained from his study and another study (Brown et al. 1989, Brown and Iverson 1992) show the necessity to develop specific BEFs for each region and forest type in the West-Iran forests. The general models of total AGB should be carefully used in specific areas or carbon projects (Noble et al. 2000).

Conclusions

The models developed in this study are recommended only when DBH is between 10 and 60cm in high forests. For the coppice shoots the functions for BEFs can be applied for woodlands with crown diameter ranging from 1.5 to 8m. The increase value for %RMSE and relative bias in leaf and twig leads to the conclusion that the BEF was not a good biomass predictor for non-woody components of individual trees with single stems. The current expansion factors, which are about foliage, can be applied for rough estimation of the biomass of these components. In summary, we developed simple predictive equations for determination of Brant's Oak biomass based on crown variables. These equations provide a useful tool for rapid estimation of AGB for two vegetation forms of Brant's Oak at the stand level.

Acknowledgements

This research is supported by Research Centre of Agriculture and Natural resources and General office of Natural Resources of Chaharmahal & Bakhtiari province. We would like to thank all of our co-workers who helped us and were instrumental in planning and executing the tedious work that laid the foundation for this research. We extend special thanks to two people who were with us all the way, Habib Akbarian for assistance in the field work and Zeinab Gholipour in the laboratory.

References

Brown S., Gillespie A.J.R., Lugo A.E. 1989 - *Biomass estimation methods for tropical forests with applications to forest inventory data.* Forest Science 35 (4): 381-902.

Brown S., Iverson L.R. 1992 - *Biomass estimates for tropical forests.* World Resources Revue 4 (3): 366-383.

Chhabra A., Palria S., Dadhwal V.K. 2002 - *Growing stock-based forest biomass estimate for India.* Biomass and Bioenergy 22: 187-194.

Chavé J., Condit R., Lao S., Caspersen J.P., Foster RB., Hubbell SP. 2003 - *Spatial and temporal variations of biomass in a tropical forest; results from a large census plot in Panama.* Journal of Ecology 91: 240-252.

Forrest Frank E. 2010 - *Crown Volume Estimates. Eastern.*

Native Tree Society. 29 p.

Guo Z., Fang J., Pan Y., Birdsey R. 2009 - *Inventory-based estimates of forest biomass carbon stocks in China: A comparison of three methods.* Forest Ecology and Management 259 (7): 1225-1231.

Hari P., Nikinama E., Holmberg M. 1990 - *Photosynthesis, transpiration and nutrient uptake in relation to tree structure.* In: Dixon, R., Meldahl, R., G.A., R., Warren, W. (Eds.), Process modelling of forest growth responses to environmental stress. Portland Timber Press: 50-57.

Heidari Safari Kouchi A., Moradian Fard F., Rostami Shahraji T., Iranmanesh Y. 2017 - *Biomass and carbon allocation of 10-year-old poplar (Populous alba L.) plantations of west Iran.* Forest Research: Open access 6 (2):1-6.

IPCC 2003 - *Good practice guidance for land use, land-use change and forestry.* Institute for Global Environmental Strategies (IGES). Hayama, Kanagawa.

Johnson W.C., Sharpe D.M. 1983 - *The ratio of total to merchantable forest biomass and its application to the global carbon budget.* Canadian Journal of Forest Research 13: 372-383.

Kantola A., Makela A. 2006 - *Development of biomass proportions in Norway spruce (Picea abies* [L.] Karst.). Trees-Structure and Function 20: 111-121.

Kauppi P.E., Kauppi E., Tomppo E., Ferm A. 1995 - *C and N storage in living trees within Finland since 1950s.* Plant Soil 168/169: 633-638.

Lacointe A. 2000 - *Carbon allocation among tree organs: a review of basic processes and representation in functional-structural tree models.* Annals of Forest Science 57: 521-533.

Lehtonen A., Makipaa R., Heikkinen J., Sievanen R., Liski J. 2004 - *Biomass expansion factors (BEFs) for Scots pine, Norway spruce and birch according to stand age for boreal forests.* Forest Ecology and Management 188: 211-224.

Lethonen A. 2005 - *Carbon stocks and flows in forest ecosystems based on forest inventory data.* Dissertation, University of Helsinki Finnish, 51 p.

Lehtonen A., Cienciala E., Tatarinov F., Makipaa R. 2007 - *Uncertainty estimation of biomass expansion factors for Norway spruce in the Czech Republic.* Annals of Forest Science 64: 133-140.

Levy P.E., Hale S.E., Nicoll B.C. 2004 - *Biomass expansion factors and root: shoot ratios for coniferous tree species in Great Britain.* Forestry 77 (5): 421-430.

Losi C.J., Siccamaa T.G., Condit R., Morales J.E. 2003 - *Analysis of alternative methods for estimating carbon stock in young tropical plantations.* Forest Ecology and Management 184: 355–368.

Muukkonen P. 2007 - *Generalised allometric volume and biomass equations for some tree species in Europe.* European Journal of Forest Research 126: 157-166.

Návar J. 2009 - *Allometric Equations and Expansion Factors for Tropical Dry Forest Trees of Eastern Sinaloa, Mexico.* Tropical and Subtropical Agroecosystems 10: 45-52.

Noble I., A.P.P.S M, Hougton R., Lashof D., Makundi W., Murdiyarso D., Murray B., Sombroek W., Valentini R. 2000 - *36 Implications of different definitions and generic issues.* In: "Land Use, Land-Use Change, and Forestry. A special report to IPCC". Watson R., Noble I., Bolin B., Ravindranath N. H., Verardo D., Dokken D. (Eds.). Cambridge University Press, Cambridge: 53-118.

Nogueira E.M., Fearnside P.M., Nelson B.W., Barbosa R.I., Keizer E.W.H. 2008 - *Estimates of forest biomass in the Brazilian Amazon: new allometric equations and adjustments to biomass from wood-volume inventories.* Forest Ecology and Management 256: 1853-1867.

Oliver C.D., Larson B.C. 1996 - *Forest Stand Dynamics.* John Wiley & Sons, New York. 518 p.

Pajtik J., Konôpka B., Lukac M. 2008 - *Biomass functions and expansion factors in young Norway spruce (Picea abies [L.] Karst) trees.* Forest Ecology and Management 256: 1096-1103.

Peichl M., Arain M.A. 2007 - *Allometry and partitioning of above and belowground tree biomass in an age-sequence of white pine forests.* Forest Ecology and Management 253: 68-80.

Sanquetta A.P., Silva F.S. 2011 - *Biomass expansion factor and root-to-shoot ratio for Pinus in Brazil.* Carbon Balance and Management 6 (6): 1-8.

Satoo T., Madgwick H.A.I. 1982 - *Forest biomass.* Forestry Sciences Martinus Nijhoff/ Dr. W. Junk Publisher, The Hague. 162 p.

Sharp D.D., Lieth H., Whigham D. 1975 - *Assessment of regional productivity in North Carolina.* In: "Primary Productivity of the Biosphere". Lieth, H., Whittaker R.H. (Eds.), Springer-Verlag, New York, USA: 131-146.

Simon L.L., Edmund V.J.T. 2000 - *Effects of above and below-ground competition on growth and survival of rain forest tree seedlings.* Ecology 81 (9): 2525-2538.

Soares P., Tome M. 2004 - *Analysis of the effectiveness of biomass expansion factors to estimate stand biomass.* In: Proceedings of the International Conference on Modeling Forest Production. Austria, April: 19 - 22.

Soares P., Tome M. 2012 - *Biomass expansion factors for Eucalyptus globules stands in Portugal. Forest Systems* 21 (1): 141-152.

Somogyi Z., Cienciala E., Makipaa R., Muukkonen P., Lehtonen A., Weiss P. 2006 - *Indirect methods of large-scale forest biomass estimation.* European Journal of Forest Research 126 (2): 197-207.

Teobaldelli M., Somogyi Z., Migliavacca M., Usoltsev V.A. 2009 - *Generalized functions of biomass expansion factors for conifers and broadleaved by stand age, growing stock and site index.* Forest Ecology and Management 257: 1004-1013.

Wang J., Zhang C., Xia F., Zhao X., Wu L., Gadow K.V. 2011 - *Biomass Structure and Allometry of Abies nephrolepis (Maxim) in Northeast China.* Silva Fennica 45(2): 211-226.

Waring R.H., Schlesinger W.H. 1985 - *Forest ecosystems: concepts and management.* Orlando FL: Academic Press. 340 p.

West P.W. 2009 - *Tree and Forest Measurement.* Verlag Berlin Heidelberg, Springer. 190 p.

Wieser G. 2007 - *Limitation by an insufficient carbon assimilation and allocation.* In: "Trees at their upper limit Netherlands". Wieser G, Tausz M (Eds.), Springer. 113 p.

PERMISSIONS

All chapters in this book were first published in ASR, by CREA Forestry and Wood; hereby published with permission under the Creative Commons Attribution License or equivalent. Every chapter published in this book has been scrutinized by our experts. Their significance has been extensively debated. The topics covered herein carry significant findings which will fuel the growth of the discipline. They may even be implemented as practical applications or may be referred to as a beginning point for another development.

The contributors of this book come from diverse backgrounds, making this book a truly international effort. This book will bring forth new frontiers with its revolutionizing research information and detailed analysis of the nascent developments around the world.

We would like to thank all the contributing authors for lending their expertise to make the book truly unique. They have played a crucial role in the development of this book. Without their invaluable contributions this book wouldn't have been possible. They have made vital efforts to compile up to date information on the varied aspects of this subject to make this book a valuable addition to the collection of many professionals and students.

This book was conceptualized with the vision of imparting up-to-date information and advanced data in this field. To ensure the same, a matchless editorial board was set up. Every individual on the board went through rigorous rounds of assessment to prove their worth. After which they invested a large part of their time researching and compiling the most relevant data for our readers.

The editorial board has been involved in producing this book since its inception. They have spent rigorous hours researching and exploring the diverse topics which have resulted in the successful publishing of this book. They have passed on their knowledge of decades through this book. To expedite this challenging task, the publisher supported the team at every step. A small team of assistant editors was also appointed to further simplify the editing procedure and attain best results for the readers.

Apart from the editorial board, the designing team has also invested a significant amount of their time in understanding the subject and creating the most relevant covers. They scrutinized every image to scout for the most suitable representation of the subject and create an appropriate cover for the book.

The publishing team has been an ardent support to the editorial, designing and production team. Their endless efforts to recruit the best for this project, has resulted in the accomplishment of this book. They are a veteran in the field of academics and their pool of knowledge is as vast as their experience in printing. Their expertise and guidance has proved useful at every step. Their uncompromising quality standards have made this book an exceptional effort. Their encouragement from time to time has been an inspiration for everyone.

The publisher and the editorial board hope that this book will prove to be a valuable piece of knowledge for researchers, students, practitioners and scholars across the globe.

LIST OF CONTRIBUTORS

Mauro Maesano, Bruno Lasserre and Marco Marchetti
Department of Biosciences and Territory, University of Molise, Italy

Francesco Chianucci, Emilio Amorini, Tessa Giannini and Andrea Cutini
Consiglio per la ricerca in agricoltura e l'analisi dell'economia agraria, Forestry Research Centre (CRA-SEL), Arezzo, Italy

Luca Mattioli
Province of Arezzo, Arezzo, Italy

Andrea Marcon, Roberta Chirichella and Marco Apollonio
Department of Sciences for Nature and Environmental Resources, University of Sassari, Sassari, Italy

Piermaria Corona, Nicola Puletti and Francesco Chianucci
Consiglio per la ricerca in agricoltura e l'analisi dell'economia agraria, Forestry Research Centre (CRA-SEL), Arezzo, Italy

Davide Ascoli and Giovanni Bovio
University of Turin, Department of Agriculture, Forest and Food Sciences, Grugliasco, Italy

Anna Barbati and Luigi Portoghesi
University of Tuscia, Department for Innovation in Biological, Agro-food and Forestry systems (DIBAF), Viterbo, Italy

Giuseppe Colangelo, Mario Elia, Raffaele Lafortezza and Giovanni Sanesi
University of Bari, Department of Agricultural and Environmental Sciences, Bari, Italy

Vittorio Garfì and Marco Marchetti
University of Molise, Department of BioScience and Territory, Pesche (IS), Italy

Francesco Iovino
University of Calabria, Department of Informatics, Modeling, Electronics and Systems Engineering (D.I.M.E.S.), Arcavacata di Rende (CS), Italy

Vittorio Leone
University of Basilicata, Department of Crop Systems, Forestry and Environmental Sciences, Potenza, Italy

Raffaella Lovreglio
University of Sassari, Department of Agriculture, Nuoro, Italy

Enrico Marchi and Susanna Nocentini
University of Florence, Department of Agricultural, Food and Forestry Systems, Firenze, Italy

Giuliano Menguzzato
University Mediterranea of Reggio Calabria, Department of Agriculture, Reggio Calabria, Italy

Rodolfo Picchio
University of Tuscia, Department of Science and Technology for Agriculture, Forests, Nature and Energy, Viterbo, Italy

Maurizio Marchi, Cristiano Castaldi, Paolo Merlini and Fulvio Ducci
Consiglio per la ricerca in agricoltura e l'analisi dell'economia agraria, Forestry Research Centre (CREA-SEL), Arezzo, Italy

Susanna Nocentini
University of Florence, Department of Agricultural, Food and Forestry Systems, Firenze, Italy

James M Roshetko
World Agroforestry Centre (ICRAF) Southeast Asia Regional Program, Bogor, West Java, Indonesia

Manuel Bertomeu
University of Extremadura, UEX, Plasencia, Cáceres, Spain

P. Corona, A. Cutini and Ugo Chiavetta
Consiglio per la ricerca in agricoltura e l'analisi dell'economia agraria, Forestry Research Centre (CREA-SEL, Arezzo, Italy)

Elena Paoletti
Consiglio Nazionale delle Ricerche, Institute for Sustainable Plant Protection (CNR-IPSP, Sesto Fiorentino, Italy)

Jonathan Sheppard, Matthias Urmes, Christopher Morhart and Heinrich Spiecker
Chair of Forest Growth and Dendroecology, Albert-Ludwigs-Universität, Freiburg, Germany

Angelo Vitone
Università degli Studi del Molise, Campobasso, Italy

Jaime Coello, Miriam Piqué and Pere Rovira
Forest Sciences Centre of Catalonia, Solsona, Spain

Francesca Giannetti and Gherardo Chirici
Department of Agricultural, Food and Forestry Systems, University of Florence, Firenze, Italy

Elena Paoletti
Institute for Sustainable Plant Protection, National Research Council (CNR), Firenze, Italy

Marco Borghetti
Scuola di Scienze Agrarie, Forestali, Alimentari e dell'Ambiente, Università della Basilicata, Potenza, Italy

Piermaria Corona
Consiglio per la ricerca in agricoltura e l'analisi dell'economia agraria, Forestry Research Centre (CREA SEL), Arezzo, Italy

Gabriele Bucci
Institute of Bioscience and Bioresources, National Research Council (CNR), Firenze, Italy

Maurizio Marchi, Ugo Chiavetta and Paolo Cantiani
Consiglio per la ricerca in agricoltura e l'analisi dell'economia agraria, Research Centre for Forestry and Wood, Arezzo (Italy)

Enoma Aigbedion and William W. Modugu
University of Benin, Department of Forestry and Wildlife, Benin City, Edo State (Nigeria)

Friday Nwabueze Ogana
University of Ibadan, Department of Forest Reaources Management Ibadan, Oyo State (Nigeria)

Carlotta Ferrara and Luca Salvati
CREA Research Centre of Forestry and Wood, Arezzo (Italy)

Pier Matteo Barone
The American University of Rome, Roma (Italy)

Most Jannatul Fardusi
Consiglio per la Ricerca in Agricoltura e l'Analisi dell'Economia Agraria - Research Centre for Forestry and Wood, Arezzo (Italy)
Università degli Studi della Tuscia - Department for the Innovation in Biological, Agro-food and Forest systems,Viterbo (Italy)

Anna Barbati
Università degli Studi della Tuscia - Department for the Innovation in Biological, Agro-food and Forest systems,Viterbo (Italy)

Francesco Chianucci
Consiglio per la Ricerca in Agricoltura e l'Analisi dell'Economia Agraria - Research Centre for Agriculture and Environment, Rome (Italy)

Giovanni Argenti
Università degli Studi di Firenze, Scienze delle Produzioni Agro-alimentari e dell'Ambiente, Firenze (Italy)

Elisa Bianchetto
CREA Research Centre for Agriculture and Environment, Firenze (Italy)

Fabrizio Ferretti
CREA Research Centre of Forestry and Wood, Arezzo (Italy)

Fabrizio Ferretti and Chiara Sboarina
CREA Research Centre for Forestry and Wood, Arezzo (Italy)

Clara Tattoni, Alfonso Vitti, Paolo Zatelli, Francesco Geri and Marco Ciolli
Dipartimento di Ingegneria Civile Ambientale e Meccanica, Università degli Studi di Trento, Trento (Italy)

Enrico Pompei
MIPAAF, Roma (Italy)

Fulvio Ducci
Chair of FPS COST Action FP1202 "Strengthening conservation: a key issue for adaptation of marginal/peripheral populations of forest trees to climate change in Europe (MaP-FGR)" CREA Research centre for Forestry and Wood, Arezzo (Italy)

Kevin Donnelly
Vice Chair Centre for Ecology and Hydrology, Edinburgh (United Kingdom)

Claudia Becagli, Nicola Puletti, Ugo Chiavetta, Paolo Cantiani and Gianfranco Fabbio
Consiglio per la Ricerca e la sperimentazione in Agricoltura, Forestry Research Centre, Arezzo (CRA-SEL), Italy

Luca Salvati
Consiglio per la Ricerca e la sperimentazione in Agricoltura, Centro di ricerca per lo studio delle Relazioni Pianta-Suolo, Roma (CRA-RPS), Italy

Maria Cristina Monteverdi
Consiglio per la Ricerca e la sperimentazione in Agricoltura, Forestry Research Centre, Arezzo (CRA-SEL), Italy

Sara Da Canal and Paolo De Angelis
Department for Innovation in Biological, Agro-food and Forest systems (DIBAF), University of Tuscia, Italy

Alberto Del Lungo
Forestry Department FAO. Forestry Officer, Project LTO, GCP/RAB/013/ITA

Salvatore Masi
School of Engineering University of Basilicata, Italy

Hocine Larbi
Université de Mascara, Algeria

Sofia Bajocco, Elisabetta Raparelli, Francesco Patriarca, Giovanni Di Matteo, Pierfrancesco Nardi and Luigi Perini
Consiglio per la Ricerca e la sperimentazione in Agricoltura, Unità di ricerca per la climatologia e la metereologia applicate all'agricoltura, Roma (CRA-CMA), Italy

Luca Salvati
Consiglio per la Ricerca e la sperimentazione in Agricoltura, Centro di ricerca per lo studio delle Relazioni Pianta-Suolo, Roma (CRA-RPS), Italy

Giuseppe Scarascia Mugnozza
University of Tuscia, Dipartimento per la Innovazione nei sistemi Biologici, Agroalimentari e Forestali, Viterbo (UNITUS -DIBAF), Italy

Dora Cimini, Antonio Tomao, Walter Mattioli and Anna Barbati
University of Tuscia, Department for Innovation in Biological, Agro-food and Forest systems (DIBAF), Viterbo, Italy

Piermaria Corona
Consiglio per la Ricerca e la sperimentazione in Agricoltura, Forestry Research Centre (CRA-SEL), Arezzo, Italy

Friday Nwabueze Ogana
University of Ibadan, Ibadan, Nigeria - Department of Social and Environmental Forestry
Opposite Agbowo Shopping Complex, Off Sango-Ojoo Expressway Ibadan 200284 Nigeria

Sabrina Raddi
DAGRI, Università degli Studi di Firenze, Florence (Italy)

Federico Magnani
DISTAL, Alma Mater - Università di Bologna, Bologna (Italy)

Jimena Forero Montaña
University of Puerto Rico Rio Piedras, Faculty of Natural Sciences, Department of Biology , San Juan Puerto Rico

Jess K. Zimmerman
University of Puerto Rico Rio Piedras, Environmental Sciences, San Juan Puerto Rico

Thrity Vakil
Tropic Ventures Research and Education Foundation, Patillas, Puerto Rico

Mark Nelson
Tropic Ventures Research and Education Foundation, Patillas, Puerto Rico
Institute of Ecotechnics (US/UK), Santa Fe, New Mexico, United States

Marìa Eglee Perez
University of Puerto Rico Rio Piedras, Mathematics, San Juan, Puerto Rico

Joaquin Medin
University of Puerto Rico in Bayamon, Physics, Bayamon, Puerto Rico

Yaghoub Iranmanesh
Research Division of Natural Resources, Chaharmalhal and Bakhtiari Agricultural and Natural Resources Research and Education Center, AREEO, Shahrekord, I.R. Iran

Hormoz Sohrabi and Sayed Mohsen Hosseini
Faculty of Natural Resources and Marine Science, University of Tarbiat Modares, Noor, I.R. Iran

Khosro Sagheb Talebi
Research Institutes of Forest and Rangelands, AREEO, Tehran, I.R. Iran

Heidari Safari Kouchi Abouzar
Faculty of Natural Resources, University of Guilan, Sowme´eh sara, I.R. Iran

Index

www.ingramcontent.com/pod-product-compliance
Lightning Source LLC
LaVergne TN
LVHW060536210926
848485LV00005B/1211